An Introduction to
Aqueous Electrolyte Solutions

An Introduction to Aqueous Electrolyte Solutions

Margaret Robson Wright
Formerly of St Andrews University, UK

John Wiley & Sons, Ltd

Email (for orders and customer service enquiries): cs-books@wiley.co.uk
Visit our Home Page on www.wileyeurope.com or www.wiley.com

Other Wiley Editorial Offices

John Wiley & Sons Inc., 111 River Street, Hoboken, NJ 07030, USA

Jossey-Bass, 989 Market Street, San Francisco, CA 94103-1741, USA

Wiley-VCH Verlag GmbH, Boschstr. 12, D-69469 Weinheim, Germany

John Wiley & Sons Australia Ltd, 42 McDougall Street, Milton, Queensland 4064, Australia

John Wiley & Sons (Asia) Pte Ltd, 2 Clementi Loop #02-01, Jin Xing Distripark, Singapore 129809

John Wiley & Sons Canada Ltd, 6045 Freemont Blvd, Mississauga, Ontario, L5R 4J3

Wiley also publishes its books in a variety of electronic formats. Some content that appears in print
may not be available in electronic books.

Library of Congress Cataloging-in-Publication Data

Wright, Margaret Robson.
    An introduction to aqueous electrolyte solutions / Margaret Robson Wright.
      p. cm.
    Includes index.
    ISBN 978-0-470-84293-5 (cloth)
    ISBN 978-0-470-84294-2 (paper)
    1. Electrolyte solutions. 2. Chemical equilibrium. 3. Solution
(Chemistry) I. Title.
    QD565.W75 2007
    541'.372–dc22   2007011329

British Library Cataloguing in Publication Data

A catalogue record for this book is available from the British Library

ISBN 978-0-470-84293-5 (cloth)
ISBN 978-0-470-84294-2 (paper)

Typeset in 10/12 pt Times by Thomson Digital Noida, India
This book is printed on acid-free paper responsibly manufactured from sustainable forestry
in which at least two trees are planted for each one used for paper production.

*Dedicated with much love, affection and gratitude
to my husband Patrick
for all his help and encouragement
throughout my research and teaching career.*

Contents

6 Neutralisation and pH Titration Curves   139

7 Ion Pairing, Complex Formation and Solubilities   177

8 Practical Applications of Thermodynamics for Electrolyte Solutions   215

10 Concepts and Theory of Non-ideality 349

Preface

This book covers a very large and important branch of physical chemistry, namely the properties, behaviour and nature of aqueous electrolyte solutions. It does not cover electrode kinetics. The preliminary chapter entitled "Guidance to Students" sets outs the aims of the book, the philosophy behind the book and the manner in which the author tries to help students to understand the topics covered. As such it is almost an extra preface. The book emphasises, through the use of many worked problems, how experimental data can be used to obtain fundamental electrochemical quantities, such as thermodynamic quantities describing the equilibrium properties and behaviour of ions in aqueous solution. Topics such as acid/base behaviour, salts, buffers, ion pairing and complexing, and solubilities of sparingly soluble salts are the main equilibrium situations discussed. It also covers non-equilibrium behaviour such as ionic molar conductivities, molar conductivities, transport numbers and mobilities. Extensive coverage is given to emf's, and the use which physical chemists can make of the information derived from such studies. All these topics are illustrated by numerous worked problems where the student is led through the arguments with the use of detailed help and guidance.

The first chapter of the book sets the stage for many of the topics dealt with later, and, in particular, is a prelude to the development of the two major theoretical topics described in the book, namely the theory of non-ideality and conductance theory. The conventional giants of these fields are Debye and Hückel with their theory of non-ideality and Debye, Hückel, Fuoss and Onsager with their various conductance equations. These topics are dealt with in Chapters 10 and 12. In addition, the author has included for both topics a qualitative account of modern work in these fields. There is much exciting work being done at present in these fields, especially in the use of statistical mechanics and computer simulations for the theory of non-ideality. Likewise some of the advances in conductance theory are indicated.

Many students veer rapidly away from topics which are quantitative, involve theoretical deductions and, in particular, where mathematical equations are involved. This book attempts to allay these fears by guiding the student throughout in a step-by-step development which explains the logic, reasoning and actual manipulation. For this reason a large fraction of the text is devoted to worked problems in which detailed and explanatory answers are given. These worked problems are not an optional extra, but are an integral and necessary part of the presentation.

As in the other sections of the book, the theoretical chapters aim to help students understand modern thinking about the behaviour and nature of electrolyte solutions by giving a slow, explanatory, "student friendly" yet rigorous introduction to such thinking. There are two major problems for students to understand before they can grasp a theory: firstly the concepts, and then the mathematics used to quantify them. In the author's experience, students rarely manage to deduce the chemical and physical concepts from the mathematical development. These concepts need to be fully and explicitly explained first before the student can start to understand the development of the theory, or even the physical limitations and approximations inherent in the equations describing non-ideality and conductance. It is the author's firm belief, based on many years of teaching students, that students can gain considerable understanding and knowledge of what is involved in a theoretical discussion of what goes on at the microscopic ionic level without delving into the mathematics first. For this reason the concepts involved in each stage of the theoretical development are fully explained before the mathematical description is given. Once the concepts have been assimilated it is then time to move on to the mathematical presentation, which in this book is developed with the same full explanatory detail.

The final chapter deals with the very important topic of solvation. Many modern experimental advances especially in the field of synchroton generated X-rays, isotopically substituted neutron diffraction, nmr and computer simulation have given a wealth of information about the microscopic behaviour of the solvent in the presence of ions. In many respects the emphasis has moved from "what happens to an ion when in water" to "what happens to water when an ionic substance is dissolved in it".

If through the written word I can help students to understand and to feel confident in their ability to learn, and to teach them in a manner which gives them the feeling of a direct contact with the teacher, then this book will not have been written in vain. It is the teacher's duty to show students how to achieve understanding, and to think scientifically. The philosophy behind this book is that this is best done by detailed explanation and guidance. It is understanding, being able to see for oneself, and confidence which help to stimulate and sustain interest. This book attempts to do precisely that.

This book is the result of the accumulated experience of forty very stimulating years of teaching students at all levels. During this period I regularly lectured to students, but more importantly, I was deeply involved in devising tutorial programmes at all levels where consolidation of lecture material was given through many problem-solving exercises. I also learned that providing detailed explanatory answers to these exercises proved very popular and successful with students of all abilities. During these years I learned that being happy to help, being prepared to give extra explanation and to spend extra time on a topic could soon clear up problems and difficulties which many students thought they would never understand. Too often teachers forget that there were times when they themselves could not understand, and when a similar explanation and preparedness to give time were welcome. To all the many students who have provided the stimulus and enjoyment of teaching I give my grateful thanks.

I am very grateful to John Wiley & Sons for giving me the opportunity to publish this book, and to indulge my love of teaching, and to the following editors with John Wiley & Sons. Firstly, I would like to thank Andy Slade who initially dealt with the commissioning of this book. In particular, I owe a great debt to Rachael Ballard who has cheerfully and with great patience guided me through the problems of preparing the manuscript for publication. Invariably, she has been extremely helpful. Finally I would like to express my thanks to Robert Hambrook for his help and patience in the final stages of preparing the book for publication. I have found all extremely friendly and helpful throughout.

Similarly, I would like to thank Professor Andy Abbott of Leicester University for his encouragement and advice after reading several chapters of this book.

To my mother, Mrs Anne Robson, I have a very deep sense of gratitude for all the help she gave me in her lifetime in furthering my academic career. I owe her an enormous debt for her invaluable, excellent and irreplaceable help with my children when they were young and I was working part-time during the teaching terms of the academic year. Without her help and her loving care of my children I would never have gained the continued experience in teaching, and I could never have written this book. My deep and most grateful thanks are due to her.

My husband, Patrick, has, throughout my teaching career and throughout the thinking about and writing of this book, been a source of constant support and help and encouragement. His very high intellectual calibre and wide-ranging knowledge and understanding have provided many fruitful and interesting discussions. He has read in detail the whole manuscript and his clarity, insight and considerable knowledge of the subject matter have been of invaluable help. I owe him many apologies for the large number of times when I have interrupted his own activities to pursue a discussion of aspects of the material presented here. It is to his very great credit that I have never been made to feel guilty about doing so. My debt to him is enormous, and my most grateful thanks are due to him.

Finally, my thanks are due to my three children who have always encouraged me in my teaching, and have encouraged me in the writing of my books. Their very fine and keen minds have always been a source of stimulus to me, and have kept my interest in young people flourishing.

Margaret Robson Wright
Formerly of Universities of Dundee and St Andrews
November 2006.

Preliminary Chapter

Guidance to Students

It is hoped that students using this book will read through this brief chapter before going on to study individual chapters.

My aim in writing this book is to guide and help students to an understanding of this large and important field of chemistry. It is aimed primarily at undergraduates studying core chemistry and this includes students of inorganic and organic aqueous solution chemistry. Chapters 1–9 and Chapter 11 cover all the main topics in general electrochemistry and equilibria in solutions, including use of the Debye-Hückel equations to correct for non-ideality. By general electrochemistry, I mean topics such as electrolysis, electrochemical cells, reversible emfs, weak acids and bases, complexing and solubilities, conductance and transport numbers, etc. Knowledge of these topics is an essential component of undergraduate studies, and this book is written for students who need help in understanding and assimilating these topics. This material is necessary knowledge for all chemistry students, not just for those studying physical chemistry. For instance, students studying solution kinetics have to be as versatile and knowledgeable in electrolyte solution chemistry as they are in kinetics. Students studying inorganic solution chemistry would likewise limit themselves drastically if they did not have electrochemical ideas at their fingertips. It has also become clear that electrochemical topics and concepts are becoming increasingly valuable and necessary in looking at many aspects of biological solutions.

This book thus attempts to help **all** chemistry students. In the core topics outlined above, the method used is to explain each topic as clearly as possible covering all the potential sources of misunderstanding. More time is therefore devoted to each topic than is typically the case in other accounts. Many students will veer rapidly away from topics which are quantitative and involve mathematical equations. This book attempts to allay these fears by guiding the student through these topics in a step-by-step development which explains the logic, reasoning and actual manipulation. For this reason, a large fraction of the text is devoted to worked examples where the answers are given in a detailed and explanatory manner. The worked problems are **not** an **optional extra** to the main text: they are an **integral and necessary** part of the presentation. Students should read through them along with the main text, so that thinking about the worked problems and reading through the explanations given in the development of the topic under consideration should be mutually instructive. Once the chapter, or section of a chapter has been studied and assimilated, I would suggest that the student should leave the section for some days, and then come back to the worked problems and work through them without looking at the answers. Once a genuine attempt has been made to work through the problem, the answer should then be consulted to correct any mistakes, or to help with difficulties which may have arisen. A student should not be dismayed if several attempts may be needed for some of the worked problems.

The book is written in a manner deliberately designed to give you, the student, the feeling of direct contact with the teacher. It is the teacher's duty to help the student to understand, and the author's philosophy and experience has always been that this is best done by detailed explanation and guidance.

Chapters 1–9 and Chapter 11 deal with the routine 'working' material of electrochemistry and the manipulation of experimental data. Chapters 10, 12 and 13 deal with the theoretical aspects of electrochemistry, viz., the theories of non-ideality and conductance, and the theoretical aspects of solvation. All of these are very important aspects of electrochemistry, but they are inherently more difficult. Nonetheless, it is important that modern students should have some knowledge of these topics.

Because you, the student, may feel less sure of the ground in theoretical electrochemistry, a similar detailed explanatory approach is taken here. There are two main problems which you, the student, have to overcome before you can understand a theory: first the concepts, and then the mathematical clothing of these concepts. The concepts need to be fully and explicitly explained before the student can hope to understand the theoretical development of these concepts. Chapter 1 gives a summary of all the 'more simple' concepts and ideas on which the theories of electrolyte solutions are based.

Chapters 10 and 12 can be thought of as in two parts, the first part of which is the basic theory of non-ideality and the basic theory of conductance. The development of these theories is explained in a step-by-step manner, with the essential concepts coming first followed by a statement of what the physical problem is which has to be overcome by the mathematical treatment, which then follows. These are difficult sections conceptually and mathematically, but the author has found that it is possible, by going sufficiently slowly and with sufficient explanation, to convey this to the 'ordinary' student, i.e. a student who is not mathematically orientated. The resultant final equations of these theories are those on which the manipulation and interpretation of experimental data such as are discussed in Chapters 1–9 and 11 are based.

The tail end of each of Chapters 10 and 12 is then devoted to the more modern aspects of non-ideality and conductance. These are given in a qualitative manner. The modern approaches to the theories of non-ideality and conductance have taken place over the years since the late 1960s by which time the basic theories of electrolytes had been formulated. The short qualitative descriptions of the advanced topics of the post late 1960s are given so that the interested student can gain some idea of where the excitement of modern work lies. This is a very active field at present, and there is a reasonable hope that final all-embracing theories of non-ideality and conductance will be forthcoming. There is great scope and excitement over the use of computer simulations based on statistical mechanical arguments, and the brief descriptions are given in the hope that students will gain some idea and appreciation of the exciting 'new' directions.

Chapter 13 is much more qualitative, and the theoretical developments here can be given in a more descriptive manner than in Chapters 10 and 12. Solvation is a very important topic not just for the physical chemist, but for both inorganic and organic chemists. It is proving to be a fundamental and important aspect of modern biochemical and biological studies.

List of Symbols

A	absorbance $\left\{= \log_{10}\left(\frac{I}{I_0}\right)\right\}$
A	constant in Debye-Hückel theory
A	area
a	activity, radius
a_i	activity of species i
$\mathring{a}$	distance of closest approach
a	coefficient in expression $S = a + b\Lambda^0$
B	constant in Debye-Hückel theory
b	coefficient of linear term in Debye-Hückel extended equation
b	quantity $\left\{= \dfrac{z^2 e^2}{4\pi\varepsilon_0\varepsilon_r kT\mathring{a}}\right\}$ appearing as upper limit of Bjerrum's integral, and in theory of dependence of Λ on concentration
b	coefficient in expression $S = a + b\Lambda^0$
C	constant of integration in solution of Poisson-Boltzmann equation
C_p	heat capacity
c	concentration
c_{actual}	actual concentration of a species
c_{stoich}	stoichiometric concentration
D	constant of integration in solution of Poisson-Boltzmann equation
E	potential difference; in particular emf
$E^{\ominus}$	standard emf
E, E_1, E_2	coefficients in expressions for dependence of Λ on concentration
e	electronic charge
F	Faraday constant
F	coefficient involved in "viscous" correction in theory of dependence of Λ on concentration
f	force
$g(r), g^{(2)}(r)$	radial distribution function
$g^{(n)}(r_{12}, r_{13}, \ldots)$	general distribution function
G	Gibbs free energy
G^{E}	excess free energy in theories of non-ideality
G_{++}, G_{--}, G_{+-}	integrals involved in Yvon/Kirkwood/Born/Green/Bogolyubov theory of non-ideality
H	enthalpy
I	ionic strength

I	intensity of light
I_0	initial intensity of light
I	current
I_+, I_-	contributions of cation and anion to current
J, J_1, J_2	coefficients in expressions for dependence of Λ on concentration
j	central reference ion of ionic atmosphere
K	equilibrium constant
K_a	acid ionisation constant
K_b	basic ionisation constant
K_s	solubility product
K_w	ionic product for water
K_n	step-wise formation constant of complex
K_{ideal}	ideal equilibrium constant
k	Boltzmann's constant
k	rate constant
l	length
m	mass
N	Avogadro's constant
N_i	total number of ions of species i
n_i	number of ions of species i per unit volume; in particular of bulk solution
n_i'	local value of n_i at given position in ionic atmosphere
n_i	amount of substance i in a given system
$\bar{n}$	mean number of ligands per metal ion
p	pressure
Q	electric charge
$Q(b)$	Bjerrum's integral
q	heat transferred
q	characteristic distance in Bjerrum's theory
R	diameter of Gurney co-sphere
R	hydrodynamic radius
R	electrical resistance
R	gas constant
r	distance; in particular radius
S	entropy
S	coefficients in expression for dependence of Λ on concentration
s	solubility
T	absolute temperature
t	time
t_+, t_-	transport numbers of cation and anion
U	energy
u_+, u_-	mobilities of cation and anion
V	volume
v	velocity
W	number of microscopically distinct arrangements
w	work done **on** system
X	electric field
z, z_+, z_-	number of charges on an ion

α	fraction dissociated
α	fraction of ion pairs which are contact ion pairs (in Fuoss theory of conductance)
α_n	fraction having n ligands attached
α_0	fraction having no ligands attached
β	fraction associated
β_n	overall formation constant of complex containing n ligands
γ	activity coefficient
γ	fraction not associated (in Fuoss theory of conductance)
γ_i	activity coefficient of a species i
$\gamma_\pm$	mean ionic activity coefficient
ΔC_p	change in heat capacity
ΔC_p^θ	standard value of the change in heat capacity
ΔG	change in Gibbs free energy
ΔG^θ	standard value of the change in Gibbs free energy
ΔH	change in enthalpy
ΔH^θ	standard value of the change in enthalpy
ΔS	change in entropy
ΔS^θ	standard value of change in entropy
ΔU	change in energy
ΔU^θ	standard value of change in energy
ΔV	change in volume
ΔV^θ	standard value of change in volume
ε	molar absorption coefficient
ε_r	relative permittivity
ε_0	universal constant
η	viscosity
θ	angle
κ	conductivity, quantity fundamental to the Debye-Hückel theory
κ_+, κ_-	contributions of cation and anion to conductivity κ
Λ	molar conductivity
Λ^0	limiting value of molar conductivity
λ_i	ionic molar conductivity of species i
λ_+, λ_-	ionic molar conductivities of cation and anion
λ_+^0, λ_-^0	limiting ionic molar conductivities for cation and anion
μ	dipole moment
μ_i	chemical potential of species i
μ_i^θ	standard value of the chemical potential of species i
ρ	charge density (i.e. charge per unit volume)
ρ	resistivity
τ	time
τ	relaxation time; in particular of ionic atmosphere
Φ	total potential energy
ϕ	potential energy for all interactions affecting one ion
ϕ_{ij}	potential energy of interaction between ions (or molecules) i and j
ψ	electrical potential

ψ_j total electrical potential at a distance r from the central j ion due to this ion and its ionic atmosphere

ψ_j' electrical potential at a distance r from the central j ion due to the ion itself

ψ_j'' electrical potential at a distance r from the central j ion due to the ionic atmosphere

ψ_j^* electrical potential at surface of j ion due to the ionic atmosphere

1
Concepts and Ideas: Setting the Stage

This book is about the nature, behaviour and properties of electrolyte solutions. The types of particles which are present in such solutions are discussed, and the experimental evidence is reviewed, from which the structure and nature of both the solute and solvent species is inferred. The development of the theories which describe the nature of electrolyte solutions is also considered.

But first of all it is useful to gather together in one place most of the facts and ideas which are pertinent and relevant to the understanding of electrolyte solutions. It also helps understanding if an explanation of how and why these facts are relevant is given, and this chapter sets out to do this.

Aims

By the end of this chapter you should be able to:

- describe what is meant by an electrolyte solution;

- list simple properties of ions and consider possible modifications to these properties;

- discuss the molecular structure of water and the effect which ions can have on it;

- discuss the difference between polar and non-polar solvents;

- explain what is meant by dipole, induced dipole and alignment of dipoles;

- list possible ion–ion interactions, ion–solvent interactions and solvent–solvent interactions;

An Introduction to Aqueous Electrolyte Solutions. By Margaret Robson Wright
© 2007 John Wiley & Sons Ltd   ISBN 978-0-470-84293-5 (cloth) ISBN 978-0-470-84294-2 (paper)

- distinguish between ideal and non-ideal solutions;

- know what is meant by an ion pair, complex, chelate and micelle;

- understand the significance of base-lines for theoretical predictions about the behaviour of solutions containing free ions only, and their use in determining equilibrium constants for processes in solution;

- appreciate the difficulties involved in distinguishing the various types of associated species in solution.

1.1   Electrolyte solutions – what are they?

Electrolyte solutions are solutions which can conduct electricity. Colligative properties such as the lowering of the vapour pressure, depression of the freezing point, elevation of the boiling point and osmotic pressure all depend on the number of individual particles present in solution. They thus give information about the number of particles **actually** present in solution. For some solutes it is found that the number of particles **actually** present in solution is greater than would be expected from the formula of the compound.

In the study of electrolyte solutions, two types of solute can be distinguished:

- (a) Those where the number of particles present is an integral number of times the number of particles expected on the basis of the stoichiometric unit, such as

  NaCl(aq): 1 stoichiometric unit $\rightarrow$ 2 particles
  $CaCl_2$(aq): 1 stoichiometric unit $\rightarrow$ 3 particles

  and **this ratio does not change with change in concentration**.

- (b) Those, such as ethanoic acid CH_3COOH(aq), or NH_3(aq), where the number is greater than that corresponding to the stoichiometric unit, but is much less than the values found in category (a). Here the ratio of the **actual number** of particles to the **stoichiometric number** of stoichiometric units **increases dramatically with decrease in concentration**.

The electrical conductance of aqueous solutions has been studied. Some are virtually non-conducting, some weakly conducting and some are highly conducting. Conduction of a current through a solution implies the existence of charged particles, and so conducting solutions must contain charged particles – ions. The highly conducting solutions correspond to solutions appearing in category (a), while the weakly conducting solutions correspond to category (b).

X-ray diffraction studies show that some solids consist of discrete molecular units, while others are giant lattices held together by strong electrostatic interactions and with no one cation specifically belonging to any particular anion, and vice versa.

The solutes whose structure in the solid is a giant ionic lattice give strongly conducting solutions whose colligative properties place them in category (a). Colours of ionic solutions are also indicative of individual charged particles being present, for instance, copper salts are always blue, dichromate salts are orange.

The conclusion to be drawn from these studies is that in solution the solute can exist as:

- **Molecular units**: non-conducting, **normal** colligative properties, X-ray structures showing discrete molecular units in the solid.

- **Molecular units plus ions**: weakly conducting, colligative properties showing **slightly** more than the expected numbers of particles present, X-ray structure showing discrete molecular units in the solid.

- **Ions**: highly conducting, colligative properties **considerably greater** than expected, X-ray structure showing a giant ionic lattice.

Solutes giving in solution:

- **Molecular units**   are called **non-electrolytes**

- **Molecular units plus ions**   are called **weak electrolytes**

- **Ions only**   are called **strong electrolytes**

For the weak electrolytes the molecules are present in equilibrium with ions derived from the molecules:

$$\text{molecules} \rightleftharpoons n \text{ ions}$$

$$K_{\text{dissoc}} = \left(\frac{[\text{ions}]^n}{[\text{molecules}]} \right)_{\text{equilibrium}} \tag{1.1}$$

The concept and meaning of equilibrium along with the equilibrium constant are discussed in Chapter 2.

For a **weak electrolyte** which is only partially dissociated into ions at moderate concentrations there is a dramatic increase in the fraction of ions present as the concentration decreases. This shows up directly in the dramatic increase in the molar conductivity as the concentration decreases.

A **strong electrolyte** consists of ions with no significant amounts of molecular species present. The molar conductivity should be independent of concentration, but is not. This will be discussed in detail in Chapter 12.

With extensive study it was soon found that anomalies existed, for instance:

- for strong electrolytes, the molar conductivity is not independent of concentration;

- electrolytes which are strong in aqueous solution are shown to behave like typical weak electrolytes in solvents such as dioxan, acetone, or methanol;

- these anomalies are reflected in other studies using strong and weak electrolytes.

This forced chemists into focusing their attention on two main conceptual points:

- What exactly is an electrolyte solution like, and how does it behave at the **molecular** level? In particular, because ions are charged particles, do electrostatic interactions play a part in the **observable** behaviour of electrolyte solutions?

- What exactly is the role of the solvent, and should its **molecular** properties be included as well as its **bulk** properties?

Attempting to answer these two fundamental questions led to considerable advances in the theory of electrolyte solutions, and in experimental methods with which to quantify and test these theories.

1.2   Ions – simple charged particles or not?

Some simple basic properties of ions are regularly used in the discussion of electrolyte solutions and in theories describing the behaviour of electrolyte solutions. These ideas are often physically naïve and must be modified before a physically realistic description of electrolyte solutions can be given.

These properties are summarised below with indications of which are naïve. A very brief indication of which modifications will be needed is given after the summary. Full discussion of these points is dealt with in later chapters.

1.2.1   Simple properties of ions

- Ions have integral positive or negative charges.

- Ions have finite definite sizes – **but** see discussion below of solvation.

- Ions are often considered to be spherically symmetrical – **but** see discussion below of shapes of ions.

- The charge is normally considered to be evenly distributed over the surface of the ion – **but** see discussion below of charge-separated ions.

- Ions are considered to be unpolarisable – **but** see discussion below of polarising power and polarisability.

- Each ion moves as an independent entity – **but** see discussion below of ion pairing and micelle clustering.

- Ions can be ordered in terms of their ease of discharge at an electrode.

- $H_3O^+(aq)$ and $OH^-(aq)$ show special properties.

1.2.2 Modifications needed to these simple ideas: a summary

Sizes

Sizes of ions in solids are found by X-ray crystallography and are termed 'crystallographic radii'. These radii are often used in discussions of properties of electrolyte solutions. But they really should not be used for ions in solution. Many studies indicate that bare ions rarely exist in solution, and their effective size is a combination of the crystallographic radii plus a contribution from solvation effects. Electrochemical experiments can yield solvation numbers, but the main evidence comes from other studies which will be discussed in Chapter 13. Knowledge about solvation is vital to the understanding of the behaviour of electrolyte solutions and has proved to be of crucial importance in determining the behaviour of biological systems. Because of the considerable current interest in this topic a full chapter will be devoted to solvation.

Shapes

Because so much of the theoretical discussions of electrolyte solutions is based on the assumption that ions are spherically symmetrical, there is the tendency to forget that many ions are certainly not symmetrical. A conscious effort should be made to think about the shape of an ion as well as its charge.

Ions such as Ca^{2+}, Mg^{2+}, Cl^-, SO_4^{2-}, PO_4^{3-}, are nearly spherically symmetrical, as are many of the complex ions found in inorganic chemistry:

$$Co(NH_3)_6^{2+} \qquad Fe(CN)_6^{4-}$$
$$Ni(CH_2(COO)_2)_2^{2-} \qquad AlCl_6^{3-}$$

but many organic ions are not, e.g.:

$$^+NH_3CH_2CH_2COOCH_2CH_2C_6H_5CHBrCHBrCOO^-$$

while complex ions, like those which are often found in biologically active solutions, generally are non-spherical.

Modern theories of electrolyte solutions using statistical mechanical ideas are now able to take cognisance of the shapes of ions (see Sections 10.17.3 and 10.19).

Distribution of charge on an ion

Even distribution of charge over the surface of an ion is a valid assumption for simple spherical ions such as Ca^{2+}, SO_4^{2-}, $AlCl_6^{3-}$, but is most probably invalid for ions such as

$$(CH_3)_3N^+CH_2CH_2CH_2CH_3 \qquad CH_3CH_2CH_2COO^-$$

Furthermore, many organic ions, often associated with biologically important substances, have a total charge which is a multiple of one, but where this charge is made up of individual charges at different sites in the molecule, as in

$$(CH_3)_3N^+CH_2CH_2N^+(CH_3)_3 \qquad SO_3^-CH_2CH_2CH_2SO_3^-$$
$$^+NH_3CH_2COO^- \qquad\qquad\quad ^+NH_3(CH_2CONH)_nCH_2COO^-$$
$$SO_3^-CH_2CH_2COO^- \qquad\qquad proteins$$

polyphosphates   nucleotides

polysaccharides

It is quite obvious that even distribution of charge over these ions simply will not occur, but a further important factor must also be considered. Do these substances behave as ions with a given overall net charge, and can they be treated as though they were equivalent to simple ions such as Ca^{2+}, SO_4^{2-}, or as though each individual charge simulated a separate individual ion? Evidence suggests that in certain solvents and above certain concentrations, effects such as these are highly pertinent to the understanding of the behaviour of these electrolytes.

The behaviour of a charge-separated ion is likely to be different from that of simple ions. Ideas and theories developed for the simple cases will have to be modified for these situations.

Unpolarisable ions

It is clearly untrue to consider ions to be unpolarisable, but theoretical treatments generally discuss ions as though they were unpolarisable.

However, even the simple I^- ion is highly polarisable. This is a well accepted fact in other branches of chemistry. Fajans' Rules in inorganic chemistry deal explicitly with this effect in bonding. In summary:

- small highly charged cations have a strong polarising power because of the intense field around them and can thus have a very strong effect in producing induced dipoles or aligning permanent dipoles in other ions;

and

- large anions with the charge dispersed over a large volume can be very easily polarised by highly polarising cations resulting in induced dipoles in the anion, or alignment of permanent dipoles if present.

These effects are present in the simple ions of standard substances such as Cs^+, Ca^{2+}, Cl^-, SO_4^{2-}, NO_3^-, CH_3COO^- and $(CH_3)_4N^+$. They are also going to be of considerable importance in electrolyte solutions where many of the ions are large and complex, for example protein ions, phospholipid ions, nucleic acids, ions of neurotransmitters and so on. Dipoles, induced dipoles and alignment of dipoles are discussed later (see Sections 1.5, 1.7.2 and 1.7.3).

Complete dissociation into ions

One of the main functions of the solvent is simply to reduce the forces of interaction between ions and thereby reduce the electrostatic potential energy of this interaction. Physically this corresponds to allowing the ions to exist as ions. This is a bulk effect.

Experiments and theoretical considerations, which will be described later, show that:

(a) Increasing the concentration of the solution increases the electrostatic interaction energy between the ions.

(b) Solvents of low polarity result in a greater electrostatic interaction energy between the ions compared with the electrostatic interaction energy between the ions when they are in solvents of high polarity.

There can be situations where the energy of interaction between two ions of opposite charge becomes so high that the ions cease to be independent of each other, and move around as a single unit which survives throughout several collisions before being able to separate. Such a unit is called an **ion pair**. Ion pairs are found in aqueous solutions as well as in low polarity solvents.

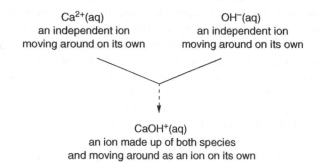

Ca^{2+}(aq)
an independent ion
moving around on its own

OH^-(aq)
an independent ion
moving around on its own

$CaOH^+$(aq)
an ion made up of both species
and moving around as an ion on its own

This is a very important topic in electrolyte studies and will be referred to often throughout this book.

Ions can also be formed into clusters called **micelles**, and this can become very important in colloidal solutions.

1.3 The solvent: structureless or not?

The solvent is the medium in which the solute exists. It is often called a dielectric. A dielectric can be thought of in terms of an insulator, which is a substance which stops or tends to stop the flow of charge, in other words to stop a current passing through it.

If a substance which acts as an insulator is placed between two charges, it reduces:

a) the field strength;

b) the force acting between the charges;

c) the electrostatic potential energy between the two charges

and the factor by which it reduces these quantities is the relative permittivity, ε_r.

But pay particular attention: what is important is that this definition of the relative permittivity is independent of any assumption that the dielectric is composed of atoms or molecules, and so requires no discussion of the medium at the microscopic level.

In effect, the relative permittivity is just a constant of proportionality characteristic of the medium.

This is precisely what is meant when the medium is described as a structureless dielectric or continuum. In particular, when discussing the role of the solvent in electrolyte solutions it is often described as a continuous medium, or a structureless medium. Most of the theoretical discussions of electrolyte solutions formulate the theoretical equations in terms of factors which involve the macroscopic quantity, ε_r. Use of this bulk quantity in the equations implicitly means a description in terms of the solvent being a structureless dielectric, with no microscopic or molecular structure.

However, the use of a macroscopic quantity in the equations does not preclude a discussion of whether or not it is reasonable, or indeed sensible, to consider the solvent as having a purely bulk macroscopic role to play. Other chemical studies show that the solvent is made up of molecules with a certain microscopic structure, so it is perfectly reasonable to expect that the microscopic structure may be of vital importance when the solvent plays its role as solvent in electrolyte solutions. Indeed, it is precisely by addressing this question that vast progress can be made in the understanding of electrolyte solutions. It is now realised that molecular details of ion–solvent interactions and consequent modifications to ion–ion and solvent–solvent interactions make a significant contribution to the behaviour of electrolyte solutions.

Studies of the microscopic structure of the solvent and its modification by the ions of the electrolyte have resulted in considerable refinements being forced onto the simple model of electrolyte solutions. Unfortunately, it is much easier to alter the model to incorporate new ideas and thought, than it is to incorporate these ideas into the mathematical framework of the theory of electrolyte solutions and its derivation. The implications of many of the topics introduced in this chapter become important in the theoretical treatments of electrolyte solutions (see Chapters 10 and 12, and for solvation, see Chapter 13).

1.4   The medium: its structure and the effect of ions on this structure

The question can be asked:

> **Does the fact that the medium (solvent) reduces the effect of one charge on another charge mean that the charges on the ions must have some effect on the solvent?**

To answer this question the details of the molecular structure of the solvent become important. At this stage some important ideas, developed in more detail in the chapter on solvation, will be introduced briefly.

The solvent is made up of molecules which are in turn made up of nuclei and electrons. In covalently bonded molecules the bonds are formed by sharing of electrons between the two atoms involved in the bond.

- A bond is termed **non-polar** if the electron distribution in the bond is **symmetrical**, i.e. the atoms have an **equal** share of the electrons of the bond.

Examples are: Cl–Cl;   H–H;   N≡N;   C–C

- A bond is termed **polar** if the electron distribution in the bond is **asymmetric**, i.e. the atoms have an **unequal** share of the electrons of the bond.

Examples are:

$$C-Cl \quad \text{written as} \quad \overset{\rightarrow}{C-Cl} \quad \text{or} \quad C^{\delta+}-Cl^{\delta-}$$

$$C-O \quad \text{written as} \quad \overset{\rightarrow}{C-O} \quad \text{or} \quad C^{\delta+}-O^{\delta-}$$

$$H-F \quad \text{written as} \quad \overset{\rightarrow}{H-F} \quad \text{or} \quad H^{\delta+}-F^{\delta-}$$

$$S-O \quad \text{written as} \quad \overset{\rightarrow}{S-O} \quad \text{or} \quad S^{\delta+}-O^{\delta-}$$

- When the bond is polar it has a **bond–dipole moment** which, on a qualitative basis, describes the degree to which the electron distribution is asymmetric. The direction in which the electron density is highest is shown by the head of an arrow, e.g. $\overset{\rightarrow}{C-Cl}$ and an indication of the relative displacement of +ve and −ve charges is given as $C^{\delta+}-Cl^{\delta-}$

- But does this tell us anything about the polarity of the molecule **as a whole?**

Answering this question means looking at the overall effect of the bond dipoles, and this requires knowledge of the symmetry of the arrangement of the atoms in the molecule.

- If **all the bonds** in the molecule are **non-polar**, or **virtually non-polar** (as in C—H), then the molecule as a whole will be **non-polar**;
  for example H_2, N_2, O_2, S_8, alkanes, benzene.

- But if **some of the bonds** are **polar**, then the molecule as such will be:

  - **non-polar** if the arrangement of the atoms is **symmetrical** and the vector bond-dipole moments cancel out, as in CCl_4.

  - **polar** if the arrangement of the atoms is **not symmetrical** enough and the vector bond-dipole moments do not cancel out, as in CH_3Cl, H_2O, CH_3OH.

1.5 How can these ideas help in understanding what might happen when an ion is put into a solvent?

Putting a charge (ion) into a
medium (solvent) results in:

partial or complete alignment
(dielectric saturation) of the
permanent dipoles if the
solvent molecule is polar.

displacement of the electrical centres
of the molecule if the solvent
molecule is non-polar, resulting in an
induced dipole.

This effect is greater the nearer the solvent molecule is to the ion, and the higher the concentration of the solution.

This effect is greater the more intense the electric field experienced by the solvent molecule, that is, the more highly polarising the ion is.

- **if the molecule is polar**: alignment of the permanent dipoles is the major effect, but polarisability effects must not be ignored;

- **if the molecule is non-polar**: the only effect comes from induced dipoles and this can be of crucial importance.

Both these effects are of great importance for ion–solvent interactions in solution. In particular, their existence implies that ions **do modify** solvent structure. This, in turn, implies that there must be **consequent modification** of solute–solute interactions in particular, and also of solvent–solvent interactions though these are of relatively less importance.

The relative permittivity, ε_r, of the solvent allows an assessment of the effect of the solvent on the strength of the ion–ion interactions to be made. Coulomb's Law (see Section 10.4.2) states that:

the force of electrostatic interaction between two ions in a solution, $f = \dfrac{z_1 z_2 e^2}{4\pi\varepsilon_0\varepsilon_r r^2}$

so that the larger the relative permittivity, ε_r, the smaller the force of interaction. The value of the relative permittivity, in principle, is measured by observing the effect of an external field on the solution. However, in practice, this experiment is difficult to carry out because it will be complicated by the effect of the external field being predominantly to cause conduction of a current through the solution (see Chapter 12).

Nonetheless it is still possible to talk about the effect of the field **on the solvent**. The ions in a solution can align the permanent dipoles of the solvent and can also induce dipoles in the solvent. The more the dipoles are aligned by the ions the smaller will be the value of the relative permittivity. If the ions have totally aligned all the dipoles in the solvent molecules, there will be none left for the external field to align. The net result is that the only effect that the external field can have will be to induce further dipoles – a much smaller effect. Hence the measured relative permittivity will be low for such a situation.

Furthermore, the possibility of alignment of the dipoles of the solvent molecules by the ions leads to the following conclusion. For the solution, the simplest model to be envisaged is one where three possible situations can be thought of. And for each situation a question can be asked:

- There is a region close to the ion where all the permanent dipoles are **completely aligned** – the region of dielectric saturation.

    Question: Can it possibly be valid to use the macroscopic relative permittivity in the theoretical treatments?

- There is a region where **partial alignment** occurs, and the situation changes through the region from **complete alignment** to one where there is **non-alignment**.

Question: What value should be assigned to the relative permittivity here; should it vary throughout the region?

● There is a region where almost **no alignment** occurs and where the solvent behaviour approximates to that of the pure solvent.

Question: Is it legitimate to use the bulk relative permittivity of the pure solvent here? – the answer here is probably 'yes'.

From this very basic and elementary discussion it should be abundantly clear that the solvent plays a crucial role in the behaviour of electrolyte solutions, quite apart from its role as a dielectric reducing the forces of interaction between ions. Such considerations are of vital importance in physical chemistry and have only been tackled rigorously in the past three decades or so. Where complex electrolytes such as are encountered in biological chemistry are concerned, they are of crucial importance and may well dominate the behaviour of such solutions.

1.6   Electrostriction

In the vicinity of each ion, a certain shrinkage of the solvent is likely to occur as a result of the attraction between the ionic charge and the polar molecules. This is called **electrostriction**, and leads to a local increase in the density of the solvent around each ion since more molecules will be packed around the ion than would be present in that volume were the ion not present.

Electrostriction is important in solvation, but has not ever been properly incorporated in any detail into electrolyte theory.

1.7   Ideal and non-ideal solutions – what are they?

Ideality is a concept which can be used for a pure substance **only if** the interactions between the particles are **negligible** as in a gas at low pressures. Since pure liquids and pure solids are condensed phases, there must be **significant** forces of interaction between their fundamental particles. It is, therefore, meaningless to talk about ideality for either the pure liquid or pure solid. But **ideality and non-ideality** are important when talking about **mixtures**.

A two component mixture can alter in composition from the situation where the amount of component A is zero to one where the amount of component B is zero. In between these two limiting situations there are mixtures with varying proportions of the two components. A mixture where the amount of **one** component **tends to zero** corresponds to the **ideal** mixture. Mixtures where there are **finite** amounts of **both** components present correspond to **non-ideal** mixtures.

Electrolyte solutions are mixtures where the components are the solute and the solvent. When the concentration of the solute **tends to zero,** called infinite dilution, the solution is regarded as **ideal**. When there are **finite** concentrations of solute the solution is regarded as **non-ideal**.

In electrolyte solutions there will be interactions between:

- solute–solute particles

- solute–solvent particles

- solvent–solvent particles

In electrolyte solutions the solute is partly or wholly in the form of ions in solution. Because ions are charged particles, there will be electrostatic interactions between the ions, and between the ions and the solvent, over and above the solvent–solvent interactions.

1.7.1   Solute–Solute interactions, i.e. ion–ion interactions

These are basically made up of five contributions:

- **Long-range coulombic interactions**, i.e. those acting over long distances between the ions. These are electrostatic interactions obeying the Coulomb inverse square law (Section 10.4.2).

$$\text{force} \propto r^{-2}$$

- **Ion-induced dipole interactions** if the ions are polarisable, as indeed most ions are (see Section 1.7.3 for a discussion of induced dipoles).

- **Short-range attractions**, i.e. those acting over short distances. These can be coulombic or non-coulombic in nature.

- **Short-range repulsions** which become significant at short distances between the ions. They are always present when two particles approach close to each other.

- **Hard spheres** where the short-range repulsive force becomes infinite when the ions come into contact. It will be seen later (Chapters 10 and 12), that the term 'into contact' is not unambiguous and is open to several interpretations.

1.7.2   Solute–solvent interactions, i.e. ion–solvent interactions – collectively known as solvation

These interactions are made up of:

- **Ion–dipole interactions**.

- **Ion-induced dipole interactions**.

(see Section 1.5 and Section 1.7.3 below for a description of these effects.)

These interactions are **attractive** when the interaction is between the end of the dipole which has an opposite charge to that of the ion with which the dipole or induced dipole is interacting.

They are **repulsive** when the ion is in close enough proximity to the end of the dipole which has a charge the same as the ion.

1.7.3 Solvent–solvent interactions

These are as described below, and are the same as are present in a pure liquid, though it must always be remembered that they may well be **altered** or **modified** as a result of the presence of the solute.

The interactions present are:

- **attractions** between solvent molecules;

- **repulsions** between solvent molecules.

The **attractions** for the solvent molecules are mainly associated with permanent or induced dipoles in the molecules.

- **dipole–dipole interactions**: dipoles in one molecule can interact with dipoles in another molecule;

- **dipole-induced dipole interactions**: dipoles in one molecule can interact with another molecule to produce an induced dipole;

- **induced dipole-induced dipole interactions or dispersion forces**: induced dipoles in one molecule can interact with induced dipoles in another molecule;

- **dipole–quadrupole or higher effects**: dipoles in one molecule can interact with quadrupoles in other molecules.

A quadrupole is a distribution of charge more complicated than a dipole. It can be exemplified as:

$$\delta^- \qquad \delta^+ \qquad \delta^+ \qquad \delta^-$$

$$\underline{\qquad\qquad} \qquad \underline{\qquad\qquad}$$

The molecule CO_2 has a quadrupole of this nature

$$\delta^- \qquad\qquad \delta^+\delta^+ \qquad\qquad \delta^-$$

$$O\!=\!=\!=\!=\!C\!=\!=\!=\!=\!O$$

The **repulsions** for solvent molecules are mainly a consequence of the molecules getting closer together, and finite sizes causing repulsions:

a) quantum mechanical repulsion associated with induced dipole-induced dipole repulsions;

b) permanent dipole–permanent dipole repulsion arising when the geometrical orientation of the molecules in space brings like charge ends of dipoles in each molecule close together.

1.8 The ideal electrolyte solution

In the ideal electrolyte solutions all three interactions are present:

- ion–ion interactions;

- ion–solvent interactions; and

- solvent–solvent interactions,

and the **ideal** electrolyte is defined as the infinitely dilute solution where the **concentration of the solute**, (i.e. the ions) $\longrightarrow$ **zero**.
Under these conditions:

- solvent–solvent interactions are significant, and are similar to those described in Section 1.7.3 above;

- solute–solvent interactions are present, but are considered to be of less significance than solvent–solvent interactions because the ratio of solute to solvent is very low;

- solute–solute interactions are present, but are considered to be relatively small because of the low concentration of solute.

Repulsions are present, and will be significant for the solvent–solvent interactions, but are considered negligible for solute–solute and solute–solvent interactions because the concentration of solute $\longrightarrow$ zero.

1.9 The non-ideal electrolyte solution

Physically this corresponds to all concentrations of solute, ions, other than that of infinite dilution where the concentration of the ions $\longrightarrow$ zero. What contributes to non-ideality can best be explained by considering in turn what happens, as the concentration of solute increases, to the three types of interaction discussed already.

- Non-ideality corresponds to all **ion–ion** interactions **over and above** those considered to be present in the ideal solution. These are any **modified** ion–ion interactions resulting from increase in the solute concentration. This is the major factor giving rise to non-ideality in electrolyte solutions. There will be **three** ion–ion interactions:

- cation and anion interactions,

- cation and cation interactions,

- anion and anion interactions.

There are a large number of modified interactions which can be considered as contributing to the non-ideality of the electrolyte solution. All of them result in increasing non-ideality as the solute concentration increases. They will be discussed in Chapters 10 and 12.

- Non-ideality also corresponds to all **ion–solvent** interactions which are **over and above** those considered to be present in the ideal solution, i.e. any **modified** ion–solvent interactions resulting from the increase in solute concentration. These interactions become more important at **high** concentration, and the contribution of ion–solvent interactions to non-ideality in electrolyte solutions becomes more important at high concentrations. An ion can interact with the solvent and can modify the solvent around it, or two ions could modify the solvent in between them. This would correspond to an ion–solvent interaction different from the ideal case and would lead to non-ideality which would increase as the solute concentration increases. It would also lead to modified solvent–solvent interactions.

- Non-ideality also corresponds to all **solvent–solvent** interactions which are **over and above** those considered to be present in the ideal solution, i.e. any **modified** solvent–solvent interactions resulting from the presence of the solute (cation and anion) at concentrations greater than infinite dilution. However, this is a relatively minor effect and of less importance than contributions to non-ideality which are discussed above. For instance, the cation or anion, or both, could disturb the solvent structure present in the pure solvent, and this in turn would lead to modified solvent–solvent interactions which would then be considered as non-ideal. This modification would become increasingly greater as the solute concentration increases, giving rise to increasing non-ideality.

1.10 Macroscopic manifestation of non-ideality

The **actual** electrostatic potential energy of a real electrolyte solution is a sum of all these possible interactions. Each one makes its own contribution to the total electrostatic potential energy, and each contribution has a different weighting depending on the physical situation being considered. The **ideal** electrostatic potential energy is the sum of all interactions present in the ideal solution where the concentration of solute tends to zero. The difference between the **real actual** potential energy and the **ideal** value represents the **non-ideal** electrostatic potential energy. This can be shown to be equivalent to the **non-ideal** part of the total free energy, G, of the solution, and this non-ideal part of G is often termed the **excess free energy**,

G^E, of the solution (see Chapter 10, Section 10.6.15). This, in turn, can be considered in the simple primitive treatment of electrolyte solutions to manifest itself as an **activity coefficient** $\gamma_\pm$, or in a less primitive version as:

$$\gamma_\pm \quad \text{plus a solvation term}.$$

In modified treatments the excess free energy can be considered to manifest itself as:

$$\gamma_\pm \quad \text{plus a solvation term plus an ion pair association term}$$

and in more advanced treatments the effects of shape, charge distribution, polarising power and polarisability will come in.

The term **mean** activity coefficient is used here, rather than an individual activity coefficient for the cation and the anion, and is defined in Sections 8.22 and 8.23 and Worked Problems 8.13 and 8.14 and below.

The activity coefficient is defined in terms of the activity and the concentration (see Section 8.21.1).

$$a = \gamma c \tag{1.2}$$

Both activity and concentration are experimental quantities. It is possible to **talk** about the activity and the activity coefficient of each type of ion making up the electrolyte, but the activity and the activity coefficient of the **individual** ions **cannot** be measured experimentally. Activity is a property of the electrolyte solution as a whole. Hence the use of **mean** activities and **mean** activity coefficients. The term mean is **not** used in its common sense of an average quantity, but is used in a rather different sense which reflects the number of ions which result from each given formula:

$$NaCl \longrightarrow 2 \text{ ions}$$
$$CaCl_2 \longrightarrow 3 \text{ ions}$$

- for symmetrical electrolytes: $AB(s) \rightarrow A^{x+}(aq) + B^{x-}(aq)$

$$\gamma_\pm = \{\gamma_{A^{x+}}\gamma_{B^{x-}}\}^{1/2} \tag{1.3}$$

e.g. for NaCl:

$$\gamma_\pm^2(NaCl) = \gamma_{Na^+}\gamma_{Cl^-} \tag{1.4}$$
$$\gamma_\pm(NaCl) = \sqrt[2]{\gamma_{Na^+}\gamma_{Cl^-}} \tag{1.5}$$

- for unsymmetrical electrolytes: $A_xB_y(s) \rightarrow xA^{y+}(aq) + yB^{x-}(aq)$

$$\gamma_\pm = \{(\gamma_{A^{y+}})^x(\gamma_{B^{x-}})^y\}^{1/(x+y)} \tag{1.6}$$

and for $CaCl_2$:

$$\gamma_\pm^3(CaCl_2) = \gamma_{Ca^{2+}}\gamma_{Cl^-}^2 \tag{1.7}$$
$$\gamma_\pm(CaCl_2) = \sqrt[3]{\gamma_{Ca^{2+}}\gamma_{Cl^-}^2} \tag{1.8}$$

This will become much clearer once the thermodynamic description of activity, activity coefficient and mean activity coefficient has been developed (see Sections 8.21 to 8.23).

1.11 Species present in solution

In a solution the solvent is the most abundant species. The solute can exist in various possible forms, but only the following will be discussed.

- Free ions

- Undissociated molecules in equilibrium with free ions

- Ion pairs in equilibrium with free ions and other complexes and chelates

- Complexes and chelates in equilibrium with free ions and other complexes and chelates

- Micelle clusters in equilibrium with free ions

The last four situations are equilibria and an equilibrium constant describes each equilibrium (see Chapter 2). Describing the equilibrium means postulating which species are involved in the equilibrium, which, in turn, requires that:

- it is possible actually to observe the species present by chemical and physical means; or

- chemical judgement suggests that the species do actually exist.

1.12 Formation of ion pairs from free ions

An ion pair is a physical entity with no specific chemical interactions between the ions. The ions of the ion pair move together as a single unit and are held together by electrostatic forces of the coulomb type acting over the short distances that the ions are apart in the ion pair. These coulombic forces impose a certain degree of cohesion on the unit and this is sufficiently great to overcome the tendency for normal thermal motion to cause the ions to move around as separate particles each with its own translational degrees of freedom.

Because the forces holding the ions together are of this physical nature, they depend on three factors:

- the charges on the ions;

- the sizes of the ions;

- the relative permittivity of the solvent in the vicinity of the ion.

It is very important to realise that these forces are therefore **independent** of the **chemical nature** of the ions. On this basis, it would be expected that electrolytes which have ions of the same charge and are of similar size would have equilibrium constants of similar magnitude **if** the associated species is an ion pair. This is found for some 2:2 sulphates, and for some cations of similar size which associate with $Cl^-(aq)$ or $I^-(aq)$.

1.12.1 Charge distribution on the free ion and the ion pair

- The **charge** on a **simple** ion is usually unambiguous, e.g. $Mg^{2+}(aq)$ or $NO_3^-(aq)$, and the charge distribution for such ions is probably approximately spherically symmetrical. But the charge and the charge distribution for some ions may not be so clearly defined, for instance:

$$CH_3CH_2CH_2CH_2NH_3^+(aq)$$
$$(CH_3)_3N^+CH_2CH_2CH_2COO^-(aq)$$

- The **charge** on an **ion pair** is the algebraic sum of the charges on the individual ions:

  - a $+2$ cation with a -2 anion gives an ion pair with an **overall** charge of zero, while

  - a $+3$ cation with a -1 anion gives an ion pair with an **overall** charge of $+2$.

However, it is imperative to think more deeply than this.

An ion pair of **zero overall** charge must **not** be treated as though it were a **neutral molecule**. At best it can be regarded as a dipolar molecule, but it is probably more like a charge-separated ion.

- a dipolar ion pair

$$Mg^{2+}(aq) + SO_4^{2-}(aq) \xrightarrow{\longrightarrow} (Mg\overset{\longrightarrow}{SO_4})(aq)$$

- a charge-separated ion pair

$$Mg^{2+}(aq) + SO_4^{2-}(aq) \rightleftharpoons (Mg^{2+}SO_4^{2-})(aq)$$

There are important implications arising from this.

If the ion pair is uncharged:

- Interionic interactions between the ion pair and free ions and other ion pairs will be set up and contribute to the non-ideality of the solution.

- The ion pair should be given a mean activity coefficient different from unity. This is rarely done, but see Section 12.17.

- The ion pair will probably be able to conduct a current, though its contribution will be small compared with that from the free ion. It is generally assumed to make a zero contribution, but see Section 12.17.

If the ion pair has an overall charge:

- It should not be treated as though it were a single charge with a spherically symmetrical distribution of charge.

- It is probably more like a dipolar charged ion, e.g. $CaOH^+(aq)$ or a charge-separated ion pair $(Ca^{2+}OH^-)(aq)$.

- For an overall charged ion pair, a mean ionic activity coefficient must always be assigned to it, generally calculated from the Debye-Hückel theory on the basis of the overall charge.

- It will also have a non-zero molar conductivity, though in practice the magnitude is very difficult to assess.

1.12.2 Size of an ion and an ion pair in solution

Most size correlations for free ions have used crystallographic radii which represent the bare ion. But there is no doubt that most ions are solvated in solution, though it is difficult to assess precisely the extent of solvation, and hence the size of the solvated ion.

Likewise, the ion pair will be solvated, and some estimate of its size is required. Furthermore, the change in solvation pattern on forming the ion pair is of crucial importance.

Three limiting situations can be envisaged, but other intermediate situations are possible:

- An ion pair is formed with no disruption of the individual solvation sheaths of the individual ions, so that in the ion pair these solvation sheaths of the individual ions are in contact and solvent is present between the ions (see Figure 1.1(a)).

- An ion pair is formed with total disruption of the individual solvation sheaths of the individual ions so that in the ion pair the bare ions are in contact and there is no solvent between the ions (see Figure 1.1(b)).

- An ion pair is formed with the partial disruption of the individual solvation sheaths of the individual ions so that in the ion pairs some solvent has been squeezed out, but there is still some solvent present between the ions (see Figure 1.1(c)).

The ion pairs which can be found are thus not necessarily identical, and there is the possibility that different experimental methods may pick out and detect only one kind of ion pair, for instance, detect contact ion pairs but not solvent separated ion pairs.

A further formal definition commonly used in inorganic chemistry can be proposed:

- an **outer-sphere** ion pair is one where one, or at most two, solvent molecules lie between the ions.

- an **inner-sphere** ion pair is one where the bare ions are in contact – all solvent sheaths have been eliminated from between the ions.

However, both inner and outer-sphere ion pairs are still solvated as the **composite** unit, as are the ion pairs described as contact or solvent separated.

Although the **definitions** can be quite unambiguous, **experimental** classification into contact or solvent separated, or inner and outer sphere ion pairs most certainly is not unambiguous, and may even, at best, be only a guess. This is exactly the same problem as is encountered when discussing the formal and experimental distinctions between complexes and ion pairs.

Attempts to distinguish experimentally between the formation of:

   a)   ion pairs from free ions, and

   b)   complexes and chelates from free ions

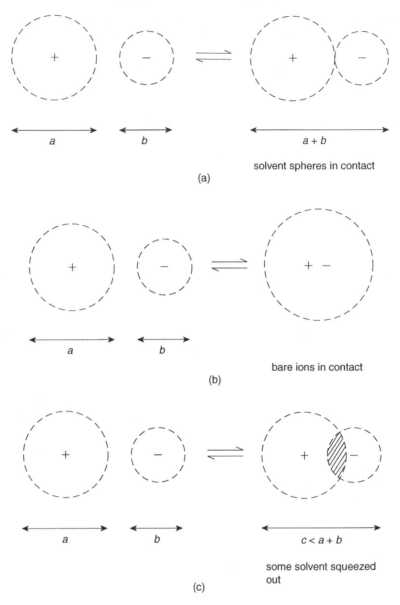

Figure 1.1   (**a**) Solvent sheaths of individual ions in contact; (**b**) bare ions in contact; (**c**) partial disruption of the solvation sheaths of ions showing some solvent being squeezed out.

often end up in deep water, and considerable scepticism must be cast on the interpretation of many experiments which purport to make a clean-cut distinction. Also, different experimental methods may pick up behaviour typical of different types of associated species, so that a comparison of results from different methods may add to the confusion. It is easy to give definitions, but more difficult to decide what sort of species is present.

1.13 Complexes from free ions

Where a complex is formed there is an intimate chemical interaction between the ions. Some electronic rearrangement is occurring resulting in covalent interactions, in contrast to the physical coulombic electrostatic interactions involved in the formation of an ion pair.

If complexes and chelates involve intimate chemical interactions, the extent of association should reflect the chemical nature of the ions involved. Equilibrium constants should be different, and possibly even very grossly different, for equilibria which superficially seem very similar and alike, for instance, association of one species with ions of similar size and charge. The situation is reminiscent of that found for the dissociation constants of weak acids and bases where the magnitude of the equilibrium constants depends on the chemical nature of the species involved. It is in stark contrast to that expected for the formation of ion pairs, where the magnitude of the association constant is expected to be independent of the chemical nature of the ions involved.

The metal ions:

$$Cu^{2+}(aq), \quad Ni^{2+}(aq), \quad Co^{2+}(aq), \quad Zn^{2+}(aq), \quad Mn^{2+}(aq)$$

have crystallographic radii which are very similar, and they all have the same charge, $+2$. When they interact with oxalate

oxalate

and with glycinate

glycinate

a wide variation in the association constants is found. The glycinates, for instance, have values ranging from 2.75×10^3 to 4.2×10^8 mol^{-1} dm^3. Complexes are assumed to be formed.

On the other hand, it is believed that the interaction of the same metal ions with $SO_4^{2-}(aq)$ results in an ion pair. Here the association constants are very similar, ranging from 1.9×10^2 to 2.9×10^2 mol^{-1} dm^3.

1.14 Complexes from ions and uncharged ligands

If an associated species is formed between an ion and an **uncharged** ligand, it is usually assumed that a complex is formed and that electronic rearrangements of a chemical nature have occurred.

When aqueous NH_3 is added to an aqueous solution containing Cu^{2+} ions an intense blue coloration indicative of electronic rearrangement occurs, and the main species formed is the complex:

$$Cu^{2+}(aq) + 4NH_3(aq) \rightleftharpoons Cu(NH_3)_4^{2+}(aq)$$

Formation of an intimate chemical species implies a fairly drastic alteration in the solvation sphere around the Cu^{2+}(aq), with the NH_3 ligands displacing the solvent molecules from around the ion.

1.15   Chelates from free ions

If the ligand ion is a simple ion, such as I^-(aq) or OH^-(aq), then there is only one possible point of attachment irrespective of whether an ion pair or a complex is formed.

   If a ligand has two charges located in different parts of the ion, then there are two possible points of attachment of the cation, and if the interactions are covalent the associated species is called a **chelate**:

$$Cd^{2+}(aq) + \quad \begin{array}{c} O \diagdown \quad CH_2 \quad \diagup O \\ C \qquad C \\ | \qquad | \quad (aq) \\ O^- \qquad O^- \end{array} \rightleftharpoons \quad \begin{array}{c} O \diagdown \quad CH_2 \quad \diagup O \\ C \qquad C \\ | \qquad | \quad (aq) \\ O^- \quad {}_{Cd}^{2+} \quad O^- \end{array}$$

   Ligands can be ions as in the example above, but they can also be ions having neutral points of attachment within the molecule, for instance amino acid anions such as glycinates. Does the metal ion then simply ion pair with the carboxylate part, or does the amino part also become involved to form a chelate?, i.e. is the product:

- an ion pair

$$Cu^{2+}(aq) + NH_2CH_2COO^-(aq) \rightleftharpoons NH_2CH_2COO^-Cu^{2+}(aq)$$

or

- a chelate?

$$Cu^{2+}(aq) + NH_2CH_2COO^-(aq) \rightleftharpoons \begin{array}{c} CH_2-C=O(aq) \\ | \qquad | \\ NH_2 \quad O^- \\ \diagdown \quad \diagup \\ Cu^{2+} \end{array}$$

Evidence, mainly spectroscopic, suggests that the associated species is a chelate in this case.

   Ligands can also be neutral, and act by virtue of their lone pairs, e.g. the standard chelating agent $NH_2CH_2CH_2NH_2$.

1.16   Micelle formation from free ions

Here clustering of ions of like charge occurs to give a cluster of colloidal size. Unambiguous detection of micelle formation is fairly easy experimentally because:

- Formation gives clusters of such a size as to be detected by standard techniques for colloidal solutions such as Tyndall's beam effect where scattering of light by the solution occurs.

- Abrupt removal of such large numbers of ions from solution gives such a dramatic change in the properties of the solution that the effect is easily detected.

Electrolytes showing clustering properties are typified by paraffin chain salts where there is a long paraffin-like chain with a cationic group at the end, as in a quaternary ammonium group:

$$CH_3CH_2CH_2CH_2CH_2CH_2N^+(CH_3)_3$$

or a paraffin-like chain with an anionic group at the end like the carboxylates in soaps, and sulphonates in detergents:

$$CH_3CH_2CH_2CH_2CH_2CH_2COO^-$$

$$CH_3CH_2CH_2CH_2C_6H_4SO_3^{2-}$$

These electrolytes behave like a non-associated electrolyte up to a certain concentration, and then alter abruptly, with the properties changing dramatically. This is attributed to the rapid onset of micelle formation at a certain critical concentration. In the micelle, the paraffin chains face inwards with the charged groups lying on the outer surface where their charge is partially neutralised by small simple ions of opposite charge, 'counter ions', fitting into the spaces of the cluster. Overall this has an effect on the properties of the solution similar to that expected if large numbers of ions are removed from solution.

Added salts can encourage the aggregation of ions to micelles, 'salting out'. This is of considerable importance in biological electrolytes, such as bile salts and phospholipids.

Micelles and polyelectrolytes must always be clearly distinguished. Many biological electrolytes are long chain species, for instance, proteins, where the **individual** molecule or electrolyte is similar in size to the micelle. Many such polyelectrolytes have positive and negative groups occurring alternately or irregularly along the chain. The biggest difference between micelles and biological polyelectrolytes is in the relative mobility of the polymer segments of the polyelectrolyte which are often coiled but can become uncoiled on addition of electrolytes which bind to the polyelectrolyte. In contrast, the micelle can break up into its individual ions, that is, it can come to pieces, whereas the polymer does not. The electrostatic and configurational energy changes are not well understood and represent a challenge to the biologist and chemist alike. Solvation effects are also of major consideration, and have been shown to be of crucial importance.

1.17 Measuring the equilibrium constant: general considerations

This is generally given by comparing the behaviour expected for a given situation with what is actually observed.

The expected or predicted behaviour of the solution is worked out on the basis of a
postulate of the solution being made up of free ions only,
and this, in turn,
requires that there is a theoretical base-line which describes the behaviour expected
for a solution made up totally of ions,
since
without this theoretical base-line of predicted behaviour it is impossible to start to

infer from the observed experimental behaviour anything about possible species actually present if any of the processes discussed in Sections 1.13 to 1.17 occur,
because
it is precisely the deviations from this base-line which are used to infer the existence of particles other than free ions.

1.18 Base-lines for theoretical predictions about the behaviour expected for a solution consisting of free ions only, Debye-Hückel and Fuoss-Onsager theories and the use of Beer's Law

1.18.1 Debye-Hückel and Fuoss-Onsager equations

Setting up base-lines presupposes that it is known exactly how a solution consisting of free ions only would behave over a range of concentrations, and it is precisely here that problems arise. The theory behind the setting up of base-lines for the Debye-Hückel and Fuoss–Onsager theories is given in Chapters 10 and 12, and only the conclusions are summarised here.

The equations are:

$$\log_{10}\gamma_{\pm} = -A|z_1z_2|\sqrt{I} \qquad \text{Debye-Hückel limiting law} \qquad (1.9)$$

$$\log_{10}\gamma_{\pm} = \frac{-A|z_1z_2|\sqrt{I}}{1 + B\dot{a}\sqrt{I}} \qquad \text{Debye-Hückel equation} \qquad (1.10)$$

$$\log_{10}\gamma_{\pm} = \frac{-A|z_1z_2|\sqrt{I}}{1 + B\dot{a}\sqrt{I}} + bI \qquad \text{Debye-Hückel extended equation} \qquad (1.11)$$

$$\Lambda = \Lambda^0 - S\sqrt{c} \qquad \text{Fuoss-Onsager limiting law} \qquad (1.12)$$

$$\Lambda = \Lambda^0 - S\sqrt{c} + Ec\,\log_{10}c + Jc \qquad \text{Fuoss-Onsager extended equation} \qquad (1.13)$$

where:

I is the ionic strength,

Λ is the molar conductivity at finite concentrations,

Λ^0 is the molar conductivity at infinite dilution.

The constants A and B are defined in Section 10.6.16 and S, E and J are defined in Section 12.10.

In these equations, with the possible exception of Equation (1.11), all quantities on the right hand side are known or calculable, and in this respect these equations should be excellent as base-lines for predicting the behaviour of electrolyte solutions. These equations account for non-ideality resulting from long range coulombic interactions, and can be used to predict the behaviour expected for such a solution. Any deviation from this predicted behaviour is then interpreted as due to some process which is removing ions from solution. This could be incomplete ionisation for a weak acid, or formation of a complex or an ion pair or other equilibria. A comparison of the actual observed behaviour of the solution with that predicted after non-ideality has been considered often leads to a determination of an equilibrium constant for the process assumed to be occurring. This all presupposes that the equations correcting for non-ideality are valid.

These two base-lines are based on theories which are conceptually very similar, and deal directly with non-ideality resulting from coulombic interionic interactions.The third base-line is totally different conceptually. It is based on Beer's law which describes the intensity of absorption of radiation by a solution as a function of concentration. It does not deal with non-ideality, and assumes that non-ideality of the coulombic interaction type has no effect on the theoretical expression.

1.18.2   Beer's Law equation

Spectroscopic and spectrophotometric methods are a standard way of identifying equilibrium processes and ascribing equilibrium constants to them. These methods are based on absorption of radiation by the species present in solution. The technique can be described qualitatively by saying that:

- The position of the lines in the spectrum gives the nature and identity of the absorbing species.

- The intensity of the lines gives the concentration of the absorbing species.

The intensity and the concentration are related by Beer's law which is the base-line for electrolyte studies by spectroscopic techniques.

$$\log_{10} \frac{I_0}{I} = A = \varepsilon c d \tag{1.14}$$

where I_0 is the incident intensity,

I is the intensity of the radiation transmitted,

A is the absorbance of the solution, and is directly measured by the instrument,

ε is a constant of proportionality – the molar absorption coefficient – and this depends on the nature of the absorbing species,

c is the concentration of the absorbing species,

d is the path length through which radiation is transmitted.

A and ε depend on the wavelength of the radiation used.

It is generally accepted that Beer's law is valid over most experimental concentrations. Deviations from the predicted values are interpreted as due to removal of the absorbing species by some equilibrium process, and a value for the equilibrium constant can be calculated.

Spectroscopic methods are also used to study equilibria in solution where **new peaks** are obtained when an equilibrium process is set up in solution. Observation of a **new peak** is indicative of a **new species** being formed, and study of that new absorption using Beer's law should generate an equilibrium constant. Spectroscopic methods can also help to distinguish between inner and outer-sphere associated species. For instance, there are theoretical and experimental reasons for assuming that:

- The visible spectrum picks up **inner-sphere** ion pairs and complexes.
- The UV spectrum picks up **outer-sphere** ion pairs and complexes.

One advantage of this method is that the deviations from expected Beer's law values are not attributable even in part to non-ideality, but are totally due to the setting up of an equilibrium process, in contrast to activity and conductance methods where deviations may be due to unaccounted for non-ideality plus association.

1.19 Ultrasonics

This technique has been used to determine association constants for equilibria in electrolyte solutions, particularly for ion association. It has also been used to study solvation effects on ion association (see Section 13.10.5).

A sound wave is passed through an electrolyte solution in which there are species present at equilibrium. This is done over a range of frequencies of the sound wave, and the absorption of sound energy by the solution, or the velocity of the sound wave is measured for each frequency. The sound wave is equivalent to a pressure wave or to a series of alternating temperatures, and these call for a corresponding **wave of new equilibrium positions to be set up**.

If the concentrations of the species present in solution can alter rapidly enough then **each new equilibrium position can be set up** as each new temperature is attained. When this happens, the equilibria and the concentrations can keep in phase with the periodic displacement of the sound wave and there is no great change in the velocity or absorption of the wave.

If reaction does not happen fast enough, then the concentrations of the species present will not alter with the periodic displacement and the **required new equilibrium positions are not set up**. Again the absorption of energy or the velocity of sound will not alter much over a range of frequencies.

If, however, the concentrations can alter rapidly enough, and the new equilibrium positions can be set up but are **out of phase** with the periodic displacement of the sound wave, then this shows up as an increase in the velocity of the sound wave, or as a dramatic increase in the absorption of sound (see Figure 1.2(a) and (b)). Each time this happens an equilibrium process is adjusting itself. The **number of times** this happens shows directly the **number of equilibrium processes** occurring in the solution which are being disturbed by the sound wave (see Figure 1.2(c) and 1.2(d)). Sometimes two or more peaks are superimposed, but these can be resolved by standard curve-fitting techniques.

The beauty of the ultrasonic method lies in this possibility of picking out how many equilibrium processes are present, and multistep equilibria can often be disentangled with relative ease. This is something which the other methods are either incapable of, or which they can only do with difficulty and ambiguity.

However, there still remains the fundamental problem of identifying the process chemically. Identification of the process in the chemical sense is ambiguous as there is no direct chemical observation of the system. It is easy to distinguish between chemical processes and physical processes, such as ion–solvent interactions or energy transfer, by the frequencies at which the maxima in the absorption of sound occurs. Identification of the chemical equilibria and the chemical species present are inferred through a fit of theory plus inference with experiment, and the data may be susceptible to more than one interpretation. One clue which has been used is that the frequency at which absorption of

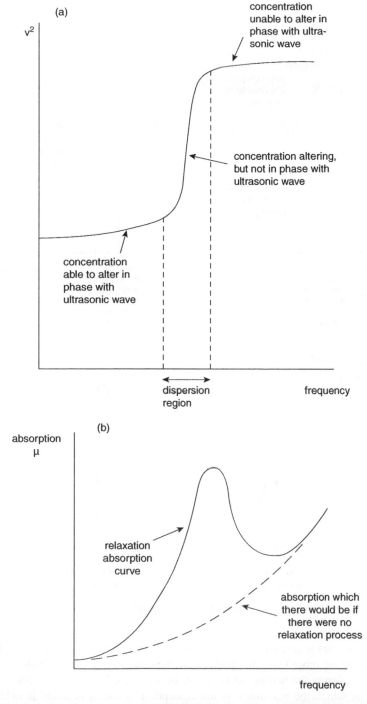

Figure 1.2 (a) Periodic relaxation diagram showing the dependence of the velocity of an ultrasonic wave on frequency for one equilibrium; (b) periodic relaxation diagram showing the dependence of the absorption of an ultrasonic wave on frequency for one equilibrium; (c) periodic relaxation for a double equilibrium: diagram for velocity of wave; (d) periodic relaxation for a double equilibrium: diagram for absorption of wave.

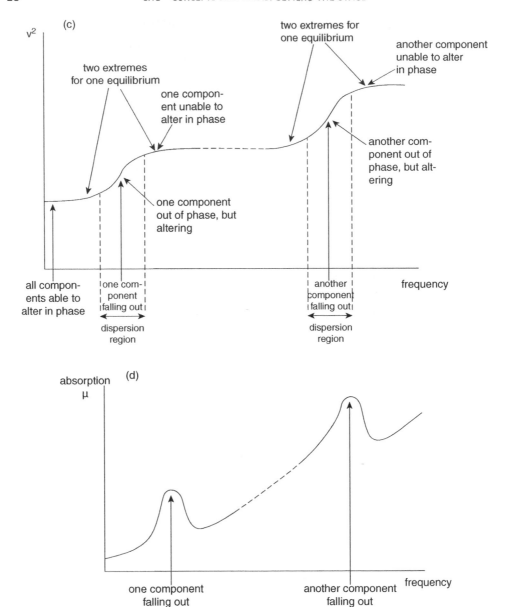

Figure 1.2   (*Continued*)

the sound wave occurs is related to the rate constants for the processes involved. This means that fast processes are more likely to involve ionic processes rather than covalent interactions. One other clue allows distinctions to be made between inner and outer-sphere species. If the frequency at which the maximum in the absorption of sound is found is independent of the nature of the cation, an outer-sphere species is involved. When solvent molecules separate ions, this means that the nature of the cation has less effect. If the frequency depends on the nature of the cation, then an inner-sphere species is involved.

However, inference and chemical knowledge are still the main basis for attributing a particular type of chemical process to a given absorption.

1.20   Possibility that specific experimental methods could distinguish between the various types of associated species

This will be discussed where appropriate throughout the book and below, but it is worthwhile issuing a word of caution here when assessing the likely success in making such distinctions. In fact, no technique has yet been devised which will enable an **unambiguous** and **categorical** distinction to be made between the various possible types of associated species. Claims that a technique can do so must have the evidence critically assessed. Ultrasonics and spectroscopic methods probably allow a distinction to be made with reasonable certainty between outer and inner-sphere ion pairs, while conductance work and scattering of light experiments allow micelle formation to be picked up very easily, but these are quite exceptional achievements in the field of association. Inferences made from thermodynamic and kinetic experiments have often been used, but they rely very heavily on being able to predict a model for the ideal solution – a topic fraught with difficulties as will be seen when the theories of electrolyte solutions are discussed.

1.21   Some examples of how chemists could go about inferring the nature of the species present

Various examples will be given throughout the book, and the following is only a brief survey with highly selective coverage, the main idea being simply to **illustrate** how to go about using experimental results to infer something about the chemical nature of the solution being studied. It would be useful to keep these examples in mind as the more theoretical aspects of electrolyte solutions are being developed.

(i)   A combined spectroscopic and K_{assoc} magnitude example

- The association constants for equilibria involving $Co^{2+}(aq)$ with $SO_4^{2-}(aq)$ and $S_2O_3^{2-}(aq)$ are very similar:

$$K_{CoSO_4} = 2.3 \times 10^2 \, mol^{-1} \, dm^3. \quad K_{CoS_2O_3} = 1.2 \times 10^2 \, mol^{-1} \, dm^3.$$

Inference: The associated species are of very similar nature and are likely to be ion pairs since formation of an ion pair is influenced mainly by the charges on the associating species which are the same here.

- Both species show changes in the UV absorption, but formation of the thiosulphate associated species gives a change in the visible spectrum while formation of the sulphate does not.

Inference: $CoSO_4(aq)$ is an outer-sphere solvent-separated ion pair; CoS_2O_3 is an inner-sphere contact ion pair.

(ii) Ultrasonic studies

Sulphates of $Ni^{2+}(aq)$, $Mg^{2+}(aq)$, $Ca^{2+}(aq)$, $Mn^{2+}(aq)$, $Co^{2+}(aq)$, $Al^{3+}(aq)$, $Zn^{2+}(aq)$, $Be^{2+}(aq)$ give two absorption maxima, with the high-frequency maxima being independent of the nature of the cation, and the low-frequency maxima being dependent on the nature of the cation.

Inference: The following equilibria are set up:

$$M^{2+}(aq) + SO_4^{2-}(aq) \underset{A}{\overset{\text{diffusion}}{\rightleftharpoons}} (M^{2+}(H_2O)_2SO_4^{2-})(aq)$$

outer-sphere solvent-separated ion pair

$$\underset{B}{\rightleftharpoons} (M^{2+}(H_2O)SO_4^{2-})(aq) \underset{C}{\rightleftharpoons} (M^{2+}SO_4^{2-})(aq)$$

outer-sphere solvent-   inner-sphere contact

separated ion pair   contact ion pair

- Two maxima mean two equilibria to be identified. A is ruled out because diffusion is unlikely to give absorption of sound at the frequencies involved.

- The high-frequency maxima which are independent of the nature of the cation suggest that both species are solvent separated. This suggests that equilibrium B is involved here.

- The low-frequency maxima which are dependent on the nature of the cation suggest that the equilibrium involves a species with no solvent between the ions. This suggests that equilibrium C is involved here, since in this equilibrium the last H_2O molecule is being expelled to give the contact ion pair.

(iii) Another spectroscopic example

When $SO_4^{2-}(aq)$ is added to a solution containing $[(Co(NH_3)_5H_2O)]^{3+}(aq)$, a slow process is observed in the visible spectrum, and an immediate very rapid change is noted in the UV spectrum.

Inference: Two equilibria are involved:

$$[(Co(NH_3)_5H_2O)]^{3+}(aq) + SO_4^{2-}(aq) \overset{A}{\rightleftharpoons} [\{Co(NH_3)_5\}^{3+}SO_4^{2-}](aq) + H_2O(l)$$

$$\text{or } [Co(NH_3)_5SO_4]^+(aq) + H_2O(l)$$

$$[(Co(NH_3)_5H_2O)]^{3+}(aq) + SO_4^{2-}(aq) \overset{B}{\rightleftharpoons} [\{Co(NH_3)_5H_2O\}^{3+}\text{solvent }SO_4^{2-}](aq)$$

- The slow change in the visible spectrum is attributed to equilibrium (A). A change in the visible spectrum suggests formation of a contact ion pair $[\{Co(NH_3)_5\}^{3+}SO_4^{2-}]$ (aq) or a complex ion $[Co(NH_3)_5SO_4]^+(aq)$. A slow change suggests formation of a complex with replacement of H_2O by SO_4^{2-}. This would mean an intimate chemical rearrangement involving covalent bonds.

- The fast change in the UV spectrum is attributed to equilibrium (B). Change in the UV spectrum suggests an outer-sphere solvent-separated species. A fast change would suggest a simple physical formation of an ion pair.

(iv) Use of Bjerrum's theory

Bjerrum's theory (Section 10.12) deals explicitly with formation of ion pairs, and can be used to calculate an expected value for the association constant for an equilibrium between two ions and an ion pair. This predicted value can be compared with the observed value.
The conclusions are:

- If the two values are similar, then it is likely that ion pairing is involved since Bjerrum's theory deals explicitly with short-range coulombic interactions which are predominantly dependent on the charges on the ions.

- If $K_{observed} \gg K_{Bjerrum}$, then it is likely that complexing is involved since a much larger K_{assoc} would suggest additional interactions over and above those involved in Bjerrum-type association which is dependent on electrostatic interactions only. The differences between the observed and predicted values would have to be sufficiently large for them not to be easily attributable to inadequacies in the theoretical base-line for analysing the data, or in the derivation of the Bjerrum equation itself.

(v) Application of thermodynamic reasoning

Most of these methods are based on correlations of the observed equilibrium constant with properties like: atomic number of the cation, type of cation, type of anion, type of ligand, pK of the ligand if it is a weak acid or weak base, crystallographic radii, solvated radii, electrostatic interaction energies, ionisation energies of the cation, symmetry of the ligand and more exotic properties.
Ion association can be described as:

$$M^{n+} + A^{m-}(aq) \rightleftharpoons MA^{(n-m)+}(aq)$$

$$K_{assoc} = \left(\frac{[MA^{(n-m)^+}][\gamma^{(n-m)^+}]}{[M^{n+}]\gamma^{n+}[A^{m-}]\gamma^{m-}} \right)_{equilib} \tag{10.15}$$

$$\text{with } \Delta G^{\theta} = -RT \log_e K_{assoc} \tag{10.16}$$

$$\text{since } \Delta G^{\theta} = \Delta H^{\theta} - T \Delta S^{\theta} \tag{10.17}$$

$$\log_{10} K_{assoc} = -\frac{\Delta H^{\theta}}{2.303RT} + \frac{\Delta S^{\theta}}{2.303R} \tag{10.18}$$

Equations 10.17 and 10.18 show up the fundamental flaw in any such correlation. ΔG^{θ} and hence K_{assoc} are composite quantities, and values of ΔG^{θ} or K_{assoc} could be very similar for different systems **simply** because variations in ΔH^{θ} and ΔS^{θ} **compensate out**. Care must be taken to ensure that this is not the case. This is especially so when the conclusions from the experimental data depend crucially on the magnitude of ΔG^{θ} (or K_{assoc}), or on an observed variation of ΔG^{θ} or K_{assoc} with properties such as those listed above. Ideally both K_{assoc} and

$\Delta H^{\theta}_{\text{assoc}}$ should be measured and from this $\Delta S^{\theta}_{\text{assoc}}$ can be found. Modern precision calorimetric data allow highly accurate values of $\Delta H^{\theta}_{\text{assoc}}$ to be found and in turn this can lead to very accurate $\Delta S^{\theta}_{\text{assoc}}$ values.

This fundamental problem is one which has also been shown to be of supreme importance when solvation is being considered.

However, unless both ΔG^{θ} (or K_{assoc}) and $\Delta H^{\theta}_{\text{assoc}}$ are measured, then the experimenter is forced into using the old correlation methods. Useful information can be obtained provided the interpretation is not pushed too far.

2

The Concept of Chemical Equilibrium: An Introduction

This chapter introduces very elementary, but fundamental, ideas and concepts of chemical equilibrium. Most of the topics mentioned will be developed in further chapters. However, it is very important that the simple concepts given in this chapter are understood before proceeding further. This chapter is more an explanatory revision exercise than a formal development of new ideas.

Aims

By the end of this chapter you should be able to:

- distinguish between irreversible and reversible reactions;

- understand the significance of the position of equilibrium, and the approach to equilibrium;

- list the species present at equilibrium, and quote the algebraic form of the equilibrium constant, K, for any reaction;

- predict the direction of reaction from experimental data;

- have a qualitative knowledge of the effect of temperature and pressure on the equilibrium constant;

- know the distinction between stoichiometric and actual concentrations, and be able to formulate the relation between them;

An Introduction to Aqueous Electrolyte Solutions.   By Margaret Robson Wright
© 2007 John Wiley & Sons Ltd   ISBN 978-0-470-84293-5 (cloth) ISBN 978-0-470-84294-2 (paper)

- see the importance of the concept of electrical neutrality for electrolyte solutions, and use it to obtain relations between the actual concentrations of the ions in the solution.

2.1   Irreversible and reversible reactions

The concept of equilibrium is one of the most fundamental ideas in chemistry, and is highly relevant to the study of electrolyte solutions. When reactions occur, two situations can result:

- reaction can either proceed to completion when the reactant which is not in excess is completely converted to product, e.g. the reaction of $S_2O_8{}^{2-}(aq)$ with $I^-(aq)$, proceeds to virtual completion. Such reactions are called **irreversible**.

$$S_2O_8^{2-}(aq) + 2I^-(aq) \longrightarrow 2SO_4^{2-}(aq) + I_2(aq)$$

or

- reaction does not proceed to completion and a constant amount of reactants and products are present finally, e.g. the reaction

$$CH_3Br(aq) + I^-(aq) \rightleftharpoons CH_3I(aq) + Br^-(aq)$$

does not go to completion and constant amounts of reactants and products are present **even** when all apparent reaction has ceased. Such reactions are termed **reversible** and the final state is called being **at equilibrium**. The progress of the reaction to the final state is described as the **approach to equilibrium**.

These two reactions are sufficiently slow for the reaction to be easily followed with time. However, this need not be the case and the final state can be set up very rapidly indeed, with the outward appearance of an instantaneous reaction. Most equilibria involving electrolytes fall into this category, and the special techniques of fast reactions would be needed if the approach to equilibrium were to be demonstrated.

2.2   Composition of equilibrium mixtures, and the approach to equilibrium

At equilibrium **all the species** involved in the equilibrium **must be present**.

The following are examples where there is a rapid approach to equilibrium:

- ethanoic acid dissolved in $H_2O(l)$ has undissociated ethanoic molecules, the ethanoate ion, $H_3O^+(aq)$, and $H_2O(l)$ all present.

$$CH_3COOH(aq) + H_2O(l) \rightleftharpoons H_3O^+(aq) + CH_3COO^-(aq)$$

- in the formation of a solution of the complex:

$$Fe^{3+}(aq) + SCN^-(aq) \rightleftharpoons FeSCN^{2+}(aq)$$

$Fe^{3+}(aq)$, $SCN^-(aq)$ and $FeSCN^{2+}(aq)$ are present.

- formation of the ion pair, $CaOH^+$:

$$Ca^{2+}(aq) + OH^-(aq) \rightleftharpoons CaOH^+(aq)$$

gives an equilibrium mixture $Ca^{2+}(aq)$, $OH^-(aq)$ and $CaOH^+(aq)$.

Heterogeneous reactions behave similarly. The solubility of sparingly soluble salts is an example of a heterogeneous reaction where equilibrium is set up rapidly.

- $BaSO_4(s)$, $Ba^{2+}(aq)$ and $SO_4{}^{2-}(aq)$ are all present at equilibrium in the reaction:

$$BaSO_4(s) \rightleftharpoons Ba^{2+}(aq) + SO_4^{2-}(aq)$$

- in the redox reaction:

$$Fe^{2+}(aq) + Ag^+(aq) \rightleftharpoons Fe^{3+}(aq) + Ag(s)$$

all four species are present.

A very important point:

These observations can be summarised in the general statement that at equilibrium **all of the species involved in the reversible reaction must be present simultaneously**. It is particularly important to remember this when a heterogeneous reaction is involved, e.g. in the solubility and redox reactions the **solid** phases $BaSO_4$ and Ag **must be present** at equilibrium.

The equilibrium which is set up is a **dynamic** state where the **forward** and **back** reactions are **still occurring**, even although it would **appear** that **reaction has ceased**. It is not possible to demonstrate this experimentally, but it is nonetheless one of the fundamental properties of equilibrium.

2.3 Meaning of the term 'position of equilibrium' and formulation of the equilibrium constant

Experimental determinations of the composition of reaction mixtures at equilibrium leads to the concept of the position of equilibrium, and to a further general principle about chemical equilibria: the **meaning** and **formulation** of the **equilibrium constant**. This is one of the most important principles in the whole of chemistry and governs many aspects of the behaviour of electrolyte solutions.

This chapter examines the **experimental** basis for the algebraic form of the equilibrium constant for a given reaction and the consequent implications. In Section 8.14 the **theoretical**

derivation of the equilibrium constant using the principles of thermodynamics is given. This will demonstrate why experimental results lead to a specific expression for the equilibrium constant and no other.

The reaction:

$$Fe^{2+}(aq) + Ag^+(aq) \rightleftharpoons Fe^{3+}(aq) + Ag(s)$$

can be followed using analytical techniques. $[Fe^{2+}]$, $[Ag^+]$ and $[Fe^{3+}]$ can be followed with time, and when the concentrations become constant, reaction is at equilibrium. The **relative amounts** of each substance in the equilibrium mixture specify the **position of equilibrium**. If the experiment is repeated with different initial conditions, it is still found that the concentrations approach constant values, but these differ according to the initial amounts of reactants and/or product taken. From this it is clear that the **position** of equilibrium **is not unique**, but depends on the circumstances of the experiment. The position of equilibrium can also be illustrated in terms of the fraction of reactant converted. This can be defined as {the amount reacted/the total amount of reactant present initially}:

$$\text{fraction of } Fe^{2+}(aq) \text{ converted} = \frac{[Fe^{2+}]_{converted}}{[Fe^{2+}]_{initially}} = \frac{([Fe^{2+}]_{initially} - [Fe^{2+}]_{equilibrium})}{[Fe^{2+}]_{initially}} \quad (2.1)$$

A similar quantity can be formulated for $Ag^+(aq)$:

$$\text{fraction of } Ag^+(aq) \text{ converted} = \frac{[Ag^+]_{converted}}{[Ag^+]_{initially}} = \frac{([Ag^+]_{initially} - [Ag^+]_{equilibrium})}{[Ag^+]_{initially}} \quad (2.2)$$

These quantities will also depend on the initial conditions.

The experimental data also allows the **algebraic relation** between these equilibrium concentrations to be investigated. It is found that when reaction has reached equilibrium the quotient:

$$\left(\frac{[Fe^{3+}]}{[Fe^{2+}][Ag^+]} \right)_{eq} \text{ takes a \textbf{constant} value at a fixed temperature, and takes a different value}$$

at other temperatures. This is called the **equilibrium constant**. In particular, it is this **specific** quotient that is constant at constant temperature and no other, e.g. the quotients

$$\left(\frac{[Fe^{3+}]^2}{[Fe^{2+}][Ag^+]} \right)_{eq} \text{ or } \left(\frac{[Fe^{3+}]^2}{[Fe^{2+}]^2[Ag^+]} \right)_{eq} \text{ do not take a constant value. Other possible algebraic}$$

relations could be tested against the experimental data and the numerical values of these quotients would also vary with the experimental conditions.

Experiment can thus furnish:

- the **composition** of the equilibrium mixture, and consequently the **position of equilibrium**, and that these vary with the experimental conditions.

- the **algebraic form** of the equilibrium constant and its numerical value. These are shown to be independent of the experimental conditions, and the equilibrium constant takes a unique value at constant temperature.

Other reactions can be considered in the same way.

- For the ionisation of ethanoic acid:

$$CH_3COOH(aq) + H_2O(l) \rightleftharpoons H_3O^+(aq) + CH_3COO^-(aq)$$

the quotient which is constant at equilibrium is: $\left(\dfrac{[H_3O^+][CH_3COO^-]}{[CH_3COOH]} \right)_{eq}$. In particular, there is **no term for [H_2O]** in this expression. That this is so becomes abundantly clear when the thermodynamic equilibrium quotient is derived.

- For the redox reaction:

$$2Cr^{2+}(aq) + Sn^{4+}(aq) \rightleftharpoons 2Cr^{3+}(aq) + Sn^{2+}(aq)$$

the quotient which is found to be constant is: $\left(\dfrac{[Cr^{3+}]^2[Sn^{2+}]}{[Cr^{2+}]^2[Sn^{4+}]} \right)_{eq}$

- For the heterogeneous solubility reaction:

$$CaF_2(s) \rightleftharpoons Ca^{2+}(aq) + 2F^-(aq)$$

the respective constant expression is the product: $\left([Ca^{2+}][F^-]^2 \right)_{eq}$

- For the complexing reaction:

$$HgI_2(s) + 2I^-(aq) \rightleftharpoons HgI_4^{2-}(aq)$$

the quotient is: $\left(\dfrac{[HgI_4^{2-}]}{[I^-]^2} \right)_{eq}$

For the heterogeneous reactions, there are no terms for the solid phases in the equilibrium expressions, and this can be easily shown to be the case experimentally. Altering the amount of solid present has no effect on the constancy of these quantities at equilibrium. Again that this is so can be demonstrated conclusively by the thermodynamic derivation of the algebraic form of the equilibrium relations (see Section 8.14).

These constant expressions are termed '**the equilibrium constant, K**', and this takes a constant value at a given temperature. The numerical value generally varies with temperature.

Each of these expressions has a subscript 'eq' to emphasise that the quantity is that found at equilibrium. This is done deliberately to distinguish the equilibrium situation from the corresponding quantity for situations when the reaction is not at equilibrium.

2.3.1   Ideal and non-ideal equilibrium expressions

All of the equilibrium relations quoted above are **ideal** expressions, where quotients of concentrations are constant. But in fact all the solutions will be **non-ideal** and the observed equilibrium constant will have to be corrected for non-ideality using activity coefficients for each species appearing in the equilibrium expression. In practice, extrapolation procedures involving the ionic strength are used, generally by plotting $\log K_{obs}$ vs. $\dfrac{\sqrt{I}}{1 + \sqrt{I}}$ where the K is a quotient of concentrations (see Sections 8.22 to 8.27). However, for a first discussion of the algebraic form of equilibrium constants, it is simpler to consider the **ideal** form of the equilibrium relations.

2.3.2 Prediction of the ideal algebraic form of the equilibrium constant from the stoichiometric equation

This is necessary if an elementary discussion of the principles of equilibrium is given **before** the full thermodynamic derivation of the algebraic form of the equilibrium constant. Much can be said about equilibrium without a thermodynamic approach, and without inclusion of the effects of non-ideality, i.e. without involving activity coefficients.

The algebraic form of K can be deduced directly from the stoichiometric equation, **with the proviso** that there are **no terms** in the algebraic expression for a pure solid, a pure liquid or the solvent if they appear in the stoichiometric equation. This can be contrasted with the fact that at equilibrium **all** the species involved in the equilibrium must be present. The general reaction:

$$aA + bB \rightleftharpoons cC + dD$$

where all the species reacting are either in the gas phase or in solution, has an equilibrium constant:

$$K = \left(\frac{[C]^c [D]^d}{[A]^a [B^b]} \right)_{eq} \tag{2.3}$$

The stoichiometric coefficients in the stoichiometric equation describing the reversible reaction appear as powers in the equilibrium relation.

Worked problem 2.1

Question

Write down the algebraic expression for the equilibrium constant for the following reactions:

- $CO(g) + 2H_2(g) \rightleftharpoons CH_3OH(g)$

- $NH_2COONH_4(s) \rightleftharpoons 2NH_3(g) + CO_2(g)$

- $Hg^{2+}(aq) + Hg(l) \rightleftharpoons Hg_2^{2+}(aq)$

- $AgCl(s) + 2NH_3(aq) \rightleftharpoons Ag(NH_3)_2^+(aq) + Cl^-(aq)$

- $La(OH)_3(s) \rightleftharpoons La^{3+}(aq) + 3OH^-(aq)$

- $[Co(NH_3)_6]^{3+}(aq) + 6H_3O^+(aq) \rightleftharpoons [Co(H_2O)_6]^{3+}(aq) + 6NH_4^+(aq)$

Answer

- $K = \left(\dfrac{[CH_3OH]}{[CO][H_2]^2} \right)_{eq}$

- $K = ([NH_3]^2[CO_2])_{eq}$

- $K = \left(\dfrac{[Hg_2^{2+}]}{[Hg^{2+}]} \right)_{eq}$

- $K = \left(\dfrac{[Ag(NH_3)_2^+][Cl^-]}{[NH_3]^2} \right)_{eq}$

- $K = ([La^{3+}][OH^-]^3)_{eq}$

- $K = \left(\dfrac{[Co(H_2O)_6^{3+}][NH_4^+]^6}{[Co(NH_3)_6^{3+}][H_3O^+]^6} \right)_{eq}$

2.4 Equilibrium and the direction of reaction

The isomerisation of glucose in aqueous solution can be used to illustrate this.

$$\alpha \text{ glucose(aq)} \rightleftharpoons \beta \text{ glucose(aq)}$$

The concentrations of the α and β forms found at various times throughout reaction are given in Table 2.1.
These data illustrate several points:

- This reaction does not proceed to completion. At very long reaction times there are significant amounts of both α and β forms present.

- When the concentrations of the α and β forms are plotted against time both approach a non-zero asymptote, i.e. a plateau is reached.

- The quotient $\dfrac{[\beta]}{[\alpha]}$ does not remain constant throughout reaction, but gradually approaches a constant value which is that value found at very large times. This constant value is the equilibrium constant.

Table 2.1 Isomerisation of α and β glucose at 15°C

$\dfrac{[\alpha]}{\text{mol dm}^{-3}}$	0.01956	0.01736	0.01570	0.01475	0.01394	0.01242
$\dfrac{[\beta]}{\text{mol dm}^{-3}}$	0	0.00220	0.00386	0.00481	0.00562	0.00714
$\dfrac{[\beta]}{[\alpha]}$	0	0.127	0.246	0.326	0.403	0.575
$\dfrac{\text{time}}{\text{s}}$	0	100	200	300	400	∞

- In particular, for the data given, the quotient gradually **increases** until it reaches the equilibrium value. The only way in which it can do so is for more β to be formed progressively, and for more α to be removed.
  This can be summarised as:

- **If the quotient of concentrations < the equilibrium quotient**, then reaction will occur in such a direction that the quotient increases up to the equilibrium value, i.e. reaction will occur in the forward direction. Once the equilibrium value is reached, the reaction is in its equilibrium state, and the quotient of concentrations remains constant. In the present context this would correspond to the situation where reaction occurs with only α initially (see Table 2.1), or with a mixture of α and β forms such that the quotient of concentrations is less than the equilibrium value.

- **If the quotient of concentrations > the equilibrium constant**, then reaction will occur in such a direction that the quotient decreases to the equilibrium value, i.e. reaction will occur in the back direction. Again, once the equilibrium value is reached, the reaction is in its equilibrium state and the quotient of concentrations remains constant. In the present context this would correspond to a situation where reaction occurs with only β initially, or with a mixture of α and β forms such that the quotient of concentrations is greater than the equilibrium value.

- **If the quotient of concentrations = the equilibrium constant**, then the reaction is at equilibrium and the quotient will remain constant. This would happen if a mixture of α and β forms of a composition which gives the equilibrium quotient is taken. No overall reaction would occur and the composition of the mixture would remain constant.

The difference between a quotient or product of concentrations for situations not at equilibrium, and that for equilibrium can be further illustrated by considering the following equilibrium:

$$SrSO_4(s) \rightleftharpoons Sr^{2+}(aq) + SO_4^{2-}(aq)$$

In this case when equilibrium is established all three species are present. The equilibrium product is $([Sr^{2+}][SO_4^{2-}])_{eq}$ and this takes a constant value at any given temperature. If solutions of $Sr(NO_3)_2(aq)$ and $K_2SO_4(aq)$ are mixed, predictions can be made as to whether equilibrium between $SrSO_4(s)$, $Sr^{2+}(aq)$ and $SO_4{}^{2-}(aq)$ can be set up for given compositions of the mixture.

- If initially the quotient, $[Sr^{2+}][SO_4^{2-}] > ([Sr^{2+}][SO_4^{2-}])_{eq}$, then reaction would occur so that the product can decrease to the equilibrium value, i.e. the back reaction would occur and precipitation of $SrSO_4$ would occur. This would continue until the concentrations had decreased sufficiently so that $[Sr^{2+}][SO_4^{2-}] = ([Sr^{2+}][SO_4^{2-}])_{eq}$, at which stage equilibrium is established.

- If initially $[Sr^{2+}][SO_4^{2-}] < ([Sr^{2+}][SO_4^{2-}])_{eq}$, then $SrSO_4(s)$ cannot be formed, and no precipitation is observed.

But note: if solid $SrSO_4$ were present initially, dissolution could occur. Provided sufficient solid is present initially, dissolution will continue until the product, $[Sr^{2+}][SO_4^{2-}]$, has increased to the equilibrium value, $([Sr^{2+}][SO_4^{2-}])_{eq}$, at which stage equilibrium is reached, and no more solid will dissolve. If insufficient solid $SrSO_4$ is present initially, and even if it all dissolves, there is no way in which the product, $[Sr^{2+}][SO_4^{2-}]$, can ever reach the equilibrium value, $([Sr^{2+}][SO_4^{2-}])_{eq}$. Equilibrium can never be reached.

- If the quotient, $[Sr^{2+}][SO_4^{2-}] = ([Sr^{2+}][SO_4^{2-}])_{eq}$, then equilibrium conditions exist and there would be no change.

These arguments are standard for all reversible reactions, and discussion of what happens when reactants and products are mixed, and, in particular, the prediction of the direction of reaction, are individual for each reaction. The basic principles, however, are identical for all cases. Typical reversible reactions for electrolyte solutions involve acid–base equilibria, ion pair, complex and chelate formation, and heterogeneous reactions such as solubility of sparingly soluble salts.

Worked problem 2.2

Question

The reversible reaction between methyl bromide and the iodide ion occurs according to the stoichiometric equation:

$$CH_3Br(aq) + I^-(aq) \rightleftharpoons CH_3I\,(aq) + Br^-(aq)$$

(i) Write down the form of the equilibrium constant.

(ii) When $CH_3Br(aq)$, $CH_3I(aq)$, $I^-(aq)$ and $Br^-(aq)$ are mixed at 25°C and the reaction has come to equilibrium, the equilibrium concentrations are:

$$[CH_3I] = 8.45 \times 10^{-3} \text{ mol dm}^{-3}, \quad [CH_3Br] = 1.21 \times 10^{-3} \text{ mol dm}^{-3},$$
$$[Br^-] = 2.46 \times 10^{-2} \text{ mol dm}^{-3}, \qquad [I^-] = 1.04 \times 10^{-2} \text{ mol dm}^{-3}$$

Calculate the equilibrium constant.

(iii) Initially $[CH_3I] = 8.89 \times 10^{-3}$ mol dm^{-3}. In what direction has reaction occurred?

(iv) Calculate the initial concentrations of $CH_3Br(aq)$, $Br^-(aq)$ and $I^-(aq)$. Find the magnitude of the quotient of initial concentrations, and from this predict the direction of reaction.

Answer

(i) $K = \left(\dfrac{[CH_3I][Br^-]}{[CH_3Br][I^-]} \right)_{eq}$

(ii) $K = \left(\dfrac{[CH_3I][Br^-]}{[CH_3Br][I^-]} \right)_{eq} = \dfrac{8.45 \times 10^{-3}\,mol\,dm^{-3} \times 2.46 \times 10^{-2}\,mol\,dm^{-3}}{1.21 \times 10^{-3}\,mol\,dm^{-3} \times 1.04 \times 10^{-2}\,mol\,dm^{-3}} = 16.5$

(iii) $[CH_3I]_{initially} = 8.89 \times 10^{-3}\,mol\,dm^{-3}$

$[CH_3I]_{finally} = 8.45 \times 10^{-2}\,mol\,dm^{-3}$

Since the CH_3I concentration has decreased, reaction has gone from right to left, i.e. reaction has occurred in the back direction.

(iv) Decrease in $[CH_3I] = 0.44 \times 10^{-3}\,mol\,dm^{-3}$

From the stoichiometric equation:

x mol of CH_3Br reacts with x mol of I^- to give x mol of CH_3I and x mol of Br^-

$\therefore$ decrease in $[CH_3I]$ = increase in $[CH_3Br] = 0.44 \times 10^{-3}\,mol\,dm^{-3}$

$\therefore [CH_3Br]_{eq} = [CH_3Br]_{initially} +$ increase in $[CH_3Br]$

$\therefore [CH_3Br]_{initially} = [CH_3Br]_{eq} -$ increase in $[CH_3Br]$

$\qquad\qquad = (1.21 \times 10^{-3} - 0.44 \times 10^{-3})\,mol\,dm^{-3}$

$\qquad\qquad = 0.77 \times 10^{-3}\,mol\,dm^{-3}$

$[I^-]_{eq} = [I^-]_{initially} +$ increase in $[I^-]$

$[I^-]_{initially} = [I^-]_{eq} -$ increase in $[I^-]$

$\qquad\qquad = [I^-]_{eq} -$ increase in $[CH_3Br]$

$\qquad\qquad = (1.04 \times 10^{-2} - 0.44 \times 10^{-3})\,mol\,dm^{-3}$

$\qquad\qquad = 9.96 \times 10^{-3}\,mol\,dm^{-3}$

$[Br^-]_{eq} = [Br^-]_{initially} -$ decrease in $[Br^-]$

$\qquad\quad = [Br^-]_{initially} -$ decrease in $[CH_3I]$

$[Br^-]_{initially} = [Br^-]_{eq} +$ decrease in $[CH_3I]$

$\qquad\qquad = (2.46 \times 10^{-2} + 0.44 \times 10^{-3})\,mol\,dm^{-3}$

$\qquad\qquad = 2.50 \times 10^{-2}\,mol\,dm^{-3}$

$\left(\dfrac{[CH_3I][Br^-]}{[CH_3Br][I^-]} \right)_{initially} = \dfrac{8.89 \times 10^{-3}\,mol\,dm^{-3} \times 2.50 \times 10^{-2}\,mol\,dm^{-3}}{0.77 \times 10^{-3}\,mol\,dm^{-3} \times 9.96 \times 10^{-3}\,mol\,dm^{-3}}$

$\qquad\qquad = 29.0$

$\left(\dfrac{[CH_3I][Br^-]}{[CH_3Br][I^-]} \right)_{initially} > \left(\dfrac{[CH_3I][Br^-]}{[CH_3Br][I^-]} \right)_{eq}$

Reaction always occurs in the direction such that the initial quotient changes until it reaches the equilibrium quotient. In this case the initial quotient must decrease until it reaches the equilibrium value. The only way that that is possible is for the $[CH_3I]$ to decrease and $[CH_3Br]$ to increase, i.e. reaction must occur from right to left, i.e. the back reaction must occur; which is the conclusion reached earlier.

Worked problem 2.3

Question

The dissolution of the sparingly soluble salt $Ag_2SO_4(s)$ in water occurs as:

$$Ag_2SO_4(s) \rightleftharpoons 2Ag^+(aq) + SO_4^{2-}(aq)$$

(a) What species are present at equilibrium? Write down the algebraic expression for the equilibrium constant.

(b) At $20°C$ $K = 1.2 \times 10^{-5}$ mol^3 dm^{-9}.

Explain whether precipitation of silver sulphate could occur spontaneously:

- On mixing equal volumes of 0.018 $mol\,dm^{-3}$ aqueous silver nitrate and 0.024 $mol\,dm^{-3}$ aqueous sodium sulphate;

- On mixing $20\,cm^3$ of $0.200\,mol\,dm^{-3}$ aqueous silver nitrate and $30\,cm^3$ of $0.016\,mol\,dm^{-3}$ aqueous sodium sulphate?

Answer

(a) All three species **must be present** at equilibrium.

$$K = ([Ag^+]^2[SO_4^{2-}])_{eq}$$

Note: Although all three species must be present at equilibrium, the equilibrium constant expression does not include a term for the solid Ag_2SO_4. This is in keeping with the general statement that the algebraic form of K can be deduced from the stoichiometric equation, **with the proviso** that there are **no terms** for any pure solid, pure liquid or solvent. As mentioned previously, that this is so becomes abundantly clear when the algebraic form is deduced from the thermodynamic argument. This will be explained in Sections 8.13 and 8.14.
Remember that on mixing dilution occurs and the stoichiometric concentrations will decrease.

- On mixing these solutions, equal volumes are used and the stoichiometric concentrations are halved.

0.009 $mol\,dm^{-3}$ $AgNO_3(aq)$ corresponds to 0.009 $mol\,dm^{-3}$ $Ag^+(aq)$ and 0.009 $mol\,dm^{-3}$ $NO_3^-(aq)$.
0.012 $mol\,dm^{-3}$ $Na_2SO_4(aq)$ corresponds to 0.024 $mol\,dm^{-3}$ $Na^+(aq)$ and 0.012 $mol\,dm^{-3}$ $SO_4^{2-}(aq)$.

Whether or not precipitation occurs depends on how the product$([Ag^+]^2[SO_4^{2-}])_{initially}$ compares with $([Ag^+]^2[SO_4^{2-}])_{eq}$.

$$([Ag^+]^2[SO_4^{2-}])_{initially} = (9 \times 10^{-3})^2 \text{ mol}^2 \text{ dm}^{-6} \times 1.2 \times 10^{-2} \text{ mol dm}^{-3}$$

$$= 1.0 \times 10^{-6} \text{ mol}^3 \text{ dm}^{-9}$$

$$([Ag^+]^2[SO_4^{2-}])_{eq} = 1.2 \times 10^{-5} \text{ mol}^3 \text{ dm}^{-9}.$$

$$([Ag^+]^2[SO_4^{2-}])_{initially} < ([Ag^+]^2[SO_4^{2-}])_{eq}$$

$$\text{i.e.} ([Ag^+]^2[SO_4^{2-}])_{initially} < K$$

The only process which can occur is one which will allow $([Ag^+]^2[SO_4^{2-}])_{initially}$ to increase until the equilibrium value is reached. This would correspond to further dissolution, i.e. reaction occurs from left to right. Precipitation cannot occur.

- On mixing the second solutions, unequal volumes are used. The final volume is 50 cm^3. The stoichiometric concentrations of $AgNO_3(aq)$ and $Na_2SO_4(aq)$ will be

$$\frac{20 \text{ cm}^3 \times 0.200 \text{ mol dm}^{-3}}{50 \text{ cm}^3} = 0.080 \text{ mol dm}^{-3}, \text{ and}$$

$$\frac{30 \text{ cm}^3 \times 0.016 \text{ mol dm}^{-3}}{50 \text{ cm}^3} = 9.6 \times 10^{-3} \text{ mol dm}^{-3} \text{ respectively.}$$

$$([Ag^+]^2[SO_4^{2-}])_{initially} = (0.080)^2 \text{ mol}^2 \text{ dm}^{-6} \times 9.6 \times 10^{-3} \text{ mol dm}^{-6}.$$

$$= 6.14 \times 10^{-5} \text{ mol}^3 \text{ dm}^{-9}.$$

$$([Ag^+]^2[SO_4^{2-}])_{initially} > ([Ag^+]^2[SO_4^{2-}])_{eq}$$

$$\text{i.e.} ([Ag^+]^2[SO_4^{2-}])_{initially} > K$$

In this case the only process which can occur is for $([Ag^+]^2[SO_4^{2-}])_{initially}$ to decrease until the equilibrium value is reached. This corresponds to the back reaction occurring, and precipitation of the solid will be observed.

2.5   A searching problem

Experiment shows that for **homogeneous** reactions in both the gas phase and solution, all species involved in the reversible reaction are present in the equilibrium mixture, and reaction **never reaches completion**, e.g.

$$NH_3(aq) + H_3O^+(aq) \rightleftharpoons NH_4^+(aq) + H_2O(l)$$

$$K = \left(\frac{[NH_4^+]}{[NH_3][H_3O^+]} \right)_{eq} \tag{2.4}$$

Here, if acid is added to a solution of $NH_3(aq)$, no matter which of $NH_3(aq)$ or $H_3O^+(aq)$ is in large excess, reaction will occur in the forward direction with the concentrations of both $NH_3(aq)$ and $H_3O^+(aq)$ decreasing until the concentrations of all the aqueous species reach their equilibrium values and the quotient $\left(\frac{[NH_4^+]}{[NH_3][H_3O^+]} \right)_{eq}$ is set up. The $[NH_3]$ and $[H_3O^+]$ can only decrease until $[NH_3]_{eq}$ and $[H_3O^+]_{eq}$ are reached, and neither can reach zero.

However, in reversible **heterogeneous** reactions there is the possibility that, under **certain** experimental conditions, reaction **can proceed to completion** and equilibrium can never be set up. This behaviour is **unique** to **heterogeneous** reactions, and can be demonstrated experimentally. The following reactions illustrate this **apparent** anomaly:

$$SrSO_4(s) \rightleftharpoons Sr^{2+}(aq) + SO_4^{2-}(aq)$$

$$K = ([Sr^{2+}][SO_4^{2-}])_{eq} \tag{2.5}$$

$$Fe^{3+}(aq) + Ag(s) \rightleftharpoons Fe^{2+}(aq) + Ag^+(aq)$$

$$K = \left(\frac{[Fe^{2+}][Ag^+]}{[Fe^{3+}]} \right)_{eq} \tag{2.6}$$

In the first reaction, the solid $SrSO_4$ is fairly soluble, and, if only a very small amount of the solid were initially present in water, it is possible that all of it could dissolve before the equilibrium product is reached. Reaction would then be observed to go to completion with no solid left. Equilibrium would not be established.

In the second reaction, if a small amount of solid Ag is added to a solution containing excess $Fe^{3+}(aq)$, then it is possible that all the Ag(s) could be used up before the equilibrium concentrations can be reached. If this is so equilibrium can never be set up, and it is possible for reaction to proceed to completion in the **forward** direction.

However, when the **back** reaction is considered, equilibrium will always be set up with all four species present. If a mixture of $Fe^{2+}(aq)$ and $Ag^+(aq)$ is taken, reaction will occur and the concentrations of both ions will decrease until they reach their equilibrium values, and the quotient $\left(\frac{[Fe^{3+}]}{[Fe^{2+}][Ag^+]} \right)_{eq}$ is set up. Since $[Fe^{2+}]$ and $[Ag^+]$ can only decrease to their **finite** equilibrium values, neither can decrease to zero.

2.6 The position of equilibrium

The position of equilibrium gives information about the composition of the mixture once equilibrium has been established in a reversible reaction, i.e. about the relative amounts of reactants and products present. The magnitude of the equilibrium constant does not, in itself, give this information since a quotient can be made up of various combinations of the quantities contributing to the equilibrium constant. For instance, the pH of a solution depends on the magnitude of K and the concentration of acid taken, and dilution will alter the position of equilibrium, but K is a constant and depends only on the identity of the acid. Consequently, the equilibrium constant and the position of equilibrium must not be taken to be synonymous.

There are situations where the magnitude of the equilibrium constant may allow inferences to be made about the position of equilibrium, but these must be made with caution.

- If the magnitude of the equilibrium constant is large, then it is often said that the position of equilibrium lies to the right;

- if the magnitude is small, it is said that the position of equilibrium lies to the left;

- while if it is close to unity, there are comparable amounts of reactants and products present.

These generalisations **must be used with caution** as there are situations where they can give the wrong predictions, e.g.:

For the reaction:

$$A + mB \rightleftharpoons AB_m \text{ where}$$

$$K = \left(\frac{[AB_m]}{[A][B]^m} \right)_{eq} \tag{2.7}$$

- If [B] is large and K is small, then there can be almost complete conversion to AB_m in direct contradiction to the prediction above.

$$\left(\frac{[AB_m]}{[A]} \right)_{eq} = K \times [B]^m \text{ and if [B] is large, then } [B]^m \text{ could dominate in the product,}$$

$K \times [B]^m$. Qualitatively this is expressed as: if [B] is large it will 'push the equilibrium to the right', in contrast to the situation when [A] and [B] are comparable when there is little conversion to product.

- If [B] is small, then even with K large it is possible that there is very little conversion, in contradiction to the prediction given earlier. In this case $\left(\frac{[AB_m]}{[A]} \right)_{eq}$ has a magnitude which again depends on the relative values of K and $[B]^m$. If [B] is small, then $[B]^m$ could dominate the product, $K \times [B]^m$.

2.7   Other generalisations about equilibrium

There are other general statements which are often made, and likewise should be discussed in terms of the principles of equilibrium:

- If products are removed continually, then the position of equilibrium is never set up.

- If products are added, then this will 'push the position of equilibrium to the left'.

- If reactants are added, then this will 'push the position of equilibrium to the right'.

- In electrolyte solution studies when an electrolyte is added which contains an ion in common with the reactant or product electrolyte, then the position of equilibrium can be altered. This is called the 'common ion effect', though often 'suppression' is the term more commonly used. The direction in which the equilibrium position moves depends on whether the common ion is a reactant or product.

2.8 K and pK

Experimental values of K can range over many powers of 10, and these can be expressed more concisely in a compressed scale by taking logarithms. For instance $K = 2.05 \times 10^{-3} \text{ mol dm}^{-3}$ can be more succinctly given as $pK = 2.69$, where $pK = -\log_{10}K$. Most tabulated data is given in terms of pK values rather than equilibrium constants, K. This is analogous to the common habit of expressing concentrations of H_3O^+ as pH values, where $pH = -\log_{10}[H_3O^+]$.

2.9   Qualitative experimental observations on the effect of temperature on the equilibrium constant, K

Equilibrium constants usually depend on the temperature and the general point can be made that:

- if ΔH^θ is positive, then K increases with increase in temperature;

- if ΔH^θ is negative, then K decreases with increase in temperature;

- if ΔH^θ is approximately zero, then K is nearly independent of temperature.

This can be regarded as 'a rule of thumb', but the exact relation between K and temperature is made clear from the thermodynamic analysis. This will be given in Section 8.16 and Worked Problems 8.5, 8.6 and 8.7.

The dependence of K on temperature can show both a linear and a non-linear relation between $\log_e K$ and T or $\log_e K$ and $1/T$. Sometimes, as in the case of some carboxylic acids, a maximum in K is found as the temperature is varied. This is discussed in Worked Problems 2.5 and 8.9.

Worked problem 2.4

Question

Pure water has a detectable conductance, indicating the presence of ions. These result from the self ionisation of water:

$$2H_2O(l) \rightleftharpoons H_3O^+(aq) + OH^-(aq)$$
$$K_w = ([H_3O^+][OH^-])_{eq}$$

Note: there is no term for H_2O in this expression.

The following table gives the temperature dependence of K_w. What is the sign of ΔH^θ for the self ionisation? Find the pH of pure water at these temperatures.

Values of K_w as a function of temperature

temp / °C	0	10	20	25	30	40	50	60	100
pK_w	14.94	14.53_5	14.17	14.00	13.83	13.53_5	13.26	13.02	12.30

Answer

$pK_w = -\log_{10} K_w$, $\therefore$ if pK_w decreases with temperature, then K_w must increase with temperature. This tells us that ΔH^θ for the self ionisation must be positive.

In pure water:

$$[H_3O^+]_{eq} = [OH^-]_{eq}$$
$$\therefore [H_3O^+]_{eq} = \sqrt{K_w}$$
$$pH = -\log_{10}[H_3O^+]$$
$$\therefore \text{for pure water } pH = -\log_{10}\sqrt{K_w} = 1/2 pK_w$$

From the table it becomes immediately clear that allowance must be made for this temperature dependence.

temp °C	0	10	20	25	30	40	50	60	100
pK_w	14.94	14.53$_5$	14.17	14.00	13.83	13.53$_5$	13.26	13.02	12.30
neutral pH	7.47	7.27	7.08$_5$	7.00	6.91$_5$	6.76$_5$	6.63	6.51	6.15

Worked problem 2.5

Question

The following table gives the temperature dependence of the equilibrium constant for the ionisation of ethanoic acid:

$$CH_3COOH(aq) + H_2O(aq) \rightleftharpoons H_3O^+(aq) + CH_3COO^-(aq)$$
$$K_a = \left(\frac{[H_3O^+][CH_3COO^-]}{[CH_3COOH]}\right)_{eq}$$

Draw a graph of K_a vs. T. Discuss the shape of the graph, and the significance of the shape of the graph.

Values of K_a for ethanoic acid as a function of temperature

temp °C	0	10	20	30	40	50
$\dfrac{10^5 K_a}{\text{mol dm}^{-3}}$	1.657	1.729	1.753	1.750	1.703	1.633

Answer

The graph shows a maximum between 20°C and 30°C. At the maximum $\Delta H^\ominus$ = zero. At lower temperatures K_a increases with increase in temperature, and so in this region $\Delta H^\ominus$ is positive, while to higher temperatures K_a decreases with increase in temperature, and so in this region $\Delta H^\ominus$ is negative.

2.10 Qualitative experimental observations on the effect of pressure on the equilibrium constant, K

If all the species involved in the reversible reaction are in the gas phase, then the equilibrium constant is independent of pressure. However, if the reaction is in solution, then K can depend on pressure as a consequence of the overall change in volume between products and reactants, ΔV^θ. Solution phase reactions are distinct from gas phase reactions in this respect, and it is only after the thermodynamic treatment is given that this becomes clear.

The general point can be made that:

- if ΔV^θ is positive, then K decreases with increase in pressure;

- if ΔV^θ is negative, then K increases with increase in pressure;

- if ΔV^θ is approximately zero, then K is nearly independent of pressure.

These considerations are always pertinent if there are some species in solution, e.g. solubilities are dependent on pressure. This is discussed in more detail in Section 8.18 and Worked Problem 8.10(ii).

The dependence of K on temperature and pressure can give a lot of information about details of the reaction involved. The kinetic analogues likewise give much information about the mechanism of the reaction involved.

2.11 Stoichiometric relations

These are relations which are constantly being used in equilibrium calculations (see Section (3.11) and can be illustrated by the following reactions.

- $2Cr^{2+}(aq) + Sn^{4+}(aq) \rightleftharpoons 2Cr^{3+}(aq) + Sn^{2+}(aq)$

If solutions of $Cr^{2+}(aq)$ and $Sn^{4+}(aq)$ are mixed then the initial amount, i.e. the stoichiometric amount of $Cr^{2+}(aq)$, will end up being distributed between the $Cr^{3+}(aq)$ formed in the redox reaction and the $Cr^{2+}(aq)$ left. Likewise, the initial amount, i.e. the stoichiometric amount of $Sn^{4+}(aq)$, will distribute itself between the $Sn^{2+}(aq)$ formed in the redox reaction and the $Sn^{4+}(aq)$ left. The stoichiometric amount is based on the number of mol of the Cr^{2+}, or Sn^{4+} salts weighed out.

Relations between these quantities can be written down, and these must distinguish between the total amount of $Cr^{2+}(aq)$ used and the actual amounts of $Cr^{2+}(aq)$ and $Cr^{3+}(aq)$ present, and likewise for $Sn^{4+}(aq)$ and $Sn^{2+}(aq)$.

$$[Cr^{2+}]_{\text{total or stoichiometric}} = [Cr^{2+}]_{\text{actual}} + [Cr^{3+}]_{\text{actual}} \tag{2.8}$$

$$[Sn^{4+}]_{\text{total or stoichiometric}} = [Sn^{4+}]_{\text{actual}} + [Sn^{2+}]_{\text{actual}} \tag{2.9}$$

It is these **actual** concentrations at equilibrium which appear in the equilibrium expression:

$$K = \left(\frac{[Cr^{3+}]^2_{actual}[Sn^{2+}]_{actual}}{[Cr^{2+}]^2_{actual}[Sn^{4+}]_{actual}} \right)_{eq} \quad (2.10)$$

- $Cu^{2+}(aq) + 4NH_3(aq) \rightleftharpoons Cu(NH_3)_4(aq)$

The initial stoichiometric amount of copper salt taken will end up distributed between the Cu present in the complex and $Cu^{2+}(aq)$ left as uncomplexed ions in the equilibrium mixture. Likewise the initial stoichiometric amount of $NH_3(aq)$ taken is distributed between the complex and the $NH_3(aq)$ left uncomplexed in the equilibrium mixture.

$$[Cu^{2+}]_{total \ or \ stoichiometric} = [Cu(NH_3)_4^{2+}]_{actual} + [Cu^{2+}]_{actual} \quad (2.11)$$

$$[NH_3]_{total \ or \ stoichiometric} = 4[Cu(NH_3)_4^{2+}]_{actual} + [NH_3]_{actual} \quad (2.12)$$

$$K = \left(\frac{[Cu(NH_3)_4^{2+}]_{actual}}{[Cu^{2+}]_{actual}[NH_3]^4_{actual}} \right)_{eq} \quad (2.13)$$

Note well: there is a factor 4 in the term for the complex when considering the distribution of $NH_3(aq)$ between complex and uncomplexed $NH_3(aq)$. This must appear since four molecules of NH_3 are present in each complex formed.

Such relations will be used many times throughout this book, and the many worked problems should give plenty of practice in formulating the expressions for any specific reaction.

2.12 A further relation essential to the description of electrolyte solutions – electrical neutrality

All electrolyte solutions must be electrically neutral, i.e. have no overall charge with the positive charges balancing the negative charges. This gives a further useful relation between the **actual** concentrations of the ions present in an electrolyte solution. For instance, in a mixture of NaCl(aq) and KBr(aq) there will be present $Na^+(aq)$, $K^+(aq)$, $Cl^-(aq)$, $Br^-(aq)$ as well as $H_3O^+(aq)$ and $OH^-(aq)$ from the self ionisation of $H_2O(l)$, and the total positive charge in the solution must balance the total negative charge. The positive charge comes from $Na^+(aq)$, $K^+(aq)$, $H_3O^+(aq)$ and each ion will contribute one positive charge. The negative charge comes from $Cl^-(aq)$, $Br^-(aq)$ and $OH^-(aq)$ and each ion will contribute one negative charge.

If each $Na^+(aq)$ contributes one positive charge, then n_{Na^+} mol from the NaCl will contribute n_{Na^+} mol of positive charge. Hence the total positive charge from the NaCl(aq) will be proportional to the concentration, $n_{Na^+}/$volume of solution, i.e. $[Na^+]_{actual}$. A similar argument will apply to the other positive charges and to the negative charges.

Summing up for the whole solution:

total positive charge = total negative charge

$$[Na^+]_{actual} + [K^+]_{actual} + [H_3O^+]_{actual} = [Cl^-]_{actual} + [Br^-]_{actual} + [OH^-]_{actual} \quad (2.14)$$

Since the solution will have neutral pH, $[H_3O^+]_{actual} = [OH^-]_{actual}$
then

$$[Na^+]_{actual} + [K^+]_{actual} = [Cl^-]_{actual} + [Br^-]_{actual} \qquad (2.15)$$

When divalent ions are present this multiple charge on the ion must be included, e.g. $Ba^{2+}(aq)$ present in a solution will contribute two positive charges, and the total positive charge contributed will be $2[Ba^{2+}]_{actual}$. Likewise if the $PO_4{}^{3-}(aq)$ ion were present it would contribute $3[PO_4{}^{3-}]_{actual}$ of negative charge. For the redox reaction, given in Section 2.3, where the chloride salts are used:

$$2Cr^{2+}(aq) + Sn^{4+}(aq) \rightleftharpoons 2Cr^{3+}(aq) + Sn^{2+}(aq)$$

electrical neutrality requires:

$$2[Cr^{2+}]_{actual} + 3[Cr^{3+}]_{actual} + 4[Sn^{4+}]_{actual} + 2[Sn^{2+}]_{actual} + [H_3O^+]_{actual} = [Cl^-]_{actual} + [OH^-]_{actual}$$
$$(2.16)$$

Since the solution has a neutral pH

$$2[Cr^{2+}]_{actual} + 3[Cr^{3+}]_{actual} + 4[Sn^{4+}]_{actual} + 2[Sn^{2+}]_{actual} = [Cl^-]_{actual} \qquad (2.17)$$

3

Acids and Bases: A First Approach

This chapter gives a qualitative description of acids and bases, and emphasises the importance of the hydrated proton in such equilibria. Acids are substances which give rise to a proton, and bases are substances which accept a proton, and thereby generate $OH^-(aq)$. This should set the stage for the quantitative descriptions given in Chapters 4 to 6. This chapter is basically revision, and where necessary gives extra detail so that a complete understanding emerges.

Aims

By the end of this chapter you should be able to:

- give a description of examples of simple and complex acids and bases, and realise that strong acids and strong bases are fully ionised in solution, whereas weak acids and weak bases are not fully ionised in solution, and are present in equilibrium with the products of their ionisation as an acid or base respectively;

- show that the self ionisation of the solvent, here $H_2O(l)$, is an example of acid/base behaviour;

- discuss the meaning of conjugate acids and conjugate bases, and give a more detailed account of acid/base behaviour;

- recognise that acid/base equilibria are set up in solutions of salts and buffers;

- formulate the algebraic expression for the equilibrium constant for a weak acid, K_a, and for a weak base, K_b;

- deduce the relation between K_a for a weak acid and K_b for its conjugate base, and the corresponding expression for a weak base and its conjugate acid;

- write down the stoichiometric relations between total and actual concentrations which govern all equilibria.

An Introduction to Aqueous Electrolyte Solutions.   By Margaret Robson Wright
© 2007 John Wiley & Sons Ltd   ISBN 978-0-470-84293-5 (cloth) ISBN 978-0-470-84294-2 (paper)

3.1   A qualitative description of acid–base equilibria

An acid is a substance which donates a proton, while a base is a substance which accepts a proton. In aqueous solution the proton, H^+, is generally written as $H_3O^+(aq)$ indicating that, although the monohydrate ion is the major species, other more highly hydrated ions are present. This is indicated by the (aq). $H_3O^+(aq)$ will be used throughout this book. When a substance acts as an acid it must give a proton to another species which thus acts as a base. Likewise when a base accepts a proton from a molecule, this molecule must be acting as an acid. And so acid–base behaviour does not take place in isolation. A molecule can only act as an acid in the presence of a molecule acting as a base. Similarly, a molecule can only act as a base in the presence of a molecule acting as an acid.

In this respect, water allows:

- acids to function as acids, and

- bases to function as bases

because it can both accept and donate protons. Because of its ability to act both as an acid or as a base depending on the conditions, $H_2O(l)$ is termed an **amphoteric** solvent.

3.1.1   Acidic behaviour

$$\text{acid(aq)} \quad + H_2O(l) \quad \longrightarrow \text{base(aq)} \quad + H_3O^+(aq)$$
$$\textbf{acid of } (\textbf{1}) + \textbf{base of } (\textbf{2}) \longrightarrow \textbf{base of } (\textbf{1}) + \textbf{acid of } (\textbf{2})$$

$H_2O(l)$ acts as a base by accepting a proton from the acidic species and in doing so is converted into an acid, $H_3O^+(aq)$, the original acid being converted into a base. The acid–base behaviour is essentially a proton transfer from one species to another.

Simple acids like HCl, H_2SO_4 and H_3PO_4, are sources of $H_3O^+(aq)$ in aqueous solution. These are classified as:

- monoacidic or monoprotic when there is one proton available, such as in HCl;

- dibasic or diprotic when there are two protons available as in H_2SO_4;

- tribasic or triprotic when there are three protons available as in H_3PO_4.

There are also more complex molecules which act as acids, and do so by virtue of the presence of an acidic functional group which can donate a proton to the water molecule in aqueous solution. The simpler ones are molecules containing the carboxylic acid group, COOH, the hydroxyl group OH, or the protonated amino group, NH_3^+. Sometimes more than one of these functional groups may be present.

Typical simple carboxylic acids are the monobasic ethanoic acid, CH_3COOH, and benzoic acid, C_6H_5COOH. Succinic acid is an example of a dibasic acid, and citric acid a tribasic acid.

Biotin is an example of a complex carboxylic acid.

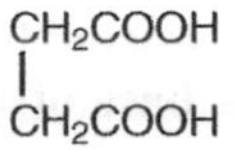

$$CH_2COOH$$
$$|$$
$$CH_2COOH$$

succinic acid

$$CH_2COOH$$
$$|\!-OH$$
$$C$$
$$|\!\setminus COOH$$
$$CH_2COOH$$

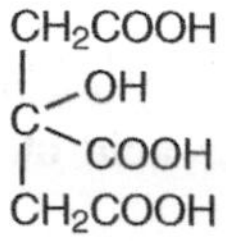

citric acid

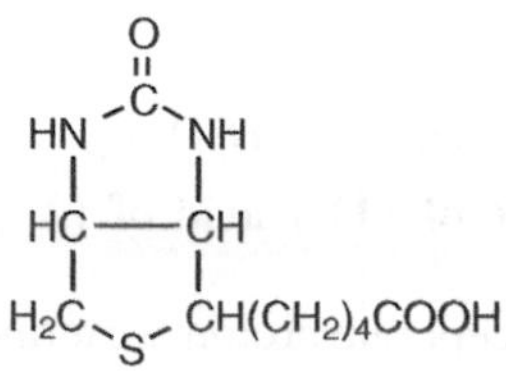

biotin

Simple alcohols containing the OH group are ethanol, C_2H_5OH, and phenol, C_6H_5OH, while carbohydrates such as glucose are complex. All function as acids.

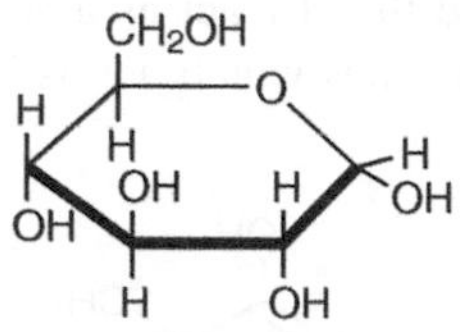

glucose

The simplest acids containing the NH_3^+ group are the protonated amino acids such as the protonated form of glycine, $^+NH_3CH_2COOH$, or the protonated form of amino acid esters such as $^+NH_3CH_2COOCH_3$. There are more complex amino acids which can have protonated amino acid groups, such as histidine, below.

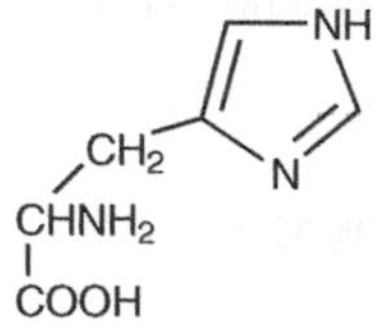

histidine

3.1.2 Basic behaviour

$$base(aq) \ + H_2O(l) \ \longrightarrow \ acid(aq) \ + OH^-(aq)$$
$$\textbf{base of}(1) + \textbf{acid of}(2) \longrightarrow \ \textbf{acid of}(1) + \textbf{base of}(2)$$

$H_2O(l)$ acts as an acid by donating a proton to the basic species and in doing so is converted into a base, $OH^-(aq)$, the original base being converted into an acid. The acid–base behaviour is again essentially a proton transfer from one species to another.

Simple bases such as NaOH and KOH act as sources of $OH^-(aq)$ in aqueous solution. The $OH^-(aq)$ ion is a base because it can accept a proton from a proton donor acid, $H_2O(l)$, in aqueous solution.

$$OH^-(aq) \ + H_2O(l) \ \longrightarrow H_2O(l) \ + OH^-(aq)$$
$$\textbf{base of } (1) + \textbf{acid of } (2) \longrightarrow \textbf{acid of } (1) + \textbf{base of } (2)$$

NH$_3$(aq) also is a simple base. It acts as a base by accepting a proton from H$_2$O(l), thereby generating OH$^-$(aq).

$$NH_3(aq) \quad + H_2O(aq) \quad \rightleftharpoons NH_4^+(aq) \quad + OH^-(aq)$$

base of (1) + acid of (2) $\rightleftharpoons$ acid of (1) + base of (2)

Bases which can accept one proton are termed monoacid, and those accepting two protons are diacidic.

Typical organic bases are the primary amines containing the NH$_2$ group, e.g. methylamine, CH$_3$NH$_2$, the secondary amines containing the group NH, e.g. dimethylamine, (CH$_3$)$_2$NH, and the tertiary amines containing the group, N, such as trimethylamine, (CH$_3$)$_3$N. Like ammonia, these accept a proton from H$_2$O and thereby act as a source of OH$^-$(aq). There are many biologically important complex molecules which act as bases through the presence of such functional groups, e.g. thymine.

thymine

Some molecules contain both acidic and basic functional groups. The amino acids such as glycine, NH$_2$CH$_2$COOH, are common examples and they act as an acid or a base, depending on the circumstances.

3.2 The self ionisation of water

This equilibrium is always set up in aqueous solution. Highly purified water has a measurable conductance corresponding to the presence of ions resulting from the ionisation of water molecules. This is another example of acid–base behaviour involving proton transfers.

$$H_2O(l) + H_2O(l) \rightleftharpoons OH^-(aq) + H_3O^+(aq)$$

acid of 1 base of 2   base of 1   acid of 2

One water molecule acts as an acid by donating a proton, the other acts as a base by accepting a proton. This equilibrium must be considered explicitly in many calculations on equilibria in aqueous solution.

3.3 Strong and weak acids and bases

Strong acids and bases are **fully ionised**, or **fully dissociated**, in water; HCl in aqueous solution exists totally as H$_3$O$^+$(aq) and Cl$^-$(aq), there is no molecular HCl present. Likewise NaOH exists totally as Na$^+$(aq) and OH$^-$(aq) and there is no molecular NaOH present.

Weak acids and bases are **incompletely ionised**, or **incompletely dissociated**, in water and there are molecular forms of the acid or base present:

- The ions of the weak acid are in equilibrium with the molecular form of the acid:

$$C_6H_5COOH(aq) + H_2O(l) \rightleftharpoons C_6H_5COO^-(aq) + H_3O^+(aq)$$

In general an acid is represented by HA and can donate a proton to $H_2O(l)$.

$$HA(aq) + H_2O(l) \rightleftharpoons A^-(aq) + H_3O^+(aq)$$

- The ions of the weak base are in equilibrium with the molecular form of the base:

$$NH_3(aq) + H_2O(l) \rightleftharpoons NH_4^+(aq) + OH^-(aq)$$

In general a base is represented by B and can accept a proton from $H_2O(l)$, thereby generating $OH^-(aq)$.

$$B(aq) + H_2O(l) \rightleftharpoons BH^+(aq) + OH^-(aq)$$

3.4 A more detailed description of acid–base behaviour

3.4.1 The weak acid, e.g. benzoic acid

$$C_6H_5COOH(aq) + H_2O(l) \rightleftharpoons C_6H_5COO^-(aq) + H_3O^+(aq)$$

- $C_6H_5COOH(aq)$ is an acid because it donates a proton to $H_2O(l)$ which is a base because it accepts the proton.

- $C_6H_5COO^-(aq)$ is a base because it accepts a proton from $H_3O^+(aq)$ which is an acid because it donates a proton.

The equation can be labelled:

$$C_6H_5COOH(aq) + H_2O(l) \rightleftharpoons C_6H_5COO^-(aq) + H_3O^+(aq)$$
acid of 1   base of 2   base of 1   acid of 2

$C_6H_5COOH(aq)$ and $C_6H_5COO^-(aq)$ are called a conjugate acid–base pair.
$H_2O(l)$ and $H_3O^+(aq)$ are called a conjugate base–acid pair

If a salt of C_6H_5COOH is dissolved in water, e.g. $C_6H_5COONa(s)$ it is fully ionised in solution to give $C_6H_5COO^-(aq)$ and $Na^+(aq)$. The anion, $C_6H_5COO^-(aq)$, acts as a base:

$$C_6H_5COO^-(aq) + H_2O(l) \rightleftharpoons C_6H_5COOH(aq) + OH^-(aq)$$

because it accepts a proton from the $H_2O(l)$ which is thus acting as an acid. The equation can again be labelled as:

$$C_6H_5COO^-(aq) + H_2O(l) \rightleftharpoons C_6H_5COOH(aq) + OH^-(aq)$$
base of 1   acid of 2   acid of 1   base of 2

$C_6H_5COO^-(aq)$ and $C_6H_5COOH(aq)$ are called a conjugate base–acid pair
$H_2O(l)$ and $OH^-(aq)$ are called a conjugate acid–base pair

There are several things to note about these two equilibria:

- when $C_6H_5COOH(aq)$ acts as an acid it generates a proton, $H_3O^+(aq)$;

- when $C_6H_5COO^-(aq)$ acts as a base it generates an $OH^-(aq)$;

- $H_2O(l)$ acts as a base in the presence of the acid $C_6H_5COOH(aq)$ because it accepts a proton;

- $H_2O(l)$ acts as an acid in the presence of the base $C_6H_5COO^-(aq)$ because it donates a proton;

- $H_2O(l)$ shows **amphoteric** behaviour.

All three equilibria are examples of proton transfer reactions.

$$C_6H_5COOH(aq) + H_2O(l) \rightleftharpoons C_6H_5COO^-(aq) + H_3O^+(aq)$$
$$C_6H_5COO^-(aq) + H_2O(l) \rightleftharpoons C_6H_5COOH(aq) + OH^-(aq)$$
$$H_2O(l) + H_2O(l) \rightleftharpoons H_3O^+(aq) + OH^-(aq)$$

i.e. the unsolvated proton, H^+, is being transferred between two species. This is probably most easily seen in the third reaction.

3.4.2   The weak base, e.g. methylamine

$$CH_3NH_2(aq) + H_2O(l) \rightleftharpoons CH_3NH_3^+(aq) + OH^-(aq)$$

- $CH_3NH_2(aq)$ is a base because it accepts a proton from $H_2O(l)$ which is an acid because it donates the proton.

- $CH_3NH_3^+(aq)$ is an acid because it donates a proton to $OH^-(aq)$ which is a base because it accepts the proton.

The equation can be labelled:

$$CH_3NH_2(aq) + H_2O(l) \rightleftharpoons CH_3NH_3^+(aq) + OH^-(aq)$$
base of 1   acid of 2   acid of 1   base of 2

$CH_3NH_2(aq)$ and $CH_3NH_3^+(aq)$ are termed a conjugate base–acid pair.
$H_2O(l)$ and $OH^-(aq)$ are called a conjugate acid–base pair

If a salt of $CH_3NH_2(aq)$ is dissolved in water, e.g. $CH_3NH_3Cl(s)$, it is fully ionised in solution to give $CH_3NH_3^+(aq)$ and $Cl^-(aq)$. $CH_3NH_3^+(aq)$ is an acid because it donates a proton to the $H_2O(l)$ which is a base because it accepts the proton. The equation can be labelled as:

$$CH_3NH_3^+(aq) + H_2O(l) \rightleftharpoons CH_3NH_2(aq) + H_3O^+(aq)$$
acid of 1   base of 2   base of 1   acid of 2

$CH_3NH_3{}^+(aq)$ and $CH_3NH_2(aq)$ are called a conjugate acid–base pair.
$H_2O(l)$ and $H_3O^+(aq)$ are called a conjugate base–acid pair.

As in the weak acid case there are several things to notice about these two equilibria:

- when $CH_3NH_2(aq)$ acts as a base it generates an $OH^-(aq)$;

- when $CH_3NH_3{}^+(aq)$ acts as an acid it generates a proton, $H_3O^+(aq)$;

- $H_2O(l)$ acts as an acid in the presence of the base $CH_3NH_2(aq)$ because it donates a proton;

- $H_2O(l)$ acts as a base in the presence of $CH_3NH_3{}^+(aq)$ because it accepts a proton;

- $H_2O(l)$ is again amphoteric.

All three equilibria are examples of proton transfer reactions:

$$CH_3NH_2(aq) + H_2O(l) \rightleftharpoons CH_3NH_3^+(aq) + OH^-(aq)$$
$$CH_3NH_3^+(aq) + H_2O(aq) \rightleftharpoons CH_3NH_2(aq) + H_3O^+(aq)$$
$$H_2O(l) + H_2O(l) \rightleftharpoons H_3O^+(aq) + OH^-(aq)$$

3.4.3 The amphoteric solvent water

This equilibrium can be described in the same way.

$$H_2O(l) + H_2O(l) \rightleftharpoons OH^-(aq) + H_3O^+(aq)$$
$$\textbf{acid of 1 base of 2} \qquad \textbf{base of 1   acid of 2}$$

Here one water molecule acts as an acid by donating a proton, and the other acts as a base by accepting a proton.

$H_2O(l)$ and $OH^-(aq)$ are a conjugate acid–base pair while $H_2O(l)$ and $H_3O^+(aq)$ are a conjugate base–acid pair.

As in the weak acid, weak base cases there are several things to notice about this equilibrum

- when $H_2O(l)$ acts as an acid it generates a proton, $H_3O^+(aq)$;

- when $H_2O(l)$ acts as base it generates an $OH^-(aq)$;

- $H_2O(l)$ acts as an acid in the presence of the base $H_2O(l)$ because it donates a proton;

- $H_2O(l)$ acts as a base in the presence of the acid $H_2O(l)$ because it accepts a proton;

- $H_2O(l)$ is again amphoteric.

This equilibrium is again an example of a proton transfer

$$H_2O(l) + H_2O(l) \rightleftharpoons OH^-(aq) + H_3O^+(aq)$$

Worked problem 3.1

Question

Write down the conjugate acid or base corresponding to the following substances:

- $C_2H_5NH_2(aq)$

- $C_6H_5O^-(aq)$

- $(CH_3)_3NH^+(aq)$

- $H_3PO_4(aq)$

- $HOOC(CH_2)_3COO^-(aq)$

Write out the chemical equations describing this behaviour.

Answer

- This is a weak base with conjugate acid, $C_2H_5NH_3{}^+(aq)$

$$C_2H_5NH_2(aq) + H_2O(l) \rightleftharpoons C_2H_5NH_3^+(aq) + OH^-(aq)$$

- This is a weak base with conjugate acid $C_6H_5OH(aq)$

$$C_6H_5O^-(aq) + H_2O(l) \rightleftharpoons C_6H_5OH(aq) + OH^-(aq)$$

- This is a weak acid with conjugate base $(CH_3)_3N(aq)$

$$(CH_3)_3NH^+(aq) + H_2O(l) \rightleftharpoons (CH_3)_3N(aq) + H_3O^+(aq)$$

- This is a weak acid with conjugate base $H_2PO_4{}^-(aq)$

$$H_3PO_4(aq) + H_2O(l) \rightleftharpoons H_2PO_4^-(aq) + H_3O^+(aq)$$

- $HOOC(CH_2)_3COO^-(aq)$ can function as both a weak acid and a weak base. Which behaviour is observed will depend on the pH.

  Functioning as a weak acid, it has the conjugate base $^-OOC(CH_2)_3COO^-(aq)$.

  $HOOC(CH_2)_3COO^-(aq) + H_2O(l) \rightleftharpoons {}^-OOC(CH_2)_3COO^-(aq) + H_3O^+(aq)$

  Functioning as a weak base, it has the conjugate acid $HOOC(CH_2)_3COOH(aq)$.

  $HOOC(CH_2)_3COO^-(aq) + H_2O(l) \rightleftharpoons HOOC(CH_2)_3COOH(aq) + OH^-(aq)$

3.5 Ampholytes

These are molecules which have a basic group and an acidic group present, and can therefore function either as an acid or a base, depending on the pH of the solution.

The most common basic group is the amino group, present either as a primary, secondary or tertiary amino grouping. Other possible basic groups are those present in heterocyclic nitrogen-containing compounds such as pyridine, quinoline and pyrrole. These bases are very weak, e.g. pyrrole has $pK_b = 14$.

Typical acidic groups are the carboxylic acid, COOH, the hydroxyl, OH, the sulphonic acid, SO_3H, and the phosphoric acid, H_2PO_4, groups.

The amino acids, both aliphatic and aromatic, and the amino phenols present some interesting behaviour. Here the species which are present depend on both the pH and the relative values of the magnitudes of K_a and K_b for the acidic and basic groups respectively.

Aliphatic amino acids

These are exemplified by the simplest, glycine, NH_2CH_2COOH. The NH_2 acts as a base by accepting a proton from water:

$$NH_2CH_2COOH(aq) + H_2O(l) \rightleftharpoons {}^+NH_3CH_2COOH(aq) + OH^-(aq)$$

and the COOH group acts as an acid by donating a proton to water:

$$NH_2CH_2COOH(aq) + H_2O(l) \rightleftharpoons NH_2CH_2COO^-(aq) + H_3O^+(aq)$$

Amino acid behaviour over a range of pH is believed to involve the charge-separated $^+NH_3CH_2COO^-(aq)$. In this case the first pK relates to the equilibrium:

$$^+NH_3CH_2COOH(aq) + H_2O(l) \rightleftharpoons {}^+NH_3CH_2COO^-(aq) + H_3O^+(aq)$$

and the second pK to:

$$^+NH_3CH_2COO^-(aq) + H_2O(l) \rightleftharpoons NH_2CH_2COO^-(aq) + H_3O^+(aq)$$

The intermediate is described as a dipolar ion, a charge-separated ion, or as a zwitterion. Evidence for this structure as opposed to the molecular species will be given in Section 6.7 and Worked Problems 6.2, 6.3 and 6.4.

Aminophenols

The aminophenols can be illustrated by *o*-aminophenol $NH_2C_6H_4OH$ where the basic group is the amino, NH_2 group, which acts by accepting a proton from water,

$$NH_2C_6H_4OH(aq) + H_2O(l) \rightleftharpoons {}^+NH_3C_6H_4OH(aq) + OH^-(aq)$$

and the hydroxyl, OH group, is the acidic group which can donate a proton to water

$$NH_2C_6H_4OH + H_2O(l) \rightleftharpoons NH_2C_6H_4O^-(aq) + H_3O^+(aq)$$

Amino phenols have two pK values. However, in complete contrast to the behaviour of the aliphatic amino acids the behaviour of the aminophenols over a range of pH involves a neutral molecule as the intermediate species.

In acidic solution the amino group is protonated, and the acidic behaviour given by the first pK value is believed to relate predominantly to the equilibrium:

$$^+NH_3C_6H_4OH(aq) + H_2O(l) \rightleftharpoons NH_2C_6H_4OH(aq) + H_3O^+(aq)$$

The second pK is believed to relate to the deprotonation of the hydroxyl group:

$$NH_2C_6H_4OH + H_2O(l) \rightleftharpoons NH_2C_6H_4O^-(aq) + H_3O^+(aq)$$

The intermediate species is believed to be the neutral molecule $NH_2C_6H_4OH$.

The aromatic amino acids

The aromatic amino acids, such as the aminobenzoic acids do not show such a clear cut distinction as the aliphatic amino acids and amino phenols, though evidence suggests that the intermediate may be molecular (see Section 6.7 and Worked Problem 6.4).

3.6 Other situations where acid/base behaviour appears

There are other types of solution which are important in acid/base behaviour, and the following is a systematic classification of the combinations which are possible.

When dealing with these acid/base equilibria the crux of the problem is often to identify the **type** of solution, e.g. consider the simple case of a solution of given concentration of NaCl. This can be prepared either by dissolving the appropriate number of mol of NaCl in a given volume of water, or by mixing the appropriate volume of a given molarity of HCl with the appropriate volume of a solution of given molarity of NaOH. This latter situation corresponds to the end point in a titration of HCl with NaOH, but the resulting solution is identical in every respect to that prepared by dissolving the appropriate amount of NaCl in water.

The first essential step is to build up a systematic classification of all possible types of solution. The next step is to use the logic of this classification to recognise in which category within the classification the given solution lies. If the solution is prepared by mixing other solutions, it is necessary to work out the number of mol of each species present and to check whether any neutralisation occurs, and, if so, to check which species remains in excess. This can only be illustrated by specific examples, as follows.

Table 3.1 works systematically through cases of:

- acids and bases,

- salts,

- salts with added acid or base,

indicating how each solution could be prepared, and quoting examples of each type. The final column in this table indicates whether each solution will be acidic, basic or neutral. This is primarily an experimental observation, but the theoretical justification can be given when calculations in acid/base equilibria are carried out (see Chapters 4, 5 and 6).

3.6.1 Salts and buffers

These will be dealt with in Chapters 5 and 6; here only the definitions will be given:

- A salt is a substance formed by the complete neutralisation of an acid with a base. They can be of several types:

Table 3.1 Classification of acid/base solutions

Type of solution	Example	Comment	Acidity/basicity
(a) strong acid	HCl		acid solution
(b) strong base	NaOH		basic solution
(c) weak acid	CH_3COOH		acid solution
(d) weak base	NH_3		basic solution
(e) salt of a strong acid/strong base	NaCl	These two are really the same situation	salt: neutral solution
(f) mixture of a strong acid + strong base with equal number of mol of each	HCl neutralised with NaOH	and describe one particular solution.	
(g) salt of a strong acid/weak base	NH_4Cl	These two are really the same situation,	salt: acidic solution
(h) mixture of strong acid + weak base with equal number of mol of each	HCl neutralised with NH_3	and describe one particular solution.	
(i) salt of a weak acid /strong base	CH_3COONa	These two are really the same situation,	salt: basic solution
(j) mixture of weak acid + strong base with equal number of mol of each	CH_3COOH neutralised with NaOH	and describe one particular solution.	
(k) salt of a weak acid/weak base	CH_3COONH_4	These two are really the same situation,	salt: solution may be acid, neutral
(l) mixture of weak acid + weak base with equal number of mol of each	CH_3COOH neutralised with NH_3	and describe one particular solution.	or basic depending on the particular salt, example here would give close to neutrality
(m) salt of a strong acid/strong base with added strong acid	NaCl + HCl	These two are really the same situation, and describe one particular solution. In effect it is	acid solution
(n) mixture of strong acid + strong base, with excess strong acid	HCl mixed with NaOH, but incompletely neutralised	the same as (a).	
(o) salt of a strong acid/strong base with added strong base	NaCl + NaOH	These two are really the same situation, and describe one	basic solution
(p) mixture of strong acid + strong base, with excess base	HCl mixed with NaOH, and taken past the end point with excess NaOH	particular solution. In effect it is the same as (b).	

(continued)

Table 3.1   (*continued*)

Type of solution	Example	Comment	Acidity/basicity
(q) salt of a strong acid/weak base with added strong acid	$NH_4Cl + HCl$	These two are really the same situation, and describe one particular solution. In effect it is the same as (a).	acid solution
(r) mixture of strong acid + weak base with excess strong acid	HCl mixed with NH_3, but incompletely neutralised		
(s) salt of a weak acid/strong base with added strong base	$CH_3COONa + NaOH$	These two are really the same situation, and describe one particular solution. In effect it is the same as (b).	basic solution,
(t) mixture of weak acid + strong base with excess strong base	CH_3COOH mixed with NaOH and taken past the end point with excess NaOH		
(u) salt of a weak acid/ strong base with added weak acid	$CH_3COONa + CH_3COOH$	These two are really the same situation, and describe one particular solution.	acid solution, and it is a **buffer solution**
(v) mixture of weak acid + strong base with excess weak acid	CH_3COOH mixed with NaOH, but incompletely neutralised		
(w) salt of a strong acid/ weak base with added weak base	$NH_4Cl + NH_3$	These two are really the same situation, and describe one particular solution.	basic solution, and it is a **buffer solution**
(x) mixture of strong acid + weak base with excess weak base	HCl mixed with NH_3 and taken past the end point with excess NH_3.		
(α) salt of a weak acid/ weak base with added weak acid	$CH_3COONH_4 + CH_3COOH$	These two are really the same situation, and describe one particular solution.	acid solution, and to a limited extent it is a **buffer solution**
(β) mixture of weak acid + weak base with excess weak acid	CH_3COOH mixed with NH_3 but incompletely neutralised		
(γ) salt of a weak acid/ weak base with added weak base	$CH_3COONH_4 + NH_3$	These two are really the same situation, and describe one particular solution.	basic solution, and to a limited extent it is a **buffer solution**
(δ) mixture of a weak acid + weak base with excess weak base	CH_3COOH mixed with NH_3, and taken past the end point with excess NH_3		

- a salt of a strong acid/strong base, e.g. NaCl,

- a salt of a strong acid/weak base, e.g. NH_4NO_3,

- a salt of a weak acid/strong base, e.g. CH_3COONa,

- a salt of a weak acid/weak base, e.g. $HCOONH_4$.

- A buffer solution consists of:

  - a weak acid + its salt with a strong base, e.g. $CH_3COOH + CH_3COONa$,

  - a weak base + its salt with a strong acid, e.g. $NH_3 + NH_4Cl$.

Worked problem 3.2

Question

Classify the following solutions into strong acid, buffer etc:

(a) $NH_3(aq)$ (b) $CH_3COOH/CH_3COONa(aq)$
(c) $KNO_3(aq)$ (d) $HCl(aq)$
(e) $CH_3CH_2NH_3Br(aq)$ (f) NH_3/excess $NaOH(aq)$
(g) $NH_4Cl/NH_3(aq)$ (h) $C_6H_5COOH(aq)$
(i) $C_6H_5OH(aq)$ (j) $C_6H_5COOK(aq)$

Answer

(a) weak base   (b) buffer of a weak acid
(c) salt of a strong acid/strong base (d) strong acid
(e) salt of a strong acid/weak base   (f) strong base,
(g) buffer of a weak base   (h) weak acid
(i) weak acid   (j) salt of a weak acid/strong base

Worked problem 3.3

Question

Classify the following solutions:

(a) 0.027 mol of $NH_3(aq) + 0.045$ mol of $HCl(aq)$

(b) 0.046 mol of $NH_3(aq) + 0.023$ mol of $HCl(aq)$

(c) 0.045 mol of $C_6H_5COOH(aq) + 0.045$ mol of $NH_3(aq)$

(d) 0.025 mol of $HCOOH(aq) + 0.015$ mol of $NaOH(aq)(d)$

(e) $20 \, cm^3$ of $0.05 \, mol \, dm^{-3}$ $CH_3CH_2COOH(aq)$ mixed with $10 \, cm^3$ of $0.02 \, mol \, dm^{-3}$ of $NaOH(aq)$

(f) $10 \, cm^3$ of $0.10 \, mol \, dm^{-3}$ $CH_3NH_2(aq)$ mixed with $10 \, cm^3$ of $0.08 \, mol \, dm^{-3}$ of $HCl(aq)$

Answer

(a) This is equivalent to $0.027 \, mol$ of $NH_4Cl(aq) + 0.018 \, mol$ of excess $HCl(aq)$, i.e. a salt of a strong acid/weak base + strong acid in excess, in effect a solution of a strong acid.

(b) This is equivalent to $0.023 \, mol$ of $NH_4Cl(aq) + 0.023 \, mol$ of $NH_3(aq)$, i.e. a salt of a strong acid/weak base + weak base, in effect a buffer solution.

(c) This is equivalent to $0.045 \, mol$ of $C_6H_5COONH_4(aq)$ i.e. a salt of a weak acid/weak base.

(d) This is equivalent to $0.015 \, mol$ of $HCOONa(aq) + 0.010 \, mol$ of $HCOOH(aq)$, i.e. a salt of a weak acid/strong base + weak acid, in effect a buffer solution.

Parts (e) and (f) require a calculation of the number of mol in each solution. Thereafter the argument is as in (a) to (d).

- concentration $= \dfrac{\text{number of mol}}{\text{volume of solution}}$

- $\therefore$ number of mol = concentration $\times$ volume

Care must be taken with the units: since concentration is usually given as $mol \, dm^{-3}$, then the volume must be given in dm^3. In this question volumes are given as, e.g. $10 \, cm^3 = 10 \times 10^{-3} \, dm^3$.

(e) The solution contains $1.0 \times 10^{-3} \, mol$ of $CH_3CH_2COOH(aq) + 0.2 \times 10^{-3} \, mol$ of $NaOH(aq)$. This is equivalent to $0.2 \times 10^{-3} \, mol$ of $CH_3CH_2COONa(aq) + 0.8 \times 10^{-3} \, mol$ of $CH_3CH_2COOH(aq)$ in excess, i.e. salt of a weak acid/strong base + excess weak acid, in effect a buffer solution.

(f) The solution contains $1.0 \times 10^{-3} \, mol$ of $CH_3NH_2(aq) + 0.8 \times 10^{-3} \, mol$ of $HCl(aq)$. This is equivalent to $0.8 \times 10^{-3} \, mol$ of $CH_3NH_3Cl(aq) + 0.2 \times 10^{-3} \, mol$ of $CH_3CH_2NH_2(aq)$ in excess, i.e. salt of a strong acid/weak base + excess weak base, in effect a buffer solution.

3.7 Formulation of equilibrium constants in acid–base equilibria

The weak acid:
$$HA(aq) + H_2O(l) \rightleftharpoons A^-(aq) + H_3O^+(aq)$$

$$K_a = \left(\frac{[A^-]_{actual}[H_3O^+]_{actual}}{[HA]_{actual}} \right)_{eq} \tag{3.1}$$

The weak base:

$$B(aq) + H_2O(l) \rightleftharpoons BH^+(aq) + OH^-(aq)$$

$$K_b = \left(\frac{[BH^+]_{actual}[OH^-]_{actual}}{[B]_{actual}} \right)_{eq} \tag{3.2}$$

There are three points to note:

- there is no term in $[H_2O]$ in either of these equilibrium quotients;

- the subscript eq refers to the equilibrium relation and emphasises that the concentrations appearing are equilibrium concentrations;

- more trivially, the subscript 'a' refers to an equilibrium constant for an acid while the subscript 'b' refers to an equilibrium constant for a base. This becomes important when considering equilibrium constants for conjugate acid–base pairs.

3.8   Magnitudes of equilibrium constants

A useful rule of thumb:

- strong acids/bases are completely ionised, and so do not have an equilibrium constant;

- moderately strong acids or bases have pK_as and pK_bs of the order of < around 3.5;

- weak acids or bases have pK_as and pK_bs of the order of around 3.5 to 10.5;

- very weak acids or bases have pK_as and pK_bs of > 10.5.

Remember: $pK = -\log_{10} K$

3.9   The self ionisation of water

$$H_2O(l) + H_2O(l) \rightleftharpoons OH^-(aq) + H_3O^+(aq)$$

$$K_w = ([H_3O^+]_{actual}[OH^-]_{actual})_{eq} \tag{3.3}$$

K_w is often called the ionic product of water.
Note: there is no term in $H_2O(l)$ in this expression. That this is so will be demonstrated in the thermodynamic derivation of the equilibrium expression (Section 8.14).

At 25°C, $K_w = 10^{-14} \, mol^2 \, dm^{-6}$ corresponding to $pK_w = 14$ at 25°C. Like most other equilibrium constants K_w varies with temperature (see Section 2.9 and Worked Problem 2.4).

The self ionisation of $H_2O(l)$ must always be considered when acid/base behaviour in aqueous solution is discussed. For instance, a weak acid in aqueous solution is a source of $H_3O^+(aq)$, but so also is the solvent, $H_2O(l)$; a weak base in aqueous solution is a source of

$OH^-(aq)$, but so also is the solvent, $H_2O(l)$.

$$HA(aq) + H_2O(l) \rightleftharpoons H_3O^+(aq) + A^-(aq)$$

$$H_2O(l) + H_2O(l) \rightleftharpoons H_3O^+(aq) + OH^-(aq)$$

'Does this matter'?

Whether it does, or does not, depends critically on the relative contributions to the concentration of $H_3O^+(aq)$ coming from each of the equilibria. If the weak acid is sufficiently ionised so that its contribution to the actual concentration of $H_3O^+(aq)$ is large compared with that from the self ionisation of water then, to a first approximation, the self ionisation of water can be ignored. The weak acid can then be taken to be the only effective source of $H_3O^+(aq)$. Provided the total concentration of the weak acid is not too small, this generally holds for acids with $pK < 10.0$.

However, when a very weak acid is considered, e.g. sucrose or glucose with pK_as of 12.7 and 12.2 respectively, the fraction ionised is very low. Here the contribution to the actual concentration of $H_3O^+(aq)$ from the self ionisation of the solvent water becomes significant and must **not** be ignored. The main source of $H_3O^+(aq)$ is now from the self-ionisation of water.

A similar situation arises for weak bases:

$$B(aq) + H_2O(l) \rightleftharpoons BH^+(aq) + OH^-(aq)$$

$$H_2O(l) + H_2O(l) \rightleftharpoons H_3O^+(aq) + OH^-(aq)$$

Again both the weak base and the solvent water are sources of $OH^-(aq)$, and the relative contributions to the actual $[OH^-]_{actual}$ from both have to assessed and an active decision has to be made as to whether the contribution from the self ionisation of water can be ignored.

These points are dealt with in detail in Chapter 4.

3.10   Relations between K_a and K_b: expressions for an acid and its conjugate base and for a base and its conjugate acid

The acid/base pair HA/A^-

There are three equilibria involved:
The acid:

$$HA(aq) + H_2O(l) \rightleftharpoons H_3O^+(aq) + A^-(aq)$$

$$K_a = \left(\frac{[H_3O^+]_{actual}[A^-]_{actual}}{[HA]_{actual}} \right)_{eq} \tag{3.1}$$

The conjugate base:

$$A^-(aq) + H_2O(l) \rightleftharpoons HA(aq) + OH^-(aq)$$

$$K_b = \left(\frac{[HA]_{actual}[OH^-]_{actual}}{[A^-]_{actual}} \right)_{eq} \tag{3.4}$$

The self ionisation of water:

$$H_2O(l) + H_2O(l) \rightleftharpoons H_3O^+(aq) + OH^-(aq)$$

$$K_w = \left([H_3O^+]_{actual}[OH^-]_{actual}\right)_{eq} \tag{3.3}$$

There are some situations for which the magnitude of K_b is required, but where only K_a is quoted. The above expressions can be used to obtain a relation between K_b, K_a, and K_w.

$$K_a \times K_b = \left(\frac{[H_3O^+]_{actual}[A^-]_{actual}}{[HA]_{actual}}\right)_{eq} \times \frac{([HA]_{actual}[OH^-]_{actual}}{[A^-]_{actual}}\right)_{eq} \tag{3.5}$$

$$= \left([H_3O^+]_{actual} \times [OH^-]_{actual}\right)_{eq} \tag{3.6}$$

$$= K_w \tag{3.7}$$

$$\therefore K_b = \frac{K_w}{K_a} \tag{3.8}$$

This can be exemplified by the acidic behaviour of propanoic acid, CH_3CH_2COOH, and the basic behaviour of the propanoate ion, $CH_3CH_2COO^-$.

$$CH_3CH_2COOH(aq) + H_2O(l) \xrightarrow{K_a} CH_3CH_2COO^-(aq) + H_3O^+(aq)$$

$$CH_3CH_2COO^-(aq) + H_2O(l) \xrightarrow{K_b} CH_3CH_2COOH(aq) + OH^-(aq)$$

The conjugate base/acid pair B/BH$^+$

There are three equilibria involved.
The base:

$$B(aq) + H_2O(l) \rightleftharpoons BH^+(aq) + OH^-(aq)$$

$$K_b = \left(\frac{[BH^+]_{actual}[OH^-]_{actual}}{[B]_{actual}}\right)_{eq} \tag{3.2}$$

The conjugate acid:

$$BH^+ + H_2O(l) \rightleftharpoons B(aq) + H_3O^+(aq)$$

$$K_a = \left(\frac{[B]_{actual}[H_3O^+]_{actual}}{[BH^+]_{actual}}\right)_{eq} \tag{3.9}$$

The self ionisation of water:

$$H_2O(l) + H_2O(l) \rightleftharpoons H_3O^+(aq) + OH^-(aq)$$

$$K_w = \left([H_3O^+]_{actual}[OH^-]_{actual}\right)_{eq} \tag{3.3}$$

Again, there is a similar relation between the three equilibrium constants.

$$K_a \times K_b = \left(\frac{[BH^+]_{actual}[OH^-]_{actual}}{[B]_{actual}}\right)_{eq} \times \left(\frac{[B]_{actual}[H_3O^+]_{actual}}{[BH^+]_{actual}}\right)_{eq} \tag{3.10}$$

$$= \left([OH^-]_{actual}[H_3O^+]_{actual}\right)_{eq} \tag{3.6}$$

$$= K_w \tag{3.7}$$

$$\therefore K_b = \frac{K_w}{K_a} \tag{3.8}$$

This can be exemplified by the basic behaviour of $NH_3(aq)$ and the acid behaviour of its conjugate acid $NH_4^+(aq)$.

$$NH_3(aq) + H_2O(l) \overset{K_b}{\rightleftharpoons} NH_4^+(aq) + OH^-(aq)$$

$$NH_4^+(aq) + H_2O(l) \overset{K_a}{\rightleftharpoons} NH_3(aq) + H_3O^+(aq)$$

Knowledge of these conversions is essential. Modern texts now compile tables of acidity constants only. In these tables, the K_a values for the conjugate acids of typical weak bases are the quantities given, and conversion of these to give the K_b values for the bases will be necessary, e.g. the acidity constants for $CH_3NH_2^+$ or $(CH_3)_2NH^+$ will be listed, from which the constants, K_b, for the bases CH_3NH_2 and $(CH_3)_2NH$ will have to be inferred.

3.11 Stoichiometric arguments in equilibria calculations

It is important to distinguish between the **actual** concentrations present in a solution, and the **total stoichiometric** concentrations (see Section 2.11).

When a solution of, for example, benzoic acid is prepared by dissolving a given number of mol in a specific volume of solution, the concentration is given as number of mol/volume of solution. When a concentration is worked out in this way, i.e. in terms of what was weighed out, it is called the **stoichiometric** concentration. But what really is of interest is what is **actually present** in solution, and a concentration based on this is called the **actual** concentration. If, on dissolution, the solute is involved in any reaction or equilibrium, then the stoichiometric concentration will not correspond to what is actually present.

When a weak acid such as ethanoic acid is dissolved in water equilibrium is established, the species present in solution are $H_3O^+(aq)$, the ethanoate ion, $CH_3COO^-(aq)$, and the **remaining** undissociated ethanoic acid, $CH_3COOH(aq)$.

$$CH_3COOH(aq) + H_2O(l) \rightleftharpoons CH_3COO^-(aq) + H_3O^+(aq)$$

and so the **total** or **stoichiometric** concentration of $CH_3COOH(aq)$ is distributed between the **actual** concentration of $CH_3COOH(aq)$ **remaining** and the **actual** concentration of $CH_3COO^-(aq)$ **formed**. This can be quantified as:

$$[CH_3COOH]_{total\ or\ stoich} = [CH_3COOH]_{actual} + [CH_3COO^-]_{actual} \tag{3.11}$$

and it is these **actual** concentrations which are related to each other by the equilibrium expression:

$$K_a = \left(\frac{[CH_3COO^-]_{actual}[H_3O^+]_{actual}}{[CH_3COOH]_{actual}} \right)_{eq} \tag{3.12}$$

These distinctions between **actual** and **total** concentrations are also important when formulating the expression for the fraction ionised:

$$\text{fraction ionised} = \left(\frac{\text{actual amount of the original substance present as ions}}{\text{total amount of the substance present originally}} \right)_{\substack{(3.13) \\ \text{at equilibrium}}}$$

$$= \left(\frac{[CH_3COO^-]_{actual}}{[CH_3COOH]_{total}} \right)_{eq} \tag{3.14}$$

A decision may have to be made as to whether it is legitimate to **approximate** actual concentrations with total concentrations. Sometimes this may be justified and sometimes it is not, and each case must be decided individually. In the present case, ethanoic acid is a weak acid, and **unless the total concentration is very low** it will be valid to approximate:

$$\text{i.e. } [CH_3COOH]_{actual} \approx [CH_3COOH]_{total} \tag{3.15}$$

When the stoichiometric concentration of the ethanoic acid is very low, it is no longer legitimate to make this approximation.

For a moderately strong weak acid such as chlorethanoic acid, $ClCH_2COOH$, there is fairly extensive ionisation forming the anion, $ClCH_2COO^-(aq)$, and a significant amount of the original $ClCH_2COOH(aq)$ will have been removed.

$$ClCH_2COOH(aq) + H_2O(l) \rightleftharpoons ClCH_2COO^-(aq) + H_3O^+(aq)$$
$$[ClCH_2COOH]_{total} = [ClCH_2COOH]_{actual} + [ClCH_2COO^-]_{actual} \tag{3.16}$$

In this case no approximation can be made.

3.12 Procedure for calculations on equilibria

The first stage in any calculation requires writing down the following:

- the chemical equation describing the equilibrium

- formulating the expression for the equilibrium constant;

- writing down the sources of the species appearing in the equilibrium constant;

- formulating the relation between the total and actual concentrations for relevant species;

- looking to see if there are any approximations possible, and noting their limitations and consequences.

These points are relevant to any equilibrium whether it be for a weak acid, a buffer solution, formation of an ion pair or a solubility problem. It will become clear that this procedure will be used time and time again in this book.

4
Equilibrium Calculations for Acids and Bases

In this chapter the arguments involved in calculations on weak acid and weak base equilibria and pH are developed with special emphasis on the reasoning and logic involved in the calculations. Appreciation and understanding of the points made in this chapter are highly relevant to all equilibrium calculations such as those for gas phase, complex formation, ion pairing and heterogeneous reactions. The reasoning involved in all these apparently disparate situations is similar, and so a full understanding of this chapter forms a strong basis for understanding in other types of equilibria. In particular, the chemical significance of the assumptions and various approximations is explained fully, as are the physical limitations imposed by these approximations. The chapter is progressive in terms of the rigour of the calculations.

The chapter ends with a reassessment and rigorous derivation of the pH of a solution of a weak acid or base where no approximations are made. This section can be ignored on a first reading, and only studied when a full assimilation of the earlier material has been achieved.

Aims

By the end of this chapter you should be able to:

- list the species present and their sources for solutions of a weak acid or weak base;

- understand the concept of electrical neutrality;

- appreciate the approximations which can be made in such calculations, and the limitations to their application, and in particular;

- discuss, apply and explain when the approximations relating to the extent of ionisation and to the self ionisation of water can, or cannot, be made for a weak acid and for a weak base;

An Introduction to Aqueous Electrolyte Solutions. By Margaret Robson Wright
© 2007 John Wiley & Sons Ltd ISBN 978-0-470-84293-5 (cloth) ISBN 978-0-470-84294-2 (paper)

- discuss the dependence of the fraction ionised on the pH, or on pK_a or pK_b, and demonstrate that this fraction reaches a limiting value which is not necessarily 100%, but can be considerably smaller for weaker acids and bases;

- explain the effect of the **stoichiometric** concentration of the weak acid or base on the fraction ionised for acids and bases.

4.1   Calculations on equilibria: weak acids

Let the weak acid be represented by HA.

$$HA(aq) + H_2O(l) \rightleftharpoons H_3O^+(aq) + A^-(aq)$$

$$K_a = \left(\frac{[H_3O^+]_{actual}[A^-]_{actual}}{[HA]_{actual}} \right)_{eq} \tag{4.1}$$

This is an equation in three unknowns, provided the magnitude of K_a is known.

The general procedure to be followed in all calculations requires consideration of the following points.

- **Species present in solution are:**

the acid, HA(aq), its conjugate base, $A^-(aq)$, $H_3O^+(aq)$, $H_2O(l)$ and a trace amount of $OH^-(aq)$ by virtue of the self ionisation of water.

- **Source of A^-(aq):**

the solution is one of the weak acid on its own, and HA(aq) is the only source of $A^-(aq)$.

- **Sources of H_3O^+(aq):**

  - $H_3O^+(aq)$ is formed, along with $A^-(aq)$, by the ionisation of the weak acid HA(aq);

  - the self ionisation of water is also a source of $H_3O^+(aq)$;

  - there are, therefore, **two** sources of $H_3O^+(aq)$;

  - a decision will have to be made as to whether the self ionisation of water can be ignored.

- **Stoichiometric relations:**

$$[HA]_{total\ or\ stoich} = [HA]_{actual} + [A^-]_{actual} \tag{4.2}$$

$$\therefore [HA]_{actual} = [HA]_{total\ or\ stoich} - [A^-]_{actual} \tag{4.3}$$

A decision will have to be made as to whether it is valid to approximate this expression to

$$[HA]_{actual} \approx [HA]_{total\ or\ stoich}$$

- **Substituting into the equilibrium expression:**

this gives:

$$K_a = \left(\frac{[H_3O^+]_{actual}[A^-]_{actual}}{[HA]_{total\ or\ stoich} - [A^-]_{actual}} \right)_{eq} \tag{4.4}$$

This reduces the expression to one in two unknowns. To solve this equation it is necessary:

- **either** to obtain another independent equation;

- **or** to make approximations.

4.1.1 Possible approximations for the weak acid

Approximation 1 relating to the extent of ionisation of the weak acid

In the ionisation:

$$HA(aq) + H_2O(l) \rightleftharpoons H_3O^+(aq) + A^-(aq)$$

if only a small amount of the weak acid is ionised, then only a small amount of HA(aq) will be removed and the concentration of the anion, A^-(aq), formed will be small compared with the stoichiometric concentration of the weak acid. The actual concentration of HA(aq) remaining will consequently be only slightly less than the initial stoichiometric concentration, i.e.:

$$[A^-]_{actual} \ll [HA]_{total\ or\ stoich} \tag{4.5}$$

$$\therefore [HA]_{actual} \approx [HA]_{total\ or\ stoich} \tag{4.6}$$

so that:

$$K_a \approx \left(\frac{[H_3O^+]_{actual}[A^-]_{actual}}{[HA]_{total\ or\ stoich}} \right)_{eq} \tag{4.7}$$

This approximation is based on the assumption that only a small amount of ionisation occurs.

But is it always justifiable to make this approximation?

The approximation generally holds for acids with pK_a values around 4.0 or greater. However, diluting a solution of a weak acid dramatically increases the fraction ionised, so that very low stoichiometric concentrations will render the approximation invalid (see Section 4.7). For stronger acids, the extent of ionisation becomes sufficiently large for the approximation not to hold (see Section 4.1.3).

The approximation holds:

provided the extent of ionisation is limited and the acid is not too dilute.

Approximation 2 relating to the self ionisation of water

In the ionisation:

$$HA(aq) + H_2O(l) \rightleftharpoons H_3O^+(aq) + A^-(aq)$$

$[H_3O^+]_{actual}$ is made up from $H_3O^+(aq)$ produced by the ionisation of HA **and** from $H_3O^+(aq)$ produced by the self ionisation of $H_2O(l)$. If the contribution from HA(aq) is large compared with that from the $H_2O(l)$, then HA(aq) is the main source of $H_3O^+(aq)$, and to a first approximation the self ionisation can be ignored as a source of $H_3O^+(aq)$. In the case of a weak acid on its own, HA(aq) is the only source of $A^-(aq)$, and, if the approximation above is made, it is also the only source of $H_3O^+(aq)$, and as they are produced in a 1:1 ratio:

$$[H_3O^+]_{actual} \approx [A^-]_{actual} \tag{4.8}$$

Substituting into the equilibrium expression:

$$K_a = \left(\frac{[H_3O^+]_{actual}[A^-]_{actual}}{[HA]_{actual}} \right)_{eq} \tag{4.1}$$

gives:

$$K_a \approx \left(\frac{[H_3O^+]^2_{actual}}{[HA]_{actual}} \right)_{eq} \tag{4.9}$$

$$\approx \left(\frac{[H_3O^+]^2_{actual}}{[HA]_{total} - [H_3O^+]_{actual}} \right)_{eq} \tag{4.10}$$

If K_a is known, then the approximation has reduced the equilibrium expression to one in one unknown, $[H_3O^+]_{actual}$.

But is it justifiable to ignore the contribution to $[H_3O^+]_{actual}$ from the self ionisation of $H_2O(l)$?

This approximation holds for acids with pK_a of around 10.0 or less, unless the concentration is very low. If the concentration of the weak acid is very low, then **even if** all the acid were ionised, the concentration of $H_3O^+(aq)$ would still approach that from the self ionisation of water. For acids which have $pK_a > 10$ and are thus very weak acids the extent of ionisation will be so low that the contribution to $[H_3O^+]$ from the self ionisation of water must not be ignored (see Section 4.1.4).

The approximation will hold:

provided the acid is not too weak and not too dilute.

4.1.2   The weak acid where both approximations can be made

$$K_a = \left(\frac{[H_3O^+]_{actual}[A^-]_{actual}}{[HA]_{actual}} \right)_{eq} \tag{4.1}$$

If both approximations hold then:

- $[HA]_{actual} \approx [HA]_{total\ or\ stoich}$ (4.6)

- $[H_3O^+]_{actual} \approx [A^-]_{actual}$ (4.8)

giving:

$$K_a \approx \left(\frac{[H_3O^+]^2_{actual}}{[HA]_{total\ or\ stoich}} \right)_{eq}$$ (4.11)

This is an equation in one unknown and can be solved directly.

$$[H_3O^+]_{actual} = \sqrt{K_a[HA]_{total}}$$ (4.12)

$$pH = -\log_{10}[H_3O^+]_{actual}$$ (4.13)

This argument will hold for acids with pK_a lying in the range 4–10, provided the solution is not too dilute.

4.1.3 The weak acid where there is extensive ionisation and approximation 1 is invalid

This will happen with moderately strong weak acids of pK_a less than about 4, or for weaker acids in dilute solution.

Where there is fairly extensive ionisation, the contribution to the **actual** concentration of $H_3O^+(aq)$ from the ionisation of $HA(aq) \gg$ than that from the self ionisation of $H_2O(l)$. The weak acid is then effectively the only source of both $A^-(aq)$ and $H_3O^+(aq)$ and the self ionisation of water can be ignored:

$$\therefore [H_3O^+]_{actual} \approx [A^-]_{actual}$$ (4.8)

i.e. approximation 2 holds.

However, in this case the first approximation does not hold, i.e.

$$[HA]_{actual} \neq [HA]_{total\ or\ stoich}$$ (4.14)

and the rigorous expression:

$$[HA]_{actual} = [HA]_{total\ or\ stoich} - [A^-]_{actual}$$ (4.3)

must be used in the equilibrium expression:

$$K_a = \left(\frac{[H_3O^+]_{actual}[A^-]_{actual}}{[HA]_{actual}} \right)_{eq}$$ (4.1)

The final expression is therefore:

$$K_a \approx \left(\frac{[H_3O^+]^2_{actual}}{[HA]_{total} - [A^-]_{actual}} \right)_{eq} \tag{4.15}$$

$$\approx \left(\frac{[H_3O^+]^2_{actual}}{[HA]_{total} - [H_3O^+]_{actual}} \right)_{eq} \tag{4.10}$$

This is an equation in one unknown, $[H_3O^+]_{actual}$, and can, therefore, be solved. It is, however, a quadratic equation, but it can be solved by the standard mathematical technique:

$$[H_3O^+]^2_{actual} + K_a[H_3O^+]_{actual} - K_a[HA]_{total} = 0 \tag{4.16}$$

which is of the form:

$$ax^2 + bx + c = 0 \tag{4.17}$$

and has solution:

$$x = \frac{-b \pm \sqrt{b^2 - 4ac}}{2a} \tag{4.18}$$

where:

$$x = [H_3O^+]_{actual} \tag{4.19}$$

$$a = 1 \tag{4.20}$$

$$b = K_a \tag{4.21}$$

$$c = -K_a[HA]_{total} \tag{4.22}$$

from which the pH of the solution can be found.

4.1.4   The weak acid is sufficiently weak so that the self ionisation of water cannot be ignored and approximation 2 is invalid

For acids with pK_a greater than around 10.0 the acid is so weak that the fraction ionised is very low. The contribution to the **actual** concentration of H_3O^+ from the weak acid is then so small that the contribution from the self ionisation of water is comparable to, or even greater than, that from the weak acid. There are now **two** sources of $H_3O^+(aq)$.

$$
\begin{aligned}
[H_3O^+]_{actual} &= \text{contribution from HA(aq)} + \text{contribution from H}_2\text{O(l)} \\
&= [A^-]_{actual} + \text{contribution from H}_2\text{O(l)}
\end{aligned}
\tag{4.23}
$$

There is only **one** source of $A^-(aq)$, the weak acid HA(aq), but since both contributions to $[H_3O^+]_{actual}$ must still be considered (Equation 4.23), then it follows that:

$$[H_3O^+]_{actual} \neq [A^-]_{actual} \tag{4.24}$$

and both concentrations must be known explicitly.

However, because the fraction ionised is so low, then the actual concentration of HA(aq) will be very close to the stoichiometric concentration of HA(aq), and the first approximation is valid, i.e.:

$$[HA]_{actual} \approx [HA]_{total\ or\ stoich} \qquad (4.6)$$

Taking the equilibrium expression:

$$K_a = \left(\frac{[H_3O^+]_{actual}[A^-]_{actual}}{[HA]_{actual}} \right)_{eq} \qquad (4.1)$$

and applying the above conclusions gives:

$$K_a \approx \left(\frac{[H_3O^+]_{actual}[A^-]_{actual}}{[HA]_{total\ or\ stoich}} \right)_{eq} \qquad (4.7)$$

This is an equation in two unknowns, and, as in Section 4.1, either a further approximation is required, or a second independent equation must be formulated. There are no further approximations possible, but electrical neutrality (see Section 4.1.5 and Section 2.12) requires that:

$$[H_3O^+]_{actual} = [A^-]_{actual} + [OH^-]_{actual} \qquad (4.25)$$

This now has three unknowns since it includes the unknown $[OH^-]_{actual}$. However, there is a further relation between the two unknowns, $[H_3O^+]_{actual}$ and $[OH^-]_{actual}$, namely the ionic product of water:

$$K_w = [H_3O^+]_{actual}[OH^-]_{actual} \qquad (4.26)$$
$$[OH^-]_{actual} = K_w/[H_3O^+]_{actual} \qquad (4.27)$$

giving:

$$[A^-]_{actual} = [H_3O^+]_{actual} - K_w/[H_3O^+]_{actual} \qquad (4.28)$$

which can be substituted into Equation (4.7) giving:

$$K_a \approx \left(\frac{[H_3O^+]_{actual}\left([H_3O^+]_{actual} - \dfrac{K_w}{[H_3O^+]_{actual}} \right)}{[HA]_{total\ or\ stoich}} \right)_{eq} \qquad (4.29)$$

$$\approx \left(\frac{[H_3O^+]^2_{actual} - K_w}{[HA]_{total\ or\ stoich}} \right)_{eq} \qquad (4.30)$$

$$[H_3O^+]^2_{actual} = K_a[HA]_{total\ or\ stoich} + K_w \qquad (4.31)$$

$$[H_3O^+]_{actual} = \sqrt{(K_a[HA]_{total\ or\ stoich} + K_w)} \qquad (4.32)$$

from which the pH can be calculated.

Take note: This calculation made use of a very important property of electrolyte solutions which is often required in calculations on solution equilibria, i.e. electrical neutrality.

4.1.5   Electrical neutrality

All electrolyte solutions are electrically neutral, i.e. the positive charges are balanced out by the negative charges:

the total positive charge = the total negative charge.

(see Section 2.12).

In a solution of a weak acid HA:

- The positive charge comes from $H_3O^+(aq)$, and each $H_3O^+(aq)$ contributes one $+ve$ charge.

- The negative charge comes from $A^-(aq)$ and $OH^-(aq)$, and each $A^-(aq)$ contributes one $-ve$ charge and each $OH^-(aq)$ contributes one $-ve$ charge.

If each $H_3O^+(aq)$ contributes one positive charge, then $n_{H_3O^+}$ mol of $H_3O^+(aq)$ will contribute $n_{H_3O^+}$ mol of positive charges. Hence the total positive charge from a solution containing $H_3O^+(aq)$ will be proportional to the concentration, $(n_{H_3O^+}/\text{volume of solution})$, i.e. $[H_3O^+]_{actual}$. Applying this argument to the anions, $A^-(aq)$, $OH^-(aq)$, electrical neutrality requires that:

$$[H_3O^+]_{actual} = [A^-]_{actual} + [OH^-]_{actual} \tag{4.25}$$

When a divalent or trivalent ion is present, this multiple charge on the ion must be taken care of. Each $SO_4^{2-}(aq)$ present in a solution will contribute two negative charges, and the total negative charge contributed would be $2[SO_4^{2-}]$.

Likewise each triply charged ion such as $Al^{3+}(aq)$ would contribute three positive charges, and the total positive charge contributed would be $3[Al^{3+}]$.

4.2   Some worked examples

Working through problems on pH is one of the best ways of checking whether the principles are fully understood. This, of course, implies that the problems are tackled from first principles and **not** by direct substitution into standard equations. It is very important that the significance and limitations of the various approximations are understood, and there is no better way of ensuring this than by **detailed** working through problems. Much of the logic and reasoning is similar to that used in other types of equilibria, e.g. solubility, ion pair and complex formation.

Worked problem 4.1

Question

Working from first principles, find the pH of a 20 cm^3 sample of a 0.05 mol dm^{-3} solution of a weak acid of p$K_a = 5.75$ at 25°C. Calculate the fraction ionised, and comment on its value. How does this correlate with any assumptions made in the calculation of the pH?

Answer

$$pK_a = 5.75 \therefore K_a = 1.778 \times 10^{-6} \, \text{mol dm}^{-3}$$

$$HA(aq) + H_2O(l) \rightleftharpoons H_3O^+(aq) + A^-(aq)$$

$$K_a = \left(\frac{[H_3O^+]_{actual}[A^-]_{actual}}{[HA]_{actual}} \right)_{eq}$$

This is a fairly weak acid present at a moderate concentration.

- The self ionisation of water can be ignored, the ionisation of the acid will be sufficiently extensive for the acid to be the main source of $H_3O^+(aq)$.

- There is one source of $H_3O^+(aq)$ and $A^-(aq)$, and they are formed in a 1:1 ratio,

$$\therefore [H_3O^+]_{actual} = [A^-]_{actual}$$

- The acid is sufficiently weak to be only slightly ionised.

$$\therefore [HA]_{actual} = [HA]_{total}$$

$$\therefore K_a = \left(\frac{[H_3O^+]^2_{actual}}{[HA]_{total}} \right)_{eq}$$

$$[H_3O^+] = (1.778 \times 10^{-6} \, \text{mol dm}^{-3} \times 0.05 \, \text{mol dm}^{-3})^{1/2}$$
$$= 2.98 \times 10^{-4} \, \text{mol dm}^{-3}$$
$$pH = 3.53$$

$$\text{Fraction ionised} = \left(\frac{[A^-]_{actual}}{[HA]_{total}} \right)_{eq}$$

Since in the approximations used: $[H_3O^+]_{actual} = [A^-]_{actual}$

$$\text{Fraction ionised} = \left(\frac{[H_3O^+]_{actual}}{[HA]_{total}} \right)_{eq} = \frac{2.98 \times 10^{-4} \, \text{mol dm}^{-3}}{0.05 \, \text{mol dm}^{-3}} = 5.96 \times 10^{-3}$$

Comments

- $[H_3O^+] = 2.98 \times 10^{-4} \, \text{mol dm}^{-3}$. This should be compared with $[H_3O^+] = 10^{-7} \, \text{mol dm}^{-3}$ in pure water at 25°C. Hence it is fully justified to assume that the self ionisation of water can be ignored.

- The fraction ionised $= 5.96 \times 10^{-3}$. Hence the weak acid is only 0.6% ionised. This fully justifies the assumption that $[HA]_{actual} \approx [HA]_{total}$.

Worked problem 4.2

Question

If the acid used had instead a pK_a value of 1.85 how would the reasoning be altered? Find the pH, and the fraction ionised. Explain the approximations made.

Answer

$$pK_a = 1.85 \therefore K_a = 1.41 \times 10^{-2}\,mol\,dm^{-3}.$$

This is a fairly strong weak acid.

- There will be extensive ionisation. The amount of H_3O^+ produced by the acid will be considerably greater than that from the self ionisation of water, which can be ignored.

$$\therefore [H_3O^+]_{actual} = [A^-]_{actual}$$

- The extensive ionisation means that a large fraction of the HA(aq) will be removed to form A^-(aq)

$$\therefore [HA]_{total} = [HA]_{actual} + [A^-]_{actual}$$

$$K_a = \left(\frac{[H_3O^+]_{actual}[A^-]_{actual}}{[HA]_{actual}} \right)_{eq}$$

$$\therefore K_a = \left(\frac{[H_3O^+]_{actual}^2}{[HA]_{total} - [A^-]_{actual}} \right)_{eq}$$

$$= \left(\frac{[H_3O^+]_{actual}^2}{[HA]_{total} - [H_3O^+]_{actual}} \right)_{eq}$$

$$1.41 \times 10^{-2}\,mol\,dm^{-3} = \frac{[H_3O^+]_{actual}^2}{0.05\,mol\,dm^{-3} - [H_3O^+]_{actual}}$$

$$[H_3O^+]_{actual}^2 + 1.41 \times 10^{-2}[H_3O^+]_{actual} - 7.05 \times 10^{-4} = 0$$

$$[H_3O^+]_{actual} = -\frac{1.41 \times 10^{-2} \pm \sqrt{2 \times 10^{-4} + 28.2 \times 10^{-4}}}{2}\,mol\,dm^{-3}$$

$$= 2.05 \times 10^{-2}\,mol\,dm^{-3}\ or\ -8.45 \times 10^{-2}\,mol\,dm^{-3}$$

Take care: The negative value is meaningless, and so only one root is taken.

$$\therefore [H_3O^+]_{actual} = 2.05 \times 10^{-2}\,mol\,dm^{-3}$$

$$pH = 1.69$$

Take note: Quadratic equations always have two roots. When solving quadratic equations a check should always be made to see whether both solutions are **physically** possible.

In this case the second root is negative and is not physically significant. Sometimes one root will give a concentration which is greater than the initial concentration and so is also non-meaningful.

$$\text{Fraction ionised} = \left(\frac{[A^-]_{\text{actual}}}{[HA]_{\text{total}}}\right)_{\text{eq}} = \left(\frac{[H_3O^+]_{\text{actual}}}{[HA]_{\text{total}}}\right) = \frac{2.05 \times 10^{-2}\,\text{mol dm}^{-3}}{5 \times 10^{-2}\,\text{mol dm}^{-3}} = 0.41$$

Comments

- $[H_3O^+] = 2.05 \times 10^{-2}\,\text{mol dm}^{-3}$. This should be compared with $[H_3O^+] = 10^{-7}\,\text{mol dm}^{-3}$ in pure water at 25°C. It is fully justified to assume that the self ionisation of water can be ignored.

- The fraction ionised $= 0.41$. Hence the weak acid is 41% ionised, and this indicates how totally unjustified it would be to make the assumption that $[HA]_{\text{actual}} = [HA]_{\text{total}}$.

- This conclusion can be derived in another way, by finding what $[H_3O^+]_{\text{actual}}$ would be **if both approximations are assumed to hold**, when:

$$K_a = \left(\frac{[H_3O^+]^2_{\text{actual}}}{[HA]_{\text{total}}}\right)_{\text{eq}}.$$

- Substitution gives $[H_3O^+]_{\text{actual}} = 2.65 \times 10^{-2}\,\text{mol dm}^{-3}$, from which the fraction ionised would be 53%, indicating unambiguously that the assumptions must be re-examined.

Worked problem 4.3

Question

Discuss the reasoning required to find the pH and the fraction ionised for an acid of concentration $5 \times 10^{-3}\,\text{mol dm}^{-3}$ and $pK_a = 12.25$ at 25°C.

Answer

$$pK_a = 12.25 \therefore K_a = 5.6 \times 10^{-13}\,\text{mol dm}^{-3}.$$

This is a very weak acid.

- There will be very little ionisation, and only a very small fraction of the HA(aq) will be removed to form A^-(aq).

$$\therefore [HA]_{\text{total}} = [HA]_{\text{actual}}$$

- However, it is such a weak acid that the self ionisation of water cannot be ignored.

$$\therefore [H_3O^+]_{actual} \neq [A^-]_{actual}$$

- Electrical neutrality will therefore have to be used.

$$\therefore [H_3O^+]_{actual} = [A^-]_{actual} + [OH^-]_{actual}$$

$$K_a = \left(\frac{[H_3O^+]_{actual}[A^-]_{actual}}{[HA]_{actual}} \right)_{eq}$$

$$\therefore K_a = \left(\frac{[H_3O^+]_{actual}\{[H_3O^+]_{actual} - K_w/[H_3O^+]_{actual}\}}{[HA]_{total}} \right)_{eq}$$

$$= \left(\frac{[H_3O^+]_{actual}^2 - K_w}{[HA]_{total}} \right)_{eq}$$

$$5.62 \times 10^{-13} \, mol \, dm^{-3} = \frac{[H_3O^+]^2 - K_w}{5.0 \times 10^{-2} mol \, dm^{-3}}$$

$$[H_3O^+]_{actual}^2 = (28.1 \times 10^{-15} + 10^{-14}) \, mol^2 \, dm^{-6}$$

$$[H_3O^+]_{actual} = 1.95 \times 10^{-7} \, mol \, dm^{-3}$$

$$pH = 6.71$$

To calculate the fraction ionised requires $[A^-]_{actual}$, and this has to be found from the equation for electrical neutrality.
Be careful here! It is **not** valid to use the approximate expression, $[H_3O^+]_{actual} = [A^-]_{actual}$; instead the following **must** be used.

$$[H_3O^+]_{actual} = [A^-]_{actual} + [OH^-]_{actual}$$

$$\therefore [A^-]_{actual} = [H_3O^+]_{actual} - K_w/[H_3O^+]_{actual}$$

$$[A^-]_{actual} = 1.95 \times 10^{-7} \, mol \, dm^{-3} - 10^{-14} \, mol^2 \, dm^{-6}/1.95 \times 10^{-7} \, mol \, dm^{-3}$$

$$= 1.44 \times 10^{-7} \, mol \, dm^{-3}$$

$$\text{Fraction ionised} = \left(\frac{[A^-]_{actual}}{[HA]_{total}} \right)_{eq} = \frac{1.44 \times 10^{-7} mol \, dm^{-3}}{5 \times 10^{-2} mol \, dm^{-3}} = 2.9 \times 10^{-6}$$

Comments

- $[H_3O^+]_{actual} = 1.95 \times 10^{-7} \, mol \, dm^{-3}$. Comparing this with $[H_3O^+] = 10^{-7} \, mol \, dm^{-3}$ in pure water at 25°C shows that there is a significant contribution to the actual $[H_3O^+]$ from the water. It would certainly not be justified to assume the self ionisation of water can be ignored.

- The fraction ionised is 2.9×10^{-6}. Hence the weak acid is only 2.9×10^{-4} % ionised, so the assumption that $[HA]_{actual} = [HA]_{total}$ is certainly fully justified.

4.3 Calculations on equilibria: weak bases

Students often have difficulty in seeing for themselves

- the analogies between the arguments used for weak acids and for weak bases

- that calculations on finding the pH of a solution of a weak base on its own follow **precisely** the same line of reasoning as those for the weak acid.

- The following attempts to rectify these problems.

Let the weak base be represented by B

$$B(aq) + H_2O(l) \rightleftharpoons BH^+(aq) + OH^-(aq)$$

$$K_b = \left(\frac{[BH^+]_{actual}[OH^-]_{actual}}{[B]_{actual}}\right)_{eq} \tag{4.33}$$

where, as before, the concentrations appearing in the equilibrium expression are the **actual** concentrations of the species in solution. Again, provided pK_b is known, this is an equation in three unknowns.

- **Species present in solution are:**

the base, B(aq), its conjugate acid, $BH^+(aq)$, $OH^-(aq)$, $H_2O(l)$ and trace amounts of $H_3O^+(aq)$ as a result of the self ionisation of water.

- **Source of $BH^+(aq)$:**

this is formed by the ionisation of the weak base, and this is the only source.

- **Sources of $OH^-(aq)$:**

  - $OH^-(aq)$ is formed along with $BH^+(aq)$ by the protonation of the weak base B(aq);

  - it is also formed from the self ionisation of $H_2O(l)$;

  - there are thus two sources which must be explicitly considered;

  - a decision will have to be made as to whether the self ionisation of water can be ignored.

- **Stoichiometric relations**

$$[B]_{total\ or\ stoich} = [B]_{actual} + [BH^+]_{actual} \tag{4.34}$$

$$\therefore [B]_{actual} = [B]_{total\ or\ stoich} - [BH^+]_{actual} \tag{4.35}$$

A decision will have to be made as to whether it is valid to approximate this expression to

$$[B]_{actual} \approx [B]_{total\ or\ stoich}$$

- **Substituting into the equilibrium expression:**

This gives:

$$K_b = \left(\frac{[OH^-]_{actual}\,[BH^+]_{actual}}{[B]_{total\ or\ stoich} - [BH^+]_{actual}} \right)_{eq} \quad (4.36)$$

This reduces the expression to one in two unknowns. To solve this equation it is necessary:

- **either** to obtain another independent equation;

- **or** to make approximations.

4.3.1 Possible approximations for the weak base

Approximation 1 relating to the extent of protonation of the weak base

If B(aq) is a **sufficiently weak base** for the fraction protonated to be **small**, then very little BH^+(aq) will be formed and very little B(aq) will have been removed.

$$[B]_{total\ or\ stoich} = [B]_{actual} + [BH^+]_{actual} \quad (4.34)$$

which, when $[BH^+]_{actual} \ll [B]_{total\ or\ stoich}$, can be approximated to:

$$[B]_{actual} \approx [B]_{total\ or\ stoich} \quad (4.37)$$

$$\therefore K_b \approx \left(\frac{[BH^+]_{actual}[OH^-]_{actual}}{[B]_{total\ or\ stoich}} \right)_{eq} \quad (4.38)$$

But is it always justifiable to make this approximation?

As with the weak acid, the approximation generally holds for pK_b values around 4 or greater. But again, diluting the solution of the weak base will dramatically increase the fraction protonated, so that the approximation will not hold for very low stoichiometric concentrations. For stronger bases, the extent of protonation becomes sufficiently high for the approximation to be invalid.

The approximation holds:

provided that the extent of protonation is limited and the base is not too dilute.

Approximation 2 relating to the self ionisation of water

If the contribution to OH^-(aq) from the self ionisation of water is **small** compared with the contribution from the protonation of B(aq), then to a first approximation it can be ignored.

There is, therefore, effectively only one source of $BH^+(aq)$ and $OH^-(aq)$, i.e. from the weak base, $B(aq)$, and as they are produced in a 1:1 ratio:

$$\therefore [BH^+]_{actual} \approx [OH^-]_{actual} \tag{4.39}$$

Substituting into the equilibrium expression:

$$K_b = \left(\frac{[BH^+]_{actual}[OH^-]_{actual}}{[B]_{actual}} \right)_{eq} \tag{4.33}$$

gives:

$$K_b \approx \left(\frac{[OH^-]^2_{actual}}{[B]_{actual}} \right)_{eq} \tag{4.40}$$

$$\therefore K_b \approx \left(\frac{[OH^-]^2_{actual}}{[B]_{total \text{ or stoich}} - [OH^-]_{actual}} \right)_{eq} \tag{4.41}$$

But is it justifiable to ignore the contribution to $[OH^-]_{actual}$ from the self ionisation of H_2O?

The approximation holds for bases with pK_b of around 10 or less, unless the concentration is very low. If the concentration of the weak base were very low, then **even if** all the base were protonated, the concentration of $OH^-(aq)$ would still approach that from the self ionisation of water. For bases with $pK_b > 10$, and thus are very weak bases, the extent of protonation will be so low that the contribution to $[OH^-]$ from the self ionisation of water must not be ignored.

The approximation will hold:

provided that the base is not too weak or too dilute.

However, if the base, $B(aq)$ is **very weak**, then the contribution to $OH^-(aq)$ will be small and similar to, or smaller than, the amount contributed from the self ionisation of water. There are two sources of $OH^-(aq)$ and neither can be ignored.

$$[OH^-]_{actual} = \text{contribution from } B(aq) + \text{contribution from } H_2O(l)$$

$$= [BH^+]_{actual} + \text{contribution from } H_2O(l) \tag{4.42}$$

$$\therefore [BH^+]_{actual} \neq [OH^-]_{actual} \tag{4.43}$$

and the equation describing electrical neutrality must be used. The sum of the positive charges must be balanced out by the sum of the negative charges, giving:

$$[H_3O^+]_{actual} + [BH^+]_{actual} = [OH^-]_{actual} \tag{4.44}$$

$$K_w/[OH^-]_{actual} + [BH^+]_{actual} = [\dot{O}H^-]_{actual} \tag{4.45}$$

$$\therefore [BH^+]_{actual} = [OH^-]_{actual} - K_w/[OH^-]_{actual} \tag{4.46}$$

4.3.2 The weak base where both approximations can be made

$$K_b = \left(\frac{[BH^+]_{actual}[OH^-]_{actual}}{[B]_{actual}} \right)_{eq} \tag{4.33}$$

If both approximations hold then :

- $[B]_{actual} \approx [B]_{total\ or\ stoich}$ \hfill (4.37)

- $[BH^+]_{actual} \approx [OH^-]_{actual}$ \hfill (4.39)

giving:

$$K_b \approx \left(\frac{[OH^-]^2_{actual}}{[B]_{total\ or\ stoich}} \right)_{eq} \tag{4.47}$$

This is an equation in one unknown and can be solved directly.

$$[OH^-]_{actual} = \sqrt{K_b[B]_{total\ stoich}} \tag{4.48}$$

$$[H_3O^+]_{actual} = K_w/[OH^-]_{actual} \tag{4.49}$$

$$pH = -\log_{10}[H_3O^+]_{actual} \tag{4.13}$$

This will be the case for bases with pK_b in the range 4 to 10, provided the base is not too dilute.

4.3.3 The weak base where there is extensive protonation and approximation 1 is invalid

This will happen with moderately strong weak bases of pK_b less than about 4, or for weaker bases in very dilute solution.

$$K_b = \left(\frac{[BH^+]_{actual}[OH^-]_{actual}}{[B]_{actual}} \right)_{eq} \tag{4.33}$$

Where there is fairly extensive protonation, the contribution to the **actual** concentration of $OH^-(aq)$ from the protonation of $B(aq) \gg$ than that from the self ionisation of $H_2O(l)$. The base is then effectively the only source of both $OH^-(aq)$ and $BH^+(aq)$ and the self ionisation of water can be ignored:

$$\therefore [OH^-]_{actual} \approx [BH^+]_{actual} \tag{4.39}$$

i.e. approximation 2 holds.

However, in this case the first approximation does not hold, i.e.

$$[B]_{actual} \neq [B]_{total\ or\ stoich} \tag{4.50}$$

and the rigorous expression:

$$[B]_{actual} = [B]_{total\ or\ stoich} - [BH^+]_{actual} \tag{4.35}$$

must be used in the equilibrium expression:

$$K_b = \left(\frac{[BH^+]_{actual}[OH^-]_{actual}}{[B]_{actual}} \right)_{eq} \tag{4.33}$$

The final expression is therefore:

$$K_b \approx \left(\frac{[OH^-]_{actual}^2}{[B]_{total\ or\ stoich} - [BH^+]_{actual}} \right)_{eq} \tag{4.51}$$

$$\approx \left(\frac{[OH^-]_{actual}^2}{[B]_{total\ or\ stoich} - [OH^-]_{actual}} \right)_{eq} \tag{4.52}$$

$$[OH^-]_{actual}^2 + K_b[OH^-]_{actual} - K_b[B]_{total\ or\ stoich} = 0 \tag{4.53}$$

As in the corresponding situation for a weak acid, this is a quadratic equation which can be solved using the standard procedure outlined earlier.

4.3.4 The weak base is sufficiently weak so that the self ionisation of water cannot be ignored and approximation 2 is invalid

For bases with pK_b greater than around 10.0 the base is so weak that the fraction protonated is very low. The contribution to the **actual** concentration of $OH^-(aq)$ from the weak base is so small that the contribution from the self ionisation of water is comparable to, or even greater than, that from the weak base. There are now **two** sources of $OH^-(aq)$.

$$[OH^-]_{actual} = \text{contribution from B(aq)} + \text{contribution from } H_2O(l)$$
$$= [BH^+]_{actual} + \text{contribution from } H_2O(l) \tag{4.42}$$

There is only **one** source of $BH^+(aq)$, the weak base B(aq), and so:

$$\therefore [OH^-]_{actual} \neq [BH^+]_{actual} \tag{4.43}$$

and both concentrations must be known explicitly, and the additional equation describing electrical neutrality must be used.

However, because the fraction protonated is so low, then the actual concentration of B(aq) will be very close to the stoichiometric concentration of B(aq), and the first approximation is valid, i.e.:

$$[B]_{actual} \approx [B]_{total\ or\ stoich} \tag{4.37}$$

Taking the equilibrium expression:

$$K_b = \left(\frac{[BH^+]_{actual}[OH^-]_{actual}}{[B]_{actual}} \right)_{eq} \tag{4.33}$$

and applying the above conclusions gives:

$$K_b \approx \left(\frac{[BH^+]_{actual}[OH^-]_{actual}}{[B]_{total\ or\ stoich}} \right)_{eq} \tag{4.38}$$

This is an equation in two unknowns and, for the reasons indicated in the weak acid case, electrical neutrality must be used to obtain a relation between $[BH^+]_{actual}$ and $[OH^-]_{actual}$.

$$[H_3O^+]_{actual} + [BH^+]_{actual} = [OH^-]_{actual} \tag{4.44}$$

$$\therefore [BH^+]_{actual} = [OH^-]_{actual} - K_w/[OH^-]_{actual} \tag{4.46}$$

which can be substituted into Equation (4.38), giving:

$$K_b \approx \left(\frac{[OH^-]_{actual}\left([OH^-]_{actual} - \dfrac{K_w}{[OH^-]_{actual}}\right)}{[B]_{total\ or\ stoich}} \right)_{eq} \tag{4.54}$$

$$\approx \left(\frac{[OH^-]^2_{actual} - K_w}{[B]_{total\ or\ stoich}} \right)_{eq} \tag{4.55}$$

$$[OH^-]^2_{actual} = K_b[B]_{total\ or\ stoich} + K_w \tag{4.56}$$

$$[OH^-]_{actual} = \sqrt{\left(K_b[B]_{total\ or\ stoich} + K_w\right)} \tag{4.57}$$

from which the pH can be found.

4.4   Some illustrative problems

Again as a test to the understanding of the theoretical description, the best procedure is to apply the arguments to an actual numerical problem. The conclusions outlined in Section 4.3 are illustrated in the worked problems below.

Worked problem 4.4

Question

NH_3 has a $pK_b = 4.75$ at 25°C. What is the pH and fraction protonated for a $0.02\ mol\ dm^{-3}$ solution of $NH_3(aq)$? How do these values change if the concentration drops to $1.00 \times 10^{-5}\ mol\ dm^{-3}$ and the same approximations are made? Comment on the result.

Answer

$$pK_b = 4.75 \therefore K_b = 1.78 \times 10^{-5}\ mol\ dm^{-3}.$$

$$NH_3(aq) + H_2O(l) \rightleftharpoons NH_4^+(aq) + OH^-(aq)$$

$$K_b = \left(\frac{[NH_4^+]_{actual}[OH^-]_{actual}}{[NH_3]_{actual}} \right)_{eq}$$

This is a fairly weak base present at moderate concentrations.

- The self ionisation of water can be ignored; the protonation of the base will be sufficiently extensive for the base to be the main source of $OH^-(aq)$.

- There is only one source of $OH^-(aq)$ and $NH_4^+(aq)$, and they are formed in a 1:1 ratio,

$$\therefore [NH_4^+]_{actual} = [OH^-]_{actual}$$

- The base is sufficiently weak to be only slightly protonated

$$\therefore [NH_3]_{actual} = [NH_3]_{total}$$

Making these approximations:

$$K_b = \left(\frac{[NH_4^+]_{actual}[OH^-]_{actual}}{[NH_3]_{actual}} \right)_{eq}$$

becomes:

$$K_b = \frac{[OH^-]^2_{actual}}{[NH_3]_{total}}$$

$$\therefore [OH^-]^2_{actual} = K_b[NH_3]_{total}$$

$$= (1.78 \times 10^{-5} \times 0.02)\, mol^2\, dm^{-6}$$

$$= 3.56 \times 10^{-7}\, mol^2\, dm^{-6}$$

$$\therefore [OH^-]_{actual} = 5.96 \times 10^{-4}\, mol\, dm^{-3}$$

But,

$$[H_3O^+]_{actual} = \frac{K_w}{[OH^-]_{actual}}$$

$$= \frac{1 \times 10^{-14}\, mol^2\, dm^{-6}}{5.96 \times 10^{-4}\, mol\, dm^{-3}}$$

$$= 1.68 \times 10^{-11}\, mol\, dm^{-6}$$

$$pH = -\log_{10}[H_3O^+] = 10.78$$

Note: An alternative way in which to work out the pH goes directly to pOH and from this:

$$pH = pK_w - pOH$$
$$pOH = 3.22$$
$$pH = 14.00 - 3.22 = 10.78$$

$$\text{Fraction protonated} = \frac{[NH_4^+]_{actual}}{[NH_3]_{total}} = \frac{5.96 \times 10^{-4}\, mol\, dm^{-3}}{2 \times 10^{-2}\, mol\, dm^{-3}} = 0.030$$

% protonation = 3.0

Comments

- $[OH^-]_{actual} = 5.96 \times 10^{-4}\, mol\, dm^{-3}$ which is sufficiently high for the contribution of $[OH^-]$ from $H_2O(l)$ to be insignificant. The approximation of ignoring the self ionisation of water is fully justified.

- The fraction protonated is so low that the assumption of only slight protonation is also fully justified.

- Both approximations are valid.

Effect of dilution

If the $[NH_3]_{total}$ now becomes $1.00 \times 10^{-5} \, mol \, dm^{-3}$, substitution into

$$K_b = \frac{[OH^-]^2_{actual}}{[NH_3]_{total}} \text{ gives :}$$

$$[OH^-]^2_{actual} = (1.78 \times 10^{-5} \times 1.00 \times 10^{-5}) \, mol^2 \, dm^{-3}$$

$$= 1.78 \times 10^{-10} \, mol^2 \, dm^{-6}$$

$$\therefore [OH^-]_{actual} = 1.33 \times 10^{-5} \, mol \, dm^{-3}$$

This concentration is **just** sufficiently large for the approximation of ignoring the self ionisation of water to be **just** valid.

The fraction protonated is also altered. Substitution into the expression for the fraction protonated now gives:

$$\text{Fraction protonated} = \frac{[NH_4^+]_{actual}}{[NH_3]_{total}} = \frac{1.33 \times 10^{-5} \, mol \, dm^{-3}}{1.00 \times 10^{-5} \, mol \, dm^{-3}} = 1.33$$

This value is impossible since the NH_3 cannot be 133% protonated, and indicates that the approximation assuming that very little of the NH_3 is protonated is invalid, and the calculation has to be altered to allow for this.

If the self ionisation of water can still be ignored, then:

$$\therefore [NH_4^+]_{actual} = [OH^-]_{actual}$$

and, because of extensive protonation, the following equation must be used:

$$[NH_3]_{actual} = [NH_3]_{total} - [NH_4^+]_{actual}$$

$$K_b = \left(\frac{[NH_4^+]_{actual}[OH^-]_{actual}}{[NH_3]_{actual}} \right)_{eq}$$

becomes:

$$K_b = \left(\frac{[OH^-]^2_{actual}}{[NH_3]_{total} - [OH^-]_{actual}} \right)_{eq}$$

$$[OH^-]^2_{actual} + K_b[OH^-]_{actual} - K_b[NH_3]_{total} = 0$$

$$[OH^-]_{actual} = \frac{-1.78 \times 10^{-5} \pm \sqrt{(1.78 \times 10^{-5})^2 + 4 \times 1.78 \times 10^{-5} \times 1.00 \times 10^{-5}}}{2} \, mol \, dm^{-3}$$

$$= \frac{-1.78 \times 10^{-5} \pm \sqrt{10.29 \times 10^{-10}}}{2} \, mol \, dm^{-3}$$

$$= \frac{-1.78 \times 10^{-5} \pm 3.21 \times 10^{-5}}{2}$$

$$= 0.714 \times 10^{-5} \, mol \, dm^{-3}$$

Note: there is an absurd negative concentration as the second root.

$$pOH = 5.15$$
$$pH = 8.85$$

$$\text{Fraction protonated} = \frac{[NH_4^+]_{actual}}{[NH_3]_{total}} = \frac{0.71 \times 10^{-5}\,mol\,dm^{-3}}{1.00 \times 10^{-5}\,mol\,dm^{-3}} = 0.71$$

The NH_3(aq) solution is now 71% protonated, a much more likely figure.

This problem illustrates for a weak base just how dramatically the fraction protonated increases with decrease in concentration. This observation is exactly analogous to the corresponding observation for the fraction ionised for a weak acid.

A value of $[OH^-]_{actual} = 0.71 \times 10^{-5}\,mol\,dm^{-3}$ suggests that the assumption that the self ionisation can be ignored is probably **just** valid. Any further decrease in concentration would require that neither assumption can be made (see Worked Problems 4.5 and 4.6).

The procedure for dealing with this situation is outlined in Worked Problem 4.6. Under these conditions the fraction protonated would appear to be equal to unity because of rounding up problems, and the fraction not protonated would then have to be calculated first. This is discussed in Section 4.5.

Worked problem 4.5

Question

The base, guanidine $NH=C(NH_2)_2$ has a $pK_b = 0.40$ at 25°C. Show that, by assuming only slight protonation, an impossible value is obtained for both the $[OH^-]_{actual}$ and the fraction protonated in a $5 \times 10^{-2}\,mol\,dm^{-3}$ solution. Find accurate values for both.

$$HN=C(NH_2)_2(aq) + H_2O(l) \rightleftharpoons H_2N^+=C(NH_2)_2(aq) + OH^-(aq)$$

$$K_b = \left(\frac{[H_2N^+=C(NH_2)_2]_{actual}[OH^-]_{actual}}{[HN=C(NH_2)_2]_{actual}} \right)$$

Assume that it is valid to:

- ignore the self ionisation of water, $\therefore [H_2N^+ = C(NH_2)_2]_{actual} = [OH^-]_{actual}$

- neglect any protonation of the weak base,

$$[HN=C(NH_2)]_{total} = [HN=C(NH_2)]_{actual} + [H_2N^+=C(NH_2)_2]_{actual}$$

$$\therefore [HN=C(NH_2)]_{actual} = [HN=C(NH_2)]_{total} - [H_2N^+=C(NH_2)_2]_{actual}$$

This can be approximated to

$$[HN=C(NH_2)]_{actual} = [HN=C(NH_2)]_{total}$$

$$\therefore K_b = \left(\frac{[OH^-]^2_{actual}}{[HN=C(NH_2)_2]_{total}}\right)_{eq}$$

$$\therefore [OH^-]^2_{actual} = K_b[NH=C(NH_2)_2]_{total}$$

$$= (0.398 \times 5 \times 10^{-2})\,mol^2\,dm^{-6}$$

$$= 1.99 \times 10^{-2}\,mol^2\,dm^{-3}$$

$$\therefore [OH^-]_{actual} = 0.141\,mol^{-1}\,dm^{-3}$$

But the concentration of weak base initially in solution was 0.05 mol dm^{-3}, which is **less** than this calculated value of [OH$^-$]$_{actual}$ produced by ionisation. This is impossible, and hence the approximation used is invalid for this base.

Likewise the calculated approximate fraction protonated is impossible:

$$\text{Approximate fraction protonated} = \frac{[H_2N^+=C(NH_2)_2]_{actual}}{[HN=C(NH_2)_2]_{total}}$$

$$= \frac{[OH^-]_{actual}}{[HN=C(NH_2)_2]_{total}}$$

$$= \frac{0.141\ mol\,dm^{-3}}{0.050\ mol\,dm^{-3}}$$

$$= 2.82$$

A more rigorous calculation assuming that it is valid to ignore the self ionisation of water, but acknowledging that there is significant protonation gives:

$$K_b = \left(\frac{[H_2N^+=C(NH_2)_2]_{actual}[OH^-]_{actual}}{[HN=C(NH_2)_2]_{actual}}\right)$$

$$= \left(\frac{[H_2N^+=C(NH_2)_2]_{actual}[OH^-]_{actual}}{[HN=C(NH_2)_2]_{total} - [H_2N^+=C(NH_2)_2]_{actual}}\right)$$

$$= \left(\frac{[OH^-]^2_{actual}}{[HN=C(NH_2)_2]_{total} - [OH^-]_{actual}}\right)$$

$$[OH^-]^2_{actual} = K_b[HN=C(NH_2)_2]_{total} - K_b[OH^-]_{actual}$$

$$[OH^-]^2_{actual} + 0.398[OH^-]_{actual} - 0.398 \times 0.05 = 0$$

$$[OH^-]^2_{actual} + 0.398[OH^-]_{actual} - 1.99 \times 10^{-2} = 0$$

$$\therefore [OH^-]_{actual} = \frac{-0.398 \pm \sqrt{0.158 + 0.0796}}{2}\ mol\,dm^{-3}$$

$$= \frac{-0.398 \pm \sqrt{0.238}}{2}\ mol\,dm^{-3}$$

$$= \frac{-0.398 \pm 0.487}{2}\ mol\,dm^{-3}$$

$$\therefore [OH^-]_{actual} = 0.045 \text{ mol dm}^{-3} \text{ and an absurd negative root.}$$

$$[H_3O^+]_{actual} = \frac{K_w}{[OH^-]_{actual}} = \frac{1 \times 10^{-14} \text{ mol}^2 \text{ dm}^{-6}}{0.045 \text{ mol dm}^{-3}} = 2.22 \times 10^{-13} \text{ mol dm}^{-3}$$

$$\therefore \text{ pH} = 12.65$$

$$\text{Fraction protonated} = \frac{0.045 \text{ mol dm}^{-3}}{0.050 \text{ mol dm}^{-3}} = 0.90$$

% protonated $= 90$

which indicates that any approximation assuming only slight protonation is totally inadequate. However, $[OH^-]_{actual} = 4.45 \times 10^{-2} \text{ mol dm}^{-3}$ which is sufficiently high that any contribution to the concentration of $OH^-(aq)$ from the self ionisation of water is insignificant.

A final example considers a weak base where it is completely unjustified to ignore the self ionisation of water.

Worked problem 4.6

Find the pH and fraction protonated of a 0.01 mol dm^{-3} solution of urea, p$K_b = 13.82$ at 25°C. Comment on the results.

Answer

Urea is a very weak base: $K_b = 1.51 \times 10^{-14} \text{ mol dm}^{-3}$.

$$O=C(NH_2)_2(aq) + H_2O(l) \rightleftharpoons O=C(NH_2)(NH_3^+)(aq) + OH^-(aq)$$

$$K_b = \left(\frac{[O=C(NH_2)(NH_3^+)]_{actual}[OH^-]_{actual}}{[O=C(NH_2)_2]_{actual}} \right)$$

The important species in solution are: $O=C(NH_2)_2(aq)$, its protonated form $O=C(NH_2)(NH_3^+)(aq)$ and $OH^-(aq)$.

Source of $O=C(NH_2)(NH_3^+)(aq)$: from protonation of urea, $O=C(NH_2)_2(aq)$.

Source of $OH^-(aq)$: from protonation of urea $O=C(NH_2)_2(aq)$,
from the self ionisation of $H_2O(l)$.

Urea is a very weak base and the amount of $OH^-(aq)$ resulting from the self ionisation of water will be significant in comparison to the amount from the protonation of urea, and:

$$\therefore [O=C(NH_2)(NH_3^+)]_{actual} \neq [OH^-]_{actual}$$

However, since urea is such a weak base then there will be only slight protonation, and:

$$[O=C(NH_2)_2]_{total} = [O=C(NH_2)_2]_{actual} + [O=C(NH_2)(NH_3^+)]_{actual}$$

$$[O=C(NH_2)_2]_{actual} = [O=C(NH_2)_2]_{total} - [O=C(NH_2)(NH_3^+)]_{actual}$$

can be approximated to:

$$[O=C(NH_2)_2]_{\text{actual}} = [O=C(NH_2)_2]_{\text{total}}$$

Since the self ionisation of water cannot be ignored for this base the approximate relation, $[O=C(NH_2)(NH_3^+)]_{\text{actual}} = [OH^-]_{\text{actual}}$ cannot be used. Another relation between $[O=C(NH_2)(NH_3^+)]_{\text{actual}}$ and $[OH^-]_{\text{actual}}$ must be found. Electrical neutrality gives the relation:

$$[O=C(NH_2)(NH_3^+)]_{\text{actual}} + [H_3O^+]_{\text{actual}} = [OH^-]_{\text{actual}}$$

$$[O=C(NH_2)(NH_3^+)]_{\text{actual}} = [OH^-]_{\text{actual}} - [H_3O^+]_{\text{actual}}$$

$$= [OH^-]_{\text{actual}} - \frac{K_w}{[OH^-]_{\text{actual}}}$$

Substitution into the equilibrium relation:

$$K_b = \left(\frac{[O=C(NH_2)(NH_3^+)]_{\text{actual}}[OH^-]_{\text{actual}}}{[O = C(NH_2)_2]_{\text{actual}}} \right)$$

gives:

$$K_b = \frac{\left([OH^-]_{\text{actual}} - \dfrac{K_w}{[OH^-]_{\text{actual}}} \right) [OH^-]_{\text{actual}}}{[O = C(NH_2)_2]_{\text{actual}}}$$

$$= \frac{\left([OH^-]_{\text{actual}} - \dfrac{K_w}{[OH^-]_{\text{actual}}} \right) [OH^-]_{\text{actual}}}{[O = C(NH_2)_2]_{\text{total}}}$$

$$\therefore 1.51 \times 10^{-14}\, \text{mol dm}^{-3} = \frac{\left([OH^-]_{\text{actual}} - \dfrac{1.00 \times 10^{-14}\, \text{mol}^2\, \text{dm}^{-6}}{[OH^-]_{\text{actual}}} \right) [OH^-]_{\text{actual}}}{0.01\, \text{mol dm}^{-3}}$$

$$1.51 \times 10^{-16}\, \text{mol dm}^{-3} = [OH^-]^2_{\text{actual}} - 1.00 \times 10^{-14}\, \text{mol}^2\, \text{dm}^{-6}$$

$$[OH^-]^2_{\text{actual}} = (1.51 \times 10^{-16} + 1.00 \times 10^{-14})\, \text{mol}^2\, \text{dm}^{-6}$$

$$= 101.5 \times 10^{-16}\, \text{mol}^2\, \text{dm}^{-6}$$

$$[OH^-]_{\text{actual}} = 1.01 \times 10^{-7}\, \text{mol dm}^{-3}$$

$$\therefore [H_3O^+]_{\text{actual}} = \frac{1.00 \times 10^{-14}}{1.01 \times 10^{-7}}\, \text{mol}^{-3}$$

$$= 0.99 \times 10^{-7}\, \text{mol dm}^{-3}$$

$$\text{pH} = 7.00$$

The $[OH^-]_{\text{actual}}$ is $1.01 \times 10^{-7}\, \text{mol dm}^{-3}$ which is **very close** to the value of $[OH^-]$ resulting from the ionisation of pure water. The self ionisation of water **must not** be ignored for this base.

The fraction protonated must be calculated taking note that the self ionisation of water must be included. Electrical neutrality gives:

$$[O{=}C(NH_2)(NH_3^+)]_{actual} = [OH^-]_{actual} - [H_3O^+]_{actual}$$

$$= (1.01 \times 10^{-7} - 0.99 \times 10^{-7})\,mol\,dm^{-3}$$

$$= 0.02 \times 10^{-7}\,mol\,dm^{-3}$$

$$Fraction\ protonated = \frac{[O{=}C(NH_2)(NH_3^+)]_{actual}}{[O{=}C(NH_2)_2]_{total}} = \frac{0.02 \times 10^{-7}\,mol\,dm^{-3}}{0.01\ mol\,dm^{-3}} = 2 \times 10^{-7}$$

There is thus very, very little protonation of the urea.

4.5 Fraction ionised and fraction not ionised for a weak acid; fraction protonated and fraction not protonated for a weak base

The fraction ionised is defined in terms of the amount of the original acid which is converted to the anion.

$$Fraction\ ionised = [A^-]_{actual}/[HA]_{total} \tag{4.58}$$

$$= [H_3O^+]_{actual}/[HA]_{total} \tag{4.59}$$

This expression holds under conditions where the self ionisation of water can be ignored. If this is not justified the expression becomes:

$$Fraction\ ionised = \frac{[H_3O^+]_{actual} - \dfrac{K_w}{[H_3O^+]_{actual}}}{[HA]_{total}} \tag{4.60}$$

Remember: if the self ionisation of water cannot be ignored, the electrical neutrality equation must be used:

$$[H_3O^+]_{actual} = [A^-]_{actual} + [OH^-]_{actual} \tag{4.25}$$

If ionisation is almost complete, then both these expressions become quotients of two approximately equal quantities so that the difference between the quotients and unity is very small. A direct calculation of the fraction ionised will give a result which is to all intents and purposes equal to unity, i.e. it is impossible to discriminate between 100% ionisation and very nearly 100% ionisation, and $[A^-]_{actual}$ **may appear to be equal** to $[HA]_{total}$. When this happens another route has to be followed to find out whether the acid is, for example, 99%, 99.5% or 99.9% ionised. Exactly analogous reasoning applies in the case of a weak base, except for the fact that the fraction calculated is more correctly described as a fraction protonated, from which the fraction not protonated can be found.

The following analysis is given for the weak acid

$$K_a = \left(\frac{[H_3O^+]_{actual}[A^-]_{actual}}{[HA]_{actual}}\right)_{eq} \tag{4.1}$$

If $[A^-]_{actual}$ is very nearly equal to $[HA]_{total}$ i.e. when the fraction ionised is approaching unity, then:

$$[HA]_{actual} = \left(\frac{[H_3O^+]_{actual}[A^-]_{actual}}{K_a}\right)_{eq} \qquad (4.61)$$

$$\approx \left(\frac{[H_3O^+]_{actual}[HA]_{total}}{K_a}\right)_{eq} \qquad (4.62)$$

$$\frac{[HA]_{actual}}{[HA]_{total}} \approx \frac{[H_3O^+]_{actual}}{K_a} \qquad (4.63)$$

$\dfrac{[HA]_{actual}}{[HA]_{total}}$ is the fraction **not** ionised and can be reliably evaluated since determination of $[H_3O^+]_{actual}$ can be highly accurate. It is straightforward to find whether this fraction comes to 0.0010 or 0.0005 or some other value.

From this the fraction ionised can be found as:

$$\text{fraction ionised} = 1 - \text{fraction not ionised} \qquad (4.64)$$

rather than the less definitive value of unity.

A similar treatment is given for the weak base:

$$K_b = \left(\frac{[BH^+]_{actual}[OH^-]_{actual}}{[B]_{actual}}\right)_{eq} \qquad (4.33)$$

When the fraction protonated becomes very close to unity, i.e. $[BH^+]_{actual}$ becomes very close to $[B]_{total}$:

$$[B]_{actual} = \frac{[BH^+]_{actual}[OH^-]_{actual}}{K_b} \qquad (4.65)$$

$$\approx \frac{[B]_{total}[OH^-]_{actual}}{K_b} \qquad (4.66)$$

$$\frac{[B]_{actual}}{[B]_{total}} \approx \frac{[OH^-]_{actual}}{K_b} \qquad (4.67)$$

$\dfrac{[B]_{actual}}{[B]_{total}}$ is the fraction **not** protonated and can be reliably evaluated since determination of $[OH^-]_{actual} = K_w/[H_3O^+]_{actual}$ can be highly accurate. It is straightforward to find whether this comes to 0.0010 or 0.0005 or some other value. This will allow an accurate calculation of the fraction protonated.

4.6   Dependence of the fraction ionised on pK_a and pH

There is an important question which should be asked.

Will all acids approach a limiting value for the fraction ionised, and is it possible that all acids can approach unity for the fraction ionised?

In this context, it is worthwhile noting that in Worked Problem 4.4 the fraction ionised increases with decrease in concentration.

The relations in Section 4.5 can be used to demonstrate that there is a limiting value to the fraction ionised for any acid, or to the fraction protonated for a weak base. Although some weak acids or bases appear to approach a value of unity, it must **not be assumed** that this is always the case. Furthermore it can be very useful to estimate what this limiting value actually is for any given acid or base.

Worked Problem 4.4 showed that the fraction of $NH_3(aq)$ protonated increases dramatically as the concentration of the weak base decreases, and this leads to the question:

What is the maximum value which the fraction ionised can reach?

The argument for a weak acid is:

$$K_a = \left(\frac{[H_3O^+]_{actual}[A^-]_{actual}}{[HA]_{actual}} \right)_{eq} \tag{4.1}$$

$$\frac{[HA]_{actual}}{[A^-]_{actual}} = \frac{[H_3O^+]_{actual}}{K_a} \tag{4.68}$$

$$\text{fraction ionised} = \frac{[A^-]_{actual}}{[HA]_{total}} \tag{4.58}$$

$$\frac{1}{\text{fraction ionised}} = \frac{[HA]_{total}}{[A^-]_{actual}} \tag{4.69}$$

$$= \frac{[HA]_{actual} + [A^-]_{actual}}{[A^-]_{actual}} = \frac{[HA]_{actual}}{[A^-]_{actual}} + 1 \tag{4.70}$$

$$= \frac{[H_3O^+]_{actual}}{K_a} + 1 = \frac{[H_3O^+]_{actual} + K_a}{K_a} \tag{4.71}$$

$$\therefore \text{fraction ionised} = \frac{K_a}{K_a + [H_3O^+]_{actual}} \tag{4.72}$$

$$\therefore \text{fraction not ionised} = 1 - \frac{K_a}{K_a + [H_3O^+]_{actual}} \tag{4.73}$$

$$= \frac{[H_3O^+]_{actual}}{[H_3O^+]_{actual} + K_a} \tag{4.74}$$

These expressions are extremely useful as they allow a calculation of the fraction ionised to be made for any acid of given pK_a at any pH. For instance an acid of $pK_a = 4.52$ can be made up as a solution of pH $= 5.60$.

$K_a = 3.02 \times 10^{-5}$ mol dm^{-3} and $[H_3O^+]_{actual} = 2.51 \times 10^{-6}$ mol dm^{-3} giving:

$$\text{fraction ionised} = \frac{3.02 \times 10^{-5} \text{mol dm}^{-3}}{3.02 \times 10^{-5} \text{mol dm}^{-3} + 2.51 \times 10^{-6} \text{mol dm}^{-3}} = 0.92$$

Likewise the pH corresponding to a fraction ionised equal to 0.25 can be found.

$$0.25 = \frac{3.02 \times 10^{-5} \text{ mol dm}^{-3}}{3.02 \times 10^{-5} \text{ mol dm}^{-3} + [H_3O^+]_{actual}}$$

$$\therefore [H_3O^+]_{actual} = 9.16 \times 10^{-5} \text{ mol dm}^{-3}$$

$$pH = 4.04$$

A corresponding derivation of the fraction not protonated can be carried out for a weak base.

$$B(aq) + H_2O(l) \rightleftharpoons BH^+(aq) + OH^-(aq)$$

$$K_b = \left(\frac{[BH^+]_{actual}[OH^-]_{actual}}{[B]_{actual}} \right)_{eq} \tag{4.33}$$

$$\therefore \frac{[BH^+]_{actual}}{[B]_{actual}} = \frac{K_b}{[OH^-]_{actual}} \tag{4.75}$$

$$\text{fraction protonated} = \frac{[BH^+]_{actual}}{[B]_{total}} \tag{4.76}$$

$$\frac{1}{\text{fraction protonated}} = \frac{[B]_{total}}{[BH^+]_{actual}} = \frac{[B]_{actual} + [BH^+]_{actual}}{[BH^+]_{actual}} = \frac{[B]_{actual}}{[BH^+]_{actual}} + 1 \tag{4.77}$$

$$= \frac{[OH^-]_{actual}}{K_b} + 1 = \frac{[OH^-]_{actual} + K_b}{K_b} \tag{4.78}$$

$$\therefore \text{ fraction protonated} = \frac{K_b}{K_b + [OH^-]_{actual}} \tag{4.79}$$

$$\text{fraction not protonated} = 1 - \frac{K_b}{K_b + [OH^-]_{actual}} \tag{4.80}$$

$$= \frac{[OH^-]_{actual}}{K_b + [OH^-]_{actual}} \tag{4.81}$$

As with the weak acid, these equations can be used to calculate the fraction protonated at a given pH for any weak base, or to find the pH at which a given weak base has a given fraction protonated.

4.6.1   Maximum % ionised for a weak acid and maximum % protonated for a weak base

Perhaps more importantly, Equations (4.72) and (4.79) can be used to calculate the maximum % ionised or the maximum % protonated for a weak acid or weak base respectively.

In very dilute **acid** solutions at 25°C, $[H_3O^+]_{actual} \rightarrow 10^{-7}\,mol\,dm^{-3}$. If $[H_3O^+]_{actual}$ less than this value is found then the solution is not acidic but **basic**. $[H_3O^+]_{actual} = 10^{-7}\,mol\,dm^{-3}$ is thus the limiting value for an acid solution at 25°C. Using this value, the limiting value to the fraction ionised, or the limiting value to the fraction not ionised, can be found for acids of stated pK_a values.

$\dfrac{K_a}{mol\,dm^{-3}}$	10^{-3}	10^{-4}	10^{-5}	10^{-6}	10^{-7}	10^{-8}	10^{-9}
$\dfrac{K_a}{K_a + 10^{-7}}$	$\dfrac{10^{-3}}{10^{-3} + 10^{-7}}$	$\dfrac{10^{-4}}{10^{-4} + 10^{-7}}$	$\dfrac{10^{-5}}{10^{-5} + 10^{-7}}$	$\dfrac{10^{-6}}{10^{-6} + 10^{-7}}$	$\dfrac{10^{-7}}{10^{-7} + 10^{-7}}$	$\dfrac{10^{-8}}{10^{-8} + 10^{-7}}$	$\dfrac{10^{-9}}{10^{-9} + 10^{-7}}$
fraction ionised	$\rightarrow 1$	$\rightarrow 1$	$\rightarrow 0.99$	$\rightarrow 0.91$	$\rightarrow 0.50$	$\rightarrow 0.09$	$\rightarrow 0.001$

From this table it can be seen that for acids weaker than $pK_a = 5$, it becomes impossible to approach 100% ionisation.

This table does show that it must always be remembered that, although there is a dramatic increase in % ionisation as an acid is diluted, this will not inevitably lead to virtually complete ionisation. This will only occur for acids stronger than around $pK_a = 5$.

Worked problem 4.7

Question

Use the equations of Section 4.6 to find the maximum % protonation for aqueous solutions of NH_3, guanidine and urea.

Answer

For a weak base the limiting value to a basic solution is given by $[OH^-]_{actual} = 10^{-7} \, mol \, dm^{-3}$

- $NH_3(aq), pK_b = 4.75, K_b = 1.78 \times 10^{-5} \, mol \, dm^{-3}$:

$$\text{Maximum fraction protonated} = \frac{1.78 \times 10^{-5} \, mol \, dm^{-3}}{1.78 \times 10^{-5} \, mol \, dm^{-3} + 1 \times 10^{-7} \, mol \, dm^{-3}}$$
$$= 0.994$$
$$\text{Maximum \% protonation} = 99.4$$

- guanidine $pK_b = 0.40, K_b = 0.398 \, mol \, dm^{-3}$:

$$\text{Maximum fraction protonated} = \frac{0.398 \, mol \, dm^{-3}}{0.398 \, mol \, dm^{-3} + 1 \times 10^{-7} mol \, dm^{-3}}$$
$$= 1$$
$$\text{Maximum \% protonation} = 100$$

- urea $pK_b = 13.82, K_b = 1.51 \times 10^{-14} \, mol \, dm^{-3}$:

$$\text{Maximum fraction protonated} = \frac{1.51 \times 10^{-14} \, mol \, dm^{-3}}{1.51 \times 10^{-14} \, mol \, dm^{-3} + 1 \times 10^{-7} \, mol \, dm^{-3}}$$
$$= 1.51 \times 10^{-7}$$
$$\text{Maximum \% protonation} = 1.51 \times 10^{-5}$$

Although both guanidine and ammonia can approach 100% protonation, these calculations show that urea can never be more than very, very slightly protonated.

These results do not include any approximations regarding the fraction ionised or the self ionisation of water. The magnitude for the limiting value of $[H_3O^+]_{actual}$ or of $[OH^-]_{actual} = 10^{-7} \, mol \, dm^{-3}$ is, of course, temperature dependent, because of the temperature dependence of K_w, K_a or K_b (see Worked Problems 2.4 and 2.5).

4.7 The effect of dilution on the fraction ionised for weak acids lying roughly in the range: $pK_a = 4.0$ to 10.0

An important point to note:

In previous sections it has been stated that, for such acids, it is valid to assume that both approximations hold, but the proviso 'provided the acid is not too dilute' has been added. This is a very important factor, and demonstrates that when approximations are made, active steps must be taken to check that they are valid and to recognise their limitations.

The following problem shows how important is an appreciation of the effects which dilution of the weak acid has on the validity of the approximations made in any calculations.

Worked problem 4.8

Question

Worked Problem 4.1 asks for a calculation of the pH and the fraction ionised for a 5.0×10^{-2} mol dm^{-3} solution of an acid with $pK_a = 5.75$ at 25°C. Repeat the calculation for the following concentrations of acid:

$\frac{[\text{HA}]_{\text{total}}}{\text{mol dm}^{-3}}$	5.00×10^{-3}	5.00×10^{-4}	1.00×10^{-4}	5.00×10^{-5}	1.00×10^{-5}	5×10^{-6}	1×10^{-6}

Comment on the results.

Answer

$$HA(aq) + H_2O(l) \rightleftharpoons H_3O^+(aq) + A^-(aq)$$

$$K_a = \left(\frac{[H_3O^+]_{\text{actual}} [A^-]_{\text{actual}}}{[HA]_{\text{actual}}} \right)_{\text{eq}}$$

- ignore self ionisation of water: $[H_3O^+]_{\text{actual}} = [A^-]_{\text{actual}}$
- assume very little ionisation: $[HA]_{\text{actual}} = [HA]_{\text{total}}$

$$K_a = \left(\frac{[H_3O^+]_{\text{actual}} [A^-]_{\text{actual}}}{[HA]_{\text{actual}}} \right)_{\text{eq}}$$

$$\therefore K_a = \left(\frac{[H_3O^+]^2_{\text{actual}}}{[HA]_{\text{total}}} \right)_{\text{eq}}$$

$$[H_3O^+]^2_{\text{actual}} = K_a [HA]_{\text{total}}$$

$\dfrac{[HA]_{total}}{mol\ dm^{-3}}$	5.00×10^{-3}	5.00×10^{-4}	1.00×10^{-4}	5.00×10^{-5}	1.00×10^{-5}	5×10^{-6}	1×10^{-6}
$\dfrac{[H_3O^+]^2_{actual}}{mol^2 dm^{-6}}$	8.89×10^{-9}	8.89×10^{-10}	1.78×10^{-10}	8.89×10^{-11}	1.78×10^{-11}	8.89×10^{-12}	1.78×10^{-12}
$\dfrac{[H_3O^+]}{mol\ dm^{-3}}$	9.43×10^{-5}	2.98×10^{-5}	1.33×10^{-5}	9.43×10^{-6}	4.22×10^{-6}	2.98×10^{-6}	1.33×10^{-6}
pH	4.03	4.53	4.88	5.03	5.38	5.53	5.88
$\dfrac{[A^-]_{actual}}{[HA]_{total}}$	0.019	0.060	0.133	0.188	0.42	0.596	1.33
% ionised	1.9	6.0	13	19	42	60	133

Comments: Pay **particular attention** to these as they are very important.

- From the table it is clear that the fraction ionised increases dramatically as the concentration of the acid decreases. This immediately raises the questions as to whether it is appropriate to use the approximation, $[HA]_{actual} \approx [HA]_{total}$, and even more important, 'at what stage does this approximation become invalid?' – remember, **this is an acid for which it would normally be assumed that the approximations are valid.**

- A closer look at the table shows that at very low concentrations the calculated % ionised becomes greater than 100, a sure sign that the approximation is invalid.

- Also, at very low concentrations $[H_3O^+]_{actual}$ becomes sufficiently small for the assumption that the self ionisation can be ignored to be queried.

To resolve this issue a further look at the approximations is necessary. It is also worthwhile to see whether there is any way in which the fraction ionised can be found without recourse to the full scale analysis.

4.8 Reassessment of the two approximations: a rigorous expression for a weak acid

If the stoichiometric concentration of a weak acid becomes too small then, even with an acid with a pK_a within the normal limits of the approximations, it may become necessary to consider both extensive ionisation and the self ionisation of water, i.e.

- significant ionisation: $[HA]_{actual} = [HA]_{total\ or\ stoich} - [A^-]_{actual}$ $\hspace{2cm}$ (4.3)

- the self ionisation of water, and make use of electrical neutrality:

$$[H_3O^+]_{actual} = [A^-]_{actual} + [OH^-]_{actual} \tag{4.25}$$

$$K_a = \left(\frac{[H_3O^+]_{actual}[A^-]_{actual}}{[HA]_{actual}}\right)_{eq} \tag{4.1}$$

$$K_a = \left(\frac{[H_3O^+]_{actual}[A^-]_{actual}}{[HA]_{total\ or\ stoich} - [A^-]_{actual}}\right)_{eq} \tag{4.4}$$

$$[H_3O^+]_{actual} = [A^-]_{actual} + [OH^-]_{actual} \tag{4.25}$$

$$= [A^-]_{\text{actual}} + \frac{K_w}{[H_3O^+]_{\text{actual}}} \tag{4.82}$$

Substituting the expression for $[A^-]_{\text{actual}}$ given by Equation (4.82) gives:

$$K_a = \frac{[H_3O^+]_{\text{actual}}\{[H_3O^+]_{\text{actual}} - K_w/[H_3O^+]\}}{[HA]_{\text{total}} - \{[H_3O^+]_{\text{actual}} - K_w/[H_3O^+]\}} \tag{4.83}$$

$$\text{Numerator}: \quad [H_3O^+]_{\text{actual}}\left\{\frac{[H_3O^+]^2_{\text{actual}} - K_w}{[H_3O^+]_{\text{actual}}}\right\} = [H_3O^+]^2_{\text{actual}} - K_w \tag{4.84}$$

$$\text{Denominator}: \quad \frac{[HA]_{\text{total}}[H_3O^+]_{\text{actual}} - [H_3O^+]^2_{\text{actual}} + K_w}{[H_3O^+]_{\text{actual}}} \tag{4.85}$$

$$\frac{\text{Numerator}}{\text{Denominator}} = \frac{\{[H_3O^+]^2_{\text{actual}} - K_w\}[H_3O^+]_{\text{actual}}}{[HA]_{\text{total}}[H_3O^+]_{\text{actual}} - [H_3O^+]^2_{\text{actual}} + K_w} \tag{4.86}$$

$$K_a = \frac{[H_3O^+]^3_{\text{actual}} - K_w[H_3O^+]_{\text{actual}}}{[HA]_{\text{total}}[H_3O^+]_{\text{actual}} - [H_3O^+]^2_{\text{actual}} + K_w} \tag{4.87}$$

$$[H_3O^+]^3_{\text{actual}} - K_w[H_3O^+]_{\text{actual}} = K_a[HA]_{\text{total}}[H_3O^+]_{\text{actual}} - K_a[H_3O^+]^2_{\text{actual}} + K_aK_w \tag{4.88}$$

$$[H_3O^+]^3_{\text{actual}} + K_a[H_3O^+]^2_{\text{actual}} - \{K_a[HA]_{\text{total}} + K_w\}[H_3O^+]_{\text{actual}} - K_aK_w = 0 \tag{4.89}$$

This is a cubic expression, and can only be solved by successive approximations.

The first approximation: this ignores the term in $[H_3O^+]^3_{\text{actual}}$ and the resulting quadratic term is solved in the usual way to give $[H_3O^+]_{\text{actual}}$.

The second approximation: this uses the approximate value to evaluate the magnitude of $[H_3O^+]^3_{\text{actual}}$ approximately, and this value is fed into Equation (4.89) as a constant term, and the quadratic equation then solved to give a second approximate $[H_3O^+]_{\text{actual}}$.

Further similar approximations are made until a constant $[H_3O^+]_{\text{actual}}$ is found.

From this the pH and the fraction ionised are found, and if the latter approximates to unity, then the fraction not ionised should be found.

However, as should be clear from the above, this fully rigorous analysis involves a lengthy and tedious set of calculations, and is avoided whenever possible. Fortunately, in most chemical situations there is rarely any need to work at the low concentrations which demand such analysis, and normally, either the fully approximate calculation, or one involving only one of the approximations, can be used.

4.9   Conjugate acids of weak bases

What has been discussed above for weak acids is also equally applicable to the conjugate acids of weak bases such as NH_3, or the various amines, e.g.:

$$NH_3(aq) + H_2O(l) \rightleftharpoons NH_4^+(aq) + OH^-(aq)$$

The cation, NH_4^+ (aq), is the conjugate acid of the weak base $NH_3(aq)$:

$$NH_4^+(aq) + H_2O(l) \longrightarrow NH_3(aq) + H_3O^+(aq)$$

and this acid equilibrium is handled in exactly the same way as for the weak acids discussed previously. These conjugate acids are found as salts of the weak bases, and will be discussed in the next chapter.

4.10   Weak bases

The behaviour of weak bases is exactly analogous to that for weak acids, and the conclusions reached for weak acids can be taken over and applied directly to weak bases.

The approximations of '**no extensive ionisation**' and '**ignoring the self ionisation of water**' apply equally to weak bases with pK_b values in the range 4.0 to 10.0, with the all-important proviso '**provided the solution is not too dilute**' again being necessary.

Weak bases exhibit the same dramatic increase in fraction protonated as the stoichiometric concentration decreases, with all cases reaching a limiting value. In the case of fairly strong weak bases this is virtually 100%, but for the weaker bases the fraction ionised becomes progressively smaller.

The quadratic equation has to be solved for fairly strong weak bases, and for weak bases at low concentration. The self ionisation of water can likewise be ignored for a large number of weak bases, but becomes progressively less able to be ignored as the base becomes weaker, or as the concentration decreases for ordinary weak bases.

The cubic equation is also necessary for very weak bases at low concentrations, but fortunately for normal situations it is rarely necessary to use this more rigorous treatment.

4.11   Effect of non-ideality

This will be dealt with in detail in Sections 8.24 to 8.28 where the theoretical background is given. The worked problems in these sections deal with specific aspects of the effect of non-ideality on acid/base behaviour.

For the present discussion it suffices to say that allowing for non-ideality would make virtually no difference in cases where either the solution of the acid or base is extremely dilute and hence the ionic strength is low, or where the acid or base is very weak.

The greatest effect will arise with a solution of a 'fairly strong weak acid or base'. The aspect which does present the greatest difficulty is that pH corresponds to an activity rather than a concentration, as has been assumed here. These problems will be discussed in Chapter 8, as indicated above.

5
Equilibrium Calculations for Salts and Buffers

This chapter continues the analysis given in Chapter 4. Here it will be shown that the treatment for **salts** of weak acids and weak bases is **identical** to that for weak acids and weak bases. This is because these salts are made up from the conjugate base of a weak acid, or the conjugate acid of a weak base. The analysis will again require discussion of whether or not the self ionisation of water can be ignored, and whether or not only slight ionisation, or protonation, can be assumed. Attention will be focused on the limitations imposed by the approximations and their physical significance. The chapter ends with a discussion of the properties, behaviour and analysis of buffer solutions. The standard buffer equation, the Henderson equation, although regularly used and simple to manipulate, is shown to be of somewhat limited application. The analytical treatment of buffer solutions is more complex than that for weak acids, bases or salts, and involves a further assumption not needed in the treatment of weak acids or bases.

Aims

By the end of this chapter you should be able to:

- identify salts of a strong acid/strong base, weak acid /strong base, weak base/strong acid, weak acid/weak base;

- compare the contributions to the equation for electrical neutrality for solutions of a weak acid or weak base with those for their salts;

- recognise that only two equilibria are required for calculations of the pH of solutions of salts of a weak acid/strong base, weak base/strong acid, but three are required for salts of a weak acid/weak base;

- calculate the pH of solutions for each type of salt;

An Introduction to Aqueous Electrolyte Solutions.   By Margaret Robson Wright
© 2007 John Wiley & Sons Ltd   ISBN 978-0-470-84293-5 (cloth) ISBN 978-0-470-84294-2 (paper)

- explain and justify the approximations involved in each type of calculation;

- recognise the different types of buffer solutions, and quote their properties;

- understand that the treatment for a buffer solution involves three simultaneous equilibria;

- derive the approximate expression for the pH of a buffer solution and explain the approximations involved and their limitations;

- follow through the rigorous calculation for the pH of a buffer solution and assess the approximations which can then be made;

- understand what is meant by buffer capacity and discuss the range of pH for which good buffer capacity is attained;

- compare the effects of adding a strong acid to water, to a solution of a weak acid, and to a buffer solution.

5.1   Aqueous solutions of salts

A look at Table 3.1 shows that there are several types of salts.

5.2   Salts of strong acids/strong bases

These are formed by the neutralisation of a strong acid with a strong base and are simply dealt with. The salt formed by the neutralisation of HCl(aq) with NaOH(aq) exists solely as the ions Na^+(aq) and Cl^-(aq). Neither of these ions is involved in any acid/base behaviour with water since HCl(aq) and NaOH(aq) are a strong acid and strong base respectively, and are fully ionised in solution. The pH of the solution is governed by the self ionisation of water and, at 25°C, is equal to 7.00.

5.3   Salts of weak acids/strong bases

In contrast, the salt formed by the neutralisation of methanoic acid, HCOOH(aq), with NaOH(aq) gives a basic solution with the pH dependent on the concentration of the salt. Sodium methanoate is the salt of a weak acid and strong base, and in the solid consists of Na^+ and $HCOO^-$ ions. When dissolved in water, Na^+(aq) ions show no acid/base behaviour with the water since NaOH(aq) is a strong base consisting of Na^+(aq) ions and OH^-(aq) ions only. However, $HCOO^-$(aq) is the conjugate base of the weak acid HCOOH(aq), and will thus accept a proton from the solvent water to form the undissociated acid HCOOH(aq) in the following equilibrium. The extent to which it does so depends on the concentration of the salt and the pK_b of the base $HCOO^-(aq)$.

$$HCOO^-(aq) + H_2O(l) \rightleftharpoons HCOOH(aq) + OH^-(aq)$$

This is a basic solution since $OH^-(aq)$ is formed, and this is why salts of a **weak acid/strong base** always have **basic** pH values.

$$K_b = \left(\frac{[HCOOH]_{actual}[OH^-]_{actual}}{[HCOO^-]_{actual}} \right)_{eq} \tag{5.1}$$

As mentioned in Section 3.10, pK_b for the methanoate ion $HCOO^-(aq)$ sometimes may not be tabulated, but instead may have to be worked out from the acid ionisation constant pK_a for the acid methanoic acid, HCOOH.

$$K_b \times K_a = \left(\frac{[HCOOH]_{actual}[OH^-]_{actual}}{[HCOO^-]_{actual}} \right)_{eq} \times \left(\frac{[HCOO^-]_{actual}[H_3O^+]_{actual}}{[HCOOH]_{actual}} \right)_{eq} \tag{5.2}$$

$$= \left([OH^-]_{actual}[H_3O^+]_{actual} \right)_{eq} \tag{5.3}$$

$$= K_w \tag{5.4}$$

$$\therefore K_b = \frac{K_w}{K_a} \tag{5.5}$$

Calculation of the pH of a solution of a salt of a **weak acid/strong base** uses **exactly** the same procedure as does the calculation of the pH of a solution of a **weak base**. The same considerations apply. For cases where the pK_b lies in the range 4.0 to 10.0, then the two standard approximations apply **unless the stoichiometric concentration is very low**. For low concentrations it is likely that the self ionisation of water has to be considered. This is also necessary when considering the salt of a weak acid/strong base where the conjugate base has a pK_b greater than around 10.0.

For the other extreme, when the conjugate base has a pK_b less than around 4.0, significant protonation will occur and the full expression, Equation (5.6), must be used:

$$[\text{conjugate base}]_{total} = [\text{conjugate base}]_{actual} + [\text{weak acid formed}]_{actual} \tag{5.6}$$

Here $[\text{conjugate base}]_{total}$ is the stoichiometric concentration of the salt of the weak acid/strong base in question.

In general, the conclusions for the weak base protonation equilibria apply here to the salt of the weak acid/strong base.

5.4 Salts of weak bases/strong acids

The salt formed by the neutralisation of the base ammonia $NH_3(aq)$ with $HCl(aq)$ gives an acid solution with the pH dependent on the concentration of the salt. Ammonium chloride is the salt of a weak base and strong acid, and in the solid consists of $NH_4^+(aq)$ and $Cl^-(aq)$ ions. When dissolved in water, $Cl^-(aq)$ ions show no acid/base behaviour with the water since $HCl(aq)$ is a strong acid consisting of $H_3O^+(aq)$ ions and $Cl^-(aq)$ ions only. However, $NH_4^+(aq)$ is the conjugate acid of the weak base $NH_3(aq)$, and will thus donate a proton to the solvent water to form the unprotonated base $NH_3(aq)$ in the following equilibrium. The extent to which it does so depends on the concentration of the salt and the pK_a of the acid $NH_4^+(aq)$.

$$NH_4^+(aq) + H_2O(l) \rightleftharpoons H_3O^+(aq) + NH_3(aq)$$

This is an acidic solution since $H_3O^+(aq)$ is formed, and this is why salts of a **weak base/strong acid** always have **acidic** pH values.

$$K_a = \left(\frac{[NH_3]_{actual}[H_3O^+]_{actual}}{[NH_4^+]_{actual}} \right)_{eq} \qquad (5.7)$$

As mentioned in Section 3.10, the pK_b values for weak bases are now generally listed in terms of pK_a for the conjugate acid, in this example $NH_4^+(aq)$. And so there is, in general, no need for cases of salts of weak base/strong acids to make the conversion between pK_b and pK_a (Equation 5.5).

The calculation of the pH of a solution of a salt of a **weak base/strong acid** uses **exactly** the same procedure as does the calculation of the pH of a solution of a **weak acid**. The same considerations apply. For cases where the pK_a lies in the range 4.0 to 10.0, then the two standard approximations apply **unless the stoichiometric concentration is very low**. For low concentrations it is likely that the self ionisation of water has to be considered. This is also necessary for the case of a salt of a weak base/strong acid where the conjugate acid has a pK_a greater than around 10.0.

For the other extreme, when the conjugate acid has a pK_a less than around 4.0, significant ionisation will occur and the full expression:

$$[\text{conjugate acid}]_{total} = [\text{conjugate acid}]_{actual} + [\text{weak base formed}]_{actual} \qquad (5.8)$$

Here $[\text{conjugate acid}]_{total}$ is the stoichiometric concentration of the salt of the weak base/strong acid in question.

In general, the conclusions for the weak acid ionisation equilibria apply here to the salt of the weak base/strong acid.

Worked problem 5.1

Question

Assuming that the self ionisation of water can be ignored and that the fraction protonated is small, calculate the pH of a 0.001 mol dm^{-3} solution of sodium benzoate. The acid ionisation constant of benzoic acid is 6.76×10^{-5} mol dm^{-3}, and the ionic product for H_2O is 1.00×10^{-14} mol^2 dm^{-6} at $25°C$. Check on the validity or otherwise of the assumptions and carry out a more rigorous calculation if necessary. What would happen if the concentration of the salt were 5×10^{-4} mol dm^{-3}?

Answer

Sodium benzoate is the salt of a weak acid/strong base, and will have a basic pH. The benzoate ion is the conjugate base of benzoic acid, as shown in the following equilibrium:

$$C_6H_5COO^-(aq) + H_2O(l) \rightleftharpoons C_6H_5COOH(aq) + OH^-(aq)$$

$$K_b = \left(\frac{[C_6H_5COOH]_{actual}[OH^-]_{actual}}{[C_6H_5COO^-]_{actual}} \right)_{eq}$$

$$K_a \times K_b = K_w$$

$$\therefore K_b = \frac{K_w}{K_a} = \left(\frac{[C_6H_5COOH]_{actual}[OH^-]_{actual}}{[C_6H_5COO^-]_{actual}} \right)_{eq}$$

Assuming that:

- the self ionisation of water can be ignored, then:

$$[C_6H_5COOH]_{actual} = [OH^-]_{actual}$$

- there is only slight conversion to $C_6H_5COOH(aq)$, then:

$$[C_6H_5COO^-]_{actual} = [C_6H_5COO^-]_{total}$$

Substitution into the expression for the equilibrium relationship gives:

$$\frac{K_w}{K_a} = \left(\frac{[OH^-]^2_{actual}}{[C_6H_5COO^-]_{total}} \right)_{eq}$$

$$\therefore [OH^-]^2_{actual} = \frac{K_w \times [C_6H_5COO^-]_{total}}{K_a}$$

$$= \frac{1.00 \times 10^{-14} mol^2\ dm^{-6} \times 0.001\ mol\ dm^{-3}}{6.76 \times 10^{-5}\ mol\ dm^{-3}}$$

$$[OH^-]_{actual} = 3.85 \times 10^{-7}\ mol\ dm^{-3}$$

$$[H_3O^+]_{actual} = \frac{K_w}{[OH^-]_{actual}} = \frac{1.00 \times 10^{-14} mol^2\ dm^{-6}}{3.85 \times 10^{-7} mol\ dm^{-3}} = 2.60 \times 10^{-8} mol\ dm^{-3}$$

$$pH = 7.59$$

Comments

- $[OH^-]$ from the self ionisation of pure water $= 1.00 \times 10^{-7}\ mol\ dm^{-3}$.

  $[OH^-]_{actual}$ found in this calculation $= 3.85 \times 10^{-7}\ mol\ dm^{-3}$

Hence it is probably unjustified to assume that the self ionisation of water can be ignored.

- Assuming that the self ionisation of water can be ignored and protonation is limited, gives:

$$[C_6H_5COOH]_{actual} = [OH^-]_{actual} = 3.85 \times 10^{-7}\ mol\ dm^{-3}$$

from which the fraction converted to benzoic acid would be:

$$\frac{[C_6H_5COOH]_{actual}}{[C_6H_5COO^-]_{total}} = \frac{3.85 \times 10^{-7} mol\ dm^{-3}}{0.001\ mol\ dm^{-3}} = 3.85 \times 10^{-4}$$

which is very small.

This suggests that the approximation that $[C_6H_5COO^-]_{actual} = [C_6H_5COO^-]_{total}$ is fully justified.

A more rigorous analysis of this problem is given in Section 5.4.1, and its application to the situation where the salt solution is at the lower concentration of 5×10^{-4} mol dm^{-3} is given.

5.4.1   A more rigorous treatment of worked problem 5.1

This replaces the statement that $[C_6H_5COOH]_{actual} = [OH^-]_{actual}$ by the electrical neutrality statement:

$$[C_6H_5COO^-]_{actual} + [OH^-]_{actual} = [H_3O^+]_{actual} + [Na^+]_{total} \qquad (5.9)$$

A very important point which must not be ignored:

Although the salt of a weak acid/strong base **behaves** as a solution of a weak base it must not be overlooked that the solution is that of a **salt**, and not that of a weak base on its own. The electrical neutrality **statement must always contain a contribution from the cation of the salt**, here Na$^+$(aq).

$$[Na^+]_{total} = [C_6H_5COO^-]_{total} = [C_6H_5COO^-]_{actual} + [C_6H_5COOH]_{actual} \qquad (5.10)$$

Substituting for $[C_6H_5COO^-]_{actual}$ from Equation (5.9) allows $[Na^+]_{total}$ to cancel out giving:

$$[C_6H_5COOH]_{actual} = [OH^-]_{actual} - [H_3O^+]_{actual} \qquad (5.11)$$

$$\therefore [C_6H_5COOH]_{actual} = [OH^-]_{actual} - \frac{K_w}{[OH^-]_{actual}} \qquad (5.12)$$

$$= \frac{[OH^-]^2_{actual} - K_w}{[OH^-]_{actual}} \qquad (5.13)$$

Substituting into the equilibrium relation,

$$K_b = \frac{K_w}{K_a} = \left(\frac{[C_6H_5COOH]_{actual}[OH^-]_{actual}}{[C_6H_5COO^-]_{actual}} \right)_{eq} \text{ gives :}$$

$$\frac{K_w}{K_a} = \left(\frac{[OH^-]^2_{actual} - K_w}{[OH^-]_{actual}} \times \frac{[OH^-]_{actual}}{[C_6H_5COO^-]_{actual}} \right) \qquad (5.14)$$

This can be approximated by assuming very slight conversion of the benzoate ion to benzoic acid, justified by the results of the first part of Worked Problem 5.1 above, so that $[C_6H_5COO^-]_{actual} = [C_6H_5COO^-]_{total}$, giving:

$$\frac{K_w}{K_a} = \left(\frac{[OH^-]^2_{actual} - K_w}{[C_6H_5COO^-]_{total}} \right)_{eq} \qquad (5.15)$$

$$\therefore [OH^-]^2_{actual} = \frac{K_w \times [C_6H_5COO^-]_{total}}{K_a} + K_w \qquad (5.16)$$

$$[OH^-]^2_{actual} = \frac{1.00 \times 10^{-14} \text{mol}^2 \text{ dm}^{-6} \times 0.001 \text{ mol dm}^{-3}}{6.76 \times 10^{-5} \text{mol dm}^{-3}} + 1.00 \times 10^{-14} \text{mol}^2 \text{ dm}^{-6}$$

$$= (14.79 \times 10^{-14} + 1.00 \times 10^{-14}) \text{ mol}^2 \text{ dm}^{-6}$$

$$= 15.79 \times 10^{-14} \text{ mol}^2 \text{ dm}^{-6}$$

$$\therefore [\text{OH}^-]_{\text{actual}} = 3.97 \times 10^{-7} \text{ mol dm}^{-3}$$

$$[\text{H}_3\text{O}^+]_{\text{actual}} = \frac{1.00 \times 10^{-14} \text{mol}^2 \text{ dm}^{-6}}{3.97 \times 10^{-7} \text{mol dm}^{-3}} = 2.52 \times 10^{-8} \text{ mol dm}^{-3}$$

$$\therefore \text{pH} = 7.60$$

This new value, $[\text{OH}^-]_{\text{actual}} = 3.97 \times 10^{-7}$ mol dm^{-3} using the more rigorous method can be contrasted with the fully approximate treatment which gave $[\text{OH}^-]_{\text{actual}} = 3.85 \times 10^{-7}$ mol dm^{-3}. The difference between the two values is 0.12×10^{-7} mol dm^{-3} showing that the fully approximate value is 3% low. This indicates it **is just possible to avoid** having to consider the self ionisation of water for this salt.

If the concentration is reduced to 5×10^{-4} mol dm^{-3} it is likely that the self ionisation of water will have to be included in the calculation. Using the rigorous expression derived above:

$$[\text{OH}^-]^2_{\text{actual}} = \frac{K_w \times [\text{C}_6\text{H}_5\text{COO}^-]_{\text{total}}}{K_a} + K_w \tag{5.16}$$

$$[\text{OH}^-]^2_{\text{actual}} = \frac{1.00 \times 10^{-14} \text{mol}^2 \text{ dm}^{-6} \times 5 \times 10^{-4} \text{mol dm}^{-3}}{6.76 \times 10^{-5} \text{mol dm}^{-3}} + 1.00 \times 10^{-14} \text{mol}^2 \text{ dm}^{-6}$$

$$= 8.40 \times 10^{-14} \text{mol}^2 \text{ dm}^{-6}$$

$$[\text{OH}^-]_{\text{actual}} = 2.90 \times 10^{-7} \text{mol dm}^{-3}$$

$$[\text{H}_3\text{O}^+]_{\text{actual}} = \frac{1.00 \times 10^{-14} \text{mol}^2 \text{ dm}^{-6}}{2.90 \times 10^{-7} \text{mol dm}^{-3}} = 3.45 \times 10^{-8} \text{ mol dm}^{-3}$$

$$\text{pH} = 7.46$$

If the self ionisation of water were ignored then the fully approximate treatment would give:

$$[\text{OH}^-]^2_{\text{actual}} = \frac{K_w \times [\text{C}_6\text{H}_5\text{COO}^-]_{\text{total}}}{K_a} \tag{5.17}$$

$$= \frac{1.00 \times 10^{-14} \text{mol}^2 \text{ dm}^{-6} \times 5.00 \times 10^{-4} \text{mol dm}^{-3}}{6.76 \times 10^{-5} \text{mol dm}^{-3}}$$

$$[\text{OH}^-]_{\text{actual}} = 2.72 \times 10^{-7} \text{ mol dm}^{-3}$$

$$[\text{H}_3\text{O}^+]_{\text{actual}} = \frac{1.00 \times 10^{-14} \text{mol}^2 \text{ dm}^{-6}}{2.72 \times 10^{-7} \text{mol dm}^{-3}} = 3.68 \times 10^{-8} \text{ mol dm}^{-3}$$

$$\text{pH} = 7.43$$

This can be contrasted with the more rigorous calculation which gave $[\text{OH}^-]_{\text{actual}} = 2.90 \times 10^{-7}$ mol dm^{-3} and pH $= 7.46$. The difference between the two values for $[\text{OH}^-]_{\text{actual}}$ is 0.18×10^{-7} mol dm^{-3} corresponding to the fully approximate calculation giving $[\text{OH}^-]_{\text{actual}}$ which is 6.2% low. It would thus not be valid to ignore the self ionisation of water for this concentration of the salt.

Take note: When comparing the fully approximate with the rigorous calculation, the difference is much more obvious if the $[\text{OH}^-]_{\text{actual}}$ values are compared, rather than the pH values. This is because the pH values correspond to a compressed scale.

Hence, provided the stoichiometric concentration is not too low, calculations on the protonation of the conjugate base anion using both approximations can be made for salts of weak acids/strong bases with pK_b values in the range of around 4.0 to 10.0. This fits in with the similar deductions for weak acids and weak bases.

Worked problem 5.2

Question

There are two approximations which can be made for solutions of salts:

- ignore the self ionisation of water;

- assume only small conversion to products in the equilibrium as set up.

(i) Which of the approximations are valid for the following solutions, and why?

- 0.001 mol dm^{-3} solution of potassium chloroethanoate. pK_a for chloroethanoic acid is 2.85.

- 0.001 mol dm^{-3} solution of 2-*ter*butylanilinium chloride. pK_a for the 2-*ter*butylanilinium ion is 3.80.

- 0.001 mol dm^{-3} solution of potassium nitrate.

- 0.001 mol dm^{-3} solution of sodium phenolate. pK_a for phenol is 10.00.

- 0.001 mol dm^{-3} solution of ethylammonium chloride. pK_a for $C_2H_5NH_3{}^+$ is 10.63.

(ii) Calculate the pH for a 5×10^{-3} mol dm^{-3} solution of sodium *o*-methyl phenolate. pK_a for *o*-methyl phenol is 10.28. What fraction of the original salt is converted?

Each pK_a in this question relates to 25°C.

Answer

(i)
- This is the salt of a moderately strong acid, $ClCH_2COOH$ and a strong base. The solid salt is composed of Na^+ ions and $ClCH_2COO^-$ ions. The anion is a base accepting a proton from water:

$$ClCH_2COO^-(aq) + H_2O(l) \rightleftharpoons ClCH_2COOH(aq) + OH^-(aq)$$

and this base will have a $pK_b = pK_w - pK_a = 14.00 - 2.85 = 11.15$, and $K_b = 7.08 \times 10^{-12}$ mol dm^{-3}. It is, therefore, a very weak base and the assumption that the self ionisation of water can be ignored will be invalid. The expression for electrical neutrality has to be used. The assumption of only very slight protonation of the base will be valid. The solution of this salt will have a basic pH.

- This is the salt of a moderately weak base, 2-*ter*butylaniline and the strong acid HCl. The pK_b for 2-*ter*butylaniline is $14.00 - 3.80 = 10.20$. The salt is composed of 2-*ter*butyl anilinium ions, $^+NH_3C_6H_4(C_4H_9)$ ions and Cl^- ions. The cation is an acid which will donate a proton to water:

$$^+NH_3C_6H_4(C_4H_9)(aq) + H_2O(l) \rightleftharpoons NH_2C_6H_4(C_4H_9)(aq) + H_3O^+ aq)$$

This acid has a $pK_a = 3.80$, and is thus a fairly strong acid. It is valid to assume that the contribution to the pH of this solution from the self ionisation of water can be ignored. However, since this is a moderately strong weak acid, the approximation of assuming only slight ionisation may not be justified, and the quadratic expression for $[H_3O^+]_{actual}$ may have to be solved. The solution of this salt will have an acidic pH.

- This is the solution of a salt of a strong acid/strong base.
Neither $K^+(aq)$ nor $NO_3^-(aq)$ will react with the solvent water, and the pH will be determined by the self ionisation of water. $[OH^-] = [H_3O^+] = 1.00 \times 10^{-7}$ mol dm^{-3}, and the pH $= 7.00$.

- This is a solution of the salt of a very weak acid and a strong base.
The solid salt is composed of Na^+ ions and phenolate ions $C_6H_5O^-$ ions. The anion is a base accepting a proton from water:

$$C_6H_5O^-(aq) + H_2O(l) \rightleftharpoons C_6H_5OH(aq) + OH^-(aq)$$

and this base will have a $pK_b = pK_w - pK_a = 14.00 - 10.00 = 4.00$, and $K_b = 1.00 \times 10^{-4}$ mol dm^{-3}. It is, therefore, a weak base with pK_b lying within the range where it is valid to assume that the self ionisation of water can be ignored. It is probably invalid to assume that there is only slight protonation and the rigorous expression taking account of this should be used. The solution of this salt will have a basic pH.

- The pK_b for ethylamine is $14.00 - 10.63 = 3.37$, therefore this is the salt of a moderately weak base, ethylamine, and the strong acid HCl. The salt is composed of ethylammonium, $C_2H_5NH_3^+$ ions and Cl^- ions. The cation is an acid which will donate a proton to water:

$$C_2H_5NH_3^+(aq) + H_2O(l) \rightleftharpoons C_2H_5NH_2(aq) + H_3O^+(aq)$$

This acid has a $pK_a = 10.63$, and is thus a very weak acid. The contribution to the pH of this solution from the self ionisation of water cannot be ignored, and the expression for electrical neutrality will have to be used. However, since this is a very weak acid, the approximation of assuming only slight ionisation will be justified. The solution of this salt will have an acidic pH.

(ii) pH for a 5×10^{-3} mol dm^{-3} solution of sodium o-methyl phenolate.

This is the salt of a weak acid/strong base, and so the anion of the salt is the conjugate base of the weak acid, here o-methyl phenol. The solution will have a basic pH as a result of the setting up of the equilibrium:

$$CH_3C_6H_4O^-(aq) + H_2O(l) \rightleftharpoons CH_3C_6H_4OH(aq) + OH^-(aq)$$

$$K_b = \left(\frac{[CH_3C_6H_4OH]_{actual}[OH^-]_{actual}}{[CH_3C_6H_4O^-]_{actual}} \right)_{eq}$$

pK_a for o-methyl phenol= 10.28, and so pK_b for the o-methyl phenolate anion is $14.00 - 10.28 = 3.72$, giving $K_b = 1.91 \times 10^{-4}$ mol dm^{-3}. This is a moderately strong weak base and as such the self ionisation of water can be ignored, but probably it is not possible to assume that only a small fraction of the anion is protonated.

- Ignoring the self ionisation of water gives:

$$[CH_3C_6H_4OH]_{actual} = [OH^-]_{actual}$$

- The rigorous expression is used:

$$[CH_3C_6H_4O^-]_{total} = [CH_3C_6H_4O^-]_{actual} + [CH_3C_6H_4OH]_{actual}$$
$$\therefore [CH_3C_6H_4O^-]_{actual} = [CH_3C_6H_4O^-]_{total} - [CH_3C_6H_4OH]_{actual}$$

Substituting into the equilibrium expression:

$$K_b = \left(\frac{[CH_3C_6H_4OH]_{actual}[OH^-]_{actual}}{[CH_3C_6H_4O^-]_{actual}} \right)_{eq}$$

gives:

$$1.91 \times 10^{-4} \text{ mol dm}^{-3} = \frac{[OH^-]^2_{actual}}{[CH_3C_6H_4O^-]_{total} - [OH^-]_{actual}}$$

$$= \frac{[OH^-]^2_{actual}}{0.005 \text{ mol dm}^{-3} - [OH^-]_{actual}}$$

$$\therefore [OH^-]^2_{actual} + 1.91 \times 10^{-4}[OH^-]_{actual} - 9.53 \times 10^{-7} = 0$$

$$[OH^-]_{actual} = \frac{-1.91 \times 10^{-4} \pm \sqrt{3.65 \times 10^{-8} + 3.81 \times 10^{-6}}}{2} \text{mol dm}^{-3}$$

$$= \frac{-1.91 \times 10^{-4} \pm \sqrt{3.85 \times 10^{-6}}}{2} \text{mol dm}^{-3}$$

$$= 8.85 \times 10^{-4} \text{mol dm}^{-3}$$

i.e. ignoring the negative root.

$$pOH = 3.05, \quad \text{and} \quad pH = 14.00 - 3.05 = 10.95$$

Since $[OH^-]_{actual} = 8.85 \times 10^{-4}$mol dm^{-3}, this justifies ignoring the self ionisation of water. However, it indicates that it would not have been justified to assume only slight ionisation.

Fraction ionised

$$= \frac{[CH_3C_6H_4OH]_{actual}}{[CH_3C_6H_4O^-]_{total}} = \frac{[OH^-]_{actual}}{[CH_3C_6H_4O^-]_{total}} = \frac{8.85 \times 10^{-4}\text{mol dm}^{-3}}{0.005 \text{ mol dm}^{-3}} = 0.177$$

%ionisation = 17.7.

5.5   Salts of weak acids/weak bases

These are salts such as ammonium methanoate, $HCOONH_4$, with anion $HCOO^-$ and cation NH_4^+, or methylammonium benzoate, $C_6H_5COOCH_3NH_3$, with anion $C_6H_5COO^-$ and cation $CH_3NH_3^+$. The cations act as acids and donate a proton to the solvent water, while the anions act as bases and accept a proton from the water.

The pH of such solutions can be easily calculated **provided** it is assumed that **the extent of conversion of both the anion and the cation are similar**. This will only be the case if the K_a and K_b for the cation and anion are similar; applicable only to a limited number of salts. For the rest the expression for $[H_3O^+]_{actual}$ is complex.

When a salt of a weak acid/weak base is dissolved in water three equilibria are set up and these must be simultaneously satisfied. This is one stage more complex than all the previous situations where only two equilibria are simultaneously satisfied; the acid or base equilibrium and the ionisation of water.

The salt formed by neutralisation of methanoic acid, $HCOOH(aq)$, and ammonia, $NH_3(aq)$ is ammonium methanoate, $HCOONH_4(aq)$. The three equilibria are:

$$HCOO^-(aq) + H_2O(l) \rightleftharpoons HCOOH(aq) + OH^-(aq) \quad (1)$$

$$K_b = \left(\frac{[HCOOH]_{actual}[OH^-]_{actual}}{[HCOO^-]_{actual}} \right)_{eq} \quad (5.18)$$

$$NH_4^+(aq) + H_2O(l) \rightleftharpoons NH_3(aq) + H_3O^+(aq) \quad (2)$$

$$K_a = \left(\frac{[NH_3]_{actual}[H_3O^+]_{actual}}{[NH_4^+]_{actual}} \right)_{eq} \quad (5.19)$$

$$H_2O(l) + H_2O(l) \rightleftharpoons H_3O^+(aq) + OH^-(aq) \quad (3)$$

$$K_w = ([H_3O^+]_{actual}[OH^-]_{actual})_{eq} \quad (5.20)$$

Adding the chemical equations (1) and (2), and subtracting chemical equation (3), gives the overall equation:

$$NH_4^+(aq) + HCOO^-(aq) \rightleftharpoons NH_3(aq) + HCOOH(aq)$$

The equilibrium constant for this reaction is found by appropriate multiplication and division of the individual equilibrium constants.

$$K_a \times K_b = \left(\frac{[NH_3]_{actual}[H_3O^+]_{actual}}{[NH_4^+]_{actual}} \right)_{eq} \times \left(\frac{[HCOOH]_{actual}[OH^-]_{actual}}{[HCOO^-]_{actual}} \right)_{eq} \quad (5.21)$$

$$= \left(\frac{[NH_3]_{actual}[HCOOH]_{actual}}{[NH_4^+]_{actual}[HCOO^-]_{actual}} \right)_{eq} \times \left([H_3O^+]_{actual}[OH^-]_{actual} \right)_{eq} \quad (5.22)$$

$$\therefore \frac{K_a K_b}{K_w} = \left(\frac{[NH_3]_{actual}[HCOOH]_{actual}}{[NH_4^+]_{actual}[HCOO^-]_{actual}} \right)_{eq} \quad (5.23)$$

Having formulated the equilibrium relationship, the two other commonly used relations must be considered:

- **The stoichiometric relations**:

$$[NH_4^+]_{total} = [NH_3]_{actual} + [NH_4^+]_{actual} \tag{5.24}$$

$$[HCOO^-]_{total} = [HCOOH]_{actual} + [HCOO^-]_{actual} \tag{5.25}$$

But $[HCOO^-]_{total} = [NH_4^+]_{total}$ since both come from the salt put in, and so:

$$[NH_3]_{actual} + [NH_4^+]_{actual} = [HCOOH]_{actual} + [HCOO^-]_{actual} \tag{5.26}$$

- **Electrical neutrality**:

$$[NH_4^+]_{actual} + [H_3O^+]_{actual} = [HCOO^-]_{actual} + [OH^-]_{actual} \tag{5.27}$$

As mentioned above, it is extremely difficult to formulate an expression for $[H_3O^+]_{actual}$ using these relations, and the normal practice uses the approximation of assuming that the extent of conversion of both acid and base is similar. Doing this means that the amount of $NH_3(aq)$ formed from the acid $NH_4^+(aq)$ is similar to the amount of $HCOOH(aq)$ formed from the base $HCOO^-(aq)$.

Thus in Equation (5.23):

$$[NH_3]_{actual} \approx [HCOOH]_{actual} \text{ and } [NH_4^+]_{actual} \approx [HCOO^-]_{actual} \text{ giving :}$$

$$\frac{K_a K_b}{K_w} = \left(\frac{[NH_3]_{actual}^2}{[NH_4^+]_{actual}^2}\right)_{eq} = \left(\frac{[HCOOH]_{actual}^2}{[HCOO^-]_{actual}^2}\right)_{eq} \tag{5.28}$$

Since $[NH_4^+]_{actual} = [NH_4^+]_{total} - [NH_3]_{actual}$ \hfill (5.29)

and $[HCOO^-]_{actual} = [HCOO^-]_{total} - [HCOOH]_{actual}$ \hfill (5.30)

$$\frac{K_a K_b}{K_w} = \left(\frac{[NH_3]_{actual}^2}{([NH_4^+]_{total} - [NH_3]_{actual})^2}\right)_{eq} \tag{5.31}$$

$$\frac{K_a K_b}{K_w} = \left(\frac{[HCOOH]_{actual}^2}{([HCOO^-]_{total} - [HCOOH]_{actual})^2}\right)_{eq} \tag{5.32}$$

Both of these equations have one unknown, $[NH_3]_{actual}$ and $[HCOOH]_{actual}$ respectively, and so can be solved to give either $[NH_3]_{actual}$ or $[HCOOH]_{actual}$. Depending on which concentration is found either the expression for K_a or K_b can be used to find $[H_3O^+]_{actual}$ or $[OH^-]_{actual}$.

This analysis **implicitly assumes** that the contribution to $[H_3O^+]_{actual}$ or $[OH^-]_{actual}$ from the self ionisation of water is being ignored. However, the rigorous treatment using electrical neutrality and making no assumptions as to the relative magnitudes of K_a and K_b ends up with an equation involving the fourth power of $[H_3O^+]_{actual}$ which has to be solved by successive approximations. Detailed calculations have shown that, **provided the stoichiometric concentration of salt is not too low**, the approximate and rigorous treatments give the same result, even when K_a and K_b differ by several powers of ten.

5.6 Buffer solutions

Buffers appear in chemical and biological systems as a means of controlling the pH. They have two important properties:

- they resist change in pH on addition of acid or base;

- they resist change in pH on dilution.

Table 5.1 gives a list of common buffers.

Buffer solutions consist of a **weak acid** and its **salt with a strong base**, or a **weak base** and its **salt with a strong acid**. Although most of the standard buffers are made up from weak acids plus their salts with strong bases (Table 5.1), there are also buffer solutions which can be prepared from the weak base plus its salt. The buffers quoted in Table 5.1 illustrate, however, that it is not necessary to rely on the weak base plus salt buffer to help to cover the whole range of pH. Some of these buffers are made from dibasic acids and make use of:

- the acidic properties of the undissociated dibasic acid; and

- the acidic salt produced by the neutralisation of the first ionisation of the dibasic acid,

e.g. phthalic acid and potassium hydrogen phthalate. Phthalic acid is a dibasic acid, and the mono potassium salt can still act as an acid by virtue of its remaining carboxylic acid group.

$$HOOCC_6H_4COOH(aq) + H_2O(l) \rightleftharpoons HOOCC_6H_4COO^-(aq) + H_3O^+(aq)$$

$$HOOCC_6H_4COO^-(aq) + H_2O(l) \rightleftharpoons {}^-OOCC_6H_4COO^-(aq) + H_3O^+(aq)$$

These equations represent the two stages of ionisation of the dibasic phthalic acid.

Table 5.1 Some common buffers

Buffer solutions	Approximate range of pH
phthalic acid/potassium hydrogen phthalate $C_6H_4(COOH)_2$/$HOOCC_6H_4COO^-$	2.2 to 3.8
ethanoic acid/sodium ethanoate CH_3COOH/CH_3COO^-	3.4 to 5.9
potassium hydrogen phthalate/dipotassium phthalate $HOOCC_6H_4COO^-$/$^-OOCC_6H_4COO^-$	4.0 to 6.2
potassium dihydrogen phosphate/dipotassium hydrogen phosphate $H_2PO_4^-$/HPO_4^{2-}	5.8 to 8.0
boric acid/monosodium borate H_3BO_3/$H_2BO_3^-$	7.8 to 10.0
glycine/sodium glycinate $^+NH_3CH_2COO^-$/$NH_2CH_2COO^-$	9.5 to 11.7
disodium hydrogen phosphate/trisodium phosphate HPO_4^{2-}/PO_4^{3-}	11.0 to 12.0

- Buffer pH 2.2 to 3.8: this corresponds to the phthalic acid/potassium hydrogen phthalate buffer

$$HOOCC_6H_4COOH(aq) + H_2O(l) \rightleftharpoons HOOCC_6H_4COO^-(aq) + H_3O^+(aq)$$

$$HOOCC_6H_4COO^-(aq) + H_2O(l) \rightleftharpoons HOOCC_6H_4COOH(aq) + OH^-(aq)$$

- Buffer 4.0 to 6.2: this corresponds to the potassium hydrogen phthalate/dipotassium phthalate buffer

$$HOOCC_6H_4COO^-(aq) + H_2O(l) \rightleftharpoons {}^-OOCC_6H_4COO^-(aq) + H_3O^+(aq)$$

$${}^-OOCC_6H_4COO^-(aq) + H_2O(l) \rightleftharpoons HOOCC_6H_4COO^-(aq) + OH^-(aq)$$

Three simultaneous equilibria are set up in a buffer solution, and all must be considered when formulating the expressions for $[H_3O^+]_{actual}$ or $[OH^-]_{actual}$. Carrying out actual calculations of the pH of a buffer solution is straightforward **provided** that **approximate** expressions for deriving the expressions for $[H_3O^+]_{actual}$ or $[OH^-]_{actual}$ are used. However, the reasoning used in the derivations and approximations is more tricky, and demonstrates the **absolute necessity** to be aware of what approximations are being made, and their implications and limitations.

5.6.1 Buffer: weak acid plus its salt with a strong base

This buffer is prepared by mixing appropriate amounts of solutions of weak acid and its salt. Depending on the chemical nature of the weak acid, the resulting solution can have a pH which is acidic or basic. There are three equilibria which are simultaneously set up and must be simultaneously satisfied.

- HA is a weak acid and donates a proton to water:

$$HA(aq) + H_2O(l) \rightleftharpoons A^-(aq) + H_3O^+(aq) \quad (A)$$

$$K_a = \left(\frac{[A^-]_{actual}[H_3O^+]_{actual}}{[HA]_{actual}} \right)_{eq} \tag{5.33}$$

- A^- is the conjugate base of the weak acid, HA, and thus will accept a proton from water:

$$A^-(aq) + H_2O(l) \rightleftharpoons HA(aq) + OH^-(aq) \quad (B)$$

$$K_b = \left(\frac{[HA]_{actual}[OH^-]_{actual}}{[A^-]_{actual}} \right)_{eq} \tag{5.34}$$

- the self ionisation of water:

$$H_2O(l) + H_2O(l) \rightleftharpoons H_3O^+(aq) + OH^-(aq) \quad (C)$$

$$K_w = ([H_3O^+]_{actual}[OH^-]_{actual})_{eq} \tag{5.20}$$

- Considering equilibrium (A), some $A^-(aq)$ is formed while some HA(aq) is removed.

- But simultaneously, in equilibrium (B), some HA(aq) is formed while some A^-(aq) is removed.

- $[HA]_{actual}$ comes from $[HA]_{total}$ present initially **minus** [HA] removed by conversion to A^-(aq) in equilibrium (A) **plus** [HA] formed by conversion of A^-(aq) acting as a base to give HA(aq) in equilibrium (B).

- $[A^-]_{actual}$ comes from $[A^-]_{total}$ present initially **minus** the $[A^-]$ which has been converted to HA(aq) by equilibrium (B) **plus** $[A^-]$ which has been formed from HA(aq) acting as an acid in equilibrium (A).

These statements can be rewritten as:

- $[HA]_{actual} = [HA]_{total} - [HA]_{removed\ in\ (A)} + [HA]_{formed\ in\ (B)}$ (5.35)

If it can be assumed that the amount of HA(aq) removed by equilibrium(A) is approximately equal to that formed by equilibrium (B), then the change in [HA] due to the simultaneous setting up of both equilibria will be approximately zero, and so the approximation can be made that:

$$[HA]_{actual} \approx [HA]_{total}$$ (5.36)

- $[A^-]_{actual} = [A^-]_{total} - [A^-]_{removed\ in\ (B)} + [A^-]_{formed\ in\ (A)}$ (5.37)

If it can be assumed that the amount of A^-(aq) removed by equilibrium (B) is approximately equal to that formed by equilibrium (A), then the change in $[A^-]$ due to the simultaneous setting up of both equilibria will be approximately zero, and so the approximation can be made that:

$$[A^-]_{actual} \approx [A^-]_{total}$$ (5.38)

- What is required is an expression for $[H_3O^+]_{actual}$. This can be found from either:

$$K_a = \left(\frac{[A^-]_{actual}[H_3O^+]_{actual}}{[HA]_{actual}} \right)_{eq}$$ (5.33)

or

$$K_b = \left(\frac{[HA]_{actual}[OH^-]_{actual}}{[A^-]_{actual}} \right)_{eq}$$ (5.34)

Using the approximate expressions (5.36) and (5.38) gives:

$$K_a = \left(\frac{[A^-]_{total}[H_3O^+]_{actual}}{[HA]_{total}} \right)_{eq}$$ (5.39)

$$K_b = \left(\frac{[HA]_{total}[OH^-]_{actual}}{[A^-]_{total}} \right)_{eq}$$ (5.40)

from which either $[H_3O^+]_{actual}$ or $[OH^-]_{actual}$ can be found.

Taking the expression for K_a gives:

$$[H_3O^+]_{actual} = \frac{K_a[HA]_{total}}{[A^-]_{total}} \tag{5.41}$$

This expression is often quoted in its logarithmic form, the Henderson equation:

$$\log_{10}[H_3O^+]_{actual} = \log_{10}K_a + \log_{10}\frac{[HA]_{total}}{[A^-]_{total}} \tag{5.42}$$

$$-\log_{10}[H_3O^+]_{actual} = -\log_{10}K_a - \log_{10}\frac{[HA]_{total}}{[A^-]_{total}} \tag{5.43}$$

$$pH = pK_a - \log_{10}\frac{[HA]_{total}}{[A^-]_{total}} \tag{5.44}$$

$$\text{or } pH = pK_a + \log_{10}\frac{[A^-]_{total}}{[HA]_{total}} \tag{5.45}$$

It is also often quoted as:

$$pH = pK_a + \log_{10}\frac{[salt]}{[acid]} \tag{5.46}$$

However, it is much better to deduce the expression for $[H_3O^+]_{actual}$ direct from the equilibrium quotient for K_a or K_b and calculate pH from this, rather than simply applying the Henderson buffer equation in any of its typical forms (Equations 5.44 to 5.46). This procedure eliminates any problems relating to the sign of the logarithmic term and the form of the ratio, and also shows the **significance of the approximations made**.

This analysis rests on three assumptions:

- that the self ionisation of water can be ignored;
- that it can be assumed that the extent of conversion in both equilibria (A) and (B) is small;
- and, **more importantly**, that the extent of conversion in each equilibrium is **approximately the same**.

Provided pK_a and pK_b are both within the normal range of 4.00 to 10.00 and provided the buffer solution is not too dilute, then the first two approximations should be valid. The validity of the third can only be assessed by carrying out calculations where the assumption is not made. These, however, are complex (see Section 5.6.2 below), but do show that, under normal conditions, it is valid to make both this assumption and the other two.

5.6.2   The rigorous calculation for the buffer of a weak acid plus its salt with a strong base

For this, the case of an acid buffer made up from a mixture of a weak acid, HA(aq), and its salt, NaA(aq), will again be considered so that direct comparisons can be made with the previous approximate treatment.

$$[NaA]_{initially} = [Na^+]_{total} = [A^-]_{total} \tag{5.47}$$

$$[HA]_{initially} = [HA]_{total} \tag{5.48}$$

When these are made up into an aqueous solution, equilibrium is set up simultaneously with respect to the three reactions:

$$HA(aq) + H_2O(l) \rightleftharpoons A^-(aq) + H_3O^+(aq)$$

$$A^-(aq) + H_2O(l) \rightleftharpoons HA(aq) + OH^-(aq)$$

$$H_2O(l) + H_2O(l) \rightleftharpoons H_3O^+(aq) + OH^-(aq)$$

As shown above:

$$[A^-]_{actual} = [A^-]_{total} + [A^-]_{formed\ in\ (A)} - [A^-]_{removed\ in\ (B)} \tag{5.37}$$

$$[HA]_{actual} = [HA]_{total} + [HA]_{formed\ in\ (B)} - [HA]_{removed\ in\ (A)} \tag{5.35}$$

$$[A^-]_{formed\ in\ (A)} = [HA]_{removed\ in\ (A)} \tag{5.49}$$

$$[HA]_{formed\ in\ (B)} = [A^-]_{removed\ in\ (B)} \tag{5.50}$$

What is wanted from this calculation?

Expressions for $[A^-]_{actual}$ and $[HA]_{actual}$, in terms of known quantities and/or $[H_3O^+]_{actual}$ and $[OH^-]_{actual}$, are required for substitution into K_a which can then be solved for $[H_3O]_{actual}$:

$$K_a = \left(\frac{[H_3O^+]_{actual}[A^-]_{actual}}{[HA]_{actual}}\right)_{eq} \tag{5.33}$$

This analysis illustrates some of the tricks used in manipulating equations to achieve the stated objective

Try adding Equations (5.37) and (5.35):

$$[A^-]_{actual} + [HA]_{actual} = [A^-]_{total} + [A^-]_{formed\ in\ (A)} - [A^-]_{removed\ in\ (B)}$$
$$+ [HA]_{total} + [HA]_{formed\ in\ (B)} - [HA]_{removed\ in\ (A)} \tag{5.51}$$

Substituting equations (5.49) and (5.50) gives:

$$[A^-]_{actual} + [HA]_{actual} = [A^-]_{total} + [HA]_{removed\ in\ (A)} - [HA]_{formed\ in\ (B)}$$
$$+ [HA]_{total} + [HA]_{formed\ in\ (B)} - [HA]_{removed\ in\ (A)} \tag{5.52}$$

$$[A^-]_{total} - [A^-]_{actual} = [HA]_{actual} - [HA]_{total} \tag{5.53}$$

This is an equation in two unknowns and so another independent equation is required.

So far **electrical neutrality has not been used**:

$$[Na^+]_{actual} + [H_3O^+]_{actual} = [A^-]_{actual} + [OH^-]_{actual} \tag{5.54}$$

$$\therefore [A^-]_{total} + [H_3O^+]_{actual} = [A^-]_{actual} + [OH^-]_{actual} \tag{5.55}$$

Remember: The salt in the buffer is NaA(aq) and $[Na^+]_{actual}$ must appear in the electrical neutrality (Equation 5.54).

This gives a second equation defining $[A^-]_{total} - [A^-]_{actual}$:

$$[A^-]_{total} - [A^-]_{actual} = [OH^-]_{actual} - [H_3O^+]_{actual} \tag{5.56}$$

which can be compared with the previous relation (5.53):

$$[A^-]_{total} - [A^-]_{actual} = [HA]_{actual} - [HA]_{total} \tag{5.53}$$

Equation (5.56) is useful in **two** ways:

• rearranged, it gives an expression:

$$[A^-]_{actual} = [A^-]_{total} + [H_3O^+]_{actual} - [OH^-]_{actual} \tag{5.57}$$

• it enables $[HA]_{actual}$ to be found by giving an expression for $[A^-]_{total} - [A^-]_{actual}$ which can be used with (5.53) to give:

$$[HA]_{actual} = [HA]_{total} - [H_3O^+]_{actual} + [OH^-]_{actual} \tag{5.58}$$

Two equations (Equations 5.57 and 5.58) have been found for $[A^-]_{actual}$ and $[HA]_{actual}$ respectively and these only involve known stoichiometric concentrations, plus $[H_3O^+]_{actual}$ and $[OH^-]_{actual}$. These latter concentrations are related through the ionic product for water. The equations for $[A^-]_{actual}$ and $[HA]_{actual}$ thus contain one unknown only, and these can be substituted into the equilibrium expression to generate an equation in one unknown.

$$K_a = \left(\frac{[H_3O^+]_{actual}[A^-]_{actual}}{[HA]_{actual}} \right)_{eq} \tag{5.33}$$

$$= \frac{[H_3O^+]_{actual}\{[A^-]_{total} + [H_3O^+]_{actual} - [OH^-]_{actual}\}}{[HA]_{total} - [H_3O^+]_{actual} + [OH^-]_{actual}} \tag{5.59}$$

$$[H_3O^+]_{actual} = \frac{K_a\{[HA]_{total} - [H_3O^+]_{actual} + [OH^-]_{actual}\}}{[A^-]_{total} + [H_3O^+]_{actual} - [OH^-]_{actual}} \tag{5.60}$$

This is a cubic equation, which has to be solved by successive approximations (Section 4.8).

However, one approximation which is certainly valid for a buffer with an **acidic** pH is that $[OH^-]_{actual} < [H_3O^+]_{actual}$ and making use of it:

$$[H_3O^+]_{actual} = \frac{K_a\{[HA]_{total} - [H_3O^+]_{actual}\}}{[A^-]_{total} + [H_3O^+]_{actual}} \tag{5.61}$$

Likewise for a buffer with a **basic** pH, $[H_3O^+]_{actual} < [OH^-]_{actual}$, giving:

$$[H_3O^+]_{actual} = \frac{K_a\{[HA]_{total} + [OH^-]_{actual}\}}{[A^-]_{total} - [OH^-]_{actual}} \tag{5.62}$$

These approximations convert the cubic into a quadratic equation in one unknown. These quadratic equations **must be used** in situations where the pH is less than 4 (Equation 5.61). or greater than 10 (Equation 5.62). In Equation (5.61) the $[H_3O^+]_{actual}$ is likely to be comparable to both $[HA]_{total}$ and $[A^-]_{total}$ and the term in $[H_3O^+]_{actual}$ in the right hand side of Equation (5.61) must not be dropped. Similarly, for pH values greater than 10 the term in $[OH^-]_{actual}$ will become comparable to both $[HA]_{total}$ and $[A^-]_{total}$ and must not be dropped, i.e. Equation (5.62) must be used.

A further approximation to Equation (5.60) can be made, for buffer solutions with pH values between 4 and 10. In Equation (5.60):

$$[H_3O^+]_{actual} = \frac{K_a\{[HA]_{total} - [H_3O^+]_{actual} + [OH^-]_{actual}\}}{[A^-]_{total} + [H_3O^+]_{actual} - [OH^-]_{actual}} \tag{5.60}$$

$[H_3O^+]_{actual}$ and $[OH^-]_{actual}$ will be small and can be ignored with respect to $[HA]_{total}$ and $[A^-]_{total}$ respectively. Most buffer solutions which are used in practice conform to this since $[HA]_{total}$ and $[A^-]_{total}$ are of the order of 0.5×10^{-3} mol dm^{-3}, or greater. The approximate equation used under these conditions is:

$$[H_3O^+]_{actual} = \frac{K_a[HA]_{total}}{[A^-]_{total}} \tag{5.63}$$

which is often quoted as the logarithmic expression:

$$pH = pK_a + \log_{10} \frac{[A^-]_{total}}{[HA]_{total}} \tag{5.64}$$

and is identical to the Henderson buffer equations, Equations (5.41) and (5.45), derived earlier which assumed that:

- the amount of A$^-$(aq) removed by equilibrium (B) is approximately equal to that formed by equilibrium (A);

- the amount of HA(aq) removed by equilibrium (A) is approximately equal to that formed by equilibrium (B).

A similar analysis can be given for a buffer made up from a weak base plus its salt with a strong acid in water, see Section 5.6.3 for the approximate analysis and Section 5.6.4 for a summary of the rigorous analysis.

Worked problem 5.3

Question

Phthalic acid is the source of two buffer solutions. Phthalic acid is a dibasic acid, and at 25°C the first $pK_a = 2.95$ and the mono potassium salt can act as an acid corresponding to the second $pK_a = 5.41$, by virtue of its remaining carboxylic acid group.

$$HOOCC_6H_4COOH(aq) + H_2O(l) \rightleftharpoons HOOCC_6H_4COO^-(aq) + H_3O^+(aq)$$
$$HOOCC_6H_4COO^-(aq) + H_2O(l) \rightleftharpoons {}^-OOCC_6H_4COO^-(aq) + H_3O^+(aq)$$

- The potassium hydrogen phthalate salt is the weak acid of the potassium hydrogen phthalate/dipotassium phthalate buffer which has pH range 4.0 to 6.2.

- The phthalic acid/potassium hydrogen phthalate salt buffers in the range 2.2 to 3.8.

Write down the chemical equations describing the two buffer regions.

- Can the approximate form of the equation for $[H_3O^+]_{actual}$ be used successfully for either, or both, of the buffers?

- Find the pH for both of the following solutions:

  (i) a solution prepared by mixing 100 cm^3 of 0.020 mol dm^{-3} potassium hydrogen phthalate with 150 cm^3 of 0.005 mol dm^{-3} dipotassium phthalate;

  (ii) a solution prepared by mixing 100 cm^3 of 0.020 mol dm^{-3} phthalic acid with 150 cm^3 of 0.005 mol dm^{-3} potassium hydrogen phthalate.

Comment on the results.

Answer

- Buffer 4.0 to 6.2

The **weak acid** equilibrium is:

$$HOOCC_6H_4COO^-(aq) + H_2O(l) \rightleftharpoons {}^-OOCC_6H_4COO^-(aq) + H_3O^+(aq)$$

The **salt** equilibrium is:

$$^-OOCC_6H_4COO^-(aq) + H_2O(l) \rightleftharpoons HOOCC_6H_4COO^-(aq) + OH^-(aq)$$

This buffer solution, i.e. the one buffering between 4.0 and 6.2 can be expressed by an expression analogous to Equation (5.41) because $[H_3O^+]_{actual}$ is considerably smaller than either $[HA^-]_{total}$ or $[A^{2-}]_{total}$ in the equation analogous to Equation (5.61):

$$[H_3O^+]_{actual} = \frac{K_a[HA^-]_{total}}{[A^{2-}]_{total}}$$

which in its logarithmic form corresponds to Equation (5.45):

$$pH = pK_a + \log_{10} \frac{[A^{2-}]_{total}}{[HA^-]_{total}}$$

The pH can be found from either equation.

- Buffer pH 2.2 to 3.8:

  The **weak acid** equilibrium is:

  $$HOOCC_6H_4COOH(aq) + H_2O(l) \rightleftharpoons HOOCC_6H_4COO^-(aq) + H_3O^+(aq)$$

  The **salt** equilibrium is:

  $$HOOCC_6H_4COO^-(aq) + H_2O(l) \rightleftharpoons HOOCC_6H_4COOH(aq) + OH^-(aq)$$

This buffer, i.e. the one buffering between pH 2.20 and 3.8, is outwith the range of validity of the approximate equations and the more rigorous expressions must be used. Here $[H_3O^+]_{actual}$ is comparable to both $[H_2A]_{total}$ and $[HA^-]_{total}$ in the equation which is analogous to Equation (5.61):

$$[H_3O^+]_{actual} = \frac{K_a\{[H_2A]_{total} - [H_3O^+]_{actual}\}}{[HA^-]_{total} + [H_3O^+]_{actual}}$$

$$pH = pK_a + \log_{10} \frac{[HA^-]_{total} + [H_3O^+]_{actual}}{[H_2A]_{total} - [H_3O^+]_{actual}}$$

Either of these two equations must be used to calculate the pH.

- **Calculation of the pH values for the two solutions**

The buffers are prepared by mixing each of the two solutions which are thus diluted. The final volumes are 250 cm^3.

(i) *KH phthalate as weak acid + diK phthalate as its salt*, i.e. the buffer in the range pH 4.0 to 6.2.

$$[^-OOCC_6H_4COOH]_{stoich} = \frac{0.020 \text{ mol dm}^{-3} \times 100 \text{ cm}^3}{250 \text{ cm}^3} = 8.0 \times 10^{-3} \text{mol dm}^{-3}$$

$$[^-OOCC_6H_4COO^-]_{stoich} = \frac{0.005 \text{ mol dm}^3 \times 150 \text{ cm}^3}{250 \text{ cm}^3} = 3.0 \times 10^{-3} \text{mol dm}^{-3}$$

pK_a for $^-OOCC_6H_4COOH = 5.41, K_a = 3.89 \times 10^{-6}$ mol dm^{-3}.

- The approximate analysis gives:

$$\begin{aligned}[H_3O^+]_{actual} &= \frac{K_a[HA^-]_{total}}{[A^{2-}]_{total}} \\ &= \frac{3.89 \times 10^{-6} \text{mol dm}^{-3} \times 8.0 \times 10^{-3} \text{mol dm}^{-3}}{3.0 \times 10^{-3} \text{mol dm}^{-3}} \\ &= 1.04 \times 10^{-5} \text{ mol dm}^{-3} \\ pH &= 4.98\end{aligned}$$

- The rigorous analysis gives:

$$[H_3O^+]_{actual} = \frac{K_a\{[HA^-]_{total} - [H_3O^+]_{actual}\}}{[A^{2-}]_{total} + [H_3O^+]_{actual}}$$

i.e. an equation analogous to Equation (5.61).

However, the approximate analysis gives $[H_3O^+]_{actual} = 1.04 \times 10^{-5}$ mol dm^{-3} which is small compared with either $[HA]_{stoich}$ or $[A^-]_{stoich}$ and so the corrections brought in by the rigorous analysis are insignificant.

(ii) *Phthalic acid as weak acid +KH phthalate as its salt*, i.e. the buffer in the range pH 2.2 to 3.8

$$[HOOCC_6H_4COOH]_{stoich} = \frac{0.020 \, mol \, dm^{-3} \times 100 \, cm^3}{250 \, cm^3} = 8.0 \times 10^{-3} mol \, dm^{-3}$$

$$[HOOCC_6H_4COO^-]_{stoich} = \frac{0.005 \, mol \, dm^{-3} \times 150 \, cm^3}{250 \, cm^3} = 3.0 \times 10^{-3} mol \, dm^{-3}$$

pK_a for $HOOCC_6H_4COOH = 2.95$, $K_a = 1.12 \times 10^{-3}$ mol dm^{-3}.

- The approximate analysis gives:

$$[H_3O^+]_{actual} = \frac{K_a[H_2A]_{total}}{[HA^-]_{total}}$$

$$= \frac{1.12 \times 10^{-3} mol \, dm^{-3} \times 8.0 \times 10^{-3} mol \, dm^{-3}}{3 \times 10^{-3} mol \, dm^{-3}}$$

$$= 2.99 \times 10^{-3} \, mol \, dm^{-3}$$

$$pH = 2.52$$

- The rigorous analysis gives:

$$[H_3O^+]_{actual} = \frac{K_a\{[H_2A]_{total} - [H_3O^+]_{actual}\}}{[HA^-]_{total} + [H_3O^+]_{actual}}$$

In the approximate treatment $[H_3O^+]_{actual} = 2.99 \times 10^{-3}$ mol dm^{-3}, and this is comparable to both $[HA]_{stoich}$ and $[A^-]_{stoich}$. The approximate analysis is, therefore, invalid.

$$[H_3O^+]_{actual} = \frac{K_a\{[H_2A]_{total} - [H_3O^+]_{actual}\}}{[HA^-]_{total} + [H_3O^+]_{actual}}$$

$$[H_3O^+]_{actual} = \frac{1.12 \times 10^{-3} mol \, dm^{-3}\{8.0 \times 10^{-3} mol \, dm^{-3} - [H_3O^+]_{actual}\}}{3 \times 10^{-3} mol \, dm^{-3} + [H_3O^+]_{actual}}$$

$$[H_3O^+]_{actual}^2 + 4.12 \times 10^{-3}[H_3O^+]_{actual} - 8.96 \times 10^{-6} = 0$$

$$\therefore [H_3O^+]_{actual} = \frac{-4.12 \times 10^{-3} \pm 7.27 \times 10^{-3}}{2} \, mol \, dm^{-3}$$

$$= 1.58 \times 10^{-3} \, mol \, dm^{-3}$$

$$pH = 2.80$$

Comments

The calculations demonstrate that the approximate treatment is valid for the buffer in the range pH 4.0 to 6.2, but totally invalid for the buffer in the range pH 2.2 to 3.8. Here the approximate calculation gives the pH to be 2.52, in contrast to the rigorous treatment which gives pH as 2.80.

5.6.3 Buffer: weak base plus its salt with a strong acid

This buffer is prepared by mixing appropriate amounts of solutions of the base, B, and its salt. The resulting solution has an acidic or basic pH depending on the chemical nature of the weak base considered. The analysis of this case is **logically identical** to that for the buffer made from a weak acid and its salt with a strong base. There are three equilibria which must be simultaneously satisfied:

- B(aq) is a weak base and accepts a proton from water:

$$B(aq) + H_2O(l) \rightleftharpoons BH^+(aq) + OH^-(aq) \quad (A)$$

$$K_b = \left(\frac{[BH^+]_{actual}[OH^-]_{actual}}{[B]_{actual}} \right)_{eq} \quad (5.65)$$

- $BH^+(aq)$ is the conjugate acid of B, and thus acts as an acid by donating a proton to the water:

$$BH^+(aq) + H_2O(l) \rightleftharpoons B(aq) + H_3O^+(aq) \quad (B)$$

$$K_a = \left(\frac{[B]_{actual}[H_3O^+]_{actual}}{[BH^+]_{actual}} \right)_{eq} \quad (5.66)$$

- the self ionisation of water:

$$H_2O(l) + H_2O(l) \rightleftharpoons H_3O^+(aq) + OH^-(aq)$$

$$K_w = ([H_3O^+]_{actual}[OH^-]_{actual})_{eq} \quad (5.20)$$

In equilibrium (A) some $BH^+(aq)$ is formed while some B(aq) is removed. But simultaneously, in equilibrium (B), some B(aq) is formed while $BH^+(aq)$ is removed.

- $[B]_{actual}$ comes from $[B]_{total}$ present initially **minus** [B] removed by conversion to $BH^+(aq)$ in equilibrium (A) **plus** [B] formed by conversion of $BH^+(aq)$ acting as an acid to B(aq) in equilibrium (B).

- $[BH^+]_{actual}$ comes from $[BH^+]_{total}$ present initially **minus** the $[BH^+]$ which has been converted to B by equilibrium (B) **plus** $[BH^+]$ which has been formed from B acting as a base in equilibrium (A).

These statements can be rewritten as:

- $[B]_{actual} = [B]_{total} - [B]_{removed\ in\ A} + [B]_{formed\ in\ B}$ \quad (5.67)

- $[BH^+]_{actual} = [BH^+]_{total} - [BH^+]_{removed\ in\ B} + [BH^+]_{formed\ in\ A}$ \quad (5.68)

The standard approximate derivation

- If it is assumed that the amount of B(aq) removed by equilibrium (A) is approximately equal to that formed by equilibrium (B), then the change in [B] due to the simultaneous setting up of both equilibria will be approximately zero, and so the approximation can be made that:

$$[B]_{actual} \approx [B]_{total} \tag{5.69}$$

- If it is assumed that the amount of BH^+(aq) removed by equilibrium (B) is approximately equal to that formed by equilibrium (A), then the change in $[BH^+]$ due to the simultaneous setting up of both equilibria will be approximately zero, and so the approximation can be made that:

$$[BH^+]_{actual} \approx [BH^+]_{total} \tag{5.70}$$

What is required is an expression for $[H_3O^+]_{actual}$. This can be found from either Equation (5.65) or (5.66):

Using the approximate expressions, Equations (5.69) and (5.70), gives:

$$K_b = \left(\frac{[BH^+]_{total}[OH^-]_{actual}}{[B]_{total}} \right)_{eq} \tag{5.71}$$

$$\text{or } K_a = \left(\frac{[B]_{total}[H_3O^+]_{actual}}{[BH^+]_{total}} \right)_{eq} \tag{5.72}$$

from which either $[OH^-]_{actual}$ or $[H_3O^+]_{actual}$ can be found.

Note: the analogy to Equations (5.39) and (5.40), and the identical nature of the derivation to that for the weak acid buffer.

Taking the expression for K_a gives:

$$[H_3O^+]_{actual} = \frac{K_a[BH^+]_{total}}{[B]_{total}} \tag{5.73}$$

This expression is often quoted in its logarithmic form:

$$\log_{10}[H_3O^+]_{actual} = \log_{10}K_a + \log_{10}\frac{[BH^+]_{total}}{[B]_{total}} \tag{5.74}$$

$$-\log_{10}[H_3O^+]_{actual} = -\log_{10}K_a - \log_{10}\frac{[BH^+]_{total}}{[B]_{total}} \tag{5.75}$$

$$pH = pK_a - \log_{10}\frac{[BH^+]_{total}}{[B]_{total}} \tag{5.76}$$

$$pH = pK_a + \log_{10}\frac{[B]_{total}}{[BH^+]_{total}} \tag{5.77}$$

Again note the analogy to Equations (5.44) and (5.45)

However, again it is much better to deduce the expression for $[H_3O^+]_{actual}$ direct from the equilibrium quotient for K_a or K_b and calculate pH from this. This procedure eliminates any problems relating to the sign of the logarithmic term and the form of the ratio, and probably more importantly **it emphasises the nature of the approximations involved.**

Comparing the approximate equation for the buffer prepared from the weak acid and its salt with a strong base and that for the buffer prepared from a weak base and its salt with a strong acid:

$$[H_3O^+]_{actual} = \frac{K_a[HA]_{total}}{[A^-]_{total}} \tag{5.41}$$

$$pH = pK_a + \log_{10} \frac{[A^-]_{total}}{[HA]_{total}} \tag{5.45}$$

$$[H_3O^+]_{actual} = \frac{K_a[BH^+]_{total}}{[B]_{total}} \tag{5.73}$$

$$pH = pK_a + \frac{[B]_{total}}{[BH^+]_{total}} \tag{5.77}$$

The ratio which appears in the logarithmic form corresponds to [conjugate base]/[acid], where the weak acid is either HA or BH^+. The ratio **must not** be read as [salt]/[acid] which is what is often quoted for the expression for a buffer made from the weak acid plus its salt with a strong base. Transferring to the weak base buffer on this basis would give the ratio the wrong way up, or the sign wrong.

5.6.4 The rigorous calculation for the buffer of a weak base plus its salt with a strong acid

The analysis can be carried out in an exactly similar fashion to that given in Section 5.6.2 for the buffer made from the weak acid/salt of a weak acid buffer, giving:

$$[H_3O^+]_{actual} = K_a \frac{\{[BH^+]_{total} - [H_3O^+]_{actual} + [OH^-]_{actual}\}}{[B]_{total} + [H_3O^+]_{actual} - [OH^-]_{actual}} \tag{5.78}$$

which is similar to that in Section 5.6.2, Equation (5.60).

Similar approximations can be made.

For an **acidic** pH, $[OH^-]_{actual} < [H_3O^+]_{actual}$ and making use of this:

$$[H_3O^+]_{actual} = \frac{K_a\{[BH^+]_{total} - [H_3O^+]_{actual}\}}{[B]_{total} + [H_3O^+]_{actual}} \tag{5.79}$$

Likewise for a buffer with a **basic** pH, $[H_3O^+]_{actual} < [OH^-]_{actual}$, giving:

$$[H_3O^+]_{actual} = \frac{K_a\{[BH^+]_{total} + [OH^-]_{actual}\}}{[B]_{total} - [OH^-]_{actual}} \tag{5.80}$$

Worked problem 5.4

Question

Using the approximate treatment given above, work out the pH of a buffer solution containing $0.10 \, \text{mol dm}^{-3}$ of $NH_3(aq)$ and $0.02 \, \text{mol dm}^{-3}$ $NH_4Cl(aq)$. pK_a for $NH_4^+(aq)$ at $25°C$ is 9.25.

Answer

This buffer is prepared by mixing appropriate amounts of solutions of $NH_3(aq)$ and $NH_4Cl(aq)$. The resulting solution has a basic pH.

$NH_3(aq)$ is a weak base and accepts a proton from water:

$$NH_3(aq) + H_2O(l) \rightleftharpoons NH_4^+(aq) + OH^-(aq)$$

$$K_b = \left(\frac{[NH_4^+]_{\text{actual}}[OH^-]_{\text{actual}}}{[NH_3]_{\text{actual}}} \right)_{\text{eq}}$$

$NH_4Cl(aq)$ in solution gives $NH_4^+(aq)$ which is the conjugate acid of $NH_3(aq)$, and thus acts as an acid by donating a proton to the water:

$$NH_4^+(aq) + H_2O(l) \rightleftharpoons NH_3(aq) + H_3O^+(aq)$$

$$K_a = \left(\frac{[NH_3]_{\text{actual}}[H_3O^+]_{\text{actual}}}{[NH_4^+]_{\text{actual}}} \right)_{\text{eq}}$$

The third equilibrium is that for the self ionisation of water:

$$H_2O(l) + H_2O(l) \rightleftharpoons H_3O^+(aq) + OH^-(aq)$$

$$K_w = ([H_3O^+]_{\text{actual}}[OH^-]_{\text{actual}})_{\text{eq}}$$

These three equilibria are set up simultaneously and must all be satisfied simultaneously. Using the argument given in Section 5.6.3 and the approximate expressions gives:

$$K_a = \left(\frac{[NH_3]_{\text{total}}[H_3O^+]_{\text{actual}}}{[NH_4^+]_{\text{total}}} \right)_{\text{eq}}$$

$$\therefore [H_3O^+]_{\text{actual}} = \frac{K_a[NH_4^+]_{\text{total}}}{[NH_3]_{\text{total}}}$$

pK_a for $NH_4^+(aq) = 9.25$, $K_a = 5.62 \times 10^{-10} \, \text{mol dm}^{-3}$.

$$[H_3O^+]_{\text{actual}} = \frac{5.62 \times 10^{-10} \text{mol dm}^{-3} \times 0.02 \, \text{mol dm}^{-3}}{0.10 \, \text{mol dm}^{-3}} = 1.12 \times 10^{-10} \text{mol dm}^{-3}$$

$$pH = 9.95$$

Worked problem 5.5

Question

Use Table 5.1 to choose a buffer with pH = 6.0. How would this buffer be prepared?

Answer

There are two possible buffers listed in the table:

- potassium hydrogen phthalate as the weak acid with the salt dipotassium phthalate. The second pK_a for phthalic acid is 5.41.

- sodium dihydrogen phosphate as the weak acid with the salt disodium hydrogen phosphate. The second pK_a for phosphoric acid is 6.20.

The equilibrium expression for the weak acid is:

$$K_a = \left(\frac{[H_3O^+]_{actual}[A^{2-}]_{actual}}{[HA^-]_{actual}} \right)_{eq}$$

Making the approximation that the actual concentrations of HA(aq) and A⁻(aq) can be taken as equal to the stoichiometric concentrations gives:

$$[H_3O^+]_{actual} = K_a \frac{[HA^-]_{total}}{[A^{2-}]_{total}}$$

$$\therefore \frac{[HA^-]_{total}}{[A^{2-}]_{total}} = \frac{[H_3O^+]_{actual}}{K_a}$$

For the phthalate buffer:

$$pK_a = 5.41, K_a = 3.90 \times 10^{-6} \text{ mol dm}^{-3}, [H_3O^+] = 1.0 \times 10^{-6} \text{ mol dm}^{-3}$$

$$\therefore \frac{[HA^-]_{total}}{[A^{2-}]_{total}} = \frac{1.0 \times 10^{-6} \text{mol dm}^{-3}}{3.90 \times 10^{-6} \text{mol dm}^{-3}} = 0.256$$

$$\therefore [HA^-]_{total} = 0.256 \times [A^{2-}]_{total}$$

It is then possible to choose a value for either $[HA^-]_{total}$, or $[A^{2-}]_{total}$ from which the other can be found.

$$\text{Let } [A^{2-}]_{total} = 2.00 \times 10^{-2} \text{ mol dm}^{-3}$$

$$\therefore [HA^-]_{total} = 0.256 \times 2.00 \times 10^{-2} \text{ mol dm}^{-3} = 0.512 \times 10^{-2} \text{ mol dm}^{-3}$$

i.e. $[^-OOCC_6H_4COOH]_{total} = 0.512 \times 10^{-2} \text{ mol dm}^{-3}$

and $[^-OOCC_6H_4COO^-]_{total} = 2.00 \times 10^{-2} \text{ mol dm}^{-3}$

For the phosphate buffer:

$$pK_a = 6.20, K_a = 6.31 \times 10^{-7} \text{ mol dm}^{-3}, [H_3O^+] = 1.0 \times 10^{-6} \text{ mol dm}^{-3}$$

$$\therefore \frac{[HA^-]_{total}}{[A^{2-}]_{total}} = \frac{1.0 \times 10^{-6} \text{mol dm}^{-3}}{6.31 \times 10^{-7} \text{mol dm}^{-3}} = 1.59$$

$$\therefore [HA^-]_{total} = 1.59 \times [A^{2-}]_{total}$$

It is then possible to choose a value for either $[HA^-]_{total}$, or $[A^{2-}]_{total}$ from which the other can be found.

$$Let [A^{2-}]_{total} = 2.00 \times 10^{-2} \text{ mol dm}^{-3}$$

$$\therefore [HA^-]_{total} = 1.59 \times 2.00 \times 10^{-2} \text{ mol dm}^{-3} = 3.18 \times 10^{-2} \text{ mol dm}^{-3}$$

$$i.e. [H_2PO_4^-]_{total} = 3.18 \times 10^{-2} \text{ mol dm}^{-3}$$

$$and [HPO_4^{2-}]_{total} = 2.00 \times 10^{-2} \text{ mol dm}^{-3}$$

5.6.5   Effect of dilution on buffering capacity

Buffer solutions have the capacity to resist change of pH on dilution. Addition of $H_2O(l)$ will dilute both the weak acid and its salt by the same amount and so the pH remains the same, e.g. in the case of the NH_3/NH_4Cl buffer:

$$K_a = \left(\frac{[NH_3]_{total}[H_3O^+]_{actual}}{[NH_4^+]_{total}} \right)_{eq} \tag{5.81}$$

$$\frac{[NH_4^+]_{total}}{[NH_3]_{total}} = \frac{[H_3O^+]_{total}}{K_a} \tag{5.82}$$

and so if the ratio $\dfrac{[NH_4^+]_{total}}{[NH_3]_{total}}$ remains the same, so does $[H_3O^+]_{actual}$. However, if the solution is diluted to a very large extent this no longer holds, and the self ionisation of the solvent must no longer be neglected.

5.6.6   Effect of addition of H_3O^+(aq) or OH^-(aq) on the pH of a buffer

Addition of a strong acid or a strong base to a buffer solution has very little effect on the pH, this being one of the main properties of a buffer. This can be contrasted with the effect of addition of either to water where there is a dramatic change in pH.

If 5 cm^3 of 0.1 mol dm^{-3} of HCl(aq) is added to 100 cm^3 of water, $[H_3O^+]$ will be 5 cm$^3 \times$ 0.1 mol dm^{-3}/100 cm^3 = 5×10^{-3}mol dm^{-3}, and pH is 2.30. This is to be contrasted with a pH of 7.0 for water at 25°C, showing the dramatic change in pH on addition of HCl(aq).

However, there is no such dramatic change when HCl(aq) is added to a buffer, as shown in the following analysis. Consider, for example, the weak acid/weak acid salt with a strong base buffer, i.e. HA/NaA.

The three equilibria set up are:

$$HA(aq) + H_2O(l) \rightleftharpoons A^-(aq) + OH^-\_(aq) \quad (A)$$
$$A^-_{(aq)} + H_2O(l) \rightleftharpoons HA(aq) + OH^-(aq) \quad (B)$$
$$H_2O(l) + H_2O(l) \rightleftharpoons H_3O^+(aq) + OH^-(aq) \quad (C)$$

If HCl(aq) is added to this buffer solution it will neutralise an equivalent number of mol of the base, $A^-(aq)$, present in the buffer solution and form an equivalent number of mol of HA(aq).

$$H_3O^+(aq) + A^-(aq) \rightarrow HA(aq) + H_2O(l)$$

As a result, there will be changes in the position of equilibrium in each of the equilibria (A), (B) and (C). Both before and after addition of the strong acid, the equilibrium relation

$$K_a = \left(\frac{[H_3O^+]_{actual}[A^-]_{actual}}{[HA]_{actual}}\right)_{eq} \tag{5.33}$$

will hold, and for many buffers this will reduce to the corresponding relation in terms of the stoichiometric concentration of $A^-(aq)$ and HA(aq). In consequence the ratio $[A^-]_{stoich}/[HA]_{stoich}$ will not change drastically, and hence the pH also will not change drastically.

5.6.7 Effect of addition of $H_3O^+(aq)$ or $OH^-(aq)$ on the pH of a weak acid on its own

If HCl is added to a weak acid there is a considerable change in pH showing that a weak acid on its own has no buffering capacity. The effect, however, is not as great as that observed when a strong acid or base is added to water. A solution of a weak acid consists of both the undissociated weak acid, HA(aq) and its conjugate base $A^-(aq)$.

$$HA(aq) + H_2O(l) \rightleftharpoons H_3^+O(aq) + A^-(aq)$$
$$K_a = \left(\frac{[H_3O^+]_{actual}[A^-]_{actual}}{[HA]_{actual}}\right)_{eq} \tag{5.33}$$

After the strong acid has been added, nearly all of the weak acid will be in the form of un-ionised HA - suppression of the ionisation - and the pH will be given approximately by the concentration of the strong acid. A large decrease in pH will result, but this will not be so large as is found when strong acid is added to water.

5.6.8 Buffer capacity

In effect, the term buffer capacity means the range of pH over which a buffer is effective. The commonly held view is that this range lies within $pK_a \pm 1$. The table and graph below illustrate the buffering capacity of the ethanoic acid/sodium ethanoate buffer, and allow a judgement to be made as to whether the accepted range, $pK_a \pm 1$, is reasonable. pK_a for ethanoic acid is 4.76 at 25°C.

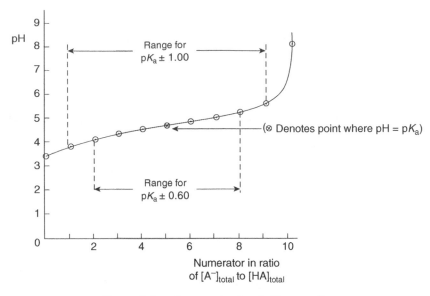

Figure 5.1 pH curve showing buffer capacity.

pH values for 1×10^{-2} mol dm^{-3} ethanoic acid and for 1×10^{-2} mol dm^{-3} sodium ethanoate are worked out using the standard approximations of ignoring the self ionisation of water and assuming only slight ionisation. The approximate buffer equation, analogous to Equations (5.39):

$$K_a = \left(\frac{[CH_3COO^-]_{total}[H_3O^+]_{actual}}{[CH_3COOH]_{total}} \right)_{eq} \tag{5.83}$$

is used in the form, analogous to Equation (5.45):

$$pH = pK_a + \log_{10} \frac{[CH_3COO^-]_{total}}{[CH_3COOH]_{total}} \tag{5.84}$$

to calculate the pH values of buffers with a total concentration, $[HA]_{total} + [A^-]_{total} = 1 \times 10^{-2}$ mol dm^{-3} but with different ratios corresponding to given % fractions of HA(aq) converted to A$^-$(aq). Figure 5.1 shows the results in graphical form, while Table 5.2 lists the actual values.

pH for 1×10^{-2} mol dm^{-3} ethanoic acid: 3.38
pH for 1×10^{-2} mol dm^{-3} sodium ethanoate: 8.38.

The graph shows that, around pH = pK$_a$, the change in pH for a given fraction of HA converted to A$^-$ is smallest and that there is good buffering capacity in this region. As the

Table 5.2 Values of pH corresponding to different values of the ratio $\dfrac{[A^-]_{total}}{[HA]_{total}}$

$\dfrac{[A^-]_{total}}{[HA]_{total}}$	$\dfrac{1}{9}$	$\dfrac{2}{8}$	$\dfrac{3}{7}$	$\dfrac{4}{6}$	$\dfrac{5}{5}$	$\dfrac{6}{4}$	$\dfrac{7}{3}$	$\dfrac{8}{2}$	$\dfrac{9}{1}$
pH	3.81	4.16	4.39	4.58	4.76	4.94	5.13	5.36	5.71

ratio, $\dfrac{[A^-]_{total}}{[HA]_{total}}$ decreases or increases from unity the buffering capacity decreases. From the graph it can be seen that the standard range of good buffer capacity, $pK_a \pm 1$, is perhaps too large, and one corresponding to a limiting ratios of $\dfrac{2}{8}$ and $\dfrac{8}{2}$ is more reasonable, i.e. $pK_a \pm 0.60$.

Note: when $\dfrac{[A^-]_{total}}{[HA]_{total}} = 1$ then $pH = pK_a$, and this corresponds to half-way to the end point in a titration of the weak acid with a strong base (see Section 6.2.3). This is a very simple method of finding a first approximate value pK_a value for a weak acid, but it **must always be remembere**d that the buffer equation to which it refers is only **approximate**.

6

Neutralisation and pH Titration Curves

The previous discussions on salts of a strong acid/strong base, weak acid/strong base, weak base/strong acid, weak acid/weak base and buffers form the basis of the interpretation of pH titration curves. Although neutralisation and acid/base behaviour have some features in common, i.e. they both involve acid/base behaviour and both are proton transfer reactions, distinctions must be drawn between them.

This chapter also considers the more complex situations involved in polybasic acid equilibria and ampholytes.

Aims

By the end of the chapter you should be able to:

- distinguish between neutralisation and acid/base behaviour;

- analyse titration curves of monobasic acids and monoacid bases;

- appreciate the distinction between the end-point and the equivalence point;

- list the acid/base equilibria for polybasic acids;

- discuss and analyse titration curves for a dibasic acid, understand the need for approximations and be able to justify them;

- follow through a brief discussion of the much more complex case of the tribasic acid;

- know what is meant by an ampholyte;

- discuss the behaviour of ampholytes.

An Introduction to Aqueous Electrolyte Solutions. By Margaret Robson Wright
© 2007 John Wiley & Sons Ltd ISBN 978-0-470-84293-5 (cloth) ISBN 978-0-470-84294-2 (paper)

6.1 Neutralisation

In neutralisation, e.g. a weak acid by a strong base, the $OH^-(aq)$ added can react in two ways:

- by neutralising the $H_3O^+(aq)$ present as a result of the dissociation of the weak acid:

$$H_3O^+(aq) + OH^-(aq) \xrightarrow{\text{neutralisation}} 2H_2O(l)$$

Removal of $H_3O^+(aq)$ will cause more of the undissociated weak acid to dissociate so as to re-establish the equilibrium:

$$HA(aq) + H_2O(l) \rightleftharpoons H_3O^+(aq) + A^-(aq)$$

$$K_a = \left(\frac{[H_3O^+]_{actual}[A^-]_{actual}}{[HA]_{actual}} \right)_{eq} \tag{6.1}$$

and progressive neutralisation by more $OH^-(aq)$ will effectively remove the weak acid until a number of mol of $OH^-(aq)$ has been added equivalent to the number of mol of weak acid present initially.

- or the $OH^-(aq)$ will react directly with the weak acid:

$$HA(aq) + OH^-(aq) \xrightarrow{\text{neutralisation}} A^-(aq) + H_2O(l)$$

and as more $OH^-(aq)$ is added progressive neutralisation will occur.

The two reactions:

$$H_3O^+(aq) + OH^-(aq) \xrightarrow{\text{neutralisation}} 2H_2O(l)$$

$$HA(aq) + OH^-(aq) \xrightarrow{\text{neutralisation}} A^-(aq) + H_2O(l)$$

are **irreversible reactions** and both are very fast. The kinetics of these reactions indicate that the rates of these reactions are comparable, and both mechanisms will, therefore, contribute to the neutralisation. Both are proton transfers, but the base added in neutralisation is the added $OH^-(aq)$, in contrast to the solvent acting as the base in the **reversible** acid dissociation of a weak acid.

$$HA(aq) + H_2O(l) \rightleftharpoons H_3O^+(aq) + A^-(aq)$$

The neutralisation of a weak base by a strong acid can be likewise described, and both neutralisation reactions will again contribute to the mechanism of neutralisation.

- the added $H_3O^+(aq)$ reacts with the $OH^-(aq)$ produced by the basic reaction of the weak base, and as a result more base will be converted to the protonated conjugate acid as the equilibrium is rapidly re-established, i.e.

$$OH^-(aq) + H_3O^+(aq) \xrightarrow{\text{neutralisation}} 2H_2O(l)$$

is followed by:

$$B(aq) + H_2O(l) \rightleftharpoons BH^+(aq) + OH^-(aq)$$

$$K_b = \left(\frac{[BH^+]_{actual}[OH^-]_{actual}}{[B]_{actual}} \right)_{eq} \tag{6.2}$$

and progressive neutralisation can occur via this mechanism as more $H_3O^+(aq)$ is added.

- or the $H_3O^+(aq)$ will react directly with the weak base:

$$B(aq) + H_3O^+(aq) \xrightarrow{\text{neutralisation}} BH^+(aq) + H_2O(l)$$

and, as more $H_3O^+(aq)$ is added, progressive neutralisation will occur.

The neutralisation of a strong acid with a strong base, or vice versa, is much more straightforward, with no acid/base equilibria involved. The neutralisation occurs via:

$$OH^-(aq) + H_3O^+(aq) \xrightarrow{\text{neutralisation}} 2H_2O(l)$$

and neutralisation continues with addition of $H_3O^+(aq)$ or $OH^-(aq)$ until an equivalent number of mol has been added.

6.2 pH titration curves

Figures 6.1 to 6.4 show the pH titration curves for the various cases given in Tables 6.1 to 6.4. These have been calculated as given below.

6.2.1 Neutralisation of a strong acid by a strong base, e.g. HCl(aq) with NaOH(aq) (Figure 6.1)

$25 \, cm^3$ of $1.00 \times 10^{-2} \, mol \, dm^{-3}$ HCl is titrated with a $5 \times 10^{-3} \, mol \, dm^{-3}$ solution of NaOH, and Table 6.1 shows how the pH of the solution alters with each addition of NaOH. The total volume of the solution increases throughout the titration. The reaction occurring is

$$H_3O^+(aq) + OH^-(aq) \xrightarrow{\text{neutralisation}} 2H_2O(l)$$

and one mol of $H_3O^+(aq)$ is neutralised by one mol of $OH^-(aq)$. The total number of mol of $OH^-(aq)$ required for complete neutralisation will be equal to the number of $H_3O^+(aq)$ present initially. Progressive neutralisation occurs throughout the titration, and from the number of mol of $OH^-(aq)$ added at any stage of neutralisation the number of mol of $H_3O^+(aq)$ neutralised can be found, from which the number of mol of $H_3O^+(aq)$ remaining is calculated. Knowing the total volume of the solution, $[H_3O^+]$ is found, giving the pH.

The initial amount of $H_3O^+(aq)$ present

$$= 25 \, cm^3 \times 1 \times 10^{-2} \, mol \, dm^{-3} = \frac{25}{1000} \, dm^3 \times 1 \times 10^{-2} \, mol \, dm^{-3} = 2.5 \times 10^{-4} \, mol$$

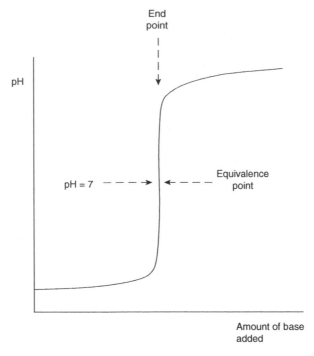

Figure 6.1   Titration of a strong acid by a strong base, e.g. HCl(aq) with NaOH(aq).

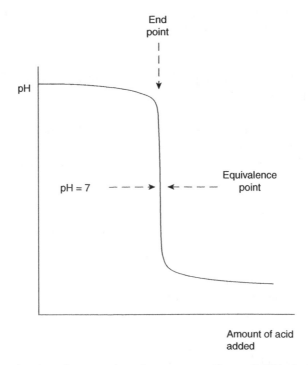

Figure 6.2   Titration of a strong base by a strong acid, e.g. NaOH(aq) with HCl(aq).

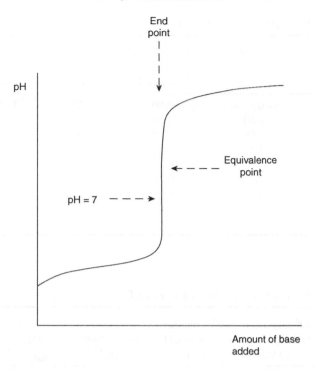

Figure 6.3 Titration of a weak acid by a strong base, e.g. $CH_3COOH(aq)$ with $NaOH(aq)$.

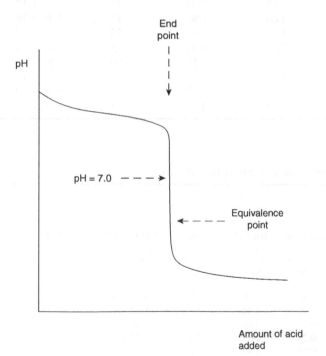

Figure 6.4 Titration of a weak base by a strong acid e.g. $NH_3(aq)$ with $HCl(aq)$.

Table 6.1 pH titration: strong acid/strong base at 25°C

volume NaOH added $\frac{}{cm^3}$	$10^4 \times$ number mol OH$^-$(aq) added	$10^4 \times$ number of mol of H$_3$O$^+$(aq) left	total volume cm^3	$\frac{[H_3O^+]}{mol\,dm^{-3}}$	pH
0	0	2.50	25	1.00×10^{-2}	2.00
10	0.50	2.00	35	5.71×10^{-3}	2.24
20	1.00	1.50	45	3.33×10^{-3}	2.48
30	1.50	1.00	55	1.82×10^{-3}	2.74
40	2.00	0.50	65	7.69×10^{-4}	3.11
45	2.25	0.25	70	3.57×10^{-4}	3.45
47.5	2.37$_5$	0.12$_5$	72.5	1.72×10^{-4}	3.76
49	2.45	0.05	74	6.76×10^{-5}	4.17
50	2.50	0	75	1.00×10^{-7}	7.00

Table 6.2 pH titration: strong base/strong acid at 25°C

volume HCl added $\frac{}{cm^3}$	$10^4 \times$ number mol H$_3$O$^+$(aq) added	$10^4 \times$ number of mol of OH$^-$(aq) left	total volume cm^3	$\frac{[OH^-]}{mol\,dm^{-3}}$	pOH	pH
0	0	2.50	25	1.00×10^{-2}	2.0	12.00
10	0.50	2.00	35	5.71×10^{-3}	2.24	11.76
20	1.00	1.50	45	3.33×10^{-3}	2.48	11.52
30	1.50	1.00	55	1.82×10^{-3}	2.74	11.26
40	2.00	0.50	65	7.69×10^{-4}	3.11	10.89
45	2.25	0.25	70	3.57×10^{-4}	3.45	10.55
47.5	2.37$_5$	0.12$_5$	72.5	1.72×10^{-4}	3.76	10.24
49	2.45	0.05	74	6.76×10^{-5}	4.17	9.83
50	2.50	0	75	1.00×10^{-7}	7.00	7.00

Table 6.3 pH titration: weak acid/strong base at 25°C

Volume NaOH added/cm^3	$10^4 \times$ number of mol OH$^-$ added	$10^4 \times$ number of mol CH$_3$COOH left	$10^4 \times$ number of mol CH$_3$COO$^-$ formed	Total volume/cm^3	[CH$_3$COOH]$_{stoich}$ left/mol dm^3	[CH$_3$COO$^-$]$_{stoich}$ formed/mol dm^3	[H$_3$O$^+$]/ mol dm^{-3}	pH
0	0.00	2.50	0.00	25	1.00×10^{-2}	0.00	4.17×10^{-4}	3.38
10	0.50	2.00	0.50	35	5.71×10^{-3}	1.43×10^{-3}	6.95×10^{-5}	4.16
20	1.00	1.50	1.00	45	3.33×10^{-3}	2.22×10^{-3}	2.61×10^{-5}	4.58
30	1.50	1.00	1.50	55	1.82×10^{-3}	2.73×10^{-3}	1.16×10^{-5}	4.94
40	2.00	0.50	2.00	65	7.69×10^{-4}	3.08×10^{-3}	4.34×10^{-6}	5.36
45	2.25	0.25	2.25	70	3.57×10^{-4}	3.21×10^{-3}	1.94×10^{-6}	5.71
49	2.45	0.05	2.45	74	6.76×10^{-5}	3.31×10^{-3}	3.55×10^{-7}	6.45
50	2.50	0	2.50	75	0.00	3.33×10^{-3}	7.24×10^{-9}	8.14

Table 6.4 pH titration: weak base/strong acid at 25°C

Volume HCl added/cm^3	$10^4 \times$ number of mol H$_3$O$^+$ added	$10^4 \times$ number of mol NH$_3$ left	$10^4 \times$ number of mol NH$_4^+$ formed	total volume/cm^3	[NH$_3$]$_{stoich}$ left/mol dm^{-3}	[NH$_4^+$]$_{stoich}$ formed/mol dm^{-3}	[H$_3$O$^+$]/ mol dm^{-3}	pH
0	0.00	2.50	0.00	25	1.00×10^{-2}	0.00	2.37×10^{-11}	10.63
10	0.50	2.00	0.50	35	5.71×10^{-3}	1.43×10^{-3}	1.41×10^{-10}	9.85
20	1.00	1.50	1.00	45	3.33×10^{-3}	2.22×10^{-3}	3.75×10^{-10}	9.43
30	1.50	1.00	1.50	55	1.82×10^{-3}	2.73×10^{-3}	8.43×10^{-10}	9.07
40	2.00	0.50	2.00	65	7.69×10^{-4}	3.08×10^{-3}	2.25×10^{-9}	8.65
45	2.25	0.25	2.25	70	3.57×10^{-4}	3.21×10^{-3}	5.04×10^{-9}	8.30
49	2.45	0.05	2.45	74	6.76×10^{-5}	3.31×10^{-3}	2.76×10^{-8}	7.56
50	2.50	0	2.50	75	0.00	3.33×10^{-3}	1.37×10^{-6}	5.86

Since HCl and NaOH are strong electrolytes, there are no acid–base equilibria to consider, other than the self ionisation of water:

$$H_3O^+(aq) + OH^-(aq) \rightleftharpoons H_2O(l)$$

$$K_w = ([H_3O^+][OH^-])_{eq} \tag{6.3}$$

6.2.2 Neutralisation of a strong base by a strong acid, e.g. NaOH(aq) with HCl(aq) (Figure 6.2)

25 cm^3 of 1.00×10^{-2} mol dm^{-3} NaOH is titrated with a 5×10^{-3} mol dm^{-3} solution of HCl, and Table 6.2 shows how the pH of the solution alters with each addition of HCl.

The calculation is essentially the same as previously. The number of mol of H$_3$O$^+$(aq) added gives the number of mol of OH$^-$(aq) neutralised, from which the number of mol of OH$^-$(aq) left and its concentration can be found.

$$pOH = -\log_{10}[OH^-] \tag{6.4}$$

$$K_w = ([H_3O^+][OH^-])_{eq} \tag{6.3}$$

$$\therefore [H_3O^+] = \frac{K_w}{[OH^-]} \tag{6.5}$$

$$pH = -\log_{10}\left(\frac{K_w}{[OH^-]}\right) \tag{6.6}$$

$$= pK_w - pOH \tag{6.7}$$

6.2.3 Neutralisation of a weak acid by a strong base, e.g. CH$_3$COOH(aq) with NaOH(aq) (Figure 6.3)

25 cm^3 of 1.00×10^{-2} mol dm^{-3} CH$_3$COOH is titrated with a 5×10^{-3} mol dm^{-3} solution of NaOH, and Table 6.3 shows how the pH of the solution alters with each addition of NaOH.

In this titration, the added $OH^-(aq)$ removes an equivalent number of mol of $CH_3COOH(aq)$ by neutralisation. From this the number of mol of the weak acid remaining is found, and from the total volume of the solution after each addition of $OH^-(aq)$ the **stoichiometric** concentration of $CH_3COOH(aq)$ left can be found. This corresponds to the **total** concentration of acid present at each stage of the titration. **But** $CH_3COOH(aq)$ is a weak acid and this stoichiometric or total concentration of $CH_3COOH(aq)$ is partly in the form of the undissociated $CH_3COOH(aq)$ molecules, and partly as $CH_3COO^-(aq)$ ions.

$$[CH_3COOH]_{stoich\ or\ total} = [CH_3COOH]_{actual} + [CH_3COO^-]_{actual} \tag{6.8}$$

$$K_a = \left(\frac{[CH_3COO^-]_{actual}[H_3O^+]_{actual}}{[CH_3COOH]_{actual}}\right)_{eq} \tag{6.9}$$

The titration shows three stages, each of which must be handled differently:

- the beginning of the titration before any $OH^-(aq)$ is added, corresponds to a solution of a weak acid on its own in water;

- after some $OH^-(aq)$ has been added, an equivalent amount of weak acid will have been removed, and an equivalent amount of its salt with NaOH has been formed, i.e. a buffer solution is formed;

- when an amount of NaOH equivalent to the number of mol of weak acid present initially has been added, all the weak acid will have been neutralised, and a corresponding amount of the salt is formed. The solution is that of a salt of a weak acid with a strong base on its own in water.

The calculations outlined above are:

- **At the beginning of the titration**: a solution of a weak acid in water.

$$CH_3COOH(aq) + H_2O(l) \rightleftharpoons CH_3COO^-(aq) + H_3O^+(aq)$$

$$K_a = \left(\frac{[CH_3COO^-]_{actual}[H_3O^+]_{actual}}{[CH_3COOH]_{actual}}\right)_{eq} \tag{6.9}$$

$pK_a = 4.76$, $K_a = 1.74 \times 10^{-4}\ mol\,dm^{-3}$.

This is a weak acid where both of the standard approximations hold, i.e. no extensive ionisation, and ignoring the self ionisation of water:

$$\therefore K_a = \frac{[H_3O^+]^2_{actual}}{[CH_3COOH]_{stoich\ or\ total}} \tag{6.10}$$

$$\therefore [H_3O^+]_{actual} = \sqrt{K_a[CH_3COOH]_{stoich\ or\ total}} \tag{6.11}$$

$$= \sqrt{1.74 \times 10^{-5}\ mol\,dm^{-3} \times 1.00 \times 10^{-2}\ mol\,dm^{-3}}$$

$$= 4.17 \times 10^{-4}\ mol\,dm^{-3}$$

$$pH = 3.38$$

- **The region of incomplete neutralisation**: the buffer region

In this region the number of mol of $CH_3COOH(aq)$ left un-neutralised is first found, from which the number of mol of $CH_3COO^-(aq)$ formed can be calculated. The total volume after each addition of $NaOH(aq)$ must be found. From this $[CH_3COOH]_{\text{stoich or total}}$ and $[CH_3COO^-]_{\text{stoich or total}}$ are calculated after each addition.

For this buffer solution, the standard approximations hold: i.e. the actual concentration equals the stoichiometric concentration:

$$[H_3O^+]_{\text{actual}} = K_a \frac{[CH_3COOH]_{\text{stoich or total}}}{[CH_3COO^-]_{\text{stoich or total}}} \tag{6.12}$$

After addition of $30\,cm^3$:

Number of mol of $OH^-(aq)$ added $= \dfrac{30}{1000}\,dm^3 \times 5 \times 10^{-3}\,mol\,dm^{-3} = 1.5 \times 10^{-4}$

Number of mol of $CH_3COOH(aq)$ initially $= \dfrac{25}{1000}\,dm^3 \times 1 \times 10^{-2}\,mol\,dm^{-3} = 2.5 \times 10^{-4}$

Number of mol of $CH_3COOH(aq)$ after neutralisation $= 1 \times 10^{-4}$

Total volume $= (25 + 30)\,cm^3 = 55\,cm^3$

$$[CH_3COOH]_{\text{left}} = \frac{1}{(55/1000)\,dm^3} \times 1 \times 10^{-4}\,mol = 1.82 \times 10^{-3}\,mol\,dm^{-3}$$

Number of mol of $CH_3COO^-(aq)$ formed $= 1.5 \times 10^{-4}$

$$[CH_3COO^-]_{\text{formed}} = \frac{1}{(55/1000)\,dm^3} \times 1.5 \times 10^{-4}\,mol = 2.73 \times 10^{-3}\,mol\,dm^{-3}$$

These are the **stoichiometric**, or **total** concentrations.

Using Equation (6.12) gives:

$$[H_3O^+]_{\text{actual}} = \frac{1.74 \times 10^{-5}\,mol\,dm^{-3} \times 1.82 \times 10^{-3}\,mol\,dm^{-3}}{2.73 \times 10^{-3}\,mol\,dm^{-3}} = 1.17 \times 10^{-5}\,mol\,dm^{-3}$$

$$pH = 4.93$$

- **Complete neutralisation**: a solution of the salt on its own in water

This corresponds to a solution of the conjugate base of the weak acid, i.e. $CH_3COO^-(aq)$ acting as a weak base.

$$CH_3COO^-(aq) + H_2O(l) \rightleftharpoons CH_3COOH(aq) + OH^-(aq)$$

$$K_b = \frac{K_w}{K_a} = \left(\frac{[CH_3COOH]_{\text{actual}}[OH^-]_{\text{actual}}}{[CH_3COO^-]_{\text{actual}}} \right)_{\text{eq}} \tag{6.13}$$

The anion of the salt is a weak base where the standard approximations hold, and:

$$[CH_3COO^-]_{\text{actual}} = [CH_3COO^-]_{\text{stoich or total}} \tag{6.14}$$

Here $[CH_3COO^-]_{stoich\ or\ total}$ must be that in the final solution at the equivalence point. It can be calculated from the number of mol of CH_3COOH present initially and the final total volume of the solution.

$$\therefore\ K_b = \frac{K_w}{K_a} = \frac{[OH^-]^2_{actual}}{[CH_3COO^-]_{stoich\ or\ total}} \tag{6.15}$$

$$[OH^-]_{actual} = \sqrt{\frac{K_w[CH_3COO^-]_{stoich\ or\ total}}{K_a}} \tag{6.16}$$

When all of the original $CH_3COOH(aq)$ has been neutralised:
number of mol of $CH_3COO^-(aq)$ = number of mol of $CH_3COOH(aq)$ present initially = 2.5×10^{-4}

Total volume of the solution $= (25 + 50)\,cm^3 = 75\,cm^3$

$$\therefore\ [CH_3COO^-]_{formed} = \frac{1}{(75/1000)\,dm^3} \times 2.5 \times 10^{-4}\,mol = 3.33 \times 10^{-3}\,mol\,dm^{-3}$$

This is the **stoichiometric** concentration of the salt formed, and using the approximate Equation (6.16) gives:

$$[OH^-]_{actual} = \sqrt{\frac{1.00 \times 10^{-14}\,mol^2\,dm^{-6} \times 3.33 \times 10^{-3}\,mol\,dm^{-3}}{1.74 \times 10^{-5}\,mol\,dm^{-3}}} = 1.38 \times 10^{-6}\,mol\,dm^{-3}$$

$$pOH = 5.86$$
$$pH = 14 - 5.86 = 8.14$$

6.2.4   Neutralisation of a weak base by a strong acid, e.g. $NH_3(aq)$ with $HCl(aq)$ (Figure 6.4)

$25\,cm^3$ of $1.00 \times 10^{-2}\,mol\,dm^{-3}$ NH_3 is titrated with a $5 \times 10^{-3}\,mol\,dm^{-3}$ solution of HCl, and Table 6.4 shows how the pH of the solution alters with each addition of HCl.
The calculations predicting the pH values for each stage in the titration are **exactly analogous** to those for the titration of a weak acid with a strong base. Again there are three separate stages:

$$pK_a \text{ for } NH_4^+(aq) = 9.25 \text{ giving } K_a = 5.62 \times 10^{-10}\,mol\,dm^{-3}$$
$$pK_b \text{ for } NH_3(aq) = 4.75 \text{ giving } K_b = 1.78 \times 10^{-5}\,mol\,dm^{-3}$$

• **at the beginning of the titration**, no HCl has been added, and the solution is that of a weak base on its own in water.

$$NH_3(aq) + H_2O(l) \rightleftharpoons NH_4^+(aq) + OH^-(aq)$$
$$K_b = \left(\frac{[NH_4^+]_{actual}[OH^-]_{actual}}{[NH_3]_{actual}}\right)_{eq} \tag{6.17}$$

Making the standard approximations of ignoring the self ionisation of water and assuming little protonation gives:

$$K_b = \frac{[OH^-]^2_{\text{actual}}}{[NH_3]_{\text{stoich or total}}} \tag{6.18}$$

$$[OH^-] = \sqrt{K_b[NH_3]_{\text{stoich or total}}} \tag{6.19}$$

- **During the titration** when HCl is being added, and there is incomplete neutralisation, the solution is a buffer. Using the standard approximation gives:

$$K_b = \left(\frac{[NH_4^+]_{\text{stoich or total}}[OH^-]_{\text{actual}}}{[NH_3]_{\text{stoich or total}}}\right)_{\text{eq}} \tag{6.20}$$

$$[OH^-] = K_b\frac{[NH_3]_{\text{stoich or total}}}{[NH_4^+]_{\text{stoich or total}}} \tag{6.21}$$

- **At the equivalence point** when all the weak base has been neutralised the solution is that of a salt of a weak base/strong acid in water, i.e. NH$_4$Cl(aq). The protonated weak base is the conjugate acid of the weak base, here NH$_3$:

$$NH_4^+(aq) + H_2O(l) \rightleftharpoons NH_3(aq) + H_3O^+(aq)$$

$$K_a = \left(\frac{[NH_3]_{\text{actual}}[H_3O^+]_{\text{actual}}}{[NH_4^+]_{\text{actual}}}\right)_{\text{eq}} \tag{6.22}$$

Making the standard approximations gives:

$$K_a = \frac{[H_3O^+]^2_{\text{actual}}}{[NH_4^+]_{\text{stoich or total}}} \tag{6.23}$$

$$[H_3O^+] = \sqrt{K_a[NH_4^+]_{\text{stoich or total}}} \tag{6.24}$$

6.3 Interpretation of pH titration curves

Figures 6.1 to 6.4 show the curves for:

- a strong acid/strong base titration,

- a strong base/strong acid titration,

- a weak acid/strong base titration,

- a weak base/strong acid titration.

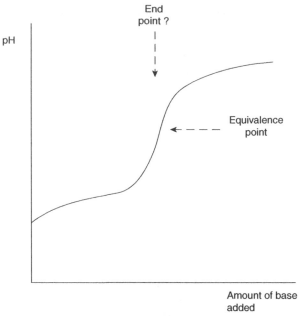

The equivalence point may be at a pH either above or
below 7.0, depending on what the acid and the base are

Figure 6.5   Titration of a weak acid by a weak base, e.g. $CH_3COOH(aq)$ with $NH_3(aq)$.

The curve for a weak acid/weak base titration is the composite of the curves for a weak acid/
strong base and weak base/strong acid titrations, as indicated (Figure 6.5).

The curves for a weak acid or weak base which are **weaker** than ethanoic acid or ammonia
respectively, are displaced upwards for the weak acid, and downwards for the weak base
respectively, as indicated (Figures 6.6(a) and 6.6(b) respectively).

The strong acid/strong base titration curve shows a smooth increase in pH with addition of
base until around 99% neutralisation when a dramatic increase in pH over several pH units is
found. Addition of excess base past the equivalence point is also shown. This is equivalent
to the titration of a corresponding amount of strong base by a strong acid as shown. If the
initial concentration of the acid, or base, is **greater** then the pH titration is displaced
downwards, or upwards, respectively, and the change in pH around the equivalence point
is even greater. If the concentrations are **lower**, then the curves are displaced upwards, or
downwards, respectively, and the change in pH around the equivalence point is not so
dramatic (Figures 6.7 and 6.8).

This affects the choice of indicator, if used, to determine the equivalence point. An
indicator should be chosen such that it changes colour in the region of rapidly changing
pH. The end-point in a titration using an indicator is thus less well defined than the
equivalence point which is the stage at which an exactly equivalent amount of acid or base
has been added to the original basic or acidic solution. It can be pinpointed exactly on a pH
titration. For the titration of a strong acid or base, the position of exact equivalence is found at
$pH = 7$. However, even a minute addition of $H_3O^+(aq)$ or $OH^-(aq)$ at this stage will cause a
dramatic change in pH.

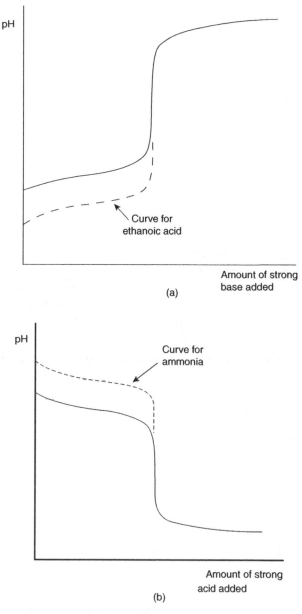

Figure 6.6(a) Titration of an acid weaker than $CH_3COOH(aq)$ with $NaOH(aq)$; **(b)** Titration of a base weaker than $NH_3(aq)$ with $HCl(aq)$.

When the weak acid/strong base or weak base/strong acid are considered several differences appear. Looking at the curve for the weak acid/ strong base titration it can be seen that there is an initial upturn in pH found at the beginning of the titration before the buffer region is reached. At this stage the buffering capacity has not set in because there is a lot of HA(aq) and very little A^-(aq). A sufficient amount of A^-(aq) has to be formed by neutralisation before the solution acts as a buffer (see Section 5.6.7).

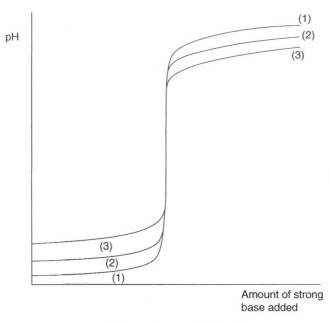

pH

(1)

(2)

(3)

(3)

(2)

(1)

Amount of strong
base added

(1) : most concentrated acid
(3) : least concentrated acid

Figure 6.7   Titration of a strong acid by a strong base with varying concentrations of acid.

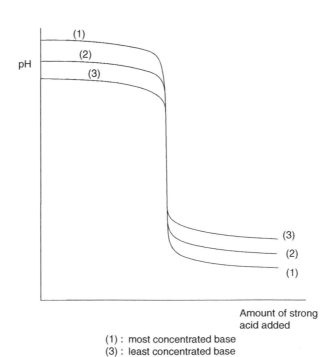

pH

(1)

(2)

(3)

(3)

(2)

(1)

Amount of strong
acid added

(1) : most concentrated base
(3) : least concentrated base

Figure 6.8   Titration of a strong base with a strong acid with varying concentrations of base.

The equivalence point for a weak acid will always occur at basic pH values, e.g. pH $= 8.14$ for the case quoted. This reflects the basic behaviour of the salt formed, in contrast to the neutral behaviour of NaCl in the strong acid/strong base titration. Furthermore, the change in pH around the equivalence point is much less dramatic, again reflecting the fact that the salt formed is involved in an acid/base equilibrium. The buffer region shows a smooth increase in pH with addition of strong base.

The graphical representation of the data obscures to some extent the true nature of the buffer region. Both the strong acid/strong base and weak acid/strong base show a very similar increase in pH with addition of base, with the slopes of the curves being very similar. However, the chemical processes occurring in each are different. For the strong acid/strong base titration, an amount of $H_3O^+(aq)$ is removed equivalent to the amount of base added. With the weak acid/strong base titration there is a difference. An amount of weak acid equivalent to the amount of base added is neutralised. This, however, is much greater than the amount of $H_3O^+(aq)$ removed. This is because the undissociated acid remaining is not fully dissociated to give an equivalent amount of $H_3O^+(aq)$, but is in equilibrium with its anion and $H_3O^+(aq)$, and the equilibrium lies well over to the left. This is in contrast with the strong acid/strong base titration where the acid remaining after the partial neutralisation exists totally as $H_3O^+(aq)$. Table 6.3 shows that whereas addition of base to the weak acid only results in a very small change in $[H_3O^+]$, this is not the case with the strong acid titration (Table 6.1).

- For the strong acid titration: addition of $10\,cm^3$ of NaOH to the **solution of total volume of $55\,cm^3$** changes $[H_3O^+]$ from $1.82 \times 10^{-3}\,mol\,dm^{-3}$ to $0.77 \times 10^{-3}\,mol\,dm^{-3}$, a change of $1.05 \times 10^{-3}\,mol\,dm^{-3}$. This does not suggest a buffer solution.

- For the weak acid titration the corresponding change in $[H_3O^+]$ is very small, i.e. from $1.16 \times 10^{-5}\,mol\,dm^{-3}$ to $4.34 \times 10^{-6}\,mol\,dm^{-3}$ corresponding to a change of $6.34 \times 10^{-6}\,mol\,dm^{-3}$, behaviour typical of a buffer solution.

6.4 Polybasic acids

These are acids where there is more than one ionisable proton per molecule of undissociated acid. The simplest examples are the dicarboxylic acids:

$$HOOC(CH_2)_nCOOH(aq) + H_2O(l) \xrightleftharpoons{K_1} HOOC(CH_2)_nCOO^-(aq) + H_3O^+(aq)$$

$$K_1 = \left(\frac{[H_3O^+]_{actual}[HOOC(CH_2)_nCOO^-]_{actual}}{[HOOC(CH_2)_nCOOH]_{actual}} \right)_{eq} \tag{6.25}$$

$$HOOC(CH_2)_nCOO^-(aq) + H_2O(l) \xrightleftharpoons{K_2} {}^-OOC(CH_2)_nCOO^-(aq) + H_3O^+(aq)$$

$$K_2 = \left(\frac{[H_3O^+]_{actual}[{}^-OOC(CH_2)_nCOO^-]_{actual}}{[HOOC(CH_2)_nCOO^-]_{actual}} \right)_{eq} \tag{6.26}$$

For all dibasic acids $pK_2 > pK_1$. This is because it is more difficult to remove the second proton from the negatively charged anion produced in the first dissociation. However, the difference between the two pK values decreases as the value of n increases, a consequence of

Table 6.5 pK values for some dibasic acids and tribasic acids at 25°C

acid	pK_1	pK_2	pK_3
oxalic HOOC—COOH	1.27	4.27	
malonic (propanedioic) HOOCCH$_2$COOH	2.86	5.70	
succinic (butanedioic) HOOC(CH$_2$)$_2$COOH	4.21	5.64	
adipic (hexanedioic) HOOC(CH$_2$)$_4$COOH	4.41	5.28	
maleic (*cis* but-2-enedioic) *cis*HOOCCH=CHCOOH	1.92	6.23	
fumaric (*trans* but-2-enedioic) *trans*HOOCH=CHCOOH	3.02	4.38	
o-phthalic *o*-C$_6$H$_4$(COOH)$_2$	2.95	5.41	
m-phthalic *m*-C$_6$H$_4$(COOH)$_2$	3.62	4.60	
p-phthalic *p*-C$_6$H$_4$(COOH)$_2$	3.54	4.46	
glycine $^+$NH$_3$CH$_2$COOH	2.22	9.86	
alanine $^+$NH$_3$CH(CH$_3$)COOH	2.22	9.97	
β alanine $^+$NH$_3$CH$_2$CH$_2$COOH	3.60	10.36	
phosphoric H$_3$PO$_4$	2.15	7.20	11.9
glutamic NH$_3$$^+$CH(COOH)CH$_2CH_2$COOH	2.30	4.51	9.95
citric HOOCC(OH)(CH$_2$COOH)$_2$	3.13	4.76	6.40

the increasing separation of charge after the second proton is removed. Table 6.5 gives pK values for some typical dibasic and tribasic acids.

Dibasic acids show two regions of **rapid change** in pH if the pKs are well separated. The middle of each of these regions is taken to be the end-point, and this may be slightly different from the point of exact equivalence. The end-point as determined by an indicator can be difficult to pin down precisely and pH titrations are more definitive. If the pKs are close together the regions of increasing pH may be poorly separated or may even merge. The end-point of the titration as determined by the middle of the region of rapid change in pH may be difficult to pin down and will be close to the second equivalence point, with the first equivalence point unable to be pinned down from the titration curve (see Figures 6.9 to 6.11).

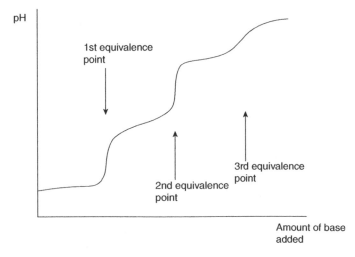

Figure 6.9 pH titration curve for phosphoric acid at 25°C. pK_1 = 2.15, pK_2 = 7.20 and pK_3 = 11.9.

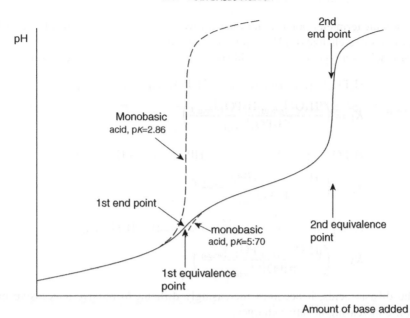

Figure 6.10 pH titration curve for malonic acid at 25°C. $pK_1 = 2.86$ and $pK_2 = 5.70$, with corresponding monobasic acids of the same pKs.

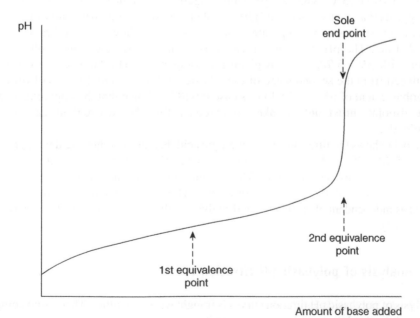

Figure 6.11 pH titration curve for hexane-1.6-dioic acid at 25°C. $pK_1 = 4.41$ and $pK_2 = 5.28$.

For this reason the term end-point rather than equivalence point will be used to describe the regions of more rapid change of pH for polybasic acids.

Phosphoric acid is a typical tribasic acid showing three dissociation equilibria:

$$H_3PO_4(aq) + H_2O(l) \xrightleftharpoons{K_1} H_2PO_4^-(aq) + H_3O^+(aq)$$

$$K_1 = \left(\frac{[H_3O^+]_{actual}[H_2PO_4^-]_{actual}}{[H_3PO_4]_{actual}}\right)_{eq} \tag{6.27}$$

$$H_2PO_4^-(aq) + H_2O(l) \xrightleftharpoons{K_2} HPO_4^{2-}(aq) + H_3O^+(aq)$$

$$K_2 = \left(\frac{[H_3O^+]_{actual}[HPO_4^{2-}]_{actual}}{[H_2PO_4^-]_{actual}}\right)_{eq} \tag{6.28}$$

$$HPO_4^{2-}(aq) + H_2O(l) \xrightleftharpoons{K_3} PO_4^{3-}(aq) + H_3O^+(aq)$$

$$K_3 = \left(\frac{[H_3O^+]_{actual}[PO_4^{3-}]_{actual}}{[HPO_4^{2-}]_{actual}}\right)_{eq} \tag{6.29}$$

As with the dibasic acids, it becomes increasingly difficult to remove successive protons because of the increasing negative charges.

pH titration curves for polybasic acids depend on the relative pK values for successive dissociations. Figure 6.9 shows the titration curve for phosphoric acid. At 25°C, p$K_1 = 2.15$, p$K_2 = 7.20$, p$K_3 = 11.9$. Here the three end-points are well separated, and are well defined, so that it **would appear as though** the curve represents the behaviour of three monobasic acids of differing strength acting independently.

Figure 6.10 shows the curve for malonic acid, propane-1,3-dioic acid, with p$K_1 = 2.86$ and p$K_2 = 5.70$ at 25°C. The two pK values again differ considerably, and the titration curve suggests that the two stages **might** be able to be treated independently of each other. The overall curve shows two separated end-points, where, however, the first end-point is not too distinct. The pH titration curves for two monobasic acids, one with p$K = 2.86$, the other with p$K = 5.70$, are also given for comparison. The latter curve can only be distinguished from the second stage of the dibasic acid in the initial stages of titration for the monobasic acid of p$K = 5.70$. In this case it will be shown that the apparent separation of the end-points **must not** be taken as evidence that the two regions can be treated independently.

Figure 6.11 shows the titration curve for adipic acid, hexane-1,6-dioic acid with p$K_1 = 4.41$ and p$K_2 = 5.28$ at 25°C. Here the curve resembles that of a monobasic acid with one extended buffer region, and only one end-point. There is thus a clear distinction between this titration curve and those for malonic and phosphoric acids. The two dissociations for this acid cannot be treated as independent of each other, and analysis of the curve must consider both equilibria simultaneously.

6.4.1 Analysis of polybasic pH titration curves

The analysis of polybasic pH titration curves is fraught with difficulties. There is no completely general analytical solution, and in these cases it is even more important to be aware of what approximations, if any, can be made, whether they are fully justified and what they correspond to physically. Assessment of the limitations imposed by these approximations must be considered explicitly. It is even more important here, in contrast to the case of a monobasic acid, to realise that each acid must be considered individually with respect to both approximations, their validity and to the consequences. Most of the analysis is complex. Calculation of the equilibrium concentrations of the species actually present in a given solution again will depend on knowing the equilibrium constants and the stoichiometric concentrations. For a polybasic acid or base they can only be found by the use of sensible approximations, or a series of successive approximations which are explicitly related to the physical situation.

6.4.2 A dibasic acid with two apparently separated pK values, e.g. malonic acid

In a solution of a dibasic acid there are several species which can be present: $H_2A(aq)$, $HA^-(aq)$, $A^{2-}(aq)$, $H_3O^+(aq)$ and $OH^-(aq)$. Which are present in significant amounts will depend crucially on the pH.

For a solution of $H_2A(aq)$ on its own in water:

- There are several equilibria which will be set up.

The first dissociation:

$$H_2A(aq) + H_2O(l) \xrightleftharpoons{K_1} HA^-(aq) + H_3O^+(aq)$$

$$K_1 = \left(\frac{[H_3O^+]_{actual}[HA^-]_{actual}}{[H_2A]_{actual}} \right)_{eq} \tag{6.30}$$

The second dissociation:

$$HA^-(aq) + H_2O(l) \xrightleftharpoons{K_2} A^{2-}(aq) + H_3O^+(aq)$$

$$K_2 = \left(\frac{[A^{2-}]_{actual}[H_3O^+]_{actual}}{[HA^-]_{actual}} \right)_{eq} \tag{6.31}$$

The self ionisation of water, $pK_w = 14$

$$2H_2O(l) \xrightleftharpoons{K_w} H_3O^+(aq) + OH^-(aq)$$

$$K_w = ([H_3O^+]_{actual}[OH^-]_{actual})_{eq} \tag{6.3}$$

The protonation of $HA^-(aq)$ as a base: $pK_{b1} = pK_w - pK_1$ $\qquad$ (6.32)

$$HA^-(aq) + H_2O(l) \xrightleftharpoons{K_{b1}} H_2A(aq) + OH^-(aq)$$

$$K_{b1} = \frac{K_w}{K_1} = \left(\frac{[H_2A]_{actual}[OH^-]_{actual}}{[HA^-]_{actual}} \right)_{eq} \tag{6.33}$$

The protonation of $A^{2-}(aq)$ as a base: $pK_{b2} = pK_w - pK_2$ $\qquad$ (6.34)

$$A^{2-}(aq) + H_2O(l) \xrightleftharpoons{K_{b2}} HA^-(aq) + OH^-(aq)$$

$$K_{b2} = \frac{K_w}{K_2} = \left(\frac{[HA^-]_{actual}[OH^-]_{actual}}{[A^{2-}]_{actual}} \right)_{eq} \tag{6.35}$$

All these equilibria will contribute to the pH of the solution, and the likely contribution of each can be assessed.

- **The stoichiometric relation is**:

$$[H_2A]_{total} = [H_2A]_{actual} + [HA^-]_{actual} + [A^{2-}]_{actual} \qquad (6.36)$$

- **Electrical neutrality gives**:

$$[H_3O^+]_{actual} = [HA^-]_{actual} + 2[A^{2-}]_{actual} + [OH^-]_{actual} \qquad (6.37)$$

Depending on the situation, it may be possible to make sensible approximations to these equations.

Making successive substitutions from the equilibrium expressions (6.30) and (6.31) into the equation for electrical neutrality (6.37) gives:

$$[H_3O^+]_{actual} = \frac{K_1[H_2A]_{actual}}{[H_3O^+]_{actual}} + \frac{2K_2[HA^-]_{actual}}{[H_3O^+]_{actual}} + \frac{K_w}{[H_3O^+]_{actual}} \qquad (6.38)$$

$$[H_3O^+]_{actual} = \frac{K_1[H_2A]_{actual}}{[H_3O^+]_{actual}} + \frac{2K_1K_2[H_2A]_{actual}}{[H_3O^+]^2_{actual}} + \frac{K_w}{[H_3O^+]_{actual}} \qquad (6.39)$$

$$\therefore \ [H_3O^+]_{actual} = K_1 \frac{[H_2A]_{actual}}{[H_3O^+]_{actual}} \left\{ 1 + \frac{2K_2}{[H_3O^+]_{actual}} \right\} + \frac{K_w}{[H_3O^+]_{actual}} \qquad (6.40)$$

Equation (6.38) states that the contributions to $[H_3O^+]_{actual}$ in succession are:

- contribution from $H_2A(aq)$ as an acid;

- contribution from $HA^-(aq)$ as an acid;

- contribution from the self ionisation of water.

Note: This analysis automatically includes a consideration of the self ionisation of water.

Equation (6.38) can be used to estimate the relative contributions to the pH for any acid at any pH, from each of the sources of $H_3O^+(aq)$. It can be shown that these depend on the values of K_1 and K_2.

Equation (6.40) is a cubic expression and can only be solved by a series of successive approximations (see Section 4.8). However, the use of **intelligent approximations** which depend on the given situation can give **approximate** values for $[H_3O^+]_{actual}$, which could then be used as a **starting point** for a **more exact** analysis. For instance, if the acid $H_2A(aq)$ is weak, then there will be only slight ionisation to $HA^-(aq)$, and any subsequent ionisation to $A^{2-}(aq)$ will be even less, and the second term probably can be ignored. If this is the case it would suggest that the second stage of the ionisation does not interfere in the ionisation of $H_2A(aq)$ in its first stage. However, it is best that each term be estimated in turn.

The following problem is an exercise in making sensible approximations, and justifying their validity.

Worked Problem 6.1

Question

Using the following data assess whether it is likely that the second stage dissociation will contribute to the pH of a $0.0100 \text{ mol dm}^{-3}$ solution of each acid on its own in water.

- malonic acid:
at 25°C, $pK_1 = 2.86$, $K_1 = 1.38 \times 10^{-3} \text{ mol dm}^{-3}$; $pK_2 = 5.70$, $K_2 = 2.00 \times 10^{-6} \text{ mol dm}^{-3}$.

- adipic acid:
at 25°C, $pK_1 = 4.41$, $K_1 = 3.89 \times 10^{-5} \text{ mol dm}^{-3}$; $pK_2 = 5.28$, $K_2 = 5.25 \times 10^{-6} \text{ mol dm}^{-3}$.

Answer

- malonic acid:
Start by using Equation (6.38)

$$[H_3O^+]_{actual} = \frac{K_1[H_2A]_{actual}}{[H_3O^+]_{actual}} + \frac{2K_2[HA^-]_{actual}}{[H_3O^+]_{actual}} + \frac{K_w}{[H_3O^+]_{actual}} \tag{6.38}$$

Possible approximations:
- assume that the second stage of ionisation can be ignored; this will have to be checked once a first approximate $[H_3O^+]_{actual}$ is found;

- malonic acid is a fairly strong weak acid in its first stage of ionisation, and so the solution will have pH < 7. Therefore, the last term in Equation (6.38), $K_w/[H_3O^+]_{actual}$ can be ignored;

- if the contribution to $[H_3O^+]_{actual}$ from the self ionisation of water and the second stage of ionisation are ignored, then:

$$[H_3O^+]_{actual} = [HA^-]_{actual}$$

- since $pK_1 = 2.86$, it cannot be assumed that there is only slight ionisation to $HA^-(aq)$, and the rigorous expression must be used:

$$[H_2A]_{total} = [H_2A]_{actual} + [HA^-]_{actual}$$

Making these assumptions, Equation (6.38) becomes:

$$[H_3O^+]_{actual} = \frac{K_1[H_2A]_{actual}}{[H_3O^+]_{actual}} = \frac{K_1\{[H_2A]_{total} - [HA^-]_{actual}\}}{[H_3O^+]_{actual}}$$

$$= \frac{K_1[H_2A]_{total} - K_1[H_3O^+]_{actual}}{[H_3O^+]_{actual}}$$

$$\therefore [H_3O^+]_{actual}^2 + K_1[H_3O^+]_{actual} - K_1[H_2A]_{total} = 0$$

This is a quadratic equation with solution:

$$[H_3O^+]_{actual} = \frac{-1.38 \times 10^{-3} \pm \sqrt{1.90_5 \times 10^{-6} + 5.52 \times 10^{-5}}}{2} \text{ mol dm}^{-3}$$

$$= \frac{-1.38 \times 10^{-3} \pm 7.56 \times 10^{-3}}{2} \text{ mol dm}^{-3}$$

$$= 3.09 \times 10^{-3} \text{ mol dm}^{-3}$$

i.e. ignoring the negative root.

This first approximate $[H_3O^+]_{actual}$ can now be used to calculate whether it is valid to ignore the second term in Equation (6.38):

Using the approximation, $[HA^-]_{actual} = [H_3O^+]_{actual}$, Equation (6.38) becomes

$$[H_3O^+]_{actual} = \frac{K_1[H_2A]_{actual}}{[H_3O^+]_{actual}} + \frac{2K_2[H_3O^+]_{actual}}{[H_3O^+]_{actual}} + \frac{K_w}{[H_3O^+]_{actual}}$$

and ignoring the last term, by approximation, this becomes:

$$[H_3O^+]_{actual} = \frac{K_1[H_2A]_{actual}}{[H_3O^+]_{actual}} + 2K_2$$

$$\text{i.e. } [H_3O^+]_{actual} - 2K_2 = \frac{K_1[H_2A]_{actual}}{[H_3O^+]_{actual}}$$

If $2K_2 \ll [H_3O^+]_{actual}$ then $[H_3O^+]_{actual} = K_1[H_2A]/[H_3O^+]_{actual}$ which corresponds to the term $2K_2[HA^-]_{actual}/[H_3O^+]_{actual}$ being dropped from Equation (6.38) as well as $K_w/[H_3O^+]_{actual}$.

This can be checked as follows:

$$2K_2 = 4 \times 10^{-6} \, \text{mol dm}^{-3}$$

first approximate $[H_3O^+]_{actual} = 3.09 \times 10^{-3} \, \text{mol dm}^{-3}$

$\therefore 2K_2 \ll [H_3O^+]_{actual}$ and the term $\dfrac{2K_2[HA^-]_{actual}}{[H_3O^+]_{actual}}$ can justifiably be dropped from

Equation (6.38) as well as $\dfrac{K_w}{[H_3O^+]_{actual}}$.

This would suggest that the second stage does not interfere with the ionisation of malonic acid in its first stage for the case of malonic acid **on its own in water** and at a stoichiometric concentration of $1 \times 10^{-2} \, \text{mol dm}^{-3}$.

- adipic acid:

at 25°C, $pK_1 = 4.41$, $K_1 = 3.89 \times 10^{-5} \, \text{mol dm}^{-3}$; $pK_2 = 5.28$, $K_2 = 5.25 \times 10^{-6} \, \text{mol dm}^{-3}$.

The same approximations as for malonic acid are used with the one possible difference that since $pK_1 = 4.41$ instead of 2.86 there is probably only slight ionisation to $HA^-(aq)$ and so $[H_2A]_{actual} \approx [H_2A]_{total}$. However, to be certain, the rigorous expression will be used as in the case of malonic acid.

$$[H_3O^+]_{actual} = \frac{K_1[H_2A]_{actual}}{[H_3O^+]_{actual}} = \frac{K_1\{[H_2A]_{total} - [HA^-]_{actual}\}}{[H_3O^+]_{actual}} = \frac{K_1[H_2A]_{total} - K_1[H_3O^+]_{actual}}{[H_3O^+]_{actual}}$$

giving the quadratic equation

$$[H_3O^+]^2_{actual} + K_1[H_3O^+]_{actual} - K_1[H_2A]_{total} = 0$$

Proceeding in this way gives a first approximate $[H_3O^+]_{actual} = 6.05 \times 10^{-4} \, mol \, dm^{-3}$. This value can now be used to calculate whether it is valid to ignore the second term as well as the third in Equation (6.38).

$2K_2 = 10.5 \times 10^{-6} \, mol \, dm^{-3}$. Thus $2K_2$ is not negligible compared with $[H_3O^+]_{actual} = 6.05 \times 10^{-4} \, mol \, dm^{-3}$. In the equation:

$$[H_3O^+]_{actual} = \frac{K_1[H_2A]_{actual}}{[H_3O^+]_{actual}} + \frac{2K_2[H_3O^+]_{actual}}{[H_3O^+]_{actual}} + \frac{K_w}{[H_3O^+]_{actual}}$$

only the last term can be ignored, and

$$[H_3O^+]_{actual} = \frac{K_1[H_2A]_{actual}}{[H_3O^+]_{actual}} + \frac{2K_2[H_3O^+]_{actual}}{[H_3O^+]_{actual}}$$

$$\therefore [H_3O^+]_{actual} = \frac{K_1[H_2A]_{actual}}{[H_3O^+]_{actual}} + 2K_2$$

This suggests that the second stage **does interfere** with the ionisation of adipic acid in its first stage to a limited extent for the case of adipic acid **on its own in water** and at a stoichiometric concentration of $1 \times 10^{-2} \, mol \, dm^{-3}$. In contrast to malonic acid, a rigorous calculation should not ignore this contribution from the second stage.

Although it is possible to make a first rough approximate estimate of $[H_3O^+]_{actual}$ for both these acids this **must not** be taken to mean that the second stage of ionisation to $HA^-(aq)$ can be ignored in the **buffer** regions or when $H_2A(aq)$ has been half neutralised, i.e. at what should be the **first equivalence-point**. Explicit expressions must be derived for these regions.

6.5   pH titrations of dibasic acids: the calculations

The complete analysis of the titration curve must include:

- the beginning of the titration when no base has been added; i.e. the weak acid on its own in water;

- the first and second buffer regions;

- the first and second end-points.

For acids like adipic acid there is only one end-point, corresponding to complete neutralisation of the dibasic acid after an amount of base has been added equivalent to twice the number of mol of acid initially present. For other acids such as malonic acid two end-points are found and there seems to be an apparent separation of the end-points. This must not, however, be taken to mean that the two stages of the titration curve can be treated independently.

6.5.1   The beginning of the titration

The analytical expression for $[H_3O^+]_{actual}$ has been given above (Equation 6.38) and approximate values for $[H_3O^+]_{actual}$ and pH given for malonic and adipic acids. The following gives a systematic, general and rigorous procedure for the successive approximations to a solution of the equation:

$$[H_3O^+]_{actual} = \frac{K_1[H_2A]_{actual}}{[H_3O^+]_{actual}} + \frac{2K_2[HA^-]_{actual}}{[H_3O^+]_{actual}} + \frac{K_w}{[H_3O^+]_{actual}} \qquad (6.38)$$

First approximation:

- Ignore $K_w/[H_3O^+]_{actual}$ and $2K_2[HA^-]_{actual}/[H_3O^+]_{actual}$.

  Assume that $[A^{2-}]_{actual}$ is negligible, and that $[H_3O^+]_{actual} = [HA^-]_{actual}$.
  Obtain a **first approximate** $[H_3O^+]_{actual}$ and hence $[HA^-]_{actual}$ by solution of the quadratic equation:

$$[H_3O^+]^2_{actual} = K_1[H_2A]_{actual} \qquad (6.41)$$
$$= K_1[H_2A]_{total} - K_1[HA^-]_{actual} \qquad (6.42)$$
$$= K_1[H_2A]_{total} - K_1[H_3O^+]_{actual} \qquad (6.43)$$
$$\text{i.e. } [H_3O^+]^2_{actual} + K_1[H_3O^+]_{actual} - K_1[H_2A]_{total} = 0 \qquad (6.44)$$

- Take these **first approximate** $[H_3O^+]_{actual}$ and $[HA^-]_{actual}$ and substitute into the expression for K_2

$$K_2 = \left(\frac{[A^{2-}]_{actual}[H_3O^+]_{actual}}{[HA^-]_{actual}} \right)_{eq} \qquad (6.31)$$

  Since it has been assumed that $[H_3O^+]_{actual} = [HA^-]_{actual}$ in the first approximation, the **first approximate** $[A^{2-}]_{actual} = K_2$.

- Take these **first approximate** $[HA^-]_{actual}$ and $[A^{2-}]_{actual}$, and the stoichiometric relation:

$$[H_2A]_{total} = [H_2A]_{actual} + [HA^-]_{actual} + [A^{2-}]_{actual} \qquad (6.36)$$

  to get a **first approximate** $[H_2A]_{actual}$.

- a set of **first approximate** concentrations has now been found.

  Using these values, a series of successive approximations are made until the concentrations have become constant.

6.5.2   The first equivalence point

At the first equivalence point half of the stoichiometric amount of $H_2A(aq)$ initially present will have been neutralised, i.e. the $H_2A(aq)$ has been converted to an **equal** stoichiometric amount of $NaHA(aq)$ which is fully ionised to $Na^+(aq)$ and $HA^-(aq)$.

There are, however, two further equilibria which must be considered, since HA^- can act as a base or as an acid:

- **acting as a base**:

$$HA^-(aq) + H_2O(l) \xrightleftharpoons{K_{b1}} H_2A(l) + OH^-(aq)$$

$$K_{b1} = \frac{K_w}{K_1} = \frac{[H_2A]_{actual}[OH^-]_{actual}}{[HA^-]_{actual}} \tag{6.33}$$

- **acting as an acid**:

$$HA^-(aq) + H_2O(l) \xrightleftharpoons{K_2} A^{2-}(aq) + H_3O^+(aq)$$

$$K_2 = \left(\frac{[A^{2-}]_{actual}[H_3O^+]_{actual}}{[HA^-]_{actual}}\right)_{eq} \tag{6.31}$$

- **The stoichiometric relation is**:

$$[H_2A]_{total} = [H_2A]_{actual} + [HA^-]_{actual} + [A^{2-}]_{actual} \tag{6.36}$$

which becomes:

$$[Na^+]_{total} = [H_2A]_{total} = [H_2A]_{actual} + [HA^-]_{actual} + [A^{2-}]_{actual} \tag{6.45}$$

- **Electrical neutrality**

This must include the contribution from the $Na^+(aq)$:

$$[Na^+]_{total} + [H_3O^+]_{actual} = [HA^-]_{actual} + 2[A^{2-}]_{actual} + [OH^-]_{actual} \tag{6.46}$$

Substituting Equation (6.46) for $[Na^+]_{total}$ into Equation(6.45)gives:

$$[H_2A]_{actual} + [HA^-]_{actual} + [A^{2-}]_{actual} + [H_3O^+]_{actual} = [HA^-]_{actual} + 2[A^{2-}]_{actual} + [OH^-]_{actual} \tag{6.47}$$

$$[H_3O^+]_{actual} = [HA^-]_{actual} + 2[A^{2-}]_{actual} + [OH^-]_{actual} - [H_2A]_{actual} - [HA^-]_{actual} - [A^{2-}]_{actual} \tag{6.48}$$

$$\therefore [H_3O^+]_{actual} = [A^{2-}]_{actual} - [H_2A]_{actual} + [OH^-]_{actual} \tag{6.49}$$

Using the equilibrium relations corresponding to K_1, K_2 and K_w gives:

$$[H_3O^+]_{actual} = \frac{K_2[HA^-]_{actual}}{[H_3O^+]_{actual}} - \frac{[HA^-]_{actual}[H_3O^+]_{actual}}{K_1} + \frac{K_w}{[H_3O^+]_{actual}} \tag{6.50}$$

$$[H_3O^+]_{actual} + \frac{[HA^-]_{actual}[H_3O^+]_{actual}}{K_1} = \frac{K_2[HA^-]_{actual}}{[H_3O^+]_{actual}} + \frac{K_w}{[H_3O^+]_{actual}} \tag{6.51}$$

Multiply both sides by $[H_3O^+]_{actual}$

$$[H_3O^+]^2_{actual}\left[1 + \frac{[HA^-]_{actual}}{K_1}\right] = K_2[HA^-]_{actual} + K_w \tag{6.52}$$

$$[H_3O^+]^2_{actual} = \frac{K_2[HA^-]_{actual} + K_w}{1 + [HA^-]_{actual}/K_1} \tag{6.53}$$

This is a quadratic equation in $[H_3O^+]_{actual}$, which ultimately involves four unknowns: $[H_2A]_{actual}$, $[HA^-]_{actual}$, and $[A^{2-}]_{actual}$ and $[H_3O^+]_{actual}$. The only easy way to obtain all unknown quantities is by successive approximations.

6.5.3   The first buffer region

In the buffer region only partial neutralisation will have occurred. The problem when polybasic acids are considered lies in the possibility that the second stage of ionisation may complicate matters. If there is no great overlap at the first end-point then it can be confidently assumed that the second ionisation will not interfere over most of the first buffer region. However, even with dibasic acids such as malonic acid where the two pK values differ by 3, it has been shown that the effects of the second ionisation cannot be ignored at the first end-point. The first buffer region, likewise, cannot be treated as if the solution were that of a monobasic acid. The closer to the end-point, the more interference will be observed, see comparison of the pH titration curve for malonic acid and that for a monobasic acid with the same pK. For acids such as adipic acid where there is no separation of the end-points, it will not be valid to assume that the two ionisation equilibria function independently. A rigorous analysis including the effects of all the possible equilibria, K_1, K_{b1}, K_2, K_{b2}, and K_w, must be found.

6.5.4   Analysis of the first buffer region

For the buffer region a series of stages of neutralisation must be considered, and this results in a series of values of the ratio, $\dfrac{[Na^+]_{total}}{[H_2A]_{total}} = x$. This ratio will be zero at the beginning of the titration, and unity at the first equivalence point. At other stages of the titration $x\%$ of the original acid has been neutralised. This ratio will be used in the equation for electrical neutrality.
The following relations are also required:

• **The equilibrium relations**:
are those given earlier, i.e. the first dissociation, the second dissociation, and the self ionisation of water (Equations 6.30, 6.31 and 6.3).

• **The stoichiometric relation**:

$$[H_2A]_{total} = [H_2A]_{actual} + [HA^-]_{actual} + [A^{2-}]_{actual} \tag{6.36}$$

Electrical neutrality:

$$[Na^+]_{total} + [H_3O^+]_{actual} = [HA^-]_{actual} + 2[A^{2-}]_{actual} + [OH^-]_{actual} \quad (6.46)$$

When a fraction x of the original acid has been neutralised:

$$[Na^+]_{total} = x[H_2A]_{total}$$
$$= x[H_2A]_{actual} + x[HA^-]_{actual} + x[A^{2-}]_{actual} \quad (6.54)$$

and the electrical neutrality equation becomes:

$$x[H_2A]_{total} + [H_3O^+]_{actual} = [HA^-]_{actual} + 2[A^{2-}]_{actual} + [OH^-]_{actual} \quad (6.55)$$

$$x[H_2A]_{actual} + x[HA^-]_{actual} + x[A^{2-}]_{actual} + [H_3O^+]_{actual}$$
$$= [HA^-]_{actual} + 2[A^{2-}]_{actual} + [OH^-]_{actual} \quad (6.56)$$

$$[H_3O^+]_{actual} = (1-x)[HA^-]_{actual} + (2-x)[A^{2-}]_{actual} - x[H_2A]_{actual} + [OH^-]_{actual} \quad (6.57)$$

$$\therefore [H_3O^+]_{actual} = (1-x)[HA^-]_{actual} + (2-x)\frac{K_2[HA^-]_{actual}}{[H_3O^+]_{actual}}$$
$$- x\frac{[HA^-]_{actual}[H_3O^+]_{actual}}{K_1} + \frac{K_w}{[H_3O^+]_{actual}} \quad (6.58)$$

$$\therefore [H_3O^+]_{actual} + x\frac{[HA^-]_{actual}[H_3O^+]_{actual}}{K_1}$$
$$= (1-x)[HA^-]_{actual} + (2-x)\frac{K_2[HA^-]_{actual}}{[H_3O^+]_{actual}} + \frac{K_w}{[H_3O^+]_{actual}} \quad (6.59)$$

Multiply by $[H_3O^+]_{actual}$

$$[H_3O^+]^2_{actual}\left[1 + \frac{x[HA^-]_{actual}}{K_1}\right]$$
$$= (1-x)[HA^-]_{actual}[H_3O^+]_{actual} + (2-x)K_2[HA^-]_{actual} + K_w \quad (6.60)$$

$$\therefore [H_3O^+]^2_{actual} = \frac{(1-x)[HA^-]_{actual}[H_3O^+]_{actual} + (2-x)K_2[HA^-]_{actual} + K_w}{1 + \frac{x[HA^-]_{actual}}{K_1}} \quad (6.61)$$

This is a quadratic equation in $[H_3O^+]_{actual}$, and is also an equation in four unknowns, and again the only easy way to obtain all unknown quantities is by successive approximations.

6.5.5 The second equivalence point

At the second equivalence point all of the stoichiometric amount of $H_2A(aq)$ initially present will have been neutralised, i.e. the $H_2A(aq)$ has been converted to an **equal** stoichiometric amount of $Na_2A(aq)$ which is fully ionised to $Na^+(aq)$ and $A^{2-}(aq)$. This occurs when a stoichiometric amount of base has been added equal to twice the stoichiometric amount of $H_2A(aq)$ originally present. This can be handled in the same way as the first equivalence point.

Depending on the values of K_1 and K_2 either the more approximate or the more rigorous methods will have to be used.

6.5.6 The second buffer region

This will be handled either by the more approximate method, or by the more rigorous method of successive approximations. In general, the more rigorous method will have to be used the closer the titration is to the first equivalence point. For most of the buffer region away from the first equivalence point, the more approximate series of successive approximations should be adequate.

6.6 Tribasic acids

Phosphoric acid is a typical tribasic acid. A similar treatment to that for dibasic acids can be given, from which it becomes clear that the expressions derived have a pattern to them.

In a solution of a tribasic acid there are several species which can be present: $H_3A(aq)$, $H_2A^-(aq)$, $HA^{2-}(aq)$, $A^{3-}(aq)$, $H_3O^+(aq)$ and $OH^-(aq)$. Which are present in significant amounts will depend crucially on the pH. For a solution of $H_3A(aq)$ on its own in water the stoichiometric relation:

$$[H_3A]_{total} = [H_3A]_{actual} + [H_2A^-]_{actual} + [HA^{2-}]_{actual} + [A^{3-}]_{actual} \qquad (6.62)$$

can be written.

The electrical neutrality relation is:

$$[H_3O^+]_{actual} = [H_2A^-]_{actual} + 2[HA^{2-}]_{actual} + 3[A^{3-}]_{actual} + [OH^-]_{actual} \qquad (6.63)$$

Depending on the situation, sensible approximations to these equations can be made. There are also several equilibria which will be set up.

The first dissociation:

$$H_3A(aq) + H_2O(l) \xrightleftharpoons{K_1} H_2A^-(aq) + H_3O^+(aq)$$

$$K_1 = \left(\frac{[H_3O^+]_{actual}[H_2A^-]_{actual}}{[H_3A]_{actual}} \right)_{eq} \qquad (6.64)$$

The second dissociation:

$$H_2A^-(aq) + H_2O(l) \xrightleftharpoons{K_2} HA^{2-}(aq) + H_3O^+(aq)$$

$$K_2 = \left(\frac{[HA^{2-}]_{actual}[H_3O^+]_{actual}}{[H_2A^-]_{actual}} \right)_{eq} \qquad (6.65)$$

The third dissociation:

$$HA^{2-}(aq) + H_2O(l) \xrightleftharpoons{K_3} A^{3-}(aq) + H_3O^+(aq)$$

$$K_3 = \left(\frac{[A^{3-}]_{actual}[H_3O^+]_{actual}}{[HA^{2-}]_{actual}} \right)_{eq} \qquad (6.66)$$

The self ionisation of water: $pK_w = 14$

$$2H_2O(l) \xrightleftharpoons{K_w} H_3O^+(aq) + OH^-(aq)$$
$$K_w = ([H_3O^+][OH^-])_{eq} \tag{6.3}$$

The protonation of H_2A^-(aq) as a base: $pK_{b1} = pK_w - pK_1$ $\hspace{2em}$ (6.67)

$$H_2A^-(aq) + H_2O(l) \xrightleftharpoons{K_{b1}} H_3A(aq) + OH^-(aq)$$
$$K_{b1} = \frac{K_w}{K_1} = \left(\frac{[H_3A]_{actual}[OH^-]_{actual}}{[H_2A^-]_{actual}}\right)_{eq} \tag{6.68}$$

The protonation of HA^{2-}(aq) as a base: $pK_{b2} = pK_w - pK_2$ $\hspace{2em}$ (6.69)

$$HA^{2-}(aq) + H_2O(l) \xrightleftharpoons{K_{b2}} H_2A^-(aq) + OH^-(aq)$$
$$K_{b2} = \frac{K_w}{K_2} = \left(\frac{[H_2A^-]_{actual}[OH^-]_{actual}}{[HA^{2-}]_{actual}}\right)_{eq} \tag{6.70}$$

The protonation of A^{3-}(aq) as a base: $pK_{b3} = pK_w - pK_3$ $\hspace{2em}$ (6.71)

$$A^{3-}(aq) + H_2O(l) \xrightleftharpoons{K_{b3}} HA^{2-}(aq) + OH^-(aq)$$
$$K_{b3} = \frac{K_w}{K_3} = \left(\frac{[HA^{2-}]_{actual}[OH^-]_{actual}}{[A^{3-}]_{actual}}\right)_{eq} \tag{6.72}$$

All these equilibria will contribute to the pH of the solution, and the likely contribution of each can be assessed.

Substituting into the equation for electrical neutrality (Equation 6.63) from the equilibrium expressions given above:

$$[H_3O^+]_{actual} = \frac{K_1[H_3A]_{actual}}{[H_3O^+]_{actual}} + \frac{2K_2[H_2A^-]_{actual}}{[H_3O^+]_{actual}} + \frac{3K_3[HA^{2-}]_{actual}}{[H_3O^+]_{actual}} + \frac{K_w}{[H_3O^+]_{actual}} \tag{6.73}$$

$$[H_3O^+]_{actual} = \frac{K_1[H_3A]_{actual}}{[H_3O^+]_{actual}} + \frac{2K_1K_2[H_3A]_{actual}}{[H_3O^+]_{actual}^2} + \frac{3K_1K_2K_3[H_3A]_{actual}}{[H_3O^+]_{actual}^3} + \frac{K_w}{[H_3O^+]_{actual}} \tag{6.74}$$

$$\therefore [H_3O^+]_{actual} = K_1\frac{[H_3A]_{actual}}{[H_3O^+]_{actual}}\left\{1 + \frac{2K_2}{[H_3O^+]_{actual}} + \frac{3K_2K_3}{[H_3O^+]_{actual}^2}\right\} + \frac{K_w}{[H_3O^+]_{actual}} \tag{6.75}$$

Equation (6.73) states that the contributions to $[H_3O^+]_{actual}$ in succession are:

- contribution from H_3A(aq) as an acid;

- contribution from H_2A^-(aq) as an acid;.

- contribution from HA^{2-} as an acid;

- contribution from the self ionisation of water.

Equation (6.73) can be used to estimate the relative contributions to the pH for any acid at any pH, from each of the sources of $H_3O^+(aq)$. It can be shown that these depend on the values of K_1, K_2 and K_3.

6.6.1   Analysis of the titration curve

This analysis is complex, but follows the pattern given for the dibasic acids. The solutions of the final equations have to be found by successive approximation, or else by computer curve fitting to the whole titration curve. In practice, this is how the analysis will be carried out.

6.6.2   An important thought

Computer curve fitting using standard programs is an **extremely good** way of coping with a **complex analysis** which otherwise would need a lot of thought in setting out the various relations required, e.g. those given in Section 6.6. It is also **very useful** for carrying out the very tedious numerical calculations which are required.

However, there are problems associated with curve fitting procedures, not least those associated with the possibility that more than one best fit is possible.

It is also **crucially important** that there is full awareness of what is going on physically and chemically in any given situation, be it a weak acid on its own in water or a buffer made up from a polybasic acid and its salt.

This is **precisely why** the analysis is set out in full for each case being considered in this chapter and others. There is no point in obtaining pKs or actual concentrations of species present in an aqueous solution if it is not known what is going on chemically in that solution.

6.7   Ampholytes

In Section 3.5, it was stated that aliphatic aminoacids and aminophenols showed strongly contrasting behaviour when studied over a range of pH. Both show acidic and basic behaviour and, for both, the titration curves show two pK values and two end-points.

Figure 6.12 shows the titration curves for glycine and m-aminophenol. Both show two separated end-points with only the first one being well defined. At 25°C the titration curves furnish the following pK values:

	pK_1	pK_2
glycine:	2.22	9.86
m-aminophenol	4.17	9.87

Glycine can be titrated with HCl and must therefore be acting as a base. When fully protonated, glycine must be $^+NH_3CH_2COOH$.

Glycine can also be titrated with NaOH and here must be acting as an acid. When fully deprotonated glycine must be $NH_2CH_2COO^-$.

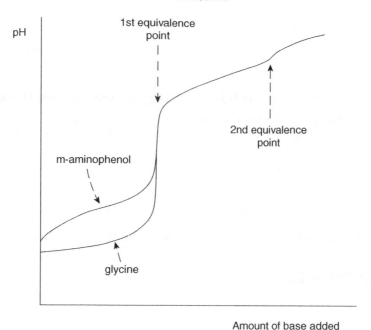

Figure 6.12 Titration curves for glycine and *m*-amino phenol at 25°C.

The titration curve has two end-points:

- one at around pH = 6 which is associated with a buffer region where half-way to the equivalence point gives pH = 2.50. This buffer region corresponds to an acid with $pK_a = 2.22$.

- one at around pH = 12 which is associated with a buffer region where half-way to the equivalence point gives pH = 9.85. This buffer region corresponds to $pK_a = 9.86$.

Remember: an approximate pK_a can be found from the buffer region of the titration curve. $pK_a \approx pH$ at half-way to the equivalence point.

The titration curve can be analysed as for a dibasic acid, with the two acid groups being the $^+NH_3$ group and the COOH group. What the titration curve **cannot tell directly** is to **which grouping** each pK refers, or **what the species** present at the first equivalence point **actually is**. This must be found from other studies, spectroscopic analysis being the most powerful. As shown in Section 3.5 there are two possible structures:

- NH_2CH_2COOH formed by neutralisation of the acidic amino group, $^+NH_3$ with OH^-.

- $^+NH_3CH_2COO^-$ formed by neutralisation of the acidic carboxyl group, COOH with OH^-.

Other studies have shown conclusively that the species present at the first equivalence point is predominantly the dipolar, charge separated, or zwitterion, $^+NH_3CH_2COO^-$.

The acid–base behaviour of glycine is thus:

$$^+NH_3CH_2COOH(aq) + H_2O(l) \xrightleftharpoons{K_1} {}^+NH_3CH_2COO^-(aq) + H_3O^+(aq)$$

and

$$^+NH_3CH_2COO^-(aq) + H_2O(l) \xrightleftharpoons{K_2} NH_2CH_2COO^-(aq) + H_3O^+(aq)$$

m-aminophenol shows a similar titration curve (Figure 6.13), again analysed as for a dibasic acid. Likewise this titration curve can only furnish values of the two pK_a values, and other studies determine the structure of the species present at the first equivalence point. They indicate that this is likely to be predominantly the molecular species $NH_2C_6H_4OH$. The acid–base behaviour is:

$$^+NH_3C_6H_4OH(aq) + H_2O(l) \xrightleftharpoons{K_1} NH_2C_6H_4OH(aq) + H_3O^+(aq)$$

$$NH_2C_6H_4OH(aq) + H_2O(l) \xrightleftharpoons{K_2} NH_2C_6H_4O^-(aq) + H_3O^+(aq)$$

Worked problem 6.2

Question

Show that the following data support the suggestion that the intermediate species, found at the first equivalence point of titrations of aliphatic amino acids with NaOH, is the charge separated zwitterion.

acid	glycine	alanine	β alanine
pK_1	2.22	2.22	3.60
pK_2	9.86	9.97	10.36

acid	$CH_3NH_3^+$	CH_3COOH
pK_a	10.62	4.76

All values relate to 25°C

Answer

pK_1 for the amino acids is similar to, though smaller than, pK_a for typical carboxylic acids such as ethanoic acid. This indicates that the process involved in these equilibria is similar:

$$CH_3COOH(aq) + H_2O(aq) \xrightleftharpoons{K_a} CH_3COO^-(aq) + H_3O^+(aq)$$

$$^+NH_3CH_2COOH(aq) + H_2O(l) \xrightleftharpoons{K_1} {}^+NH_3CH_2COO^-(aq) + H_3O^+(aq)$$

This conclusion is further supported by the observation that $pK_1 < pK_a$. This is what would be expected for the deprotonation of a –COOH in a positively charged species. The presence of the positive charge would repel the approach of a proton to the negatively charged carboxylate group $-COO^-$ and decrease the chance of protonation on the $-COO^-$, or put differently the positive charge would encourage the removal of a proton from the carboxyl group $-COOH$.

The fact that pK_2 for the amino acids is very similar to pK_a for a typical protonated amine suggests that the process involved in all the equilibria is similar, i.e. the acid–base behaviour relates to the amino group.

$$CH_3NH_3^+(aq) + H_2O(l) \xrightleftharpoons{K_a} CH_3NH_2(aq) + H_3O^+(aq)$$

$$^+NH_3CH_2COO^-(aq) + H_2O(l) \xrightleftharpoons{K_2} NH_2CH_2COO^-(aq) + H_3O^+(aq)$$

Worked problem 6.3

Question

Show that the following data support the suggestion that the intermediate species found at the first equivalence point of titration of aminophenols with NaOH is a molecular species rather than a zwitterion.

- aminophenols

aminophenol	pK_1	pK_2
o-aminophenol	4.72	9.71
m-aminophenol	4.17	9.87
p-aminophenol	5.50	10.30

- protonated aromatic amines and phenols

aromatic amines (protonated)	pK_a	phenols	pK_a
aniline	4.58	phenol	9.98
o-methyl aniline	4.39	o-methyl phenol	10.28
m-methyl aniline	4.69	m-methyl phenol	10.08
p-methyl aniline	5.12	p-methyl phenol	10.14

All values relate to 25°C

Answer

For the aminophenols $pK_1 \approx pK_a$ for typical protonated aromatic amines, while $pK_2 \approx pK_a$ for typical phenols. This can be taken to indicate that K_1 relates to a process which is at least predominantly:

$$^+NH_3C_6H_4OH(aq) + H_2O(l) \xrightleftharpoons{} NH_2C_6H_4OH(aq) + H_3O^+(aq)$$

i.e. that K_1 refers to the acid-base behaviour of the amino group.

The fact that $pK_2 \approx pK_a$ for typical phenols suggests that K_2 refers to acid–base behaviour of the phenolic group, and that K_2 relates to a process which is at least predominantly:

$$NH_2C_6H_4OH(aq) + H_2O(l) \rightleftharpoons NH_2C_6H_4O^-(aq) + H_3O^+(aq)$$

These conclusions indicate that the processes involved in the deprotonation of aminophenols is different from that for the aliphatic amino acids in that the intermediate species formed at the first equivalence point of a titration against NaOH is probably the molecular species rather than the charge separated zwitterion.

Worked problem 6.4

Question

Suggest a possible structure for the species found at the first equivalence point for the following aminobenzoic acids.

- protonated amino benzoic acids and their methyl esters

amino benzoic acids (protonated)	pK_1	pK_2	amino benzoates (protonated)	pK_a
o-amino benzoic acid	2.11	4.90	methyl o-amino benzoate	2.23
m-amino benzoic acid	3.12	4.74	methyl m-amino benzoate	3.64
p-amino benzoic acid	2.41	4.85	methyl p-amino benzoate	2.38

- protonated aromatic amines and substituted benzoic acids

protonated aromatic amines	pK_a	substituted benzoic acid	pK_a
aniline	4.58	benzoic acid	4.17
o-methyl aniline	4.39	o-methyl benzoic acid	3.91
m-methyl aniline	4.69	m-methyl benzoic acid	4.27
p-methyl aniline	5.12	p-methyl benzoic acid	4.37

All values relate to 25°C

Answer

The situation is not so clear cut with the aromatic amino acids, and it becomes more difficult to assign K_1 and K_2. This is because benzoic acid and aniline have very similar pK_as, and, on the basis of magnitude alone, pK_2 could be assigned to the acid–base behaviour of either the amino or the carboxyl group. The pK_1 values for the aminobenzoic acids are significantly lower than the pK_as for aniline and substituted anilines and for benzoic acid and substituted benzoic acids. However, the pK_a values for the methyl amino benzoates do give a clue. In these molecules there is only one group which can exhibit acid–base behaviour, i.e. the amino group. The fact that these pK_a values are very similar to pK_1 for the aminobenzoic acids could possibly suggest that the acid–base behaviour might be associated with the amino group and that the equilibrium involved is more likely to be:

$$^+NH_3C_6H_4COOH(aq) + H_2O(l) \xrightleftharpoons{K_1} NH_2C_6H_4COOH(aq) + H_3O^+(aq)$$

rather than being associated with the carboxyl group:

$$^+NH_3C_6H_4COOH(aq) + H_2O(l) \xrightleftharpoons{K_1} {}^+NH_3C_6H_4COO^-(aq) + H_3O^+(aq)$$

On this basis the lowered pK_1 for the aminobenzoic acids as compared with pK_a for the anilines could be rationalised as the effect of the conjugation of the carboxyl group to the benzene causing an increased likelihood for the protonated amino group to lose the proton. This would increase its strength as an acid relative to aniline.

On this basis pK_2 would be assigned to the acid–base behaviour of the carboxyl group:

$$NH_2C_6H_4COOH(aq) + H_2O(l) \xrightleftharpoons{K_2} NH_2C_6H_4COO^-(aq) + H_3O^+(l)$$

The alternative interpretation is that both the molecular species and the zwitterion are present at the first equivalence point, and that both equilibria contribute to the observed pK_1.

(i) $^+NH_3C_6H_4COOH(aq) + H_2O(l) \rightleftharpoons NH_2C_6H_4COOH(aq) + H_2O(l)$

This equilibrium corresponds to a K_a for the protonated amino end of the fully protonated aminobenzoic acid going to the molecular species.

(ii) $^+NH_3C_6H_4COOH(aq) + H_2O(l) \rightleftharpoons {}^+NH_3C_6H_4COO^-(aq) + H_2O(l)$

This equilibrium corresponds to a K_a for the 'acid' end of the fully protonated aminobenzoic acid going to the zwitterion.

This would mean that the observed pK_2 would also be composite:

(iii) $NH_2C_6H_4COOH(aq) + H_2O(l) \rightleftharpoons NH_2C_6H_4COO^-(aq) + H_3O^+(aq)$

This equilibrium corresponds to a K_a for the totally uncharged molecular species formed at the first equivalence point going to the fully deprotonated amino benzoate.

(iv) $^+NH_3C_6H_4COO^-(aq) + H_2O(l) \rightleftharpoons NH_2C_6H_4COO^-(aq) + H_3O^+(aq)$

This equilibrium corresponds to a K_a for the zwitterion formed at the first equivalence point going to the fully deprotonated amino benzoate.

There are, therefore, four possible equilibria involved in the deprotonation of amino benzoic acids. Spectroscopic observations indicate that, in fact, the more complex composite situation is closer to reality.

6.7.1   Analysis of the titration curves for aliphatic and aromatic amino acids, and amino phenols

This analysis will cover the possibility that both the molecular species and the zwitterion are present at the first equivalence point of a titration of the acid with base. This means that the observed K_{1obs} is a composite quantity comprised of two equilibrium constants, viz.:

- a K_a for the protonated amino end of the fully protonated amino acid or amino phenol going to the molecular species;

- a K_a for the 'acid' end of the fully protonated amino acid or amino phenol going to the zwitterion.

The observed K_{2obs} is also a composite quantity comprised of two equilibrium constants:

- a K_a for the totally uncharged molecular species formed at the first equivalence point going to the fully deprotonated amino acid or amino phenol;

- a K_a for the zwitterion formed at the first equivalence point going to the fully deprotonated amino acid or phenol.

The species present throughout the titration are:

- initially the fully protonated amino acid or aminophenol is present, represented as: $H_2X^+(aq)$;

- at the first equivalence point both the molecular species, HX(aq), (totally uncharged) and the zwitterion, $HX^{\pm}(aq)$ (charge separated, but overall neutral) may be present;

- at the second equivalence point the fully deprotonated amino acid or aminophenol is present, $X^-(aq)$.

The observed K_1

$$H_2X^+ + H_2O(l) \rightleftharpoons HX^{\pm}(aq) + H_3O^+(aq)$$
$$H_2X^+ + H_2O(l) \rightleftharpoons HX(aq) + H_3O^+(aq)$$

$$K_{1obs} = \left(\frac{[H_3O^+]([HX^{\pm}] + [HX])}{[H_2X^+]}\right)_{eq} = \left(\frac{[H_3O^+][HX^{\pm}]}{[H_2X^+]}\right)_{eq} + \left(\frac{[H_3O^+][HX]}{[H_2X^+]}\right)_{eq} \quad (6.76)$$

$$= K_a \text{ (for the 'acid' end of } H_2X^+) + K_a \text{ (for the protonated amino end of } H_2X^+)$$

where the 'acid' refers to the carboxyl in the amino acid or the hydroxyl in the aminophenol.

Analysis of the first buffer region will give a value for an observed K_1. However, it cannot be used to infer where the deprotonation equilibrium is occurring, and depending on the species it may refer to either the 'acid' end, or to the protonated amino end, or to both simultaneously. What is necessary is an independent determination of the group, or groups involved. This cannot be done from a pH titration, and often comparison of observed pK_{1obs} with pK_a values for substances where the deprotonation site is unambiguous is used. Spectroscopic observations could furnish a way out. If the relative amounts of $HX^{\pm}$ and HX could be found spectroscopically, then it would be possible to work out the relative contributions in

Equation (6.76) from the 'acid' end and the protonated amino end to the overall observed, K_{1obs}. In the absence of such evidence only sensible guesses can be made.

The observed K_2

$$K_{2obs} = \left(\frac{[H_3O^+][X^-]}{[HX^\pm] + [HX]}\right)_{eq} \tag{6.77}$$

$$\frac{1}{K_{2obs}} = \left(\frac{[HX^\pm] + [HX]}{[H_3O^+][X^-]}\right)_{eq} = \left(\frac{[HX^\pm]}{[H_3O^+][X^-]}\right)_{eq} + \left(\frac{[HX]}{[H_3O^+][X^-]}\right)_{eq} \tag{6.78}$$

$$= \frac{1}{K_{a \text{ for the zwitterion } HX^\pm}} + \frac{1}{K_{a \text{ for the totally uncharged } HX}} \tag{6.79}$$

$$K_{2obs} = \frac{\left(K_{a \text{ for the zwitterion } HX^\pm}\right) \times \left(K_{a \text{ for the totally uncharged } HX}\right)}{\left(K_{a \text{ for the zwitterion } HX^\pm}\right) + \left(K_{a \text{ for the totally uncharged } HX}\right)} \tag{6.80}$$

$$= \frac{\left(K_{a \text{ for the protonated amino end of the zwitterion } HX^\pm}\right) \times \left(K_{a \text{ for the ''acid''end of the totally uncharged } HX}\right)}{\left(K_{a \text{ for the protonated amino end of the zwitterion } HX^\pm}\right) + \left(K_{a \text{ for the ''acid'' end of the totally uncharged } HX}\right)}$$

$$\tag{6.81}$$

This expression for K_{2obs} is more complex than that for $1/K_{2obs}$ which is the expression which would be used to manipulate the experimental results. As with K_{1obs}, K_{2obs} is a composite quantity, and the relative amounts of $HX^\pm$ and HX are required before the individual K_as can be found.

It is important to realise that the pK_1 and pK_2 values for aliphatic and aromatic amino acids, and for amino phenols are likely to be composite quantities, and that four deprotonation equilibrium constants may be involved in the acid–base behaviour. It may turn out that if the species at the first equivalence point of the pH titration is predominantly the zwitterion or the molecular species. If this is so, then the observed pKs will approximate to a single equilibrium constant. But it must not be assumed that this is always the case.

7

Ion Pairing, Complex Formation and Solubilities

The three topics of ion pairing, complex formation and solubilities are typical aspects of equilibrium in electrolyte solutions, and are handled in precisely the same manner as acid–base equilibria. As in the calculations on acid–base equilibria, only the ideal case is considered. Discussion of corrections for non-ideality are deferred until Sections 8.22 to 8.28.

In this chapter pay **special attention** to:

- noticing how often **precisely** the same logic and reasoning is used as in previous chapters on equilibrium, e.g.

  - formulating the equilibrium constant relation from the chemical equations;

  - writing down both the **stoichiometric relations** and the expression for **electrical neutrality**;

  - comparing the number of unknowns with the number of equations;

and of **vital importance**,

- considering the approximations which can be made, the physical conditions under which they are valid and, once made, checking that the approximations are valid.

Aims

By the end of this chapter you should be able to:

- discuss all the above points as they are applied to equilibrium calculations in:

- ion pair formation;

An Introduction to Aqueous Electrolyte Solutions. By Margaret Robson Wright
© 2007 John Wiley & Sons Ltd   ISBN 978-0-470-84293-5 (cloth) ISBN 978-0-470-84294-2 (paper)

- complex formation;

- solubility equilibria;

- and appreciate the distinction between ion pairs, complexes and chelates.

7.1   Ion pair formation

In any solution of an electrolyte there is always the possibility that the ions of the electrolyte might not be fully dissociated in solution. As shown in Section 1.12, ion pairing results when the electrostatic interactions between two oppositely charged ions become sufficiently large for the two ions to move around as one entity, the ion pair. The extent of association into ion pairs depends on many factors, with the most important being the nature, charges and sizes of the ions, the characteristics of the solvent and the temperature. For example, in solutions of Na_2SO_4, the following occurs:

$$Na^+(aq) + SO_4^{2-}(aq) \overset{K_{assoc}}{\rightleftharpoons} NaSO_4^-(aq)$$

$$K_{assoc} = \left(\frac{[NaSO_4^-]_{actual}}{[Na^+]_{actual}[SO_4^{2-}]_{actual}} \right)_{eq} \qquad (7.1)$$

where K_{assoc} is called the association constant, in its ideal form. This is an equation in three unknowns.

Sometimes a dissociation constant is quoted. This expresses the equilibrium in terms of the dissociation of the ion pair:

$$NaSO_4^-(aq) \overset{K_{dissoc}}{\rightleftharpoons} Na^+(aq) + SO_4^{2-}(aq)$$

where:

$$K_{dissoc} = \left(\frac{[Na^+]_{actual}[SO_4^{2-}]_{actual}}{[NaSO_4^-]_{actual}} \right)_{eq} \qquad (7.2)$$

$$K_{dissoc} = 1/K_{assoc} \qquad (7.3)$$

Most early work was given in terms of the dissociation constant, mainly because of the analogy with acid–base equilibria where acid behaviour was considered in terms of the dissociation of the weak acid, HA(aq), into ions, $H_3O^+(aq)$ and $A^-(aq)$. However, this analogy must not be taken too far. In the weak acid situation, dissociation of a **molecular species**, HA(aq), occurs, but in the ion pair situation dissociation is from a species which is not a molecule; rather it is a **species which is held together by simple coulombic electrostatic interactions**. It is vital that this distinction is made clear.

In modern work the association constant is usually quoted, where:

$$K_{assoc} = \frac{1}{K_{dissoc}} \qquad (7.4)$$

$$\therefore \log_{10} K_{assoc} = \log_{10} 1 - \log_{10} K_{dissoc} \qquad (7.5)$$

$$-\log_{10} K_{assoc} = -\log_{10} 1 + \log_{10} K_{dissoc} \qquad (7.6)$$

$$pK_{assoc} = -pK_{dissoc} \qquad (7.7)$$

and so interconversion between the two formulations is straightforward.

In the example quoted above, the species present are the free ions, $Na^+(aq)$ and $SO_4^{2-}(aq)$ and the ion pair $NaSO_4^-$ (aq), and the $Na^+(aq)$ is distributed between free $Na^+(aq)$ and ion pair, $NaSO_4^-(aq)$. The stoichiometric or total concentration of $Na^+(aq)$ is given by:

$$[Na^+]_{\text{stoich or total}} = [Na^+]_{\text{actual}} + [NaSO_4^-]_{\text{actual}} \qquad (7.8)$$

A similar expression can be written for the total concentration of $SO_4^{2-}(aq)$:

$$[SO_4^{2-}]_{\text{stoich or total}} = [SO_4^{2-}]_{\text{actual}} + [NaSO_4^-]_{\text{actual}} \qquad (7.9)$$

As with acid–base equilibria, a further equation describing electrical neutrality can be written:

$$[Na^+]_{\text{actual}} + [H_3O^+]_{\text{actual}} = 2[SO_4^{2-}]_{\text{actual}} + [NaSO_4^-]_{\text{actual}} + [OH^-]_{\text{actual}} \qquad (7.10)$$

Remember: the sulphate ion is doubly charged, hence the factor two.

A solution of $Na_2SO_4(aq)$ is one of a salt of a strong acid/strong base and is neutral, and so:

$$[H_3O^+]_{\text{actual}} = [OH^-]_{\text{actual}} \qquad (7.11)$$

giving:

$$[Na^+]_{\text{actual}} = 2[SO_4^{2-}]_{\text{actual}} + [NaSO_4^-]_{\text{actual}} \qquad (7.12)$$

If K_{assoc} and the stoichiometric concentration are known then there are now sufficient equations to enable all the actual concentrations to be found.

The fraction associated, β, for each ion is given by:

$$\text{for the } Na^+(aq): \qquad \beta = \left(\frac{[NaSO_4^-]_{\text{actual}}}{[Na^+]_{\text{total}}} \right)_{\text{eq}} \qquad (7.13)$$

$$\text{for the } SO_4^{2-}(aq): \qquad \beta = \left(\frac{[NaSO_4^-]_{\text{actual}}}{[SO_4^{2-}]_{\text{total}}} \right)_{\text{eq}} \qquad (7.14)$$

Be careful: this β describing the fraction associated must **not** be confused with the β used by inorganic chemists to describe overall equilibrium constants in complex formation (see Section 7.2).

All ion pairing equilibria can be handled in **exactly** the same manner as is illustrated by the following worked problem. This is a lengthy problem, but it should be studied carefully as it illustrates several important points in the argument. These will be highlighted to bring them to your notice. The problem could be taken to be four separate exercises, but it is useful to compare the **differences in detail** which result by virtue of the differing formulae for the salts considered, **even though exactly the same overall reasoning** is being applied. It is also useful to compare the differing interpretations for each salt.

Worked problem 7.1

Question

Ion pairs are formed in the following solutions. Compare the fraction associated for both cations and anions in each of the following solutions, suggesting a possible explanation for the spread in values.

$$0.05 \, \text{mol dm}^{-3} \text{Mg}(IO_3)_2$$

$$0.05 \, \text{mol dm}^{-3} \text{Mg} \, SO_4$$

$$0.025 \, \text{mol dm}^{-3} \, (Mg)_2 Fe(CN)_6$$

Formation of the ion pairs at 25°C:

$$MgIO_3^+ : \qquad pK_{assoc} = -0.71; \qquad MgSO_4 : \quad pK_{assoc} = -2.23;$$

$$MgFe(CN)_6^{2-} : \qquad pK_{assoc} = -3.87$$

Answer

These pK_{assoc} correspond to K_{assoc} values:

$$MgIO_3^+ : \qquad K_{assoc} = 5.1 \, \text{mol}^{-1} \, \text{dm}^3; \qquad MgSO_4 : \qquad K_{assoc} = 170 \, \text{mol}^{-1} \, \text{dm}^3;$$

$$MgFe(CN)_6^{2-} : \qquad K_{assoc} = 7400 \, \text{mol}^{-1} \, \text{dm}^3.$$

- Considering the ion pair, $MgIO_3^+(aq)$:

$$Mg^{2+}(aq) + IO_3^-(aq) \xrightleftharpoons{K_{assoc}} MgIO_3^+(aq)$$

$$K_{assoc} = \left(\frac{[MgIO_3^+]_{actual}}{[Mg^{2+}]_{actual}[IO_3^-]_{actual}} \right)_{eq} \qquad (7.15)$$

Note: this is an equation in three unknowns, and so two more independent equations are needed.

The stoichiometric relations are:

$$[Mg^{2+}]_{total} = [Mg^{2+}]_{actual} + [MgIO_3^+]_{actual}$$

$$\therefore [Mg^{2+}]_{actual} = [Mg^{2+}]_{total} - [MgIO_3^+]_{actual} \qquad (7.16)$$

$$[IO_3^-]_{total} = 2[Mg^{2+}]_{total} = [IO_3^-]_{actual} + [MgIO_3^+]_{actual}$$

$$\therefore [IO_3^-]_{actual} = 2[Mg^{2+}]_{total} - [MgIO_3^+]_{actual} \qquad (7.17)$$

There are two points to watch here:

- the factor 2, each mol of $Mg(IO_3)_2$ gives one mol $Mg^{2+}(aq)$ and 2 mol $IO_3^-(aq)$;

- Equations (7.16) and (7.17) give expressions for $[Mg^{2+}]_{actual}$ and $[IO_3^-]_{actual}$ in terms of the total concentrations and the concentration of the ion pair.

Electrical neutrality gives:

$$2[Mg^{2+}]_{actual} + [MgIO_3^+]_{actual} + [H_3O^+]_{actual} = [IO_3^-]_{actual} + [OH^-]_{actual} \qquad (7.18)$$

Since the solution is of neutral pH; salt of a strong base, $Mg(OH)_2$ and the relatively strong acid, HIO_3:

$$[H_3O^+]_{actual} = [OH^-]_{actual}$$

$$\therefore 2[Mg^{2+}]_{actual} + [MgIO_3^+]_{actual} = [IO_3^-]_{actual}$$

From Equation (7.15) and substituting Equations (7.16 and 7.17):

$$K_{assoc} = \frac{[MgIO_3^+]_{actual}}{\left\{[Mg^{2+}]_{total} - [MgIO_3^+]_{actual}\right\}\left\{2[Mg^{2+}]_{total} - [MgIO_3^+]_{actual}\right\}}$$

$$[MgIO_3^+]_{actual} = K_{assoc}\left\{2[Mg^{2+}]_{total}^2 - 3[MgIO_3^+]_{actual}[Mg^{2+}]_{total} + [MgIO_3^+]_{actual}^2\right\}$$

$$\therefore K_{assoc}[MgIO_3^+]_{actual}^2 - [MgIO_3^+]_{actual}\left\{3K_{assoc}[Mg^{2+}]_{total} + 1\right\} + 2K_{assoc}[Mg^{2+}]_{total}^2 = 0$$

$$(7.19)$$

This is a quadratic equation which can be solved in the normal way. Substituting for K_{assoc} and $[Mg^{2+}]_{total}$ gives:

$$5.1[MgIO_3^+]_{actual}^2 - 1.765[MgIO_3^+]_{actual} + 0.0255 = 0$$

$$[MgIO_3^+]_{actual}^2 - 0.346[MgIO_3^+]_{actual} + 0.005 = 0$$

giving $[MgIO_3^+]_{actual} = 0.331$ mol dm^{-3} (**impossible** since $[Mg^{2+}]_{total} = 0.05$ mol dm^{-3}) or 0.015 mol dm^{-3}.

Take note: when solving quadratic equations a check must be made to see whether both solutions are **physically** possible. Often it will be found that one solution will give either a concentration which is negative or, as here, one which is greater than the initial concentration.

The analogue of Equation (7.13) for $Mg^{2+}(aq)$ is:

$$\beta = \left(\frac{[MgIO_3^+]_{actual}}{[Mg^{2+}]_{total}}\right)_{eq} = \frac{0.015\,\text{mol dm}^{-3}}{0.05\,\text{mol dm}^{-3}} = 0.30$$

i.e. Mg^{2+} is 30% associated to the ion pair $MgIO_3^+(aq)$.

The analogue of Equation (7.14) for $IO_3^-(aq)$ is:

$$\beta = \left(\frac{[MgIO_3^+]_{actual}}{[IO_3^-]_{total}}\right)_{eq} = \frac{0.015\,\text{mol dm}^{-3}}{0.10\,\text{mol dm}^{-3}} = 0.15$$

i.e. $IO_3^-(aq)$ is 15% associated to the ion pair $MgIO_3^+(aq)$.

Be careful here: Having found β for $Mg^{2+}(aq)$, it is not correct to assume that β for $IO_3^-(aq)$ is the same. This is a consequence of the stoichiometric formula where there are two IO_3^- for each Mg^{2+}.

- Considering the ion pair, $MgSO_4(aq)$:

$$Mg^{2+}(aq) + SO_4^{2-}(aq) \underset{}{\overset{K_{assoc}}{\rightleftharpoons}} MgSO_4(aq)$$

$$K_{assoc} = \left(\frac{[MgSO_4]_{actual}}{[Mg^{2+}]_{actual}[SO_4^{2-}]_{actual}}\right)_{eq} \qquad (7.20)$$

The stoichiometric relations are:

$$[Mg^{2+}]_{total} = [Mg^{2+}]_{actual} + [MgSO_4]_{actual}$$

$$\therefore [Mg^{2+}]_{actual} = [Mg^{2+}]_{total} - [MgSO_4]_{actual} \qquad (7.21)$$

$$[SO_4^{2-}]_{total} = [Mg^{2+}]_{total} = [SO_4^{2-}]_{actual} + [MgSO_4]_{actual}$$

$$\therefore [SO_4^{2-}]_{actual} = [Mg^{2+}]_{total} - [MgSO_4]_{actual} \qquad (7.22)$$

Again note: equations for $[Mg^{2+}]_{actual}$ and $[SO_4^{2-}]_{actual}$ in terms of the total concentrations and the concentration of the ion pair have been found.

Electrical neutrality gives:

$$2[Mg^{2+}]_{actual} + [H_3O^+]_{actual} = 2[SO_4^{2-}]_{actual} + [OH^-]_{actual} \qquad (7.23)$$

Since the solution is of neutral pH, salt of a strong base/strong acid:

$$[H_3O^+]_{actual} = [OH^-]_{actual}$$

$$\therefore [Mg^{2+}]_{actual} = [SO_4^{2-}]_{actual}$$

From Equation (7.20) and substituting Equations (7.21) and (7.22):

$$K_{assoc} = \frac{[MgSO_4]_{actual}}{\left\{[Mg^{2+}]_{total} - [MgSO_4]_{actual}\right\}\left\{[Mg^{2+}]_{total} - [MgSO_4]_{actual}\right\}}$$

$$[MgSO_4]_{actual} = K_{assoc}\left\{[Mg^{2+}]_{total}^2 - 2[MgSO_4]_{actual}[Mg^{2+}]_{total} + [MgSO_4]_{actual}^2\right\}$$

$$\therefore K_{assoc}[MgSO_4]_{actual}^2 - [MgSO_4]_{actual}\left\{2K_{assoc}[Mg^{2+}]_{total} + 1\right\} + K_{assoc}[Mg^{2+}]_{total}^2 = 0$$
$$(7.24)$$

This is a quadratic equation which can be solved in the normal way. Substituting for K_{assoc} and $[Mg^{2+}]_{total}$ gives:

$$170[MgSO_4]_{actual}^2 - 18[MgSO_4]_{actual} + 0.425 = 0$$

$$[MgSO_4]_{actual}^2 - 0.1059[MgSO_4]_{actual} + 0.0025 = 0$$

giving $[MgSO_4]_{actual} = 0.070$ mol dm^{-3} (**impossible** since $[Mg^{2+}]_{total} = 0.05$ mol dm^{-3}) or 0.0355 mol dm^{-3}.

The analogue of Equation (7.13) for $Mg^{2+}(aq)$ is:

$$\beta = \left(\frac{[MgSO_4]_{actual}}{[Mg^{2+}]_{total}}\right)_{eq} = \frac{0.0355 \text{ mol dm}^{-3}}{0.05 \text{ mol dm}^{-3}} = 0.71$$

i.e. $Mg^{2+}(aq)$ is 71% associated to the ion pair $MgSO_4(aq)$

The analogue of Equation (7.14) for $SO_4^-(aq)$ is:

$$\beta = \left(\frac{[MgSO_4]_{actual}}{[SO_4^{2-}]_{total}}\right)_{eq} = \frac{0.0355 \text{ mol dm}^{-3}}{0.05 \text{ mol dm}^{-3}} = 0.71$$

i.e. $SO_4^{2-}(aq)$ is 71% associated to the ion pair $MgSO_4(aq)$

Note: since $MgSO_4(aq)$ is a symmetrical electrolyte $\beta_{Mg^{2+}} = \beta_{SO_4^{2-}}$.

- Considering the ion pair, $MgFe(CN)_6^{2-}(aq)$:

$$Mg^{2+}(aq) + Fe(CN)_6^{4-}(aq) \xrightleftharpoons{K_{assoc}} MgFe(CN)_6^{2-}(aq)$$

$$K_{assoc} = \left(\frac{[MgFe(CN)_6^{2-}]_{actual}}{[Mg^{2+}]_{actual}[Fe(CN)_6^{4-}]_{actual}} \right)_{eq} \qquad (7.25)$$

The stoichiometric relations are:

$$[Mg^{2+}]_{total} = [Mg^{2+}]_{actual} + [MgFe(CN)_6^{2-}]_{actual}$$

$$\therefore [Mg^{2+}]_{actual} = [Mg^{2+}]_{total} - [MgFe(CN)_6^{2-}]_{actual} \qquad (7.26)$$

$$[Fe(CN)_6^{4-}]_{total} = 1/2[Mg^{2+}]_{total} = [Fe(CN)_6^{4-}]_{actual} + [MgFe(CN)_6^{2-}]_{actual}$$

Watch the factor ½: each mol $Mg_2Fe(CN)_6^{2-}(aq)$ gives 2 mol $Mg^{2+}(aq)$ and 1 mol $Fe(CN)_6^{2-}(aq)$:

$$\therefore [Fe(CN)_6^{4-}]_{actual} = 1/2[Mg^{2+}]_{total} - [MgFe(CN)_6^{2-}]_{actual} \qquad (7.27)$$

Electrical neutrality gives:

$$2[Mg^{2+}]_{actual} + [H_3O^+]_{actual} = 4[Fe(CN)_6^{4-}]_{actual} + 2[MgFe(CN)_6^{2-}]_{actual} + [OH^-]_{actual} \quad (7.28)$$

Remember: factor of 2 for the doubly charged ions; factor of 4 for the -4 charged ion. Since the solution is of neutral pH:

$$[H_3O^+]_{actual} = [OH^-]_{actual}$$

$$\therefore [Mg^{2+}]_{actual} = 2[Fe(CN)_6^{4-}]_{actual} + [MgFe(CN)_6^{2-}]_{actual}$$

From Equation (7.25) and substituting Equations (7.26) and (7.27):

$$K_{assoc} = \frac{[MgFe(CN)_6^{2-}]_{actual}}{\left\{[Mg^{2+}]_{total} - [MgFe(CN)_6^{2-}]_{actual}\right\}\left\{1/2[Mg^{2+}]_{total} - [MgFe(CN)_6^{2-}]_{actual}\right\}}$$

$$[MgFe(CN)_6^{2-}]_{actual} = K_{assoc}\left\{1/2[Mg^{2+}]_{total}^2 - 3/2[MgFe(CN)_6^{2-}]_{actual}[Mg^{2+}]_{total}\right.$$
$$\left. +[MgFe(CN)_6^{2-}]_{actual}^2\right\}$$

$$\therefore K_{assoc}[MgFe(CN)_6^{2-}]_{actual}^2 - [MgFe(CN)_6^{2-}]_{actual}\left\{3/2K_{assoc}[Mg^{2+}]_{total} + 1\right\}$$
$$+ 1/2K_{assoc}[Mg^{2+}]_{total}^2 = 0 \qquad (7.29)$$

This is a quadratic equation which can be solved in the normal way. Substituting for K_{assoc} and $[Mg^{2+}]_{total}$ gives:

$$7400[MgFe(CN)_6^{2-}]_{actual}^2 - 556[MgFe(CN)_6^{2-}]_{actual} + 9.25 = 0$$

$$[MgFe(CN)_6^{2-}]_{actual}^2 - 0.0751[MgFe(CN)_6^{2-}]_{actual} + 0.00125 = 1$$

giving $[MgFe(CN)_6^{2-}]_{actual} = 0.0502$ mol dm^{-3} (**impossible** since $[Fe(CN)_6^{4-}]_{total} = 0.025$mol dm^{-3}]) or 0.0249 mol dm^{-3}.

The analogue of Equation (7.13) for $Mg^{2+}(aq)$ is:

$$\beta = \left(\frac{[MgFe(CN)_6^{2-}]_{actual}}{[Mg^{2+}]_{total}}\right)_{eq} = \frac{0.0249 \text{ mol dm}^{-3}}{0.05 \text{ mol dm}^{-3}} = 0.498$$

i.e. $Mg^{2+}(aq)$ is 50% associated to the associated species $MgFe(CN)_6^{2-}(aq)$.
The analogue of Equation (7.14) for $Fe(CN)_6^{4-}(aq)$ is:

$$\beta = \left(\frac{[MgFe(CN_6^{2-})]_{actual}}{[Fe(CN)_6^{4-}]_{total}}\right)_{eq} = \frac{0.0249 \text{ mol dm}^{-3}}{0.025 \text{ mol dm}^{-3}} = 0.996$$

showing that virtually all of the anion has been removed to form the associated species.
Note: this is another case where the two βs are not equal; here there are two Mg^{2+} for each
$Fe(CN)_6^{4-}$ in $(Mg)_2Fe(CN)_6$
• The degree of association follows the pattern of the pK_{assoc} values with the order being:

$$MgFe(CN)_6^{2-} > MgSO_4 > MgIO_3^-$$

This is what would be expected on electrostatic grounds:

$$MgFe(CN)_6^{2-}(aq) \quad \text{is formed from a} + 2 \text{ ion and a} -4 \text{ ion;}$$

$$MgSO_4(aq) \quad \text{is formed from a} + 2 \text{ ion and a} -2 \text{ ion;}$$

$$MgIO_3^+(aq) \quad \text{is formed from a} + 2 \text{ ion and a} -1 \text{ ion;}$$

and the electrostatic energy is greater the larger the product of the charges. The $Fe(CN)_6^{4-}(aq)$
is a large flabby ion and will be highly polarisable compared with the smaller $SO_4^{2-}(aq)$ and
IO_3^- ions. This would be expected to lead to even stronger attractive forces (see Section 1.2.2).
Note: in this question there are sufficient equations in the equilibrium relation and the
stoichiometric relations to solve for the concentration of ion pairs and not to need electrical
neutrality. This may not always be the case and it is as well to get into the habit of always
formulating the equation describing electrical neutrality.

7.2 Complex formation

When a complex is formed there is an intimate chemical interaction between the ions, or
between an ion and an uncharged ligand. This is in contrast to the purely electrostatic
interaction which results in the formation of an ion pair. There are situations when an
unambiguous assignment of a species as an ion pair or as a complex may be difficult. These
matters are discussed in Sections 1.20 and 1.21.

However, no matter which species are considered, the equilibrium calculations involved are
identical.

But, be aware: in complex formation two formulations are typically used for the equilibrium expression, and it is crucial to be aware of the distinctions. Equilibria are either given in
terms of **step-wise formation** equilibrium constants of **successive** complexes, Ks, or in terms
of the **overall formation** constant of a **given** complex **from the metal ion and the ligand, β s.**

Direct chemical interaction between a metal ion and a ligand often results in a series of
complexes. In complex formation the ligand replaces the water coordinated to the metal ion.

Depending on the geometry of the complex formed, e.g. linear, planar, tetrahedral or octahedral up to six ligands may become progressively attached, with a few examples where larger numbers of ligands can be coordinated. If the metal ion is represented as M, where the charge on the metal ion is unspecified, and the ligand as L, then the following schematic reactions can occur:

- **step-wise formation**

$$M(aq) + L(aq) \xrightleftharpoons{K_1} ML(aq)$$

$$K_1 = \left(\frac{[ML]_{actual}}{[M]_{actual}[L]_{actual}} \right)_{eq} \tag{7.30}$$

$$ML(aq) + L(aq) \xrightleftharpoons{K_2} ML_2(aq)$$

$$K_2 = \left(\frac{[ML_2]_{actual}}{[ML]_{actual}[L]_{actual}} \right)_{eq} \tag{7.31}$$

$$ML_2(aq) + L(aq) \xrightleftharpoons{K_3} ML_3(aq)$$

$$K_3 = \left(\frac{[ML_3]_{actual}}{[ML_2]_{actual}[L]_{actual}} \right)_{eq} \tag{7.32}$$

$$\cdots\cdots\cdots\cdots\cdots\cdots\cdots\cdots\cdots\cdots$$

$$ML_{n-1}(aq) + L(aq) \xrightleftharpoons{K_n} ML_n(aq)$$

$$K_n = \left(\frac{[ML_n]_{actual}}{[ML_{n-1}]_{actual}[L]_{actual}} \right)_{eq} \tag{7.33}$$

The overall reaction is:

$$M(aq) + nL(aq) \xrightleftharpoons{K_{overall}} ML_n(aq)$$

$$K_{overall} = \left(\frac{[ML_n]_{actual}}{[M]_{actual}[L]_{actual}^n} \right)_{actual} = K_1 K_2 K_3 \dots K_n \tag{7.34}$$

where n normally can take values up to six.

- **overall formation**

In standard treatments of complex formation equilibria, instead of individual, or step-wise, Ks defined as above, overall equilibrium constants are commonly used. The symbol β is used for such equilibrium constants. In the present context, the sequence of reactions are defined as:

$$M(aq) + L(aq) \xrightleftharpoons{\beta_1} ML(aq)$$

$$\beta_1 = \left(\frac{[ML]_{actual}}{[M]_{actual}[L]_{actual}} \right)_{eq} \tag{7.35}$$

$$M(aq) + 2L(aq) \xrightleftharpoons{\beta_2} ML_2(aq)$$

$$\beta_2 = \left(\frac{[ML_2]_{actual}}{[M]_{actual}[L]_{actual}^2} \right)_{eq} \tag{7.36}$$

$$M(aq) + 3L(aq) \underset{\beta_3}{\rightleftharpoons} ML_3(aq)$$

$$\beta_3 = \left(\frac{[ML_3]_{actual}}{[M]_{actual}[L]_{actual}^3} \right)_{eq} \tag{7.37}$$

. . . .

$$M(aq) + nL(aq) \underset{\beta_n}{\rightleftharpoons} ML_n(aq)$$

$$\beta_n = \left(\frac{[ML_n]_{actual}}{[M]_{actual}[L]_{actual}^n} \right)_{actual} \tag{7.38}$$

where:

$$\beta_n = K_{overall} = K_1 K_2 K_3 \ldots K_n \tag{7.39}$$

These βs must be distinguished from the βs in ion pair calculations where β is used to describe the fraction of metal ion or ligand associated to the ion pair.

These are not the only equilibria which may have to be considered. There is the possibility that there may also be present complexes of the type $M_2L_n(aq)$, $M_3L_n(aq)$. There is also the possibility that protonation/deprotonation of the ligand has to be considered. The chemistry of complex formation is an extensive field of study, and reference to specialist inorganic texts is necessary. The following only indicates some of the typical equilibrium calculations which are involved.

7.2.1   Fractions associated

As with acid–base equilibria, fractions can be defined, e.g. the fraction, α_n, of the total M which is present as ML_n:
the fraction of M which is present as ML is:

$$\alpha_1 = \frac{[ML]_{actual}}{[M]_{total}} \tag{7.40}$$

the fraction of M present as ML_2 is:

$$\alpha_2 = \frac{[ML_2]_{actual}}{[M]_{total}} \tag{7.41}$$

while the fraction present as ML_n is:

$$\alpha_n = \frac{[ML_n]_{actual}}{[M]_{total}} \tag{7.42}$$

As well as $\alpha_1, \alpha_2, \alpha_3, \ldots \alpha_n$, it is useful to consider the fraction of the total M which is **not** complexed, α_0, where:

$$\alpha_0 = \frac{[M]_{actual}}{[M]_{total}} \tag{7.43}$$

7.2.2 Mean number of ligands bound

The experimental data is sometimes also analysed in terms of the mean number of ligands, L, bound per M:

$$\bar{n} = \frac{[L]_{\text{total bound}}}{[M]_{\text{total}}} = \frac{[L]_{\text{total}} - [L]_{\text{actual}}}{[M]_{\text{total}}} \tag{7.44}$$

This definition is analogous to the definition of the mean number of bound protons on a polybasic acid, a quantity much used by biochemists, though less often used by chemists.

7.2.3 Equilibria calculations

The various equilibrium constants in terms of step-wise or overall constants have already been defined. In the following analysis the more common practice in complex formation is used, i.e. express everything in terms of the overall equilibrium constants, i.e. β s.

Stoichiometric relations:

These are analogous to those in acid–base equilibria:

$$[M]_{\text{total}} = [M]_{\text{actual}} + [ML]_{\text{actual}} + [ML_2]_{\text{actual}} + [ML_3]_{\text{actual}} + \cdots\cdots + [ML_n]_{\text{actual}} \tag{7.45}$$

For the general case:

$$M(aq) + nL(aq) \underset{}{\overset{\beta_n}{\rightleftharpoons}} ML_n(aq)$$

$$\beta_n = \left(\frac{[ML_n]_{\text{actual}}}{[M]_{\text{actual}}[L]_{\text{actual}}^n} \right)_{\text{eq}} \tag{7.46}$$

Substitution into Equation (7.45) for each complex in terms of the corresponding β gives:

$$[M]_{\text{total}} = [M]_{\text{actual}} + \beta_1 [M]_{\text{actual}}[L]_{\text{actual}} + \beta_2 [M]_{\text{actual}}[L]_{\text{actual}}^2 + \cdots\cdots \tag{7.47}$$

$$= [M]_{\text{actual}} \left\{ 1 + \beta_1 [L]_{\text{actual}} + \beta_2 [L]_{\text{actual}}^2 + \cdots \right\} \tag{7.48}$$

A corresponding set of equations could be written in terms of $[L]_{\text{total}}$.

$$[L]_{\text{total}} = [L]_{\text{actual}} + [ML]_{\text{actual}} + 2[ML_2]_{\text{actual}} + 3[ML_3]_{\text{actual}} + \cdots\cdots\cdots \tag{7.49}$$

$$= [L]_{\text{actual}} + \beta_1 [M]_{\text{actual}}[L]_{\text{actual}} + 2\beta_2 [M]_{\text{actual}}[L]_{\text{actual}}^2 + \cdots\cdots \tag{7.50}$$

$$[L]_{\text{total}} - [L]_{\text{actual}} = \beta_1 [M]_{\text{actual}}[L]_{\text{actual}} + 2\beta_2 [M]_{\text{actual}}[L]_{\text{actual}}^2 + \cdots\cdots \tag{7.51}$$

$$= [M]_{\text{actual}} \left\{ \beta_1 [L]_{\text{actual}} + 2\beta_2 [L]_{\text{actual}}^2 + \cdots \right\} \tag{7.52}$$

Equation (7.52) is required for calculation of $\bar{n}$, Equation (7.44).

The next step considers the quantities which can be measured experimentally.

- $[M]_{\text{total}}$ and $[L]_{\text{total}}$ are known from the experimental conditions.

- $[M]_{\text{actual}}$ can often be measured experimentally from reversible emf data.

- $[L]_{\text{actual}}$ can often be measured spectroscopically, or in cases like $Cl^-(aq)$ from emfs.

- From these two quantities an experimental value for $\alpha_0 = \dfrac{[M]_{actual}}{[M]_{total}}$ and $\bar{n} = \dfrac{[L]_{total} - [L]_{actual}}{[M]_{total}}$ can be found.

- The determination of values for equilibrium constants is often a matter of interpreting the behaviour of α_0 or $\bar{n}$.

- Using the definition of α_0, and Equation (7.48), gives:

$$\alpha_0 = \frac{[M]_{actual}}{[M]_{total}} = \frac{1}{1 + \beta_1[L]_{actual} + \beta_2[L]_{actual}^2 + \cdots} \tag{7.53}$$

- Using the definition of $\bar{n}$, and Equations (7.48) and (7.52) gives:

$$\bar{n} = \frac{[L]_{total} - [L]_{actual}}{[M]_{total}} = \frac{[M]_{actual}\left\{\beta_1[L]_{actual} + 2\beta_2[L]_{actual}^2 + \cdots\cdots\right\}}{[M]_{actual}\left\{1 + \beta_1[L]_{actual} + \beta_2[L]_{actual}^2 \cdots\cdots\right\}} \tag{7.54}$$

$$= \frac{\left\{\beta_1[L]_{actual} + 2\beta_2[L]_{actual}^2 + \cdots\cdots\right\}}{\left\{1 + \beta_1[L]_{actual} + \beta_2[L]_{actual}^2 \cdots\cdots\right\}} \tag{7.55}$$

7.2.4   Determination of $\beta_1, \beta_2, \ldots\ldots$ from the dependence of α_0 on $[L]_{actual}$

Taking the reciprocal of Equation (7.53) gives:

$$\frac{1}{\alpha_0} = 1 + \beta_1[L]_{actual} + \beta_2[L]_{actual}^2 + \cdots\cdots \tag{7.56}$$

$$\therefore \left(\frac{1}{\alpha_0} - 1\right)/[L]_{actual} = \frac{1 - \alpha_0}{\alpha_0[L]_{actual}} = \beta_1 + \beta_2[L]_{actual} + \cdots\cdots \tag{7.57}$$

If there are three or more distinct complexes present, Equation (7.57) is a complicated non-linear equation, and the graph, $\dfrac{1 - \alpha_0}{\alpha_0[L]_{actual}}$ vs $[L]_{actual}$ will be a curve. Determining the values of the constants, $\beta_1, \beta_2, \beta_3, \beta_4 \ldots\ldots$ then becomes an exercise in computer curve fitting, with all the problems associated with the ambiguities which might arise from the curve fitting procedures.

If the only complexes present are ML and ML$_2$, Equation (7.57) reduces to a linear expression, and the values of β_1 and β_2 will follow from the intercept and slope of a plot of $\left\{\dfrac{1 - \alpha_0}{\alpha_0[L]_{actual}}\right\}$ against $[L]_{actual}$.

7.2.5   Determination of $\beta_1, \beta_2, \ldots\ldots$ from the dependence of $\bar{n}$ on $[L]_{actual}$

$$\bar{n} = \frac{[L]_{total} - [L]_{actual}}{[M]_{total}} = \frac{\left\{\beta_1[L]_{actual} + 2\beta_2[L]_{actual}^2 + 3\beta_3[L]_{actual}^3 + 4\beta_4[L]_{actual}^4 + \cdots\cdots\right\}}{\left\{1 + \beta_1[L]_{actual} + \beta_2[L]_{actual}^2 + \beta_3[L]_{actual}^3 + \beta_4[L]_{actual}^4 + \cdots\cdots\right\}} \tag{7.58}$$

The total concentration of bound L is $[L]_{total} - [L]_{actual}$. This quantity can be altered by varying $[L]_{total}$, and finding the corresponding values of $[L]_{actual}$ experimentally. From these

$[L]_{total} - [L]_{actual}$ values, values of $\bar{n}$ for **any given** $[L]_{total}$ and $[M]_{total}$ can be calculated and a graph of $\bar{n}$ vs. $[L]_{actual}$ drawn. Values of the βs appearing in Equation (7.58) can be found by computer curve fitting techniques, or by successive approximations.

The number of bound L can vary from zero to a value given by n, e.g. if four complexes are formed then the number of bound L could vary from zero to four. If the concentration of bound L, i.e. $[L]_{total} - [L]_{actual} = [M]_{total}$, then $\bar{n} = 1$. Simplification of the expression for $\bar{n}$ (Equation 7.58) can be made by finding the value of $[L]_{actual}$ at which $\bar{n} = 1$. This will enable the various βs to be found.

When $\bar{n} = 1$

$$\beta_1[L]_{actual} + 2\beta_2[L]_{actual}^2 + 3\beta_3[L]_{actual}^3 + 4\beta_4[L]_{actual}^4 + \cdots\cdots$$
$$= 1 + \beta_1[L]_{actual} + \beta_2[L]_{actual}^2 + \beta_3[L]_{actual}^3 + \beta_4[L]_{actual}^4 + \cdots\cdots \quad (7.59)$$

$$\text{and} \quad \beta_2[L]_{actual}^2 + 2\beta_3[L]_{actual}^3 + 3\beta_4[L]_{actual}^4 + \cdots\cdots\cdots = 1 \quad (7.60)$$

If three or more complexes are present computer curve fitting, or successive approximations, is necessary.

However, if the only complexes present are ML and ML_2, the experimental curve can be used to find the value of $[L]_{actual}$ at which $\bar{n} = 1$.

If ML and ML_2 are the only complexes present, Equation (7.59) reduces to:

$$\beta_2[L]_{actual}^2 = 1 \quad \text{when} \quad \bar{n} = 1 \quad (7.61)$$

If β_2 can be found in this way, a value for β_1 can then be inferred from the experimental curve.

Worked problem 7.2

Question

Solutions of $Cu^{2+}(aq)$ in aqueous NH_3 form a series of complexes where four (of the six) water molecules coordinated in the square planar arrangement around the central metal ion are progressively replaced by NH_3 as ligands.

The equilibria set up are:

$$Cu^{2+}(aq) + NH_3(aq) \xrightleftharpoons{K_1} Cu(NH_3)^{2+}(aq)$$

$$K_1 = \left(\frac{[Cu(NH_3)^{2+}]_{actual}}{[Cu^{2+}]_{actual}[NH_3]_{actual}}\right)_{eq} = 9770\,mol^{-1}\,dm^3 \quad \text{at } 25°C$$

$$Cu(NH_3)^{2+}(aq) + NH_3(aq) \xrightleftharpoons{K_2} Cu(NH_3)_2^{2+}(aq)$$

$$K_2 = \left(\frac{[Cu(NH_3)_2^{2+}]_{actual}}{[Cu(NH_3)^{2+}]_{actual}[NH_3]_{actual}}\right)_{eq} = 2190\,mol^{-1}\,dm^3 \quad \text{at } 25°C$$

$$Cu(NH_3)_2^{2+}(aq) + NH_3(aq) \xrightleftharpoons{K_3} Cu(NH_3)_3^{2+}(aq)$$

$$K_3 = \left(\frac{[Cu(NH_3)_3^{2+}]_{actual}}{[Cu(NH_3)_2^{2+}]_{actual}[NH_3]_{actual}}\right)_{eq} = 537\,mol^{-1}\,dm^3 \quad \text{at } 25°C$$

$$Cu(NH_3)_3^{2+}(aq) + NH_3(aq) \xrightleftharpoons{K_4} Cu(NH_3)_4^{2+}(aq)$$

$$K_4 = \left(\frac{[Cu(NH_3)_4^{2+}]_{actual}}{[Cu(NH_3)_3^{2+}]_{actual}[NH_3]_{actual}} \right)_{eq} = 93 \text{ mol}^{-1} \text{ dm}^3 \quad \text{at} \quad 25°C$$

a) Find the products, $K_1[NH_3]_{actual}$, $K_2[NH_3]_{actual}$, $K_3[NH_3]_{actual}$, $K_4[NH_3]_{actual}$, for each of the four complexes present when $[NH_3]_{actual} = 0.100 \text{ mol dm}^{-3}$, and comment on the values.

b) Find the values for β_1, β_2, β_3 and β_4.

c) Using the methods outlined in Section (7.2.1) calculate α_0, α_1, α_2, α_3, and α_4. Comment on the values found.

Answer

a) The equilibrium constant expressions give:

$$\frac{[Cu(NH_3)^{2+}]_{actual}}{[Cu^{2+}]_{actual}} = K_1[NH_3]_{actual} = 980$$

$$\frac{[Cu(NH_3)_2^{2+}]_{actual}}{[Cu(NH_3)^{2+}]_{actual}} = K_2[NH_3]_{actual} = 220$$

$$\frac{[Cu(NH_3)_3^{2+}]_{actual}}{[Cu(NH_3)_2^{2+}]_{actual}} = K_3[NH_3]_{actual} = 54$$

$$\frac{[Cu(NH_3)_4^{2+}]_{actual}}{[Cu(NH_3)_3^{2+}]_{actual}} = K_4[NH_3]_{actual} = 9$$

All of these ratios are greater than unity, which shows that for **each equilibrium** at **this particular** $[NH_3]_{actual}$ there is more of the species with the greater number of NH_3 complexed. However, these conclusions depend on the value of $[NH_3]_{actual}$, and could alter, and even reverse, as the $[NH_3]_{actual}$ decreases. For instance, when $[NH_3]_{actual} = 0.0100 \text{ mol dm}^{-3}$ the respective ratios are 98, 22, 5.4 and 0.9. The conclusion to be drawn is that **each case must be worked out individually**.

The fact that these ratios decrease as higher complexes are formed merely reflects the magnitudes of the various Ks; the greater the magnitude of K, the greater the value of the ratio.

b) As shown in Section (7.2) values of β s can be found:

$\beta_1 = K_1 = 9770 \text{ mol}^{-1} \text{ dm}^3$

$\beta_2 = K_1K_2 = 9770 \times 2190 \text{ mol}^{-2} \text{ dm}^6 = 2.14 \times 10^7 \text{ mol}^{-2} \text{ dm}^6$

$\beta_3 = K_1K_2K_3 = 9770 \times 2190 \times 537 \text{ mol}^{-3} \text{ dm}^9 = 1.15 \times 10^{10} \text{ mol}^{-3} \text{ dm}^9$

$\beta_4 = K_1K_2K_3K_4 = 9770 \times 2190 \times 537 \times 93 \text{ mol}^{-4} \text{ dm}^{12} = 1.07 \times 10^{12} \text{ mol}^{-4} \text{ dm}^{12}$

Remember: the βs are overall equilibrium constants of the type:

$$Cu^{2+}(aq) + nNH_3(aq) \xrightarrow{\beta_n} Cu(NH_3)_n^{2+}(aq)$$

$$\beta_n = \left(\frac{[Cu(NH_3)_n^{2+}]_{actual}}{[Cu^{2+}]_{actual}[NH_3]_{actual}^n} \right)_{eq}$$

Formation of the complex with the greatest number of ligands attached has the greatest overall equilibrium constant. But note, the corresponding Ks decrease.

c)
$$Cu^{2+}(aq) + NH_3(aq) \overset{\beta_1}{\rightleftharpoons} Cu(NH_3)^{2+}(aq)$$

$$\beta_1 = \left(\frac{[Cu(NH_3)^{2+}]_{actual}}{[Cu^{2+}]_{actual}[NH_3]_{actual}} \right)_{eq}$$

$$Cu^{2+}(aq) + 2NH_3(aq) \overset{\beta_2}{\rightleftharpoons} Cu(NH_3)_2^{2+}(aq)$$

$$\beta_2 = \left(\frac{[Cu(NH_3)_2^{2+}]_{actual}}{[Cu^{2+}]_{actual}[NH_3]_{actual}^2} \right)_{eq}$$

$$Cu^{2+}(aq) + 3NH_3(aq) \overset{\beta_3}{\rightleftharpoons} Cu(NH_3)_3^{2+}(aq)$$

$$\beta_3 = \left(\frac{[Cu(NH_3)_3^{2+}]_{actual}}{[Cu^{2+}]_{actual}[NH_3]_{actual}^3} \right)_{eq}$$

$$Cu^{2+}(aq) + 4NH_3(aq) \overset{\beta_4}{\rightleftharpoons} Cu(NH_3)_4^{2+}(aq)$$

$$\beta_4 = \left(\frac{[Cu(NH_3)_4^{2+}]_{actual}}{[Cu^{2+}]_{actual}[NH_3]_{actual}^4} \right)_{eq}$$

$$[Cu(NH_3)^{2+}]_{actual} = \beta_1[Cu^{2+}]_{actual}[NH_3]_{actual}$$

$$[Cu(NH_3)_2^{2+}]_{actual} = \beta_2[Cu^{2+}]_{actual}[NH_3]_{actual}^2$$

$$[Cu(NH_3)_3^{2+}]_{actual} = \beta_3[Cu^{2+}]_{actual}[NH_3]_{actual}^3$$

$$[Cu(NH_3)_4^{2+}]_{actual} = \beta_4[Cu^{2+}]_{actual}[NH_3]_{actual}^4$$

$$[Cu^{2+}]_{total} = [Cu^{2+}]_{actual} + [Cu(NH_3)^{2+}]_{actual} + [Cu(NH_3)_2^{2+}]_{actual}$$
$$+ [Cu(NH_3)_3^{2+}]_{actual} + [Cu(NH_3)_4^{2+}]_{actual}$$

$$= [Cu^{2+}]_{actual} + \beta_1[Cu^{2+}]_{actual}[NH_3]_{actual} + \beta_2[Cu^{2+}]_{actual}[NH_3]_{actual}^2 + \cdots\cdots$$

$$= [Cu^{2+}]_{actual}\left\{ 1 + \beta_1[NH_3]_{actual} + \beta_2[NH_3]_{actual}^2 + \cdots\cdots\cdots \right\}$$

$$\alpha_1 = \frac{[Cu(NH_3)^{2+}]_{actual}}{[Cu^{2+}]_{total}} = \frac{\beta_1[NH_3]_{actual}}{1 + \beta_1[NH_3]_{actual} + \beta_2[NH_3]_{actual}^2 + \cdots}$$

$$\alpha_2 = \frac{[Cu(NH_3)_2^{2+}]_{actual}}{[Cu^{2+}]_{total}} = \frac{\beta_2[NH_3]_{actual}^2}{1 + \beta_1[NH_3]_{actual} + \beta_2[NH_3]_{actual}^2 + \cdots}$$

$$\alpha_3 = \frac{[Cu(NH_3)_3^{2+}]_{actual}}{[Cu^{2+}]_{total}} = \frac{\beta_3[NH_3]_{actual}^3}{1 + \beta_1[NH_3]_{actual} + \beta_2[NH_3]_{actual}^2 + \cdots}$$

$$\alpha_4 = \frac{[Cu(NH_3)_4^{2+}]_{actual}}{[Cu^{2+}]_{total}} = \frac{\beta_4[NH_3]_{actual}^4}{1 + \beta_1[NH_3]_{actual} + \beta_2[NH_3]_{actual}^2 + \cdots}$$

$$\alpha_0 = \frac{[Cu^{2+}]_{actual}}{[Cu^{2+}]_{total}}$$

$$= \frac{[Cu^{2+}]_{actual}}{[Cu^{2+}]_{actual} + \beta_1[Cu^{2+}]_{actual}[NH_3]_{actual} + \beta_2[Cu^{2+}]_{actual}[NH_3]_{actual}^2 + \cdots}$$

$$= \frac{1}{1 + \beta_1[NH_3]_{actual} + \beta_2[NH_3]_{actual}^2 + \cdots}$$

If $[NH_3]_{actual} = 0.100 \text{ mol dm}^{-3}$:

$$\beta_1[NH_3]_{actual} = 977, \qquad \beta_2[NH_3]_{actual}^2 = 2.14 \times 10^5,$$

$$\beta_3[NH_3]_{actual}^3 = 1.15 \times 10^7, \qquad \beta_4[NH_3]_{actual}^4 = 1.07 \times 10^8$$

$$\therefore 1 + \beta_1[NH_3]_{actual} + \beta_2[NH_3]_{actual}^2 + \cdots\cdots = 1 + 977 + 2.14 \times 10^5 + 1.15 \times 10^7$$
$$+ 1.07 \times 10^8$$
$$= 1.19 \times 10^8$$

$$\therefore \alpha_0 = \frac{1}{1.19 \times 10^8} = 8.4 \times 10^{-9}, \qquad \alpha_1 = \frac{977}{1.19 \times 10^8} = 8.2 \times 10^{-6},$$

$$\alpha_2 = \frac{2.14 \times 10^5}{1.19 \times 10^8} = 1.8 \times 10^{-3}, \qquad \alpha_3 = \frac{1.15 \times 10^7}{1.19 \times 10^8} = 9.7 \times 10^{-2}, \qquad \alpha_4 = \frac{1.07 \times 10^8}{1.19 \times 10^8} = 0.90$$

These values show that the most predominant complex is that one with the greatest number of ligands attached to the central metal ion, $Cu(NH_3)_4^{2+}(aq)$, present at 90%, with the next being $Cu(NH_3)_3^{2+}(aq)$, present at $\approx 10\%$, with the other complexes present in minute amounts.

Worked problem 7.3

Question

In a set of experiments at 25°C on the formation of complexes of $Cd^{2+}(aq)$ with pyridine(aq), the **total** concentration of $Cd^{2+}(aq)$ was 0.0200 mol dm^{-3} throughout. The following results were obtained:

$\dfrac{[pyr]_{total}}{\text{mol dm}^{-3}}$	0.053	0.103	0.130	0.158	0.182
$\dfrac{[pyr]_{actual}}{\text{mol dm}^{-3}}$	0.042	0.085	0.110	0.135	0.157
$\dfrac{[Cd^{2+}]_{actual}}{\text{mol dm}^{-3}}$	0.0106	0.00645	0.00508	0.00405	0.00334

Find the values for:

- $\bar{n}$ (the mean number of pyridine bound per Cd^{2+}),

- α_0 (the fraction of the Cd^{2+} which is uncomplexed). Ignore the possibility of any association of the $Cd^{2+}(aq)$ with anions present.

- Plot a graph of: $\left(\dfrac{1}{\alpha_0} - 1 \right) \Big/ [\text{pyr}]_{\text{actual}}$ versus $[\text{pyr}]_{\text{actual}}$

Examine this graph to see whether it gives any evidence on whether complexes containing more than two pyridine are formed, and obtain values for the overall equilibrium constants β_1 and β_2.

Answer

$$\bar{n} = \frac{[\text{pyr}]_{\text{total}} - [\text{pyr}]_{\text{actual}}}{[\text{Cd}^{2+}]_{\text{total}}} \tag{7.44}$$

The values obtained are:

$([\text{pyr}]_{\text{total}} - [\text{pyr}]_{\text{actual}})/\text{mol dm}^{-3}$	0.011	0.018	0.020	0.023	0.025
$\bar{n}$		0.55	0.90	1.00	1.15	1.25

The value for the mean number of pyridine bound per Cd^{2+} eventually exceeds unity and so there must be at least one species of complex which contains more than one bound pyridine molecule.

$$\alpha_0 = \frac{[\text{Cd}^{2+}]_{\text{actual}}}{[\text{Cd}^{2+}]_{\text{total}}} \tag{7.43}$$

and the values of α_0 for each experiment are given in the table below.

The values required to plot the graph:

$$\left(\frac{1}{\alpha_0} - 1 \right) \Big/ [\text{pyr}]_{\text{actual}} \text{ vs. } [\text{pyr}]_{\text{actual}}$$

are also given in the table below.

α_0	0.530	0.322$_5$	0.254	0.202$_5$	0.167
$\dfrac{1}{\alpha_0} - 1$	0.89	2.10	2.94	3.94	4.99
$\dfrac{(\frac{1}{\alpha_0} - 1)/[\text{pyr}]_{\text{actual}}}{\text{mol}^{-1}\,\text{dm}^3}$	21.1	24.7	26.7	29.2	31.8

The graph (Figure 7.1) is slightly curved, though this would be hardly detected if the last point were absent. If the graph had been extended to higher pyridine concentrations, and had there been more points instead of only 5, a more definite conclusion would have been possible. **This shows how important it is to cover a wide enough range of experimental conditions and to have enough experimental points.**

Values of β_1 and β_2 can be found from a limited form of Equation (7.57)

$$\left(\frac{1}{\alpha_0} - 1 \right) \Big/ [\text{L}]_{\text{actual}} = \frac{1 - \alpha_0}{\alpha_0 [\text{L}]_{\text{actual}}} = \beta_1 + \beta_2 [\text{L}]_{\text{actual}} + \cdots \cdots \tag{7.57}$$

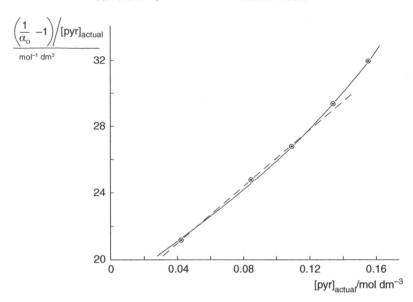

Figure 7.1   Graph of $\left(\frac{1}{\alpha_0} - 1\right)/[\text{pyridine}]_{\text{actual}}$ vs $[\text{pyridine}]_{\text{actual}}$.

If it is assumed that there are only two complexes present, this equation reduces to:

$$\left(\frac{1}{\alpha_0} - 1\right)/[\text{L}]_{\text{actual}} = \frac{1 - \alpha_0}{\alpha_0[\text{L}]_{\text{actual}}} = \beta_1 + \beta_2[\text{L}]_{\text{actual}}. \tag{7.62}$$

and so a plot of:

$$\left(\frac{1}{\alpha_0} - 1\right)/[\text{pyr}]_{\text{actual}} \text{versus } [\text{pyr}]_{\text{actual}}$$

should have intercept $= \beta_1$ and slope $= \beta_2$.

Taking the approximately linear earlier part of the graph to approximate to Equation (7.62) gives:

$$\beta_1 = 17.5 \, \text{mol}^{-1} \, \text{dm}^3$$
$$\beta_2 = 84 \, \text{mol}^{-2} \, \text{dm}^6$$

where:

$$\text{Cd}^{2+}(\text{aq}) + \text{pyr}(\text{aq}) \underset{}{\overset{\beta_1}{\rightleftharpoons}} \text{Cd}(\text{pyr})^{2+}(\text{aq})$$

$$\beta_1 = \left(\frac{[\text{Cd}(\text{pyr})^{2+}]_{\text{actual}}}{[\text{Cd}^{2+}]_{\text{actual}}[\text{pyr}]_{\text{actual}}}\right)_{\text{eq}}$$

$$\text{Cd}^{2+}(\text{aq}) + 2\text{pyr}(\text{aq}) \underset{}{\overset{\beta_2}{\rightleftharpoons}} \text{Cd}(\text{pyr})_2^{2+}(\text{aq})$$

$$\beta_2 = \left(\frac{[\text{Cd}(\text{pyr})_2^{2+}]_{\text{actual}}}{[\text{Cd}^{2+}]_{\text{actual}}[\text{pyr}]_{\text{actual}}^2}\right)_{\text{eq}}$$

7.3 Solubilities of sparingly soluble salts

There are a number of electrolytes which are only sparingly soluble in water at room temperatures. Typical examples are the silver halides, $AgCl(s)$, $AgBr(s)$, $AgI(s)$, but not $AgF(s)$ which is soluble in water. Many hydroxides are only sparingly soluble, as are fluorides such as $CaF_2(s)$ and $PbF_2(s)$. Most carbonates are sparingly soluble, as are the sulphates, $SrSO_4$, $BaSO_4(s)$, and $PbSO_4(s)$. The mercury(I) halides, $Hg_2Cl_2(s)$, $Hg_2Br_2(s)$, $Hg_2I_2(s)$ and the corresponding lead halides, e.g. $PbCl_2(s)$, are also sparingly soluble.

In the saturated solution of a sparingly soluble salt, equilibrium is set up between the solid salt and its ions in water, e.g. when $AgCl(s)$ is dissolved in water, the following equilibrium is set up:

$$AgCl(s) \xrightleftharpoons{K_s} Ag^+(aq) + Cl^-(aq)$$

where:

$$K_s = \left([Ag^+]_{actual}[Cl^-]_{actual}\right)_{eq} \tag{7.63}$$

and K_s is called the solubility product. It is important to realise that there is no term for the solid $AgCl$ in the equilibrium expression (see Section 8.14). All three species must be present for equilibrium to be set up, but only the ions appear in the equilibrium constant expression.

When a saturated solution of CaF_2 is prepared, the equilibrium set up is:

$$CaF_2(s) \xrightleftharpoons{K_s} Ca^{2+}(aq) + 2F^-(aq)$$

where:

$$\therefore K_s = \left([Ca^{2+}]_{actual}[F^-]_{actual}^2\right)_{eq} \tag{7.64}$$

and as with all the other equilibrium constants previously quoted, this is for the ideal case.

7.3.1 Formulation of the solubility product in terms of the solubility

Data on the solubility of sparingly soluble salts can be listed either in terms of the solubility product, K_s, as defined above, or as the raw experimental quantity, the solubility, s. The solubility is expressed as a number of mol per dm^3.

If the solubility in water of $AgCl(s)$ is s, then this means that s mol dm^{-3} of $AgCl$ are dissolved in the solution. Since one mol of $AgCl(s)$ gives rise to one mol of $Ag^+(aq)$ and one mol of $Cl^-(aq)$ then:

$$[Ag^+] = [Cl^-] = s \tag{7.65}$$

$$\therefore K_s = s \times s = s^2 \tag{7.66}$$

and:

$$s = \sqrt{K_s} \tag{7.67}$$

Using this relation the solubility product can be found from the solubility in water, or vice-versa.

Care must be taken when dealing with unsymmetrical electrolytes. For instance, if the solubility in water for $CaF_2(s)$ is s, then s mol dm^{-3} of $CaF_2(s)$ gives rise to s mol dm^{-3} of

$Ca^{2+}(aq)$ and $2s$ mol dm^{-3} of $F^-(aq)$, giving:

$$K_s = \left([Ca^{2+}]_{actual}[F^-]_{actual}^2\right)_{eq} \tag{7.68}$$

$$\therefore K_s = s \times (2s)^2$$

$$= 4s^3 \tag{7.69}$$

and:

$$s = \left(\frac{1}{4}K_s\right)^{1/3} \tag{7.70}$$

7.3.2   Solubility relations when a sparingly soluble salt is dissolved in a solution containing one of the ions of the solid: the common ion effect

Note: as in previous chapters it is important

- to check on the sources of the species involved in the equilibrium;

- to formulate relations between stoichiometric and actual concentrations;

- to judge when it is appropriate to make approximations;

- to see what are the physical conditions are under which the approximations are made;

- and then to look at the results to see if they validate the approximations.

The **solubilities** of a sparingly soluble salt in water and in a solution containing one of its ions will be **different**, but **both are governed by the solubility product**. For example, the solubility of $BaSO_4$ in an aqueous solution of Na_2SO_4 is governed by:

$$BaSO_4(s) \underset{K_s}{\rightleftharpoons} Ba^{2+}(aq) + SO_4^{2-}(aq)$$

$$K_s = \left([Ba^{2+}]_{actual}[SO_4^{2-}]_{actual}\right)_{eq} \tag{7.71}$$

But:

- the source of $Ba^{2+}(aq)$: dissolution of the solid only;

- the source of $SO_4^{2-}(aq)$: dissolution of the solid and the $Na_2SO_4(aq)$.

Let the solubility of the $BaSO_4$ be s',
Let the concentration of Na_2SO_4 be c

$$\therefore [Ba^{2+}]_{actual} = s' \tag{7.72}$$

$$[SO_4^{2-}]_{actual} = c + s' \tag{7.73}$$

$$\therefore K_s = s' \times (c + s') = s'c + (s')^2 \tag{7.74}$$

This is a quadratic equation in one unknown, and can be solved to give s'.
But are there **any sensible approximations** which can be made?
In the equilibrium:

$$BaSO_4(s) \underset{}{\overset{K_s}{\rightleftharpoons}} Ba^{2+}(aq) + SO_4^{2-}(aq)$$

if there is an added source of $SO_4^{2-}(aq)$, then this will push the equilibrium to the left, as compared with that set up in water alone. This will mean that $[Ba^{2+}]_{actual}$ in a solution of $BaSO_4(aq)$ and $Na_2SO_{4(aq)}$ will be less than that of $BaSO_4(s)$ dissolved in water alone.

Since the solubility of $BaSO_4(s)$ in water is low, then the solubility in $Na_2SO_4(aq)$ s' will be even lower and so:

$$s' \ll c \tag{7.75}$$
$$\therefore c + s' \approx c \tag{7.76}$$
$$\therefore K_s \approx cs' \tag{7.77}$$

and

$$s' \approx \frac{K_s}{c} \tag{7.78}$$

Worked problem 7.4

Question

Calculate the solubility of $BaSO_4(s)$ in a $0.0100 \, mol \, dm^{-3}$ solution of $Na_2SO_4(aq)$. $K_s = 1.2 \times 10^{-10} \, mol^2 \, dm^{-6}$ at $25°C$.

Answer

Let the solubility of $BaSO_4(s)$ be s'

$$\therefore [Ba^{2+}]_{actual} = s'$$
$$[SO_4^{2-}]_{actual} = s' + 0.0100 \, mol \, dm^{-3}$$
$$\therefore K_s = s' \times (0.0100 + s') = 0.0100s' + (s')^2$$

if $s' \ll 0.0100$

$$\therefore K_s = 0.0100s'$$

and

$$s' \approx \frac{K_s}{0.0100} = \frac{1.2 \times 10^{-10} \, mol^2 \, dm^{-6}}{0.0100 \, mol \, dm^{-3}} = 1.2 \times 10^{-8} \, mol \, dm^{-3}$$

The validity of the approximation can be checked by calculating the solubility, s, of $BaSO_4(s)$ on its own in water.

$$K_s = s^2 \therefore s = \sqrt{K_s} = \sqrt{1.2 \times 10^{-10}} = 1.1 \times 10^{-5} \, mol \, dm^{-3}$$

This can be contrasted with the much smaller solubility in the Na_2SO_4 solution, $s' = 1.2 \times 10^{-8}$ mol dm^{-3}, and shows that the approximation was justified.

This behaviour is very similar to the common ion effect in acid–base equilibria where the addition of a strong acid to a weak acid will suppress the dissociation of the weak acid. Here addition of a common ion will suppress the solubility of a sparingly soluble salt.

7.3.3 Possibility or otherwise of precipitation of a sparingly soluble salt when two solutions containing the relevant ions are mixed

The principles given above can be used to determine whether precipitation of solid will occur when two solutions are mixed together. $BaCl_2(s)$ and $Na_2SO_4(s)$ are both highly soluble in water, but if they are mixed together there is the possibility that precipitation of the sparingly soluble $BaSO_4$ could occur. The stoichiometric concentrations of the ions present in the mixture can be found on dilution. A decision on whether precipitation will occur can be made from the following:

- if $[Ba^{2+}]_{\text{stoich}} \times [SO_4^{2-}]_{\text{stoich}} > K_s$, then precipitation will occur to reduce the product $[Ba^{2+}]_{\text{stoich}} \times [SO_4^{2-}]_{\text{stoich}}$ until it reaches K_s. When this happens precipitation ceases, and equilibrium is established between $BaSO_4(s)$, $Ba^{2+}(aq)$ and $SO_4^{2-}(aq)$,

- if $[Ba^{2+}]_{\text{stoich}} \times [SO_4^{2-}]_{\text{stoich}} < K_s$, precipitation cannot occur. If, however, solid $BaSO_4$ is added to the solution, it will dissolve until the product, $[Ba^{2+}]_{\text{stoich}} \times [SO_4^{2-}]_{\text{stoich}}$, reaches K_s, and equilibrium is established between $BaSO_4(s)$, $Ba^{2+}(aq)$ and $SO_4^{2-}(aq)$.

- if $[Ba^{2+}]_{\text{stoich}} \times [SO_4^{2-}]_{\text{stoich}} = K_s$, then the solution is **just** saturated, with no precipitate apparent.

Worked problem 7.5

Question

Would precipitation occur when 0.020 mol dm^{-3} $BaCl_2(aq)$ is mixed with an equal volume of 0.012 mol dm^{-3} $Na_2SO_4(aq)$? K_s for $BaSO_4 = 1.2 \times 10^{-10}$ mol^2 dm^{-6} at 25°C.

Answer

Dilution will occur on mixing, and since equal volumes were used the concentrations will be halved.

$$[Ba^{2+}]_{\text{stoich}} = 0.010 \, \text{mol dm}^{-3},$$
$$[SO_4^{2-}]_{\text{stoich}} = 0.0060 \, \text{mol dm}^{-3}$$
$$[Ba^{2+}]_{\text{stoich}} \times [SO_4^{2-}]_{\text{stoich}} = (0.010 \times 0.0060) \, \text{mol}^2 \, \text{dm}^{-6} = 6 \times 10^{-5} \, \text{mol}^2 \, \text{dm}^{-6}$$
$$6 \times 10^{-5} \, \text{mol}^2 \, \text{dm}^{-6} \gg K_s, \text{ and precipitation will occur.}$$

Since $Ba^{2+}(aq)$ is in excess over $SO_4^{2-}(aq)$, nearly all the $SO_4^{2-}(aq)$ will be removed from solution. The maximum concentration of $Ba^{2+}(aq)$ which can be removed is given by the initial concentration of $SO_4^{2-}(aq)$, and is equal to $0.0060\,mol\,dm^{-3}$. $[Ba^{2+}]_{left} = (0.010 - 0.0060)\,mol\,dm^{-3} = 4.0 \times 10^{-3}\,mol\,dm^{-3}$.

The actual value of $[SO_4^{2-}]$ remaining in solution can then be found from the equilibrium relation:

$$[Ba^{2+}]_{actual} \times [SO_4^{2-}]_{actual} = K_s = 1.2 \times 10^{-10}\,mol^2\,dm^{-6}$$

$\therefore [SO_4^{2-}]_{actual} = 1.2 \times 10^{-10}\,mol^2\,dm^{-6}/4.0 \times 10^{-3}\,mol\,dm^{-3} = 3.0 \times 10^{-8}\,mol\,dm^{-3}$.

This is a very low value for $[SO_4^{2-}]_{actual}$ remaining, and justifies the assumption that virtually all of the sulphate ions will be removed as the precipitate, $BaSO_4(s)$.

The following two worked problems give further practice at handling aspects of solubilities.

Worked problem 7.6

Question

SrF_2 is sparingly soluble in water. The equilibrium constant governing the process:

$$SrF_2(s) \rightleftharpoons Sr^{2+}(aq) + 2F^-(aq)$$

has a value $K_s = 3.0 \times 10^{-9}\,mol^3\,dm^{-9}$ at $20°C$. Calculate the solubility of SrF_2 in:

a) water alone;

b) a solution which is $0.030\,mol\,dm^{-3}$ with respect to NaF;

c) a solution which is $0.030\,mol\,dm^{-3}$ with respect to $Sr(NO_3)_2$;

d) would precipitation occur if $100\,cm^3$ of a $0.050\,mol\,dm^{-3}$ solution of $Sr(NO_3)_2$ is mixed with $200\,cm^3$ of a $0.020\,mol\,dm^{-3}$ solution of NaF?

Answer

a) $SrF_2(s) \rightleftharpoons Sr^{2+}(aq) + 2F^-(aq)$

$$K_s = \left([Sr^{2+}]_{actual}[F^-]^2 \right)_{eq}$$

Let the solubility in water be s

$$\therefore [Sr^{2+}]_{actual} = s \qquad [F^-]_{actual} = 2s$$

$$K_s = s \times (2s)^2$$

$$= 4s^3$$

$$\therefore s = \sqrt[3]{\frac{K_s}{4}} = \sqrt[3]{\frac{3.0 \times 10^{-9} \, \text{mol}^3 \, \text{dm}^{-9}}{4}} = 9.1 \times 10^{-4} \, \text{mol} \, \text{dm}^{-3}$$

$$\therefore [Sr^{2+}]_{actual} = 9.1 \times 10^{-4} \, \text{mol} \, \text{dm}^{-3}$$

$$[F^-]_{actual} = 1.82 \times 10^{-3} \, \text{mol} \, \text{dm}^{-3}$$

b) This is no longer the simple case of a solubility in water. Here there is an added source of $F^-(aq)$.

Let the solubility of SrF_2 in the solution be s'.
$\therefore$ s' mol dm^{-3} of $Sr^{2+}(aq)$ and $2s'$ mol dm^{-3} of $F^-(aq)$ come from the solid SrF_2.
But 0.030 mol dm^{-3} of $F^-(aq)$ also come from NaF(aq).

$$\therefore [Sr^{2+}]_{actual} = s' \qquad [F^-]_{actual} = 2s' + 0.030$$

$$\therefore K_s = s' \times (2s' + 0.030)^2$$

This is a cubic expression which can be solved by successive approximations. But possible sensible approximations should be considered first. The solubility in water is low; $[F^-]_{actual} = 1.82 \times 10^{-3}$ mol dm^{-3} and the presence of a second source of $F^-(aq)$ will suppress the dissolution of the solid SrF_2, and so $2s' < 1.82 \times 10^{-3}$ mol dm^{-3}.

$$\therefore 2s' + 0.030 \text{ approximates to } 0.030$$

$$K_s = s' \times (2s' + 0.030)^2$$

which can be approximated to:

$$K_s = s' \times (0.030)^2$$

$$\therefore s' = \frac{3.0 \times 10^{-9} \, \text{mol}^3 \, \text{dm}^{-9}}{(0.030)^2 \, \text{mol}^2 \, \text{dm}^{-6}} = 3.3 \times 10^{-6} \, \text{mol} \, \text{dm}^{-3}$$

$$\therefore [Sr^{2+}]_{actual} = 3.3 \times 10^{-6} \, \text{mol} \, \text{dm}^{-3}$$

$$[F^-]_{actual} = 0.030 \, \text{mol} \, \text{dm}^{-3}$$

In water $[Sr^{2+}]_{actual} = 9.1 \times 10^{-4}$ mol dm^{-3}. This is much higher than 3.3×10^{-6} mol dm^{-3}, and shows conclusively that added NaF(aq) decreases the solubility of $SrF_2(s)$ dramatically.

c) In this case there is an added source of $Sr^{2+}(aq)$, and the treatment is analogous to that for part b).

Let the solubility of SrF_2 in the solution be s'
$\therefore$ s' mol dm^{-3} of $Sr^{2+}(aq)$ and $2s'$ mol dm^{-3} of $F^-(aq)$ come from the solid SrF_2.
But 0.030 mol dm^{-3} of $Sr^{2+}(aq)$ also comes from $Sr(NO_3)_2(aq)$.

$$\therefore [Sr^{2+}]_{actual} = s' + 0.030 \qquad [F^-]_{actual} = 2s'$$

$$\therefore K_s = (s' + 0.0300) \times (2s')^2$$

Again this is a cubic expression, and can be approximated in the same manner as in part b). Since s' is low, $s' + 0.030$ approximates to 0.030

$$K_s = (0.0300) \times (2s')^2$$

$$= 0.120 \times (s')^2$$

$$(s') = \sqrt{\frac{K_s}{0.120}} = \sqrt{\frac{3.0 \times 10^{-9}\,\text{mol}^3\,\text{dm}^{-9}}{0.120\,\text{mol}\,\text{dm}^{-3}}} = 1.6 \times 10^{-4}\,\text{mol}\,\text{dm}^{-3}$$

$$\therefore [Sr^{2+}]_{\text{actual}} = 0.030\,\text{mol}\,\text{dm}^{-3}$$

$$[F^-]_{\text{actual}} = 3.2 \times 10^{-4}\,\text{mol}\,\text{dm}^{-3}$$

Again, addition of a common ion decreases the solubility, but not so dramatically as in part b).

e) On mixing:

$100\,\text{cm}^3$ of $0.050\,\text{mol}\,\text{dm}^{-3}$ of $Sr^{2+}(aq)$ solution is diluted to $300\,\text{cm}^3$

$$\therefore [Sr^{2+}]_{\text{actual}} = 0.0167\,\text{mol}\,\text{dm}^{-3}$$

$200\,\text{cm}^3$ of $0.020\,\text{mol}\,\text{dm}^{-3}$ of $F^-(aq)$ solution is diluted to $300\,\text{cm}^3$

$$\therefore [F^-]_{\text{actual}} = 0.0133\,\text{mol}\,\text{dm}^{-3}$$

The solubility product, K_s, given by the product of the actual concentrations when equilibrium is established is:

$$K_s = ([Sr^{2+}]_{\text{actual}} \times [F^-]^2_{\text{actual}})_{\text{eq}} = 3.0 \times 10^{-9}\,\text{mol}\,\text{dm}^{-9}$$

The corresponding product of the actual concentrations present in the solution can be formulated and compared with the equilibrium product.

$$[Sr^{2+}]_{\text{actual}} \times [F^-]^2_{\text{actual}} = 0.0167 \times (0.0133)^2\,\text{mol}^3\,\text{dm}^{-9}$$

$$= 3.0 \times 10^{-6}\,\text{mol}^3\,\text{dm}^{-9}$$

$$[Sr^{2+}]_{\text{actual}} \times [F^-]^2_{\text{actual}} > ([Sr^{2+}]_{\text{actual}} \times [F^-]^2_{\text{actual}})_{\text{eq}}$$

and so precipitation will occur until the concentration decreases sufficiently for the equilibrium product to be attained.

So far, there has been no need to use electrical neutrality. But, as with acid–base equilibria and ion pair formation, sometimes it is needed when the equilibrium relation(s) and stoichiometric relations are fewer in number than the number of unknowns which have to be found. The following illustrative problem cannot be solved easily without invoking the relation describing electrical neutrality.

Worked problem 7.7

Question

A mixture of solid calcium sulphate and strontium sulphate is equilibrated in distilled water at $25°C$ so that an excess of each solid remains. If the solubility products are 1.2×10^{-6} and $3.2 \times 10^{-7}\,\text{mol}^2\,\text{dm}^{-6}$ respectively, calculate the concentrations of $Ca^{2+}(aq)$ and $Sr^{2+}(aq)$ in the resulting solution.

Answer

$$CaSO_4(s) \rightleftharpoons Ca^{2+}(aq) + SO_4^{2-}(aq)$$

$$K_s = ([Ca^{2+}]_{actual}[SO_4^{2-}]_{actual})_{eq}$$

$$SrSO_4(s) \rightleftharpoons Sr^{2+}(aq) + SO_4^{2+}(aq)$$

$$K_s = ([Sr^{2+}]_{actual}[SO_4^{2-}]_{actual})_{eq}$$

There are three unknowns in these two equations. To be able to obtain $[Sr^{2+}]_{actual}$, $[Ca^{2+}]_{actual}$ and $[SO_4^{2-}]_{actual}$ individually, a further independent equation relating these two quantities is needed.

One relation can be obtained from the ratio of the solubility products:

$$\left(\frac{[Ca^{2+}]_{actual}[SO_4^{2-}]_{actual}}{[Sr^{2+}]_{actual}[SO_4^{2-}]_{actual}}\right)_{eq} = \frac{1.2 \times 10^{-6}\,mol^2\,dm^{-6}}{3.2 \times 10^{-7}\,mol^2\,dm^{-6}} = 3.75$$

$$\therefore [Ca^{2+}]_{actual} = 3.75[Sr^{2+}]_{actual}$$

However, this is not an independent equation. It is just rewriting the relationships for the equilibrium constants for dissolution of $CaSO_4(s)$ and $SrSO_4(s)$, given above, in another form and, as such, does not give the third independent equation required.

Be careful with this point: always check that any extra equations are truly independent.

One further relation is still needed. The most obvious one is that dealing with electrical neutrality – the sum of the positive charges in solution must be balanced by the sum of the negative charges.

$$2[Ca^{2+}]_{actual} + 2[Sr^{2+}]_{actual} + [H_3O^+]_{actual} = 2[SO_4^{2-}]_{actual} + [OH^-]_{actual}$$

Remember: factor of 2 for the divalent ions.

Since the solution is one of two salts each formed from a strong acid/strong base, the solution will be neutral, and so:

$$[Ca^{2+}]_{actual} + [Sr^{2+}]_{actual} = [SO_4^{2-}]_{actual}$$

Substituting $[Ca^{2+}]_{actual} = 3.75[Sr^{2+}]_{actual}$ gives:

$$4.75[Sr^{2+}]_{actual} = [SO_4^{2-}]_{actual}$$

This equation taken with the solubility product relation for $SrSO_4(s)$ gives two simultaneous equations, which give:

$$([Sr^{2+}]_{actual} \times 4.75[Sr^{2+}]_{actual})_{eq} = 3.2 \times 10^{-7}\,mol^2\,dm^{-6}$$

$$\therefore [Sr^{2+}]^2 = 6.74 \times 10^{-8}\,mol^2\,dm^{-6}$$

$$\therefore ([Sr^{2+}]_{actual})_{eq} = 2.60 \times 10^{-4}\,mol\,dm^{-3}$$

But $[Ca^{2+}]_{actual} = 3.75[Sr^{2+}]_{actual}$

$$\therefore ([Ca^{2+}]_{actual})_{eq} = 9.75 \times 10^{-4}\,mol\,dm^{-3}$$

7.3.4 The effect of complexing on solubility equilibria

Ion pair formation, or complex formation, is indicated when the solubility of a sparingly soluble salt increases in the presence of an ion which could form an ion pair, or with a ligand which could form a complex with one of the ions of the salt. In these cases there are at least two equilibria, and all must be satisfied simultaneously.

Worked problem 7.8

Question

Taking the action of $edta^{4-}(aq)$ in complexing with $Ca^{2+}(aq)$ to be:

$$Ca^{2+}(aq) + edta^{4-}(aq) \xrightleftharpoons{K_1} Caedta^{2-}(aq)$$

and the value of the equilibrium constant at 20°C to be $6 \times 10^{10} \, mol^{-1} \, dm^3$, obtain a value for the equilibrium constant for the reaction:

$$CaCO_3(calcite) + edta^{4-}(aq) \xrightleftharpoons{K_2} Caedta^{2-}(aq) + CO_3^{2-}(aq)$$

The solubility product for calcite, K_s, is $4.8 \times 10^{-9} \, mol^2 \, dm^{-6}$ at 20°C.
   Calcite was added to a solution of $edta^{4-}(aq)$ until no more solid would dissolve. The concentration of the complex was found to be $0.042 \, mol \, dm^{-3}$. Find the concentrations of $edta^{4-}(aq)$ remaining and of the free $Ca^{2+}(aq)$ in solution.

Answer

There are three equilibria involved:

$$CaCO_3(calcite) \xrightleftharpoons{K_s} Ca^{2+}(aq) + CO_3^{2-}(aq)$$

$$K_s = \left([Ca^{2+}]_{actual}[CO_3^{2-}]_{actual}\right)_{eq}$$

$$Ca^{2+}(aq) + edta^{4-}(aq) \xrightleftharpoons{K_1} Caedta^{2-}(aq)$$

$$K_1 = \left(\frac{[Caedta^{2-}]_{actual}}{[Ca^{2+}]_{actual}[edta^{4-}]_{actual}}\right)_{eq}$$

$$CaCO_3(calcite) + edta^{4-}(aq) \xrightleftharpoons{K_2} Caedta^{2-}(aq) + CO_3^{2-}(aq)$$

$$K_2 = \left(\frac{[Caedta^{2-}]_{actual}[CO_3^{2-}]_{actual}}{[edta^{4-}]_{actual}}\right)_{eq}$$

Addition of the chemical equations for the first two equilibria gives the third. This is the reaction for which the equilibrium constant is required. Addition of the chemical equations corresponds to multiplication of the equilibrium constant relations:

$$K_s \times K_1 = \left([Ca^{2+}]_{actual}[CO_3^{2-}]_{actual}\right)_{eq} \times \left(\frac{[Caedta^{2-}]_{actual}}{[Ca^{2+}]_{actual}[edta^{4-}]_{actual}}\right)_{eq}$$

$$= \left(\frac{[Caedta^{2-}]_{actual}[CO_3^{2-}]_{actual}}{[edta^{4-}]_{actual}}\right)_{eq}$$

$$= K_2$$

$$\therefore K_2 = 4.8 \times 10^{-9}\ mol^2\ dm^{-6} \times 6 \times 10^{10}\ mol^{-1}\ dm^3$$

$$= 288\ mol\ dm^{-3}$$

The concentration of the complex is given. In the reaction:

$$CaCO_3(calcite) + edta^{4-}(aq) \underset{}{\overset{K_2}{\rightleftharpoons}} Caedta^{2-}(aq) + CO_3^{2-}(aq)$$

the complex and $CO_3^{2-}(aq)$ are formed in a 1:1 ratio, hence $[CO_3^{2-}]_{actual} = 0.042\ mol\ dm^{-3}$. Since $[CO_3^{2-}]_{actual}$ and $[Ca^{2+}]_{actual}$ are related to each other by the solubility product of $CaCO_3(s)$ there is now a route to $[Ca^{2+}]_{actual}$.

$$K_s = \left([Ca^{2+}]_{actual}[CO_3^{2-}]_{actual}\right)_{eq}$$

$$\therefore [Ca^{2+}]_{actual} = \frac{K_s}{[CO_3^{2-}]_{actual}} = \frac{4.8 \times 10^{-9}\ mol^2\ dm^{-6}}{0.042\ mol\ dm^{-3}} = 1.14 \times 10^{-7}\ mol\ dm^{-3}$$

There are two ways to determine $[edta^{4-}]_{actual}$.

- $[CO_3^{2-}]_{actual}$, $[Caedta^{2-}]_{actual}$ and $[edta^{4-}]_{actual}$ are all related through K_2.

$$K_2 = \left(\frac{[Caedta^{2-}]_{actual}[CO_3^{2-}]_{actual}}{[edta^{4-}]_{actual}}\right)_{eq}$$

$$\therefore [edta^{4-}]_{actual} = \frac{\left([Caedta^{2-}]_{actual}[CO_3^{2-}]_{actual}\right)_{eq}}{K_2}$$

$$= \frac{0.042\ mol\ dm^{-3} \times 0.042\ mol\ dm^{-3}}{288\ mol\ dm^{-3}}$$

$$= 6.1 \times 10^{-6}\ mol\ dm^{-3}$$

- $[Ca^{2+}]_{actual}$, $[Caedta^{2-}]_{actual}$ and $[edta^{4-}]_{actual}$ are also all related through K_1.

$$K_1 = \left(\frac{[Caedta^{2-}]_{actual}}{[Ca^{2+}]_{actual}[edta^{4-}]_{actual}}\right)_{eq}$$

$$\therefore [edta^{4-}]_{eq} = \frac{[Caedta^{2-}]_{eq}}{[Ca^{2+}]_{eq}K_1}$$

$$= \frac{0.042\ mol\ dm^{-3}}{1.14 \times 10^{-7}\ mol\ dm^{-3} \times 6 \times 10^{10}\ mol^{-1}\ dm^{-3}}$$

$$= 6.1 \times 10^{-6}\ mol\ dm^{-3}$$

7.3.5 Another interesting example

The behaviour of AgCl(s) in aqueous ammonia solutions is interesting. It is found that AgCl(s) dissolves extensively in aqueous solutions of ammonia, silver bromide dissolves quite appreciably and silver iodide dissolves only to a slight extent. In water AgCl(s) is only very sparingly soluble. This suggests that the known complexing behaviour of $NH_3(aq)$ with $Ag^+(aq)$ may be having a significant effect on the solubilities. Any $Ag^+(aq)$ formed by dissolution of AgCl(s) will be immediately mopped up by formation of the complex with $NH_3(aq)$, with consequent further dissolution of AgCl(s) occurring to maintain the solubility equilibrium.

When AgCl(s) dissolves in aqueous NH_3, two equilibria are set up and both will be simultaneously satisfied. They are:

$$AgCl(s) \underset{}{\overset{K_s}{\rightleftharpoons}} Ag^+(aq) + Cl^-(aq)$$

$$K_s = ([Ag^+]_{actual}[Cl^-]_{actual})_{eq} \tag{7.79}$$

$$= 1.8 \times 10^{-10}\,mol^2\,dm^{-6} \quad at \quad 25°C$$

$$Ag^+(aq) + 2NH_3 \underset{}{\overset{K_{assoc}}{\rightleftharpoons}} Ag(NH_3)_2^+(aq)$$

$$K_{assoc} = \left(\frac{[Ag(NH_3)_2^+]_{actual}}{[Ag^+]_{actual}[NH_3]_{actual}^2} \right)_{eq} \tag{7.80}$$

$$= 1.7 \times 10^7\,mol^{-2}\,dm^6 \quad at \quad 25°C$$

This is a very large value for the equilibrium constant for the complexing effect of $NH_3(aq)$ on $Ag^+(aq)$, and consequently the ratio:

$$\left(\frac{[Ag(NH_3)_2^+]_{actual}}{[Ag^+]_{actual}} \right)_{eq}$$

is also very large. This implies that the total concentration of **dissolved** silver, present predominantly as the complex with ammonia, can be high **even though** $[Ag^+]_{actual}$ is small.

The experimental observation is the dissolution of the solid AgCl in the $NH_3(aq)$ which continues until equilibrium is established. This overall process is:

$$AgCl(s) + 2NH_3(aq) \underset{}{\overset{K'}{\rightleftharpoons}} Ag(NH_3)_2^+(aq) + Cl^-(aq)$$

$$K' = \left(\frac{[Ag(NH_3)_2^+]_{actual}[Cl^-]_{actual}}{[NH_3]_{actual}^2} \right)_{eq} \tag{7.81}$$

K' is a composite equilibrium constant describing the simultaneous setting up of the two equilibria defined above.

$$K' = K_{assoc} \times K_s \tag{7.82}$$

$$K' = 1.7 \times 10^7\,mol^{-2}\,dm^6 \times 1.8 \times 10^{-10}\,mol^2\,dm^{-6}$$

$$= 3.1 \times 10^{-3}$$

The standard procedures can now be used to enable the solubility of AgCl(s) in aqueous NH_3 to be found for any given stoichiometric concentration of $NH_3(aq)$.

Stoichiometric relations

- $[Ag^+]_{total} = [Ag^+]_{actual} + [Ag(NH_3)_2^+]_{actual} = s$ (7.83)

where s is the solubility in $NH_3(aq)$
Since $Ag^+(aq)$ and $Cl^-(aq)$ will be formed in a 1:1 ratio:

$$[Ag^+]_{total} = [Cl^-]_{actual} = s \tag{7.84}$$

If $[Ag^+]_{actual}$ is very small, as it must be since $AgCl(s)$ is highly insoluble in water, then:

$$[Ag^+]_{total} \approx [Ag(NH_3)_2^+]_{actual} \tag{7.85}$$

- $[NH_3]_{total} = [NH_3]_{actual} + 2[Ag(NH_3)_2^+]_{actual}$ (7.86)

Be careful here: remember to include the factor 2; for each complex ion formed two NH_3 molecules are taken up.

Electrical neutrality

$$[Ag^+]_{actual} + [Ag(NH_3)_2^+]_{actual} + [H_3O^+]_{actual} = [Cl^-]_{actual} + [OH^-]_{actual} \tag{7.87}$$

But be careful: this equation can be contrasted with that given in the analysis of the solubility of $BaSO_4(s)$ in water where it could be assumed that the solution is neutral and $[H_3O^+]_{actual} = [OH^-]_{actual}$. This is not the case here; $NH_3(aq)$ is a weak base, and the pH will be >7. Fortunately, use of the electrical neutrality relation will not be required.

Equilibrium relations

$$K_s = ([Ag^+]_{actual}[Cl^-]_{actual})_{eq} \tag{7.79}$$

$$K_{assoc} = \left(\frac{[Ag(NH_3)_2^+]_{actual}}{[Ag^+]_{actual}[NH_3]_{actual}^2} \right)_{eq} \tag{7.80}$$

$$K' = \left(\frac{[Ag(NH_3)_2^+]_{actual}[Cl^-]_{actual}}{[NH_3]_{actual}^2} \right)_{eq} \tag{7.81}$$

$$K' = K_{assoc} \times K_s \tag{7.82}$$

Substituting Equation (7.86), $[NH_3]_{total} = [NH_3]_{actual} + 2[Ag(NH_3)_2^+]_{actual}$, into the expression given for K' results in:

$$K' = \left(\frac{[Ag(NH_3)_2^+]_{actual}[Cl^-]_{actual}}{([NH_3]_{total} - 2[Ag(NH_3)_2^+]_{actual})^2} \right)_{eq} \tag{7.88}$$

Since in this equilibrium $Ag(NH_3)_2^+(aq)$ and $Cl^-(aq)$ are produced in a 1:1 ratio, and since the amount of $Cl^-(aq)$ dissolved is equal to the total amount of $Ag^+(aq)$ dissolved and is thus equal to s:

$$[Ag(NH_3)_2^+]_{actual} = [Cl^-]_{actual} = s \tag{7.84}$$

$$\therefore K' = \frac{s^2}{([NH_3]_{total} - 2s)^2} \tag{7.89}$$

$$\sqrt{K'} = \sqrt{K_s K_{assoc}} = \frac{s}{[NH_3]_{total} - 2s} \qquad (7.90)$$

from which s can be found for any given stoichiometric concentration of $NH_3(aq)$.

Worked problem 7.9

Question

Use the following data to explain the observation that $AgCl(s)$ dissolves extensively in $1.6\ mol\ dm^{-3}$ aqueous NH_3, $AgBr(s)$ dissolves appreciably, but $AgI(s)$ dissolves only to a slight extent.

The solubility products, K_s, for each solid at 25°C are:

$$1.8 \times 10^{-10}\ mol^2\ dm^{-6} \quad \text{for} \quad AgCl(s)$$
$$5.0 \times 10^{-13}\ mol^2\ dm^{-6} \quad \text{for} \quad AgBr(s)$$
$$8.5 \times 10^{-17}\ mol^2\ dm^{-6} \quad \text{for} \quad AgI(s)$$

The association constant for formation of the complex of $Ag^+(aq)$ with $NH_3(aq)$, K_{assoc} is $1.7 \times 10^7\ mol^{-2}\ dm^6$.

Answer

As shown in Section 7.3.5 the overall reaction when silver halides dissolve in aqueous NH_3 is:

$$AgX(s) + 2NH_3(aq) \underset{}{\overset{K'}{\rightleftharpoons}} Ag(NH_3)_2^+(aq) + X^-(aq)$$

$$K' = \left(\frac{[Ag(NH_3)_2^+]_{actual}[X^-]_{actual}}{[NH_3]_{actual}^2} \right)_{eq}$$

and:

$$K' = K_s \times K_{assoc}$$

Using the data quoted gives the following values for K':

$$1.8 \times 10^{-10}\ mol^2\ dm^{-6} \times 1.7 \times 10^7\ mol^{-2}\ dm^6 = 3.1 \times 10^{-3} \quad \text{for } AgCl(s)$$
$$5.0 \times 10^{-13}\ mol^2\ dm^{-6} \times 1.7 \times 10^7\ mol^{-2}\ dm^6 = 8.5 \times 10^{-6} \quad \text{for } AgBr(s)$$
$$8.5 \times 10^{-17}\ mol^2\ dm^{-6} \times 1.7 \times 10^7\ mol^{-2}\ dm^6 = 1.5 \times 10^{-9} \quad \text{for } AgI(s)$$

Using the expression for K' given in Section 7.3.5 (Equation 7.90), i.e.

$$\sqrt{K'} = \sqrt{K_s K_{assoc}} = \frac{s}{[NH_3]_{total} - 2s}$$

gives:

$$s = [NH_3]_{total} \times \sqrt{K'} - 2s\sqrt{K'}$$

$$s + 2s\sqrt{K'} = [NH_3]_{total}\sqrt{K'}$$

$$s = \frac{\sqrt{K'}[NH_3]_{total}}{1 + 2\sqrt{K'}}$$

where

$$s = [X^-]_{actual} = [Ag(NH_3)_2^+]_{actual}$$

and

$$[Ag^+]_{actual} = \frac{K_s}{[X^-]_{actual}}$$

Solving this equation for each halide in turn gives the following solubilities, s, and $[Ag^+]_{actual}$:

	K'	$\sqrt{K'}$	$1 + 2\sqrt{K'}$	$\sqrt{K'}[NH_3]_{total}$	$s = [X^-]_{actual}$	K_s	$K_s/[X^-]_{actual}$ $= [Ag^+]_{actual}$
AgCl	3.1×10^{-3}	5.6×10^{-2}	1.112	8.9×10^{-2}	8.0×10^{-2}	1.8×10^{-10}	2.2×10^{-9}
AgBr	8.5×10^{-6}	2.9×10^{-3}	1.0058	4.6×10^{-3}	4.6×10^{-3}	5.0×10^{-13}	1.1×10^{-10}
AgI	1.5×10^{-9}	3.8×10^{-5}	1.00008	6.1×10^{-5}	6.1×10^{-5}	8.5×10^{-17}	1.4×10^{-12}

These calculations fall very clearly in line with the experimental observations, and indicate that the complexing behaviour of the $Ag^+(aq)$ ion with $NH_3(aq)$ along with the relevant solubility products predicts that:

AgCl(s) goes into solution to a considerable extent,

AgBr(s) goes into solution to an appreciable extent,

but AgI(s) is not much more soluble in NH_3 (aq) than in water.

7.3.6   Further examples of the effect of complexing on solubility

The complex can have an overall positive, negative or neutral charge.

Silver halides are only sparingly soluble in water. Their solubility can be enhanced, as in photographic techniques, where the complex with thiosulphate is used.

$$AgX(s) + 2S_2O_3^{2-}(aq) \rightleftharpoons Ag(S_2O_3)_2^{3-}(aq) + X^-(aq)$$

This behaviour is very similar to that with the neutral ligand $NH_3(aq)$.

When sparingly soluble salts are dissolved in a solution of the anion of the salt, it would be expected that the solubility would be decreased.

$$AgCl(s) \rightleftharpoons Ag^+(aq) + Cl^-(aq)$$

$$K_s = ([Ag^+]_{actual}[Cl^-]_{actual})_{eq} \tag{7.79}$$

Addition of $Cl^-(aq)$ will cause the equilibrium to move to the left until the product of concentrations has decreased to the equilibrium value, at which stage further precipitation of AgCl(s) will cease. Further addition of $Cl^-(aq)$ would be expected to result in further precipitation (provided sufficient $Ag^+(aq)$ is available), but it is found that when the chloride ion reaches a concentration of around 3×10^{-3} mol dm^{-3} AgCl(s) will start to re-dissolve. This is due to the formation of the complex ion $AgCl_2^-(aq)$, though others, such as $Ag^+Cl^-(aq)$, may also be involved.

A further variant on this is the observation that the highly insoluble HgI_2(s) will pass into solution if enough iodide is present. Again, this is the result of formation of a complex ion.

$$HgI_2(s) + 2I^-(aq) \rightleftharpoons HgI_4^{2-}(aq)$$

Worked problem 7.10

Question

AgCl dissolves only sparingly in water, $K_s = 1.75 \times 10^{-10}$ mol^2 dm^{-6} at 25°C. In the presence of excess $Cl^-(aq)$ the solubility at first decreases and then increases, and the following species are formed:

$$Ag^+(aq) + Cl^-(aq) \xrightleftharpoons{K_{1(assoc)}} Ag^+Cl^-(aq)$$

$$K_{1(assoc)} = \left(\frac{[Ag^+Cl^-]_{actual}}{[Ag^+]_{actual}[Cl^-]_{actual}} \right)_{eq}$$

$$Ag^+Cl^-(aq) + Cl^-(aq) \xrightleftharpoons{K_{2(assoc)}} AgCl_2^-(aq)$$

$$K_{2(assoc)} = \left(\frac{[AgCl_2^-]_{actual}}{[Ag^+Cl^-]_{actual}[Cl^-]_{actual}} \right)_{eq}$$

The following are values of the solubility of AgCl(s) in aqueous solutions of KCl:

$\dfrac{10^3[Cl^-]}{mol\,dm^{-3}}$	0.50	1.00	2.00	3.00	5.00	7.00	9.00
$\dfrac{10^7 s'}{mol\,dm^{-3}}$	6.75	5.1	4.5	4.5	4.9	5.4	5.9

a) Calculate the concentration of free $Ag^+(aq)$ in water.

b) From the value of the solubility product, calculate the expected **actual** concentrations of $Ag^+(aq)$ in each solution. Compare these values with the **total** concentration of $Ag^+(aq)$ as given by the solubility, and indicate whether these results are compatible with the presence of $Ag^+Cl^-(aq)$ and $AgCl_2^-(aq)$.

c) How would the concentrations of these species be expected to vary with increasing $Cl^-(aq)$ concentration?

d) Would this be compatible with the statement that: 'only an **anionic** species' would be expected to give a solubility which **increases** with increasing $[Cl^-]_{total}$?

e) Use the data given to find values for $K_{1(assoc)}$ and $K_{2(assoc)}$.

Answer

a) The solubility of $AgCl(s)$ in water is governed by the equilibrium:

$$AgCl(s) \xrightleftharpoons{K_s} Ag^+(aq) + Cl^-(aq)$$

$$K_s = ([Ag^+]_{actual}[Cl^-]_{actual})_{eq}$$

Let the solubility of $AgCl(s)$ in water be s, then:

$$[Ag^+]_{actual} = [Cl^-]_{actual} = s \text{ and } K_s = s^2$$

This expression applies **only if** there are **no** ion pairs present and so throughout the presence of the associated species $Ag^+Cl^-(aq)$ in the saturated solution in water is being ignored.

$$s = \sqrt{K_s} = \sqrt{1.75 \times 10^{-10} \, mol^2 \, dm^{-6}} = 1.32 \times 10^{-5} \, mol \, dm^{-3}$$

b) Let the solubility of $AgCl(s)$ in $KCl(aq)$ solution be s'.

$$K_s = ([Ag^+]_{actual}[Cl^-]_{actual})_{eq}$$

$$[Ag^+]_{actual} = \frac{K_s}{[Cl^-]_{actual}}$$

Now watch: there may be more than one source of each ion.
Sources of $Ag^+(aq)$ and $Cl^-(aq)$:

- $Ag^+(aq)$ comes from one source only, i.e. from the dissolution of the solid AgCl.

- Cl^- comes from two sources: the dissolution of the solid AgCl and the $Cl^-(aq)$ from the KCl(aq).

If the expected solubility in the KCl(aq) solution is s', then:

$$[Cl^-]_{actual} = s' + [Cl^-]_{from\ KCl}$$

Since AgCl(s) is highly insoluble in water, the $[Cl^-]_{from\ KCl} \gg s'$

$$[Cl^-]_{actual} \approx [Cl^-]_{from\ KCl}$$

$$\therefore [Ag^+]_{actual} = \frac{K_s}{[Cl^-]_{from\ KCl}}$$

and so the following table can be drawn up:

$\dfrac{10^3[Cl^-]}{mol\,dm^{-3}}$	0.50	1.00	2.00	3.00	5.00	7.00	9.00
$\dfrac{10^7[Ag^+]_{actual}}{mol\,dm^{-3}}$	3.5	1.75	0.88	0.58	0.35	0.25	0.19

The value of $[Ag^+]_{actual}$ in a saturated solution in water is $1.32 \times 10^{-5}\ mol\,dm^{-3}$. Addition of $Cl^-(aq)$ would be expected to suppress the dissolution of AgCl(s) by pushing the equilibrium to the left, the common ion effect. On a first look at the expected $[Ag^+]_{actual}$ given in the table this has occurred, but when these figures are compared with the actual observed solubilities, it becomes evident that the solubilities cannot be explained in terms of the expected common ion effect. At relatively low [KCl] the observed solubility does decrease with increasing $[Cl^-]$, but it soon reaches a minimum and then starts to increase. Furthermore the calculated $[Ag^+]_{actual}$, found assuming the common ion effect to be the sole cause of the decrease in solubility, is always less than the actual observed solubilities. Something else must be happening to cause the solubility to increase in the manner given in the table of the question.

If the species, $Ag^+Cl^-(aq)$ and $AgCl_2^-(aq)$ are also present in solution these would be expected to increase the solubility. Removal of $Ag^+(aq)$ from solution through formation of the associated species would cause the solubility equilibrium to move to the right with more solid AgCl dissolving, until the equilibrium value for $[Ag^+]_{actual}$ is set up. This value will be dictated by the simultaneous setting up of the three equilibria involved.

$$AgCl(s) \underset{}{\overset{K_s}{\rightleftharpoons}} Ag^+(aq) + Cl^-(aq)$$

$$K_s = ([Ag^+]_{actual}[Cl^-]_{actual})_{eq}$$

$$Ag^+(aq) + Cl^-(aq) \underset{}{\overset{K_{1(assoc)}}{\rightleftharpoons}} Ag^+Cl^-(aq)$$

$$K_{1(assoc)} = \left(\frac{[Ag^+Cl^-]_{actual}}{[Ag^+]_{actual}[Cl^-]_{actual}} \right)_{eq}$$

$$Ag^+Cl^-(aq) + Cl^-(aq) \underset{}{\overset{K_{2(assoc)}}{\rightleftharpoons}} AgCl_2^-(aq)$$

$$K_{2(assoc)} = \left(\frac{[AgCl_2^-]_{actual}}{[Ag^+Cl^-]_{actual}[Cl^-]_{actual}} \right)_{eq}$$

The observed solubilities are compatible with the presence of the two associated species.

c) Considering the $Ag^+Cl^-(aq)$:

$$[Ag^+Cl^-]_{actual} = K_{1\,assoc}[Ag^+]_{actual}[Cl^-]_{actual}$$
$$= K_{1\,assoc} \times K_s$$
$$= a\,constant$$

Hence, the concentration of the associated species, $Ag^+Cl^-(aq)$, would not change with increasing $[Cl^-]$, and the presence of this species would not give rise to a solubility which increases with increasing $[Cl^-]$.

On the other hand, considering the $AgCl_2^-$:

$$[AgCl_2^-]_{actual} = K_{2\,assoc}[Ag^+Cl^-]_{actual}[Cl^-]_{actual}$$
$$= K_{2\,assoc}K_{1\,assoc}[Ag^+]_{actual}[Cl^-]_{actual}[Cl^-]_{actual}$$
$$= K_{2\,assoc} \times K_{1\,assoc} \times K_s \times [Cl^-]_{actual}$$

Hence, the concentration of the associated species, $AgCl_2^-(aq)$ will increase as $[Cl^-]$ increases, and this in turn will cause more solid AgCl to dissolve.

d) These equilibrium arguments are consistent with the statement that: 'only an **anionic** species' would be expected to give a solubility which **increases** with increasing $[Cl^-]_{total}$.

e) The solubility of AgCl(s) in KCl solution is s'. The total amount of $Ag^+(aq)$ in solution is made up of contributions from the free $Ag^+(aq)$, and the two associated species, i.e.

$$[Ag^+]_{total} = s' = [Ag^+]_{actual} + [Ag^+Cl^-]_{actual} + [AgCl_2^-]_{actual}$$
$$\therefore s' - [Ag^+]_{actual} = [Ag^+Cl^-]_{actual} + K_{2\,assoc}[Ag^+Cl^-]_{actual}[Cl^-]_{actual}$$
$$= K' + K_{2\,assoc}K'[Cl^-]_{actual}$$

where $[Ag^+Cl^-]_{actual} = K' = K_{1\,assoc}K_s = a\,constant$

$\therefore$ a plot of $s' - [Ag^+]_{actual}$ against $[Cl^-]_{total}$ should be linear with :

$$intercept = K' = [Ag^+Cl^-]_{actual}$$

where $[Ag^+Cl^-]_{actual} = K_{1\,assoc} \times [Ag^+]_{actual} \times [Cl^-]_{actual}$

$$= K_{1\,assoc} \times K_s$$

$$slope = K_{2\,assoc}[Ag^+Cl^-]_{actual} = K_{2\,assoc}K_{1assoc}K_s = K_{2\,assoc}K'$$

$\therefore$ both K_{1assoc} and K_{2assoc} can be found from the graph.

$\dfrac{10^3[Cl^-]}{mol\,dm^{-3}}$	0.50	1.00	2.00	3.00	5.00	7.00	9.00
$\dfrac{10^7 s'}{mol\,dm^{-3}}$	6.75	5.1	4.5	4.5	4.9	5.4	5.9
$\dfrac{10^7[Ag^+]_{actual}}{mol\,dm^{-3}}$	3.5	1.75	0.88	0.58	0.35	0.25	0.19
$\dfrac{10^7\{s' - [Ag^+]_{actual}\}}{mol\,dm^{-3}}$	3.25	3.35	3.62	3.92	4.55	5.15	5.71

The graph, see Figure 7.2, of a plot of $(s' - [Ag^+]_{actual})$ against $[Cl^-]_{total}$ is linear with:

$$\text{intercept} = [Ag^+Cl^-]_{actual} = K_{1assoc} \times K_s = 3.1 \times 10^{-7}\,mol\,dm^{-3}$$

$$\therefore K_{1\,assoc} = \frac{3.1 \times 10^{-7}\,mol\,dm^{-3}}{K_s} = \frac{3.1 \times 10^{-7}\,mol\,dm^{-3}}{1.75 \times 10^{-10}\,mol^2\,dm^{-6}} = 1.77 \times 10^3\,mol^{-1}\,dm^3$$

$$\text{slope} = K_{2\,assoc} \times [Ag^+Cl^-]_{actual} = 2.9 \times 10^{-5}$$

$$\therefore K_{2\,assoc} = \frac{2.9 \times 10^{-5}}{[Ag^+Cl^-]_{actual}} = \frac{2.9 \times 10^{-5}}{3.1 \times 10^{-7}\,mol\,dm^{-3}} = 94\,mol^{-1}\,dm^3$$

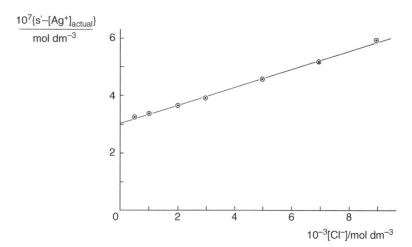

Figure 7.2 Graph of $s' - [Ag^+]_{actual}$ vs $[Cl^-]_{total}$.

8

Practical Applications of Thermodynamics for Electrolyte Solutions

In this chapter the focus will be on K, the equilibrium constant, and the following thermodynamic quantities, U, the energy, H, the enthalpy, G, the free energy, S, the entropy, V, the volume, C the heat capacity, and μ, the chemical potential. The significance of standard changes in the values of these quantities, ΔU^θ, ΔH^θ, ΔG^θ, ΔS^θ, ΔC_p^θ, and ΔV^θ for the study of electrolyte solutions will be discussed.

Attention will be directed to carrying out calculations on the use of these quantities in the study of electrolyte solutions as well as explaining the principles involved.

Aims

By the end of this chapter you should be able to:

- be aware of the importance of thermodynamic concepts and expressions in the study of electrolyte solutions and solution equilibria;

- derive the algebraic form of the equilibrium constant for a wide variety of reactions in aqueous solution from the expression $dG = \sum_i \mu_i dn_i$;

- follow through arguments leading to an interpretation of entropy in terms of order and disorder;

- discuss the effect of temperature and pressure (where relevant) on ΔH^θ, ΔG^θ, ΔS^θ and K;

- give an interpretation of ΔS^θ and ΔV^θ for reactions in electrolyte solutions;

An Introduction to Aqueous Electrolyte Solutions. By Margaret Robson Wright
© 2007 John Wiley & Sons Ltd   ISBN 978-0-470-84293-5 (cloth) ISBN 978-0-470-84294-2 (paper)

- understand the need to take account of non-ideality in solutions of electrolytes;

- formulate the relation between the mean activity coefficient and the individual activity coefficients for any electrolyte, and use the various Debye-Hückel expressions to calculate activity coefficients;

- predict the effect of ionic strength on a variety of reactions;

- make corrections for non-ideality graphically and by calculation;

- carry out calculations on all aspects of this chapter.

A very brief description of the essential thermodynamic background is given before applications to electrolyte solutions are discussed.

8.1 The first law of thermodynamics

The first law discusses **changes** in the energy of a closed system and states that there is a quantity U such that for small changes, denoted by a small d:

$$dU = q + w \tag{8.1}$$

or for large changes, denoted by a Δ, $\Delta U = q + w$.

   This reads that a small change in U can result from heat being transferred, or work being done on the system, or both occurring simultaneously.

- q can be negative, positive or zero corresponding to heat given out, heat taken in by the system, or no heat transferred.

- w is the work done **on** the system.

For experimental chemistry it is easy to visualise heat being transferred to a reaction flask, e.g. by heating. The concept of work done **on** a chemical system is more difficult to visualise. However, for the purposes of this book there are **three** types of work which have to be considered:

- **work done on the system as a result of a change in volume**, where

$$w = -p dV \tag{8.2}.$$

This will be used when setting up the thermodynamic criteria for, and equations describing, chemical equilibrium.

- **electrical work done on the system due to the passage of a charge through the system**

where

$$w = -E dQ \tag{8.3}$$

This will be used when discussing electrochemical cells.

- **electrostatic work done when charges are brought together or moved apart** where

$$w = \frac{Q_1 Q_2}{4\pi\varepsilon_0\varepsilon_r} \left(\frac{1}{r_2} - \frac{1}{r_1}\right) \tag{8.4}$$

The charges are Q_1 and Q_2, and they are initially at a distance r_1 apart and finally at a distance r_2 apart.

This will be relevant to Debye-Hückel theory where Q_1 and Q_2 become the charges z_1e and z_2e of two ions.

8.2 The enthalpy, *H*

This is a quantity whose meaning is explicable only in terms of the **defining equation**:

$$H = U + pV \tag{8.5}$$
$$dH = dU + pdV + Vdp + dVdp \tag{8.6}$$

A small change in H, dH, can result from a small change in U, dU, or a small change in p, dp, or a small change in V, dV, either individually or simultaneously. If only small changes are considered, the $dVdp$ term can be ignored to a first approximation, giving:

$$dH = dU + pdV + Vdp \tag{8.7}$$

8.3 The reversible process

A **reversible process** is one which can be made to go in reverse by a very small change in external conditions. A subsequent very small change in external conditions of an opposite kind can cause the system to return to its original state. All changes must be very small, and all changes in both directions correspond to a very small change **across a position of equilibrium**. This process is crucial to the discussion of thermodynamic aspects of equilibrium, electrolyte solution behaviour and reversible emfs. An **irreversible process** does not conform to these statements: there are no small changes across a position of equilibrium.

8.4 The second law of thermodynamics

This introduces the important quantity, S, the entropy which with the definition of G, the free energy, allows discussion of equilibria in thermodynamic terms.

The second law states that there is a quantity, S, such that for small changes in a closed system:

$$TdS \geq q \tag{8.8}$$

$$\text{or} \quad dS \geq \frac{q}{T} \tag{8.9}$$

A **reversible change** corresponds to the **equality** in the above statement of the second law. An **irreversible change** corresponds to the **inequality** in the statement of the second law.

8.5   Relations between q, w and thermodynamic quantities

When it is appropriate to do so, q can be replaced by:

- TdS for a **reversible process**, which is the type of change discussed in this book; or

- ΔU for a process at constant volume;

- ΔH for a process at constant pressure;

and w can be replaced by:

- $-pdV$ when the work done on the system is that due to a volume change;

- $-EdQ$ when the work done on the system is due to passage of a charge;

- $-pdV - EdQ$ when both contribute to the work done on the system;

- $\dfrac{Q_1 Q_2}{4\pi\varepsilon_0\varepsilon_r}\left(\dfrac{1}{r_2} - \dfrac{1}{r_1}\right)$ when electrostatic work is considered.

8.6   Some other definitions of important thermodynamic functions

The free energy, G, can be defined by the relation:

$$G = H - TS \tag{8.10}$$

A small change in G, dG, can result from a small change in H, dH, a small change in T, dT, or a small change in S, dS, and these can occur either individually or simultaneously.

$$dG = dH - TdS - SdT - dSdT \tag{8.11}$$

For small changes at constant T, $dT = 0$, and:

$$dG = dH - TdS \tag{8.12}$$

8.7   A very important equation which can now be derived

$$H = U + pV \tag{8.5}$$
$$G = H - TS \tag{8.10}$$
$$= U + pV - TS \tag{8.13}$$

taken together with the first and second laws result in a very useful thermodynamic equation required later, applicable to reversible changes where the only work done is $w = -p\mathrm{d}V$

$$\mathrm{d}G = V\mathrm{d}p - S\mathrm{d}T \tag{8.14}$$

At constant T:

$$\left(\frac{\partial G}{\partial p}\right)_T = V \tag{8.15}$$

and at constant p:

$$\left(\frac{\partial G}{\partial T}\right)_p = -S \tag{8.16}$$

where ∂ represents a partial derivative relevant to conditions where the derivative of one quantity is found with respect to another when all other variables are kept constant.

This equation is fundamental when dealing with chemical equilibria, chemical potentials and non-ideality.

8.8 Relation of emfs to thermodynamic quantities

Equation (8.14) corresponds to reversible changes where the only work done is $w = -p\mathrm{d}V$. When the work done is also due to the passage of charge, $-E\mathrm{d}Q$, it follows that

$$\mathrm{d}G = V\mathrm{d}p - S\mathrm{d}T - E\mathrm{d}Q \tag{8.17}$$

Experiments on reversible cells are normally carried out at constant T and p.

$$\therefore \mathrm{d}G = -E\mathrm{d}Q \tag{8.18}$$

An expression for $\mathrm{d}Q$ is needed, and this can be illustrated by reference to the redox reaction:

$$\mathrm{Au}^{3+}(\mathrm{aq}) + 2\mathrm{Au}(\mathrm{s}) \rightleftharpoons 3\mathrm{Au}^+(\mathrm{aq})$$

This equation reads as:

1 mol of $\mathrm{Au}^{3+}(\mathrm{aq})$ reacts with 2 mol of $\mathrm{Au}(\mathrm{s})$ to form 3 mol of $\mathrm{Au}^+(\mathrm{aq})$

But to find the charge passed when this occurs it is necessary to write out the individual redox reactions showing the number of electrons, z, involved for the reaction **with stoichiometry as written**:

$$\mathrm{Au}^{3+}(\mathrm{aq}) + 2\mathrm{e} \longrightarrow \mathrm{Au}^+(\mathrm{aq})$$
$$2\mathrm{Au}(\mathrm{s}) \longrightarrow 2\mathrm{Au}^+(\mathrm{aq}) + 2\mathrm{e}$$

The number of electrons involved, $z = 2$.

The charge passed, $\mathrm{d}Q$ for the reaction as written is:

$$\mathrm{d}Q = zF\,\mathrm{d}\left(n_{\mathrm{Au}^{3+}\text{ reacted}}\right) \tag{8.19}$$
$$\therefore \mathrm{d}G = -zFE\mathrm{d}\left(n_{\mathrm{Au}^{3+}\text{ reacted}}\right) \tag{8.20}$$

For large changes:

$$\Delta G = -zFE \tag{8.21}$$

and for the standard state, see Section 8.10:

$$\Delta G^{\theta} = -zFE^{\theta} \tag{8.22}$$

These equations are fundamental to the study of reversible cells (see Chapter 9).

8.9 The thermodynamic criterion of equilibrium

It can be shown that for **constant T and p**:

$$dG = q - TdS \tag{8.23}$$

But from the second law
$$TdS \geq q \tag{8.8}$$

$$\therefore dG \leq 0 \tag{8.24}$$

which requires that, for **constant T and p**, G can only decrease or stay the same. Hence the final state, i.e. equilibrium, can only be where G for the whole system is a minimum for a given T and p (see Figure 8.1). The slopes of tangents to this curve define the quantity, ΔG.

Equilibrium corresponds to the minimum of the curve, and at the minimum:

$$\text{the gradient of the tangent to the curve } \Delta G = 0 \tag{8.25}$$

which is the criterion of equilibrium. This requires that:

- approach to the position of equilibrium occurs with decreasing G and so conforms to the second law;

- spontaneous reaction occurs in a direction approaching equilibrium;

- moving away from the position of equilibrium would occur with increasing G and would break the second law;

- movement away from the position of equilibrium is impossible;

- at equilibrium, $\Delta G = 0$. (8.25)

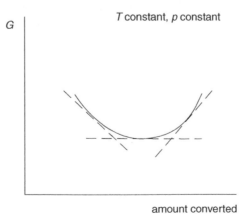

Figure 8.1   Schematic graph of G vs. amount converted.

8.10 Some further definitions: standard states and standard values

Changes in the values of thermodynamic functions such as ΔG and ΔS can be measured for various conditions, but there are conditions called 'standard conditions' where these quantities take specific values. Because these conditions are known and are standard for all species, then the measured quantities can be directly compared because they have been measured under the same standard conditions.

Standard values and conditions relate to:

- the conditions under which each substance is present,

- the phases of the substances present.

- For a **solid**, the standard state is the **pure solid**.

- For a **liquid**, the standard state is the **pure liquid**.

- For a **gas**, the standard state is **gas at a pressure of 1 bar**. Originally the standard state was taken to be the gas at 1 atm pressure, and unfortunately most of the tabulated data still relates to a standard state of 1 atm. This is something to watch out for.

- For a **solute**, the standard state is **unit activity** often **approximated** to a **concentration of 1 mol dm^{-3}**, i.e. $\gamma = 1$ where $a = \gamma c$ and $\gamma \to 1$ when $c \to 0$ (see Section 1.10 and 8.13).

- The standard states and conditions are denoted by θ, e.g. ΔG^θ is the **standard** value of ΔG and ΔH^θ is the **standard** value of ΔH.

Standard values for thermodynamic quantities such as ΔG^θ relate to substances, and to reactions between substances, present in their standard states. Generally reactions are not carried out under standard conditions, and for some reactions it may be impossible to do so. Standard thermodynamic quantities are directly related to conditions where the reaction is at equilibrium, and, in particular, can be determined by measurement of the equilibrium constant (see Section 8.14).

8.11 The chemical potential of a substance

For chemical substances, chemical reactions or phase changes, the free energy, G, depends on:

- the temperature, T,

- the pressure, p, and

- the amount of each substance, i, present, n_i

If there is an amount of one substance, n_1, present in a mixture when the amounts of all the other substances, $n_2, n_3 \ldots$, are kept constant, and where the temperature and pressure are also constant, a small change in G, dG, will result from the small change in the amount of the substance, dn_1. This can be written as:

$$(dG)_{T,p,n_2,n_3...} \propto dn_1 \qquad (8.26)$$

$$\therefore dG = \mu_1 dn_1 \qquad (8.27)$$

where μ_1 is the constant of proportionality called the chemical potential.

$$\therefore \mu_1 = \left(\frac{\partial G}{\partial n_1}\right)_{T,p,n_2,n_3...} \qquad (8.28)$$

In general:

$$\mu_i = \left(\frac{\partial G}{\partial n_i}\right)_{T,p,\text{all other } n} \qquad (8.29)$$

If the amounts of all the substances are allowed to change, but T and p are still kept constant, then the change in G will be a sum of terms for each change of amount for each substance:

$$dG = \mu_1 dn_1 + \mu_2 dn_2 + \mu_3 dn_3 + \ldots \qquad (8.30)$$

$$= \sum_i \mu_i dn_i \qquad (8.31)$$

If both T and p are allowed to vary as well:

$$dG = Vdp - SdT + \sum_i \mu_i dn_i \qquad (8.32)$$

8.12 Criterion of equilibrium in terms of chemical potentials

In a chemical reaction, reactants are removed in the forward reaction and products are formed by reaction. This is equivalent to matter being removed in the forward reaction and added by forming products. The change in G, dG, at constant T and p resulting from this has been shown to be:

$$dG = \mu_1 dn_1 + \mu_2 dn_2 + \mu_3 dn_3 + \ldots \qquad (8.30)$$

$$= \sum_i \mu_i dn_i \qquad (8.31)$$

For the general reaction occurring at constant T and p:

$$aA + bB \rightleftharpoons cC + dD$$

when the forward reaction occurs:
a mol of A react with b mol of B to form c mol of C and d mol of D
and the overall change in G is given as:

$$\Delta G = +c\mu_C + d\mu_D - a\mu_A - b\mu_B \qquad (8.33)$$

This is the change in G for a reaction which **need not necessarily be at equilibrium**.

The second law of thermodynamics requires that the **criterion for equilibrium** is that G is a minimum for constant T and p (see Section 8.9), and so at equilibrium:

$$\Delta G = 0 \tag{8.25}$$

$$\text{and } +c\mu_C + d\mu_D - a\mu_A - b\mu_B = 0 \tag{8.34}$$

or in general terms, **at equilibrium**:

$$dG = \sum_i \mu_i dn_i = 0 \tag{8.35}$$

Explicit expressions for the chemical potentials for the substances involved in the reaction are now required.

8.13 Chemical potentials for solids, liquids, gases and solutes

These can be defined by the following:

- **Solids and liquids**

$$\mu = \mu^\theta \tag{8.36}$$

where μ^θ is a constant dependent on T and slightly on p.
Note: there is no $RT\log_e$ term in the expression for μ for solids and liquids.

- **A gas**:

This is a very different situation from the solid or liquid case, and if non-identity is neglected

$$\mu = \mu^\theta + RT\log_e p \tag{8.37}$$

where μ^θ is a constant dependent on T only.
For a mixture of gases, a corresponding equation will hold for each constituent gas:

$$\mu_i = \mu_i^\theta + RT\log_e p_i \tag{8.38}$$

where μ_i^θ is a constant dependent on T only.

- **A solute in solution**

There is no explicit thermodynamic derivation of the expression for the chemical potential of a solute in solution. However, experimentally it is found that the expression for the emf of a reversible cell has an $nRT\log_e$ term in it. Since the emf is directly related to ΔG for the cell reaction, $\Delta G = -zFE$, then the expression for μ for a solute is given **by analogy** as:

$$\mu = \mu^\theta + RT\log_e a \tag{8.39}$$

where μ^θ is a constant dependent on T and slightly on p.

This equation can be written in terms of the concentration and a factor, γ, the activity coefficient:

$$a = \gamma c \quad \text{where} \quad \gamma \to 1 \quad \text{as} \quad c \to 0 \tag{8.40}$$

$$\mu = \mu^{\theta} + RT \log_e \gamma c \tag{8.41}$$

For a mixture of solutes a corresponding equation can be written for each solute:

$$\mu_i = \mu_i{}^{\theta} + RT \log_e \gamma_i c_i \tag{8.42}$$

$$a_i = \gamma_i c_i \quad \text{where} \quad \gamma_i \to 1 \text{ as } \sum_i c_i \to 0 \tag{8.43}$$

8.14 Use of the thermodynamic criterion of equilibrium in the derivation of the algebraic form of the equilibrium constant

This is a very important aspect of thermodynamics and one of great relevance to the quantitative description of electrolyte solutions. It rests directly on the quantities described above.

Consider the following reaction which involves species in the gas, solid and aqueous solution phases:

$$Pb(s) + 2CH_3COOH(aq) \rightleftharpoons 2CH_3COO^-(aq) + Pb^{2+}(aq) + H_2(g)$$

Following the argument given in Section 8.12:
At constant T and p

$$\Delta G = 2\mu_{CH_3COO^-(aq)} + \mu_{Pb^{2+}(aq)} + \mu_{H_2(g)} - \mu_{Pb(s)} - 2\mu_{CH_3COOH(aq)} \tag{8.44}$$

$$= 2\mu_{CH_3COO^-(aq)}^{\theta} + 2RT \log_e a_{CH_3COO^-(aq)} + \mu_{Pb^{2+}(aq)}^{\theta} + RT \log_e a_{Pb^{2+}(aq)}$$

$$+ \mu_{H_2(g)}^{\theta} + RT \log_e p_{H_2(g)} - \mu_{Pb(s)}^{\theta} - 2\mu_{CH_3COOH(aq)}^{\theta} - 2RT \log_e a_{CH_3COOH(aq)} \tag{8.45}$$

- for the solid, Pb(s), $\mu = \mu^{\theta}$ where μ^{θ} depends on T and slightly on p,

- for the gas, $H_2(g)$, $\mu = \mu^{\theta} + RT \log_e p_{H_2}$ where μ^{θ} depends on T only,

- for the solutes, $Pb^{2+}(aq)$, $CH_3COO^-(aq)$, $CH_3COOH(aq)$, $\mu = \mu^{\theta} + RT \log_e a$ where μ^{θ} depends on T and slightly on p.

Gathering together the standard terms:

$$\Delta G^{\theta} = 2\mu_{CH_3COO^-(aq)}^{\theta} + \mu_{Pb^{2+}(aq)}^{\theta} + \mu_{H_2(g)}^{\theta} - \mu_{Pb(s)}^{\theta} - 2\mu_{CH_3COOH(aq)}^{\theta} \tag{8.46}$$

where ΔG^{θ} is a constant dependent on T and slightly on p.

$$\therefore \Delta G = \Delta G^{\theta} + RT \log_e \frac{a_{CH_3COO^-(aq)}^2 a_{Pb^{2+}(aq)} p_{H_2(g)}}{a_{CH_3COOH(aq)}^2} \tag{8.47}$$

Note: $2RT \log_e x = RT \log_e x^2$

Pay particular attention to the following:

The quotient in Equation (8.47) is one where the values of the activities and partial pressure are ones relating to situations when the reaction is **not necessarily at equilibrium**

But: at equilibrium $\Delta G = 0$ $\therefore \Delta G^\theta = -RT \log_e \left(\dfrac{a^2_{CH_3COO^-(aq)} a_{Pb^{2+}(aq)} p_{H_2(g)}}{a^2_{CH_3COOH(aq)}} \right)_{eq}$ (8.48)

Take note:

Equation (8.48) involves a quotient which relates to activities and partial pressure when the reaction **is at equilibrium**.

Since ΔG^θ is a constant dependent on T and slightly on p, so the quotient of activities and partial pressure is also a constant dependent on T and slightly on p. This is the equilibrium constant K.

$$K = \left(\frac{a^2_{CH_3COO^-(aq)} a_{Pb^{2+}(aq)} p_{H_2(g)}}{a^2_{CH_3COOH(aq)}} \right)_{eq}$$ (8.49)

$$\Delta G^\theta = -RT \log_e K$$ (8.50)

This is one of the most important thermodynamic relations, and it allows the algebraic expression for the equilibrium constant to be identified with the **standard** free energy for the reaction.

If the analysis is **limited** to the **ideal case**, all activities can be approximated to concentrations. This is the form in which all equilibrium constants discussed in earlier chapters have been given.

Note, and the following is of crucial importance:

- there is no logarithmic term for the solid in the expression for ΔG, and correspondingly no term for the solid in the expression for K;

- likewise there will be no logarithmic term for any liquid in the expression for ΔG, and correspondingly no term for the liquid in the expression for K;

- this derivation gives a thermodynamic proof as to why in earlier chapters it has been stressed that, e.g. for the dissolution of a sparingly soluble salt in water, there is no term relating to the solid in the equilibrium constant, or for the case of the ionisation of a weak acid there is no term for the solvent H_2O;

- the algebraic expression for any reaction can be derived in a similar manner, and the pressure and temperature dependencies assessed from the dependencies of the various μ^θ values;

- The quotient of activities and partial pressure which appears as a logarithmic term in the expression for ΔG is one which is of the **same algebraic form** as that appearing in the equilibrium constant. It, however, relates to a situation when the reaction is **not** at equilibrium. It must be clearly distinguished from the quotient defining the equilibrium constant which describes a situation when reaction is at equilibrium. It is for this reason that a subscript eq is always added to the quotient describing equilibrium values.

Worked problem 8.1

Question

Explain the distinction between ΔG and ΔG^{θ}, and summarise the information contained in each quantity.

Answer

It is absolutely vital that the distinction between these two quantities is fully appreciated, and that the two are not confused.

- ΔG^{θ} is related to the equilibrium constant and to activities and partial pressures for reactions **at equilibrium**.

$$\Delta G^{\theta} = -RT \log_e K$$

Knowing K enables the value of ΔG^{θ} to be found, and vice-versa.

- ΔG^{θ} **gives information about the magnitude of the equilibrium constant, and about the position of equilibrium**:

  - If ΔG^{θ} is positive, then $K < 1$ and the equilibrium lies to the left.

  - If ΔG^{θ} is negative, then $K > 1$ and the equilibrium lies to the right.

  - If $\Delta G^{\theta} = 0$, then $K = 1$.

- ΔG relates to processes which are **not at equilibrium**

- **The sign of ΔG gives information about the direction of reaction**

  - If ΔG is positive, then reaction cannot occur in the forward direction.

  - If ΔG is negative, then reaction can occur in the forward direction.

  - If ΔG is zero, then reaction is at equilibrium.

Worked problem 8.2

Question

Deduce the algebraic form for the equilibrium constant for each of the following reactions:

(i) The ionisation of a weak acid, HA(aq),

(ii) The solubility product for dissolution of the sparingly soluble salt $LaF_3(s)$.

Answer

(i) $HA(aq) + H_2O(l) \rightleftharpoons H_3O^+(aq) + A^-(aq)$

$$\Delta G = \mu_{H_3O^+(aq)} + \mu_{A^-(aq)} - \mu_{H_2O(l)} - \mu_{HA(aq)}$$

$$= \mu^{\theta}_{H_3O^+(aq)} + RT \log_e a_{H_3O^+(aq)} + \mu^{\theta}_{A^-(aq)} + RT \log_e a_{A^-(aq)}$$

$$- \mu^{\theta}_{H_2O(l)} - \mu^{\theta}_{HA(aq)} - RT \log_e a_{HA(aq)}$$

$$= \Delta G^{\theta} + RT \log_e \frac{a_{H_3O^+(aq)} a_{A^-(aq)}}{a_{HA(aq)}}$$

where $\Delta G^{\theta} = \mu^{\theta}_{H_3O^+(aq)} + \mu^{\theta}_{A^-(aq)} - \mu^{\theta}_{H_2O(l)} - \mu^{\theta}_{HA(aq)}$ and ΔG^{θ} is a constant dependent on T and slightly on p.

Note: there is no logarithmic term involving $H_2O(l)$. This is because $\mu = \mu^{\theta}$ for a liquid.

At equilibrium $\Delta G = 0$,

$$\therefore \Delta G^{\theta} = -RT \log_e \left(\frac{a_{H_3O^+(aq)} a_{A^-(aq)}}{a_{HA(aq)}} \right)_{eq}$$

Since ΔG^{θ} is a constant dependent on T and slightly on p, $RT \log_e \left(\dfrac{a_{H_3O^+(aq)} a_{A^-(aq)}}{a_{HA(aq)}} \right)_{eq}$ is a constant dependent on T and slightly on p from which $\left(\dfrac{a_{H_3O^+(aq)} a_{A^-(aq)}}{a_{HA(aq)}} \right)_{eq}$ is a constant dependent on T and slightly on p, and is the equilibrium constant K, i.e.:

$$K = \left(\frac{a_{H_3O^+(aq)} a_{A^-(aq)}}{a_{HA(aq)}} \right)_{eq}$$

where K is a constant dependent on T and slightly on p.

(ii) $LaF_3(s) \rightleftharpoons La^{3+}(aq) + 3F^-(aq)$

$$\Delta G = \mu_{La^{3+}(aq)} + 3\mu_{F^-(aq)} - \mu_{LaF_3(s)}$$

$$= \mu^{\theta}_{La^{3+}(aq)} + RT \log_e a_{La^{3+}(aq)} + 3\mu^{\theta}_{F^-(aq)} + 3RT \log_e a_{F^-(aq)} - \mu^{\theta}_{LaF_3(s)}$$

$$= \Delta G^{\theta} + RT \log_e a_{La^{3+}(aq)} a^3_{F^-(aq)}$$

where ΔG^{θ} is a constant dependent on T and slightly on p.

Note: there is no $RT \log_e a$ term for the solid LaF_3. This is because $\mu = \mu^{\theta}$ for a solid.

Remember: $3RT \log_e x = RT \log_e x^3$

At equilibrium $\Delta G = 0$,

$$\therefore \Delta G^{\theta} = -RT \log_e \left(a_{La^{3+}(aq)} a^3_{F^-(aq)} \right)_{eq}$$

Since ΔG^{θ} is a constant dependent on T and slightly on p, $RT \log_e \left(a_{La^{3+}(aq)} a^3_{F^-(aq)} \right)_{eq}$ is a constant dependent on T and slightly on p from which $\left(a_{La^{3+}(aq)} a^3_{F^-(aq)} \right)_{eq}$ is a constant dependent on T and slightly on p, and is the equilibrium constant K, i.e.:

$$K = \left(a_{La^{3+}(aq)} a^3_{F^-(aq)} \right)_{eq}$$

where K is a constant dependent on T and slightly on p.

Note well because this is important:

- The quotients or products which appear in the expressions for ΔG, involve a quotient of activities for situations where the reaction **has not reached equilibrium**.

- The quotients or products which appear in the expressions for ΔG^θ, involve a quotient of activities for situations where the reaction **is at equilibrium**, and this is specified by the subscript eq.

The quotients or products which appear in the expressions for ΔG, ΔG^θ and K are of activities, and are therefore applicable to all conditions, ideal or non-ideal. So far in this book, such quantities have been given in terms of concentrations, and, as such, are non-ideal quantities unless they are for infinite dilution where all concentrations $\rightarrow 0$. However, many experimental determinations are based on concentrations and, in principle, should be corrected for non-ideality. This will be dealt with in Sections 8.21, 8.24 to 8.27.

Worked problem 8.3

Question

Solubility products of calcite and aragonite at 25°C are 8.9×10^{-9} mol^2 dm^{-6} and 1.22×10^{-8} mol^2 dm^{-6} respectively.

$$\text{CaCO}_3(\text{calcite}) \rightleftharpoons \text{Ca}^{2+}(\text{aq}) + \text{CO}_3^{2-}(\text{aq})$$
$$\text{CaCO}_3(\text{aragonite}) \rightleftharpoons \text{Ca}^{2+}(\text{aq}) + \text{CO}_3^{2-}(\text{aq})$$

(i) Find ΔG^θ for each of these processes.

(ii) Assuming that non-ideality can be ignored, explain whether it would be possible to dissolve

   (A) calcite

   (B) aragonite

in a solution in which both $[\text{Ca}^{2+}]$ and $\text{CO}_3^{2-}]$ are 1.00×10^{-4} mol dm^{-3}. Do this

   (a) by using an equilibrium argument;

   (b) by determining ΔG for the two possibilities.

Answer

(i) $\text{CaCO}_3(\text{calcite}) \rightleftharpoons \text{Ca}^{2+}(\text{aq}) + \text{CO}_3^{2-}(\text{aq})$
$$K_{\text{calcite}} = ([\text{Ca}^{2+}][\text{CO}_3^{2-}])_{\text{eq}}$$

$$\Delta G^\theta_{\text{calcite}} = -RT \, \log_e K_{\text{calcite}}$$
$$= -8.3145 \times 10^{-3} \, \text{kJ mol}^{-1} \, \text{K}^{-1} \times 298.15 \, \text{K} \times \log_e 8.9 \times 10^{-9}$$
$$= (-8.3145 \times 10^{-3} \times 298.15 \times -18.537) \, \text{kJ mol}^{-1} = 45.95 \, \text{kJ mol}^{-1}$$

$$CaCO_3(aragonite) \rightleftharpoons Ca^{2+}(aq) + CO_3^{2-}(aq)$$

$$K_{aragonite} = ([Ca^{2+}][CO_3^{2-}])_{eq}$$

$$\Delta G^\theta_{aragonite} = -RT\log_e K_{aragonite}$$

$$= -8.3145 \times 10^{-3}\,kJ\,mol^{-1}\,K^{-1} \times 298.15\,K \times \log_e 1.22 \times 10^{-8}$$

$$= (-8.3145 \times 10^{-3} \times 298.15 \times -18.222)\,kJ\,mol^{-1}$$

$$= 45.17\,kJ\,mol^{-1}$$

(ii) (a) In the solution $[Ca^{2+}] = [CO_3^{2-}] = 1.00 \times 10^{-4}\,mol\,dm^{-3}$, and so $[Ca^{2+}]\,[CO_3^{2-}] = 1.00 \times 10^{-8}\,mol^2\,dm^{-6}$

This can be compared with the equilibrium product given by $K_{calcite}$

$$[Ca^{2+}][CO_3^{2-}] > 8.9 \times 10^{-9}\,mol^2\,dm^{-6}$$

∴ it is impossible for calcite to dissolve in the solution.
The product can also be compared with the equilibrium product $K_{aragonite}$.

$$[Ca^{2+}][CO_3^{2-}] < 1.22 \times 10^{-8}\,mol^2\,dm^{-6}$$

∴ some aragonite will dissolve until the equilibrium product, $([Ca^{2+}][CO_3^{2-}])_{eq} = 1.22 \times 10^{-8}$ $mol^2\,dm^{-6}$, is reached.

$$(b)\Delta G = \mu_{Ca^{2+}(aq)} + \mu_{CO_3^{2-}(aq)} - \mu_{CaCO_3(s)}$$

$$= \mu^\theta_{Ca^{2+}(aq)} + \mu^\theta_{CO_3^{2-}(aq)} - \mu^\theta_{CaCO_3(aq)} + RT\log_e a_{Ca^{2+}(aq)} + RT\log_e a_{CO_3^{2-}(aq)}$$

$$= \Delta G^\theta + RT\log_e a_{Ca^{2+}(aq)} a_{CO_3^{2-}(aq)}$$

where ΔG^θ is a constant dependent on T and slightly on p. **Approximating** this expression to concentrations i.e. ignoring non-ideality, gives:

$$\Delta G = \Delta G^\theta + RT\log_e [Ca^{2+}][CO_3^{2-}]$$

For **calcite**:

$$\Delta G = 45.95\,kJ\,mol^{-1} + (8.3145 \times 10^{-3}\,kJ\,mol^{-1}\,K^{-1} \times 298.15\,K \times \log_e 1.00 \times 10^{-8})$$

$$= (45.95 - 45.66)\,kJ\,mol^{-1} = +0.29\,kJ\,mol^{-1}$$

For **aragonite**:

$$\Delta G = 45.17\,kJ\,mol^{-1} + (8.3145 \times 10^{-3}\,kJ\,mol^{-1}\,K^{-1} \times 298.15\,K \times \log_e 1.00 \times 10^{-8})$$

$$= (45.17 - 45.66)\,kJ\,mol^{-1} = -0.49\,kJ\,mol^{-1}$$

• ΔG for the process $CaCO_3(calcite) \rightleftharpoons Ca^{2+}(aq) + CO_3^{2-}(aq)$

is positive and so this process cannot occur, i.e. calcite cannot dissolve in the solution.

• ΔG for the process $CaCO_3(aragonite) \rightleftharpoons Ca^{2+}(aq) + CO_3^{2-}(aq)$

is negative and so this process can occur, i.e. aragonite can dissolve in the solution. This conclusion can be compared with that reached when non-ideality is considered (see Worked Problem 8.16).

8.15 The temperature dependence of ΔH^θ

Modern calorimetry allows highly accurate estimates of ΔH to be made, and, from these, extrapolation to zero ionic strength gives ΔH^θ, see Sections 8.24 onwards.

ΔH^θ can vary with temperature, and plots of ΔH^θ vs. T can be linear or non-linear with slopes which can be positive, negative or zero. The dependence of ΔH^θ on temperature defines another thermodynamic quantity, the change in heat capacity at constant pressure, ΔC_p^θ. Since a slope defines a derivative:

$$\left(\frac{\partial \Delta H^\theta}{\partial T}\right)_p = \Delta C_p^\theta \tag{8.51}$$

and values of ΔC_p^θ can be found experimentally from such graphs. This is the main method for determining ΔC_p^θ for reactions in solution in contrast to the possibility of using statistical mechanics for reactions of gases. It has proved extremely difficult to give significant interpretations of ΔC_p^θ values for electrolyte solutions.

Worked problem 8.4

Question

Data on the ionisation of the trimethylammonium ion show that ΔH^θ varies with temperature. Use the following data to determine $\Delta C_p{}^\theta$. Why is it not necessary to convert temperatures to kelvin in this problem?

$\dfrac{\text{temperature}}{{}^\circ\text{C}}$	15	25	35	45
$\dfrac{\Delta H^\theta}{kJ\,mol^{-1}}$	35.0	36.8	38.6	40.4

Answer

$\Delta C_p^\theta = \left(\partial \Delta H^\theta / \partial T\right)_p$. Hence a plot of ΔH^θ vs. T should be linear if ΔC_p^θ is independent of temperature and non-linear if ΔC_p^θ varies with temperature. In this case the graph is linear and the gradient gives $\Delta C_p^\theta = +180\,J\,mol^{-1}\,K^{-1}$. In this problem there is no need to convert to kelvin because all that is required is a difference in temperature, and the numerical magnitude of this difference is the same in ${}^\circ$C as in K.

8.16 The dependence of the equilibrium constant, K, on temperature

The most direct way, and possibly the most accurate method, for determining ΔH and ΔH^θ for reactions in solution is a direct calorimetric determination. However, the dependence of K on temperature also enables ΔH^θ to be found.

8.16.1 Calculation of ΔH^θ from two values of K

$\Delta G^\theta = -RT \log_e K$ and $\Delta G^\theta = \Delta H^\theta - T\Delta S^\theta$. Assuming that ΔH^θ and ΔS^0 are both independent of temperature, then:

$$\log_e \frac{K_{T_2}}{K_{T_1}} = -\frac{\Delta H^\theta}{R}\left[\frac{1}{T_2} - \frac{1}{T_1}\right] \tag{8.52}$$

However, it is highly unlikely that ΔH^θ and ΔS^0 are independent of temperature, and so the calculation will only be approximate. The smaller the range of temperature involved the smaller will be the error resulting from treating ΔH^θ as constant. This simple calculation using data limited to two temperatures only is often used to get a first approximate value.

8.16.2 Determination of ΔH^θ from values of K over a range of temperatures

Here what is required is a relation between K and T which lends itself to a plot where either an intercept or a gradient of a tangent leads to a value of ΔH^θ. Two approaches are possible.

(i) Substituting $\Delta G^\theta = -RT \log_e K$ into $\Delta G^\theta = \Delta H^\theta - T\Delta S^\theta$ and differentiating with respect to temperature and for constant p gives:

$$\left(\frac{\partial \log_e K}{\partial T}\right)_p = \frac{\Delta H^\theta}{RT^2} \tag{8.53}$$

This corresponds to a plot of $\log_e K$ vs. T where the gradient at any value of T is equal to $\Delta H^\theta/RT^2$. Because of the $1/T^2$ in the denominator, the plot will be a curve which can have either:

- a negative slope, corresponding to a negative ΔH^θ, where K decreases with temperature;

- a positive slope, corresponding to a positive ΔH^θ, where K increases with temperature;

- have a zero slope corresponding to zero ΔH^θ.

However, it would be much better to have a graph where the slope is independent of temperature, or nearly so.

(ii) What is needed is a plot of $\log_e K$ vs. $1/T$.

$$\left(\frac{\partial \log_e K}{\partial \left(\frac{1}{T}\right)}\right)_p = -T^2\left(\frac{\partial \log_e K}{\partial T}\right)_p = -T^2 \frac{\Delta H^\theta}{RT^2} = -\frac{\Delta H^\theta}{R} \tag{8.54}$$

This corresponds to a graph of $\log_e K$ vs. $1/T$. This graph can be linear or curved, but the important feature is that the gradient of the linear plot, or the gradient of a tangent to the curved plot is equal to $-\Delta H^\theta/R$, and does not involve a term in T.

- A linear graph implies that ΔH^θ is independent of temperature over the temperature range of the graph. This, in turn, implies that ΔC_p^θ is zero over the range.

- A curved plot implies that ΔH^θ is varying with temperature, and so ΔC_p^θ is non-zero over the range of the graph. Hence finding the gradients of tangents to the graph of ΔH^θ vs. T and their sign at a given temperature gives the value and sign of ΔC_p^θ at that temperature.

Worked problem 8.5

Question

The pK_a of benzoic acid at 20°C is 4.213 and at 45°C is 4.241. Calculate ΔH^θ and ΔS^θ. Explain any approximations made in the calculation and state what this implies about ΔC_p^θ.

Answer

Two temperatures only are given and so a calculation of ΔH^θ must be made from the equation:

$$\log_e \frac{K_{T_2}}{K_{T_1}} = -\frac{\Delta H^\theta}{R}\left[\frac{1}{T_2} - \frac{1}{T_1}\right]$$

which assumes that ΔH^θ does not vary with temperature over the temperature range studied.

$$pK_a = 4.213, \text{ i.e. } K_a = 6.12 \times 10^{-5}\,\text{mol dm}^{-3} \text{ at } 293.15\,\text{K},$$
$$pK_a = 4.241, \text{ i.e. } K_a = 5.74 \times 10^{-5}\,\text{mol dm}^{-3} \text{ at } 318.15\,\text{K}$$

$$\log_e \frac{5.74 \times 10^{-5}}{6.12 \times 10^{-5}} = -\frac{\Delta H^\theta}{8.3415 \times 10^{-3}\,\text{kJ mol}^{-1}\,\text{K}^{-1}}\left[\frac{1}{318.15} - \frac{1}{293.15}\right]\frac{1}{\text{K}}$$

$$0.0644 = -\frac{\Delta H^\theta}{8.3145 \times 10^{-3}\,\text{kJ mol}^{-1}\,\text{K}^{-1}}\left[\frac{293.15 - 318.15}{318.15 \times 293.15}\right]\frac{1}{\text{K}}$$

$$= \frac{\Delta H^\theta}{8.3145 \times 10^{-3}\,\text{kJ mol}^{-1}\,\text{K}^{-1}} \times 2.681 \times 10^{-4}\,\text{K}^{-1}$$

$$\therefore \Delta H^\theta = -\frac{0.0644 \times 8.3145 \times 10^{-3}}{2.681 \times 10^{-4}}\,\text{kJ mol}^{-1} = -2.00\,\text{kJ mol}^{-1}$$

Remember: R must have the units kJ mol^{-1} K^{-1} for consistency of units with those of ΔH^θ which has units kJ mol^{-1}.

If ΔH^θ is assumed to be independent of temperature between 20°C and 45°C, then so is ΔS^θ. Hence ΔS^θ can be estimated from data at either temperature.

At 20°C: $K_a = 6.12 \times 10^{-5}\,\text{mol dm}^{-3}$.

$$\Delta G^\theta = -RT \log_e K_a = -8.3145 \times 10^{-3}\,\text{kJ}\,\text{mol}^{-1}\,\text{K} \times 293.15\,\text{K} \times (-9.701)$$

$$= +23.65\,\text{kJ}\,\text{mol}^{-1}$$

$$\Delta G^\theta = \Delta H^\theta - T\Delta S^\theta$$

$$\therefore \Delta S^\theta = \frac{\Delta H^\theta - \Delta G^\theta}{T} = \frac{(-2.0 - 23.65)\,\text{kJ}\,\text{mol}^{-1}}{293.15K}$$

$$= -87.5 \times 10^{-3}\,\text{kJ}\,\text{mol}^{-1}\,\text{K}^{-1} = -87.5\,\text{J}\,\text{mol}^{-1}\,\text{K}^{-1}$$

If ΔH^θ and ΔS^θ are independent of temperature, this would imply that ΔC_p^θ is zero over the range of temperature covered.

Worked problem 8.6

Question

(i) The acid ionisation constant of the trimethylammonium ion in aqueous solution is found to be as follows:

temperature °C	10	20	30	40
pK_a	10.128	9.907	9.692	9.477

Determine ΔH^θ for the acid ionisation graphically.

(ii) More complete data is given below. Use this data to draw a more complete graph, and compare it with the graph obtained in part (i). What further information can be obtained from this extended graph?

temperature °C	0	10	20	30	40	50
pK_a	10.354	10.128	9.907	9.692	9.477	9.270

Answer

(i) This requires use of Equation (8.54) (see Section 8.16.2):

$$\left(\frac{\partial \log_e K_a}{\partial \left(\frac{1}{T}\right)}\right)_p = -\frac{\Delta H^\theta}{R}$$

A plot of $\log_e K_a$ vs. $1/T$, where the temperature is in kelvin, should have a gradient $= -\dfrac{\Delta H^\theta}{R}$ if the graph is linear.

$\dfrac{\text{temperature}}{°C}$	10	20	30	40
pK_a	10.128	9.907	9.692	9.477
$\dfrac{10^{10}K_a}{\text{mol dm}^{-3}}$	0.745	1.240	2.03	3.34
$\log_e K_a$	-23.32	-22.81	-22.32	-21.82
$\dfrac{T}{K}$	283.15	293.15	303.15	313.15
$\dfrac{10^3 \times \left(\dfrac{1}{T}\right)}{K^{-1}}$	3.532	3.411	3.299	3.193

The graph is approximately linear within the experimental error and has gradient $-4.395 \times 10^3 \text{K} = -\Delta H^\theta / R$.

$$\therefore \Delta H^\theta = +4.395 \times 10^3 \text{K} \times 8.3145 \, \text{J mol}^{-1} \text{K}^{-1} = +36.5 \, \text{kJ mol}^{-1}.$$

Watch: units of R.

ΔH^θ is thus approximately independent of temperature over the range 10°C–40°C, and so ΔC_p^θ is approximately zero over the same range. If ΔH^θ is independent of temperature, then ΔS^θ is also independent of temperature over the same range of temperature.

(ii) More complete data are given in the table.

$\dfrac{\text{temperature}}{°C}$	0	10	20	30	40	50
pK_a	10.354	10.128	9.907	9.692	9.477	9.270
$\dfrac{10^{10}K_a}{\text{mol dm}^{-3}}$	0.442	0.745	1.240	2.03	3.34	5.37
$\log_e K_a$	-23.84	-23.32	-22.81	-22.82	-21.82	-21.35
$\dfrac{T}{K}$	273.15	283.15	293.15	303.15	313.15	323.15
$\dfrac{10^3 \times \left(\dfrac{1}{T}\right)}{K^{-1}}$	3.661	3.532	3.411	3.299	3.193	3.095

(a) Using this data, the graph of $\log_e K_a$ vs. $1/T$ is now a curve, indicating that ΔH^θ and ΔS^θ both vary with temperature, leading to a non-zero ΔC_p^θ. These two graphs show how vital it is to carry out experiments over as wide a range of conditions as possible. In this case, if the more limited data is used a completely erroneous interpretation of the results is obtained.

(b) The gradient of tangents to the graph will give $-\Delta H^\theta / R$ at the particular temperature at which the tangent is drawn, hence ΔH^θ can be found at these temperatures. ΔG^θ at these temperatures can be found since $\Delta G^\theta = -RT \log_e K_a$. Since $\Delta G^\theta = \Delta H^\theta - T\Delta S^\theta$, once ΔH^θ and ΔG^θ have been found from the graph of $\log_e K_a$ vs. $1/T$ values of ΔS^θ can be found at these temperatures.

Worked problem 8.7

Question

At 25°C, the equilibrium constant, K_2, for the second stage of the acid ionisation of glycine:

$$^+NH_3CH_2COO^-(aq) + H_2O(l) \rightleftharpoons H_3O^+(aq) + H_2NCH_2COO^-(aq)$$

is 1.66×10^{-10} mol dm^{-3}, and ΔH^θ for this process is $+44.1$ kJ mol^{-1}. Estimate the acid ionisation constant, K_2, at 40°C. State any assumptions which are made. Calculate ΔS^θ at each temperature and comment.

Answer

In this question the values of ΔH^θ and K_2 at one temperature are given. Calculation of K_2 at the second temperature requires use of the equation:

$$\log_e \frac{K_{T_2}}{K_{T_1}} = -\frac{\Delta H^\theta}{R}\left[\frac{1}{T_2} - \frac{1}{T_1}\right]$$

where the temperatures must be in kelvin, and ΔH^θ is assumed to be independent of temperature which implies that ΔC_p^θ is zero.

$$\log_e \frac{K_{at\,313}}{K_{at\,298}} = -\frac{44.1\,\text{kJ mol}^{-1}}{8.3145 \times 10^{-3}\,\text{kJ mol}^{-1}\,\text{K}^{-1}}\left[\frac{1}{313.15} - \frac{1}{298.15}\right]\frac{1}{\text{K}} = +0.852$$

Remember: R must have the units kJ mol^{-1} K^{-1} for consistency of units with those of ΔH^θ which has units kJ mol^{-1}.

$$\therefore \log_e K_{at\,313} = \log_e K_{at\,298} + 0.852 = -22.519 + 0.852 = -21.667$$
$$K_{at\,313} = 3.9 \times 10^{-10}\,\text{mol dm}^{-3}$$

- Calculating values of ΔG^θ and ΔS^θ at 298.15 K:

$$\Delta G^\theta_{298} = -RT\log_e K_{298} = -8.3145 \times 10^{-3}\,\text{kJ mol}^{-1}\,\text{K}^{-1} \times 298.15\text{K}\log_e 1.66 \times 10^{-10}$$
$$= -8.3145 \times 10^{-3}\,\text{kJ mol}^{-1}\,\text{K}^{-1} \times 298.15\,\text{K} \times -22.519$$
$$= +55.82\,\text{kJ mol}^{-1}$$
$$\Delta G^\theta = \Delta H^\theta - T\Delta S^\theta$$
$$\therefore \Delta S^\theta = \frac{\Delta H^\theta - \Delta G^\theta}{T} = \frac{+(44.1 - 55.82)\,\text{kJ mol}^{-1}}{298.15\,\text{K}}$$
$$= \frac{-11.72}{298.15}\,\text{kJ mol}^{-1}\,\text{K}^{-1} = -39.3\,\text{J mol}^{-1}\,\text{K}^{-1}$$

Watch: the conversion between kJ and J when dealing with ΔS and ΔS^θ

- Calculating values of ΔG^θ and ΔS^θ at 313.15 K:

$$\Delta G_{313}^\theta = -8.3145 \times 10^{-3}\,\text{kJ}\,\text{mol}^{-1}\,\text{K}^{-1} \times 313.15\,\text{K} \times (-21.667) = +56.41\,\text{kJ}\,\text{mol}^{-1}$$

$$\Delta S_{313}^\theta = \frac{+(44.1 - 56.41)\,\text{kJ}\,\text{mol}^{-1}}{313.15\,\text{K}} = -39.3\,\text{J}\,\text{mol}^{-1}\,\text{K}^{-1}$$

This calculation indicates that ΔS^θ does not vary with temperature, which is what would be expected for a calculation which assumes that ΔH^θ is independent of temperature. **Remember**: If ΔH^θ is independent of temperature, then so must ΔS^θ.

8.17   The microscopic statistical interpretation of entropy

There are two microscopic interpretations of entropy:

- in terms of statistical mechanics,

- in terms of order and disorder.

8.17.1   The statistical mechanical interpretation of entropy

Entropy can be interpreted in terms of the number of distinct ways in which a given overall situation can occur, or the number of ways of arranging the system:

$$S = k \log_e W \tag{8.55}$$

where W is the number of distinct ways,
and k is the Boltzmann constant.

If W increases, then the number of ways of arranging the system increases, i.e. the randomness increases, and from the equation:

- an increase in randomness is equivalent to an increase in entropy;

- a decrease in randomness is equivalent to a decrease in entropy.

8.17.2   The order/disorder interpretation of entropy

Although there are some crucial distinctions, particularly for gas phase systems where the correlation is only partial, there is for electrolyte solutions a correlation between order/disorder and entropy. It is this correlation which has proved to be very fruitful for the interpretation of ΔS^θ:

- increase in order is equivalent to a decrease in randomness, which is equivalent to a decrease in entropy;

- decrease in order is equivalent to an increase in randomness, which is equivalent to an increase in entropy.

If a system changes from one which is ordered to a less ordered system, then the number of ways of arranging the system must increase, and so the randomness will increase leading to an increase in entropy.

Typical examples for electrolyte solutions are:

- Breaking down of an ionic crystal lattice to an aqueous solution of ions. Here the crystal is highly ordered and will dissolve to give at least two ions which are solvated. Forming more than one solute species will give an increase in entropy, but this may be offset by solvation increasing the order of the solvent, $H_2O(l)$, and causing a decrease in entropy. ΔS^θ for dissolution of a sparingly soluble salt can be positive or negative.

- When ions in the gas phase give ions in aqueous solution there is usually a large decrease in entropy. Here the number of solute species is the same as the number of species in the gas phase. However, the solute species will be solvated leading to an increase in the order of the solvent, and ΔS^θ will be negative.

8.18   Dependence of K on pressure

The standard chemical potential for substances other than gases depends on pressure, and so equilibrium constants for reactions in solution depend on pressure. Often a considerable dependence is found. This is highly relevant to the study of electrolyte solutions.

$$dG = Vdp - SdT \tag{8.14}$$

$$d\Delta G^\theta = \Delta V^\theta dp - \Delta S^\theta dT \tag{8.56}$$

From this it follows that, at constant temperature:

$$\left(\frac{\partial \Delta G^\theta}{\partial p}\right)_T = \Delta V^\theta \tag{8.57}$$

$$\therefore \left(\frac{\partial \log_e K}{\partial p}\right)_T = -\frac{\Delta V^\theta}{RT} \tag{8.58}$$

This corresponds to a plot of $\log_e K$ vs. p at constant T, and there are several situations which can occur.

- If the plot is linear then the gradient of the straight line is equal to $-\Delta V^\theta/RT$. If the gradient of the line is negative, then ΔV^θ is positive; if positive then ΔV^θ is negative. Linear plots are often found unless the solution is inherently highly compressible, or if the pressures are so high that the compressibility of the solution has to be considered.

- If the plot is a curve then the gradients of tangents to the graph at various pressures will give ΔV^θ at these pressures.

Worked problem 8.8

Question

The following table brings together data for propanoic acid and butanoic acid and the results, for 25°C, obtained in Worked Problems 8.5 to 8.7. Comment on and interpret these results.

Acid	$\dfrac{\Delta H^\theta}{\text{kJ mol}^{-1}}$	$\dfrac{\Delta S^\theta}{\text{J mol}^{-1}\,\text{K}^{-1}}$
C_6H_5COOH	-2.00	-87.5
CH_3CH_2COOH	-1.0	-96.2
$CH_3CH_2CH_2COOH$	-2.95	-102
$(CH_3)_3NH^+$	$+36.5$	-65.2
$^+NH_3CH_2COO^-$	$+44.1$	-39.3

Answer

ΔH^θ values:

- The carboxylic acids, C_6H_5COOH, CH_3CH_2COOH and $CH_3CH_2CH_2COOH$:

$$RCOOH(aq) + H_2O(l) \rightleftharpoons RCOO^-(aq) + H_3O^+(aq)$$

Contributions to ΔH^θ would be expected to be large and positive for bond breaking. Formation of ions from molecular species would be expected to give a large negative ΔH^θ due to solvation. The observed values are close to zero suggesting that the contributions from bond breaking and solvation approximately cancel out.

- $(CH_3)_3NH^+(aq)$:

$$(CH_3)_3NH^+(aq) + H_2O(l) \rightleftharpoons H_3O^+(aq) + (CH_3)_3N(aq)$$

Again bond breaking would be expected to have a large positive ΔH^θ, but here one ion is being replaced by another ion. A first thought would be that there would thus be no great change in solvation, so that for ΔH^θ bond breaking would predominate. The results indicate this to be the case, especially in comparison with the carboxylic acids where there is likely to be a large increase in solvation.

- The second stage acid ionisation of glycine:

$$^+NH_3CH_2COO^-(aq) + H_2O(l) \rightleftharpoons H_3O^+(aq) + NH_2CH_2COO^-(aq)$$

In effect this reaction replaces the positive charge on the glycine by the positive charge on $H_3O^+(aq)$, and is charge-wise similar to that of $(CH_3)_3NH^+(aq)$. Similar reasoning would indicate a ΔH^θ of approximately the same value as for $(CH_3)_3NH^+(aq)$, and this is observed.

ΔS^θ values:

- The carboxylic acids:

Bond breaking would be expected to give a positive ΔS^θ, but because ions are formed there will be a considerable increase in solvation. This means an increase in the order of the solvent and would lead to a large negative ΔS^θ which is observed for all the carboxylic acids.

- $(CH_3)_3NH^+(aq)$

A first thought would be that there would be no great change in entropy, but it is likely that $H_3O^+(aq)$ would be more highly solvated than $(CH_3)_3NH^+(aq)$. This would give a decrease in entropy, but one which is not as great as that for the carboxylic acids where ions are produced from molecular species. The values quoted bear this out.

- The second stage acid ionisation of glycine:

As in the case of $(CH_3)_3NH^+(aq)$, this reaction effectively replaces the positive charge on the glycine by the positive charge on $H_3O^+(aq)$, and the value of ΔS^θ could be expected to be similar to that for $(CH_3)_3NH^+(aq)$, but considerably less than for the carboxylic acids. Qualitatively this is observed.

Worked problem 8.9

Question

The acid ionisation constant for propanoic acid has been studied over a wide range of temperatures and has been shown to give a maximum in K_a at 20°C. What information does this furnish about ΔH^θ, ΔS^θ and ΔC_p^θ over this range of temperatures. What information would be required to calculate ΔV^θ?

For propanoic acid the following values are found at 25°C

$$\Delta H^\theta = -1.0 \, \text{kJ mol}^{-1}, \quad \Delta S^\theta = -96.2 \, \text{J mol}^{-1} \, \text{K}^{-1}, \quad \Delta V^\theta = -13.5 \, \text{cm}^3 \, \text{mol}^{-1}$$

Give a brief interpretation of these values.

Answer

$$\Delta G^\theta = -RT \log_e K_a$$
$$\Delta H^\theta - T\Delta S^\theta = -RT \log_e K_a$$
$$\left(\frac{\partial \log_e K_a}{\partial T} \right)_p = +\frac{\Delta H^\theta}{RT^2}$$

Since K_a shows a maximum in value at 20°C, then so will a plot of $\log_e K_a$ vs. T show a maximum at 20°C, i.e. 293.15 K.

- At the maximum i.e. 20°C, $(\partial \log_e K_a / \partial T)_p = 0$ and so ΔH^θ is also zero.

- Below the maximum, i.e. below 20°C, $(\partial \log_e K_a / \partial T)_p$ is positive, and so ΔH^θ is also positive.

- Above the maximum, i.e. above 20°C, $(\partial \log_e K_a / \partial T)_p$ is negative, and so ΔH^θ is also negative.

- The graph of ΔH^θ vs. T has a negative gradient and so ΔC_p^θ is also negative since $\Delta C_p^\theta = (\partial \Delta H^\theta / \partial T)_p$.

- If ΔH^{θ} varies with temperature, then so also will ΔS^{θ}. Since $(\partial \Delta S^{\theta}/\partial T)_p = \Delta C_p^{\theta}/T$ and ΔC_p^{θ} is negative, then ΔS^{θ} decreases with increase in temperature.

If K_a is determined over a range of pressures ΔV^{θ} can be found. Since: $(\partial \Delta G^{\theta}/\partial p)_T = \Delta V^{\theta}$, it follows that $(\partial \log_e K_a/\partial p)_T = -\Delta V^{\theta}/RT$ and so a plot of $\log_e K_a$ vs. p has slope equal to $-\Delta V^{\theta}/RT$.
- Interpretation

$$HA(aq) + H_2O(l) \rightleftharpoons H_3O^+(aq) + A^-(aq)$$

ΔH^{θ}: see Worked Problem 8.8 where bond breaking and solvation appear to balance each other.

ΔS^{θ}: see Worked Problem 8.8 where the conclusion was that solvation more than compensates for bond breaking.

ΔV^{θ}: The interpretation is very similar to that for ΔS^{θ}. Bond breaking is expected to give a positive ΔV^{θ}. Solvation results in water going from the tetrahedral array of water molecules in the bulk solvent to a more closely packed arrangement around the ions and this corresponds to a decrease in the overall volume occupied by the water molecules. The observed ΔV^{θ} is negative and so solvation is the dominant factor.

Worked problem 8.10

Question

Interpret the magnitudes of ΔS^{θ} and ΔV^{θ} for the following reactions:

(i) $\Delta S^{\theta}/\mathrm{J\,mol^{-1}\,K^{-1}}$

(a) $Ca^{2+}(aq) + SO_4^{2-}(aq) \longrightarrow Ca^{2+}SO_4^{2-}(aq)$ 67

(b) $CH_3COOH(aq) + H_2O(l) \longrightarrow CH_3COO^-(aq) + H_3O^+(aq)$ -92

(c) $C_6H_5NH_3^+(aq) + H_2O(l) \longrightarrow C_6H_5NH_2(aq) + H_3O^+(aq)$ 11.7

(d) $CaCO_3(calcite) \longrightarrow Ca^{2+}(aq) + CO_3^{2-}(aq)$ -202.9

(e) $Ni^{2+}(aq) + C_2O_4^{2-}(aq) \longrightarrow NiC_2O_4(aq)$ $+101.3$

(ii) $\Delta V^{\theta}/\mathrm{cm^3\,mol^{-1}}$

(a) $CH_3COOH(aq) + H_2O(l) \longrightarrow CH_3COO^-(aq) + H_3O^+(aq)$ -11.3

(b) $Cr(H_2O)_6^{3+}(aq) + H_2O(l) \longrightarrow H_3O^+(aq) + Cr(H_2O)_5OH^{2+}(aq)$ -3.8

(c) $C_6H_5NH_3^+(aq) + H_2O(l) \longrightarrow C_6H_5NH_2(aq) + H_3O^+(aq)$ $+4.5$

(d) $Ni^{2+}(aq) + SO_4^{2-}(aq) \longrightarrow Ni^{2+}SO_4^{2-}(aq)$ $+8.6$

(e) $SrSO_4(s) \longrightarrow Sr^{2+}(aq) + SO_4^{2-}(aq)$ -52

Answer

Only a relatively qualitative interpretation can be given in terms of the obvious gross changes occurring on reaction, e.g. consideration of change in the number of species, change in the charges and consequent changes in solvation.

(i) ΔS^{θ} **values**

(a) Formation of an ion pair from two ions corresponds to a decrease in the number of species which should result in an increase in order corresponding to a decrease in entropy. There will also be a change in solvation and the ion pair (a neutral species, but perhaps a dipole) would be expected to be less solvated than the individual ions. This would give an increase in disorder created by release of water molecules of solvation from the free ions. Observation corresponds to an increase in S and a decrease in order indicating that the change in the solvation pattern is the dominant effect.

(b) Formation of ions which are solvated gives an increase in order of the solvent and more than compensates for the decrease in order on forming two ions from one solute molecule. Therefore, overall a decrease in entropy.

(c) Here there is a small increase in entropy. A small increase in solvation and a small decrease in entropy is expected because $H_3O^+(aq)$, being smaller, will be more highly solvated than $C_6H_5NH_3^+(aq)$. This decrease is more than compensated by the increase in entropy resulting from one solute species going to two.

(d) Here two ions are formed from the solid lattice which is highly ordered and this would give an increase in entropy. However, the divalent ions will be considerably hydrated leading to a large increase in order of the solvent. The data show that solvation is predominant.

(e) In this reaction two species form one species which would result in an increase in order and a decrease in entropy. But reaction involves two charged species which will be hydrated giving an overall uncharged species which will be considerably less hydrated, resulting in a large increase in disorder for the solvent, and hence a large increase in entropy. The data shows that this latter process predominates.

(ii) ΔV^{θ} **values**

(a) There is a decrease in volume of the solvent because of solvation of the ionic products, and this dominates over the increase in volume resulting from one solute molecule forming two ions; see note at end of question.

(b) This reaction forms two ions which will be solvated, but they are formed from one highly charged, highly solvated ion. Hence there is a much smaller change in volume of the solvent compared with reaction (a). Also one ion giving two ions will give a contribution corresponding to an increase in volume.

(c) The increase in volume here reflects the increase in the number of solute species which is not compensated by the increase in solvation on forming $H_3O^+(aq)$ from $C_6H_5NH_3^+(aq)$.

(d) Formation of an ion pair of overall zero charge from two charged $(+2)$, (-2) ions will release water of solvation which will give a large increase in volume of the solvent which is not compensated by the decrease in volume due to the decrease in number of solute species.

(e) The solid ionic lattice gives solvated ions, giving a large decrease in volume of the solvent compared with the increase in volume expected for formation of two species. Solvation is dominant.

A close look at the values for ΔS^θ and ΔV^θ for given reactions shows that, in the main, there is a close correlation between them. This is not unexpected since the physical process involved is basically the same. However, there are cases where the correlation does not hold e.g. for $(CH_3)_3NH^+(aq)$ the signs of ΔS^θ and ΔV^θ are different: $\Delta S^\theta = -65\,K\,mol^{-1}\,K^{-1}$ and $\Delta V^\theta = +6.0\,cm^3\,mol^{-1}$.
Note well: in this question the argument deals with changes in the number of **solute** particles; the solvent molecule must therefore **not** be included.

8.19 Dependence of ΔG^θ on temperature

$$\Delta G^\theta = \Delta H^\theta - T\Delta S^\theta \tag{8.59}$$

The dependence of ΔG^θ on temperature is complicated by the possible dependence of both ΔH^θ and ΔS^θ on temperature. Two situations are possible:

- ΔH^θ and ΔS^θ are independent of temperature, in which case a graph of ΔG^θ vs. T would be linear with intercept $= \Delta H^\theta$ and gradient equal to $-\Delta S^\theta$. It is very unusual for this assumption to be valid.

- ΔH^θ and ΔS^θ depend on temperature, in which case the plot is now curved. The gradient of a tangent to the curve at any given T corresponds to $(\partial \Delta G^\theta / \partial T)_p$, where $(\partial \Delta G^\theta / \partial T)_p = -\Delta S^\theta$. ΔS^θ values at various temperatures can thus be found from plots of ΔG^θ vs. T. ΔH^θ values at various temperatures can then be calculated from the corresponding ΔG^θ and ΔS^θ. The standard emf of a cell is closely related to ΔG^θ and so these relations can often be used to obtain ΔH^θ and ΔS^θ from E^θ and its temperature dependence.

8.20 Dependence of ΔS^θ on temperature

For most reactions ΔS^θ, like ΔH^θ, depends on temperature, and these dependencies reflect the non-zero ΔC_p^θ found for such reactions.

For reversible processes at constant p:

$$dS = \frac{q}{T} = \frac{dH}{T} = \frac{C_p dT}{T} \tag{8.60}$$

$$\text{giving} \left(\frac{\partial \Delta S^\theta}{\partial T}\right)_p = \frac{\Delta C_p^\theta}{T} \tag{8.61}$$

This corresponds to a graph of ΔS^θ vs. T where the gradient of a tangent to the graph at any temperature equals $\Delta C_p^\theta/T$ giving ΔC_p^θ at that temperature. For many reactions which can be carried out easily under normal laboratory conditions, a direct calorimetric determination of ΔH^θ over a range of temperature will be preferable. However, for reactions carried out in an electrochemical cell, ΔC_p^θ is easily found at a given temperature from sufficiently accurate measurements of the dependence of either ΔH^θ or ΔS^θ on temperature.

Worked problem 8.11

Question

The standard emf of the cell:

$$Zn(s)|Zn^{2+}(aq) \mathbin{\vdots} Fe^{2+}(aq), Fe^{3+}(aq)|Pt(s)$$

is 1.534 V at 25°C and 1.576 V at 65°C. Find K, ΔH^θ, ΔG^θ, and ΔS^θ for this reaction at 25°C. The overall reaction is:

$$Zn(s) + 2Fe^{3+}(aq) \rightleftharpoons Zn^{2+}(aq) + 2Fe^{2+}(aq)$$

and $z = 2$, so that for the reaction as written 2 faraday is the charge passed.

Answer

$$\Delta G^\theta = -zFE^\theta$$

At 298.15K : $\Delta G^\theta = -2 \times 96485 \times 1.534$ coulomb volt mol^{-1} = -296.0 kJ mol^{-1}

At 338.15K : $\Delta G^\theta = -2 \times 96485 \times 1.576$ coulomb volt mol^{-1} = -304.1 kJ mol^{-1}

Note : coulomb volt = joule

$$\Delta G^\theta = -RT \log_e K \quad \therefore \log_e K = -\frac{\Delta G^\theta}{RT}$$

At 298.15 K:

$$\log_e K = -\frac{-296.0 \text{ kJ mol}^{-1}}{8.3145 \times 10^{-3} \text{ kJ mol}^{-1} \text{ K}^{-1} \times 298.15 \text{ K}} = 119.4$$

$$\therefore K = 7.3 \times 10^{51} \text{ mol dm}^{-3}$$

$$\Delta S^\theta = -\left(\frac{\partial \Delta G^\theta}{\partial T}\right)_p$$

$$= \frac{\{-304.1 - (-296.0)\} \text{ kJ mol}^{-1}}{40 \text{ K}} = -\frac{8.1}{40} \text{ kJ mol}^{-1} \text{ K}^{-1}$$

$$= -0.2025 \text{ kJ K}^{-1} \text{ mol}^{-1} = -202.5 \text{ J K}^{-1} \text{ mol}^{-1}$$

$$\Delta G^\theta = \Delta H^\theta - T\Delta S^\theta$$

$$\therefore \Delta H^\theta = \Delta G^\theta + T\Delta S^\theta = -296.0 \text{ kJ mol}^{-1} - 298.15 \text{ K} \times 0.2025 \text{ kJ mol}^{-1} \text{ K}^{-1}$$

$$= -356.4 \text{ kJ mol}^{-1}$$

8.21 The non-ideal case

So far the thermodynamic analysis relates only to the ideal solution, or to reaction carried out under ideal conditions. This is only approximated to in very dilute solutions, and most solutions and reactions are studied under non-ideal conditions.

All the equilibrium constants discussed earlier have been quoted in terms of concentrations. There are two reasons for developing equilibria relations in terms of concentrations:

- many actual experimental measurements are made in terms of concentrations, not activities, and the equilibrium constants are therefore concentration constants;

- it is generally simpler to give an elementary account of equilibrium constants and their related thermodynamic quantities, and then to introduce corrections for non-ideality afterwards. These corrections generally involve extrapolation of the concentration equilibrium quotient over a range of ionic strengths, rather than correcting each concentration term individually.

8.21.1 Non-ideality in electrolyte solutions

Non-ideality in electrolyte solutions is primarily due to electrostatic interactions between the ions and is taken care of by the activity coefficient, γ_i, for any species, i, defined by:

$$a_i = \gamma_i c_i \qquad \text{where} \qquad \gamma_i \to 1 \quad \text{as} \quad c_i \to 0.$$

The Debye-Hückel theory, developed in Chapter 10, gives theoretical expressions for calculating the activity coefficient of any species in solution.

Non-ideality in electrolyte solutions manifests itself in experimental studies in the following ways:

- concentration equilibrium constants are found to vary systematically with concentration;

- molar conductivities for strong electrolytes vary with concentration;

- emf expressions are not satisfactory when concentrations appear in the log[ ] terms.

In this chapter ways of modifying the theoretical expressions for the equilibrium constant for a reaction in solution to take account of non-ideality will be explained, while emf and conductance studies will be described in Chapters 9 and 11 respectively.

8.21.2 The ionic strength and non-ideality

The ionic strength takes account of both the ionic concentration and the ionic charges. For any given concentration the ionic strength will be higher the higher the charges on the ions of the electrolyte, and the larger the ionic strength the greater is the non-ideality.

The ionic strength is a special type of summation over all charged species in solution, and is defined as:

$$I = 1/2 \sum_i c_i z_i^2 \tag{8.62}$$

where c_i is the concentration of each type, i, of ion in solution, and
z_i is the algebraic charge on each type, i, of ion in solution, i.e. it includes the sign of the charge. The Debye-Hückel theory (see Sections 10.7, 10.10.1 and 10.10.2) allows a calculation of the activity coefficient, γ_i, for any ion from the known ionic strength.

- The Debye-Hückel limiting law: $\log_{10} \gamma_i = -Az_i^2 \sqrt{I}$ is valid in very dilute solutions corresponding to ionic strengths $< 1 \times 10^{-3} \, \text{mol dm}^{-3}$, and $A = 0.510 \, \text{mol}^{-1/2} \, \text{dm}^{3/2}$ for water at 25°C.

- The Debye-Hückel equation: $\log_{10} \gamma_i = -Az_i^2 \sqrt{I}/(1 + \sqrt{I})$ holds for moderate concentrations corresponding to ionic strengths in the range 1×10^{-3} to $0.1 \, \text{mol dm}^{-3}$.

- The Debye-Hückel extended equation: $\log_{10} \gamma_i = -Az_i^2 \sqrt{I}/(1 + \sqrt{I}) + bI$ is used for ionic strengths $>0.1 \, \text{mol dm}^{-3}$ where b is an empirical factor. ·

- Deviations from each of these equations are observed, and are greater the higher the charge on the ion and the greater the ionic strength.

- The activity coefficient for any uncharged species present in an aqueous electrolyte solution is, to a first approximation, taken to be unity since Az_i^2 will be zero.

Worked problem 8.12

Question

(i)   Calculate the ionic strength, I, for $0.05 \, \text{mol dm}^{-3}$ solutions of KCl(aq), MgSO₄(aq) and LaFe(CN)₆(aq).

(ii) Using the equation, $\log_{10} \gamma_i = -Az_i^2 \sqrt{I}/(1 + \sqrt{I})$ where $A = 0.510 \, \text{mol}^{-1/2} \, \text{dm}^{3/2}$ for water at 25°C, calculate the activity coefficient for the K⁺(aq), Mg²⁺(aq) and La³⁺(aq) ions for solutions of ionic strength equal to $0.1 \, \text{mol dm}^{-3}$. Comment on these values and give a brief explanation for the difference in the values.

(iii) Now calculate the activity coefficient for the K⁺(aq), Mg²⁺(aq) and La³⁺(aq) ions in the **original solutions** of concentration equal to $0.05 \, \text{mol dm}^{-3}$, and compare the results with those from part (ii).

Answer

(i) $I = 1/2 \sum_i c_i z_i^2$

- KCl(aq) solution: $I = 1/2[0.05 \times 1^2 + 0.05 \times 1^2] = 0.05\,\text{mol dm}^{-3}$

- MgSO$_4$(aq) solution: $I = 1/2[0.05 \times 2^2 + 0.05 \times 2^2] = 0.20\,\text{mol dm}^{-3}$

- LaFe(CN)$_6$(aq) solution: $I = 1/2[0.05 \times 3^2 + 0.05 \times 3^2] = 0.45\,\text{mol dm}^{-3}$

This calculation shows the dramatic increase in the ionic strength which occurs as the charge on the ions increases, **despite** all the solutions being of the **same** concentration. Non-ideality will be expected to increase in line with increasing ionic strength.

(ii) $\log_{10}\gamma_i = -Az_i^2 \dfrac{\sqrt{I}}{1+\sqrt{I}}$

$$\log_{10}\gamma_{K^+} = -0.510\,\text{mol}^{-1/2}\,\text{dm}^{3/2} \times 1^2 \times \frac{\sqrt{0.10}}{1+\sqrt{0.10}}\,\text{mol}^{1/2}\,\text{dm}^{-3/2}$$
$$= -0.510 \times 1 \times 0.240 = -0.1225$$
$$\therefore \gamma_{K^+} = 0.754$$

$$\log_{10}\gamma_{Mg^{2+}} = -0.510\,\text{mol}^{-1/2}\,\text{dm}^{3/2} \times 2^2 \times \frac{\sqrt{0.10}}{1+\sqrt{0.10}}\,\text{mol}^{1/2}\,\text{dm}^{-3/2}$$
$$= -0.510 \times 4 \times 0.240 = -0.490$$
$$\therefore \gamma_{Mg^{2+}} = 0.323$$

$$\log_{10}\gamma_{La^{3+}} = -0.510\,\text{mol}^{-1/2}\,\text{dm}^{3/2} \times 3^2 \times \frac{\sqrt{0.10}}{1+\sqrt{0.10}}\,\text{mol}^{1/2}\,\text{dm}^{-3/2}$$
$$= -0.510 \times 9 \times 0.240 = -1.102$$
$$\therefore \gamma_{La^{3+}} = 0.079$$

These are activity coefficients calculated for the **same** value of the ionic strength. The decreasing values

$$\gamma_{K^+} > \gamma_{Mg^{2+}} > \gamma_{La^{3+}}$$

reflect the increasing non-ideality as a result of increasing ionic interactions which are a consequence of increasing charge on the ion.

(iii) $\log_{10}\gamma_i = -Az_i^2 \dfrac{\sqrt{I}}{1+\sqrt{I}}$

- 0.05 mol dm^{-3} KCl has an ionic strength of 0.05 mol dm^{-3}.

$$\log_{10}\gamma_{K^+} = -0.510\,\text{mol}^{-1/2}\,\text{dm}^{3/2} \times 1^2 \times \frac{\sqrt{0.05}}{1+\sqrt{0.05}}\,\text{mol}^{1/2}\,\text{dm}^{-3/2}$$
$$= -0.510 \times 1 \times 0.1827 = -0.093$$
$$\therefore \gamma_{K^+} = 0.807$$

- 0.05 mol dm^{-3} MgSO$_4$ has an ionic strength of 0.20 mol dm^{-3}.

$$\log_{10}\gamma_{Mg^{2+}} = -0.510\,\text{mol}^{-1/2}\,\text{dm}^{3/2} \times 2^2 \times \frac{\sqrt{0.20}}{1 + \sqrt{0.20}}\,\text{mol}^{1/2}\,\text{dm}^{-3/2}$$

$$= -0.510 \times 4 \times 0.309 = -0.630$$

$$\therefore \gamma_{Mg^{2+}} = 0.234$$

- 0.05 mol dm^{-3} LaFe(CN)$_6$ has an ionic strength of 0.45 mol dm^{-3}.

$$\log_{10}\gamma_{La^{3+}} = -0.510\,\text{mol}^{-1/2}\,\text{dm}^{3/2} \times 3^2 \times \frac{\sqrt{0.45}}{1 + \sqrt{0.45}}\,\text{mol}^{1/2}\,\text{dm}^{-3/2}$$

$$= -0.510 \times 9 \times 0.401 = -1.843$$

$$\gamma_{La^{3+}} = 0.014$$

Gathering all the calculations together gives the following table:

γ_{K^+}	0.754	0.807
$I/\text{mol dm}^{-3}$	0.10	0.05
$\gamma_{Mg^{2+}}$	0.323	0.234
$I/\text{mol dm}^{-3}$	0.10	0.20
$\gamma_{La^{3+}}$	0.079	0.014
$I/\text{mol dm}^{-3}$	0.10	0.45

These results shows that increasing both the ionic strength and the charge results in a dramatic decrease in the activity coefficient, with the decrease being greatest for the most highly charged ion. This reflects the increasing non-ideality as both the charge and the ionic strength increase.

For the ideal solution the activity coefficient will be unity. Comparing this with the results obtained shows how necessary it is to take account of non-ideality in electrolyte solutions.

8.22 Chemical potentials and mean activity coefficients

The Debye-Hückel theory gives a calculation of the activity coefficients of individual ions. However, although the individual concentrations of the ions of an electrolyte solution can be measured, experiment cannot measure the individual activity coefficients. It does, however, furnish a sort of average value of the activity coefficient, called the **mean activity coefficient**, $\gamma_{\pm}$, for the **electrolyte as a whole**. The term 'mean' is not used in its common sense of an average quantity, but is used in a different sense which reflects the number of ions which result from each given formula. Such mean activity coefficients are related to the individual activity coefficients in a manner dictated by the stoichiometry of the electrolyte.

Demonstrating the thermodynamic arguments leading up to such relations requires use of the chemical potential for the solid electrolyte and the aqueous ions.

In Section 8.11 the chemical potential was defined as $\mu_i = (\partial G / \partial n_i)_{T,p,\text{all other } n}$ Consider the case of the dissolution of NaCl(s) in water:

$$\text{NaCl(s)} \rightleftharpoons \text{Na}^+(\text{aq}) + \text{Cl}^-(\text{aq})$$

For constant temperature and pressure, and using the arguments developed in Section 8.12. and 8.14:

$$\Delta G = \mu_{\text{Na}^+(\text{aq})} + \mu_{\text{Cl}^-(\text{aq})} - \mu_{\text{NaCl(s)}} \tag{8.63}$$

$$= \mu^\theta_{\text{Na}^+(\text{aq})} + \mu^\theta_{\text{Cl}^-(\text{aq})} - \mu^\theta_{\text{NaCl(s)}} + RT \log_e a_{\text{Na}^+} + RT \log_e a_{\text{Cl}^-} \tag{8.64}$$

where $\mu^\theta_{\text{Na}^+(\text{aq})}$, $\mu^\theta_{\text{Cl}^-(\text{aq})}$ and $\mu^\theta_{\text{NaCl(s)}}$ are constants dependent on T and slightly on p,

$$\therefore \Delta G = \Delta G^\theta + RT \log_e a_{\text{Na}^+} + RT \log_e a_{\text{Cl}^-} \tag{8.65}$$

where ΔG^θ is a constant dependent on T and slightly on p.

$$\therefore \Delta G = \Delta G^\theta + RT \log_e [\text{Na}^+] + RT \log_e [\text{Cl}^-] + RT \log_e \gamma_{\text{Na}^+} + RT \log_e \gamma_{\text{Cl}^-} \tag{8.66}$$

$$\therefore \Delta G = \Delta G^\theta + RT \log_e [\text{Na}^+][\text{Cl}^-] + RT \log_e \gamma_{\text{Na}^+} \gamma_{\text{Cl}^-} \tag{8.67}$$

If the concentration of the aqueous solution is c, then $[\text{Na}^+] = [\text{Cl}^-] = c$ and:

$$\Delta G = \Delta G^\theta + RT \log_e c^2 + RT \log_e \gamma_{\text{Na}^+} \gamma_{\text{Cl}^-} \tag{8.68}$$

$$= \Delta G^\theta + 2RT \log_e c + RT \log_e \gamma_{\text{Na}^+} \gamma_{\text{Cl}^-} \tag{8.69}$$

$$= \Delta G^\theta + 2RT \log_e c + 2RT \log_e (\gamma_{\text{Na}^+} \gamma_{\text{Cl}^-})^{1/2} \tag{8.70}$$

The individual ion activity coefficients cannot be measured, but an analogous equation to Equation (8.70) can be written involving the mean activity coefficient $\gamma_\pm$ for the electrolyte as a whole. This is a quantity which can be measured (see emf measurements, Section 9.20.5, Worked Problems 9.28 and 9.29 and Sections 10.6.15, 10.7, 10.10 and 10.11).

$$\Delta G = \Delta G^\theta + 2RT \log_e c + 2RT \log_e \gamma_\pm \tag{8.71}$$

Comparing this equation with:

$$\Delta G = \Delta G^\theta + 2RT \log_e c + 2RT \log_e (\gamma_{\text{Na}^+} \gamma_{\text{Cl}^-})^{1/2} \tag{8.70}$$

it follows that, since they are both describing the same quantity, ΔG for the dissolution of NaCl in water:

$$2RT \log_e \gamma_\pm = 2RT \log_e (\gamma_{\text{Na}^+} \gamma_{\text{Cl}^-})^{1/2} \tag{8.72}$$

$$\therefore \gamma_\pm = (\gamma_{\text{Na}^+} \gamma_{\text{Cl}^-})^{1/2} \tag{8.73}$$

The Debye-Hückel equation can be written in terms of a mean activity coefficient (see Section 10.6.15) $\log_{10} \gamma_\pm = -A|z_1 z_2| \sqrt{I}/(1 + \sqrt{I})$ where $|z_1 z_2|$ reads 'mod' $z_1 z_2$ and means that only the charges appear and **not** the sign. The limiting law can be written similarly.

Worked Problem 8.13

Question

(i) Deduce the relation between the mean ionic activity coefficient for $CaCl_2(aq)$ and the individual ionic activity coefficients for $Ca^{2+}(aq)$ and $Cl^-(aq)$.

(ii) Do likewise for $Fe_2(SO_4)_3(aq)$.

Answer

(i) $CaCl_2(s) \rightleftharpoons Ca^{2+}(aq) + 2Cl^-(aq)$

$$\Delta G = \mu_{Ca^{2+}(aq)} + 2\mu_{Cl^-(aq)} - \mu_{CaCl_2(s)}$$

$$\Delta G = \mu^\theta_{Ca^{2+}(aq)} + 2\mu^\theta_{Cl^-(aq)} - \mu^\theta_{CaCl_2(s)} + RT \log_e a_{Ca^{2+}} + 2RT \log_e a_{Cl^-}$$

where $\mu^\theta_{Ca^{2+}(aq)}$, $\mu^\theta_{Cl^-(aq)}$ and $\mu^\theta_{CaCl_2(s)}$ are constants dependent on T and slightly on p,

$$\Delta G = \Delta G^\theta + RT \log_e a_{Ca^{2+}} + 2RT \log_e a_{Cl^-}$$

where ΔG^θ is a constant dependent on T and slightly on p.

$$\Delta G = \Delta G^\theta + RT \log_e [Ca^{2+}] + 2RT \log_e [Cl^-] + RT \log_e \gamma_{Ca^{2+}} + 2RT \log_e \gamma_{Cl^-}$$

$$\therefore \Delta G = \Delta G^\theta + RT \log_e [Ca^{2+}][Cl^-]^2 + RT \log_e \gamma_{Ca^{2+}} \gamma^2_{Cl^-}$$

If the concentration of the aqueous solution is c, then $[Cl^-] = 2[Ca^{2+}] = 2c$ and:

$$\Delta G = \Delta G^\theta + RT \log_e (4c^3) + RT \log_e \gamma_{Ca^{2+}} \gamma^2_{Cl^-}$$
$$= \Delta G^\theta + RT \log_e 4 + 3RT \log_e c + 3RT \log_e (\gamma_{Ca^{2+}} \gamma^2_{Cl^-})^{1/3}$$

The analogous equation involving the mean activity coefficient $\gamma_\pm$ for the electrolyte as a whole is:

$$\Delta G = \Delta G^\theta + RT \log_e 4 + 3RT \log_e c + 3RT \log_e \gamma_\pm$$

Comparing these equations gives:

$$3RT \log_e \gamma_\pm = 3RT \log_e (\gamma_{Ca^{2+}} \gamma^2_{Cl^-})^{1/3}$$

$$\therefore \gamma_\pm = (\gamma_{Ca^{2+}} \gamma^2_{Cl^-})^{1/3}$$

(ii) $Fe_2(SO_4)_3(s) \rightleftharpoons 2Fe^{3+}(aq) + 3SO_4^{2-}(aq)$

Following the argument given previously,

$$\Delta G = 2\mu_{Fe^{3+}(aq)} + 3\mu_{SO_4^{2-}}(aq) - \mu_{Fe_2(SO_4)_3(s)}$$

$$\Delta G = 2\mu^{\theta}_{Fe^{3+}(aq)} + 3\mu^{\theta}_{SO_4^{2-}(aq)} - \mu^{\theta}_{Fe_2(SO_4)_3(s)} + 2RT \log_e a_{Fe^{3+}} + 3RT \log_e a_{SO_4^{2-}}$$

$$\Delta G = \Delta G^{\theta} + 2RT \log_e a_{Fe^{3+}} + 3RT \log_e a_{SO_4^{2-}}$$

$$\Delta G = \Delta G^{\theta} + 2RT \log_e [Fe^{3+}] + 3RT \log_e [SO_4^{2-}] + 2RT \log_e \gamma_{Fe^{3+}} + 3RT \log_e \gamma_{SO_4^{2-}}$$

$$\therefore \Delta G = \Delta G^{\theta} + RT \log_e [Fe^{3+}]^2 [SO_4^{2-}]^3 + RT \log_e \gamma^2_{Fe^{3+}} \gamma^3_{SO_4^{2-}}$$

where ΔG^{θ} is a constant dependent on T and slightly on p.

If the concentration of the aqueous solution is c, then $[Fe^{3+}] = 2c$ and $[SO_4^{2-}] = 3c$ and:

$$\Delta G = \Delta G^{\theta} + RT \log_e(108c^5) + RT \log_e \gamma^2_{Fe^{3+}} \gamma^3_{SO_4^{2-}}$$

$$= \Delta G^{\theta} + RT \log_e 108 + 5RT \log_e c + 5RT \log_e \{\gamma^2_{Fe^{3+}} \gamma^3_{SO_4^{2-}}\}^{1/5}$$

The analogous equation involving the mean activity coefficient $\gamma_{\pm}$ for the electrolyte as a whole is:

$$\Delta G = \Delta G^{\theta} + RT \log_e 108 + 5RT \log_e c + 5RT \log_e \gamma_{\pm}$$

Comparing these equations gives:

$$5RT \log_e \gamma_{\pm} = 5RT \log_e \{\gamma^2_{Fe^{3+}} \gamma^3_{SO_4^{2-}}\}^{1/5}$$

$$\therefore \gamma_{\pm} = \{\gamma^2_{Fe^{3+}} \gamma^3_{SO_4^{2-}}\}^{1/5}$$

Worked problem 8.14

Question

(i)   Calculate the individual activity coefficients for a 0.100 mol dm^{-3} solution of $MgCl_2(aq)$.

(ii) Use these to calculate the mean activity coefficient for the electrolyte.

(iii) Use the Debye-Hückel expression for the mean activity coefficient to calculate the mean activity coefficient.

(iv) Comment on this result.

Answer

(i)   The Debye-Hückel expression for the individual activity coefficient, γ_i is:

$$\log_{10}\gamma_i = -\frac{Az_i^2\sqrt{I}}{1+\sqrt{I}}$$

$$I = 1/2\sum_i c_i z_i^2 = 1/2\{0.100 \times (2)^2 + 0.200 \times (1)^2\} = 0.300\,\text{mol dm}^{-3}$$

$$\sqrt{I} = 0.548\,\text{mol}^{1/2}\,\text{dm}^{-3/2}$$

$$\therefore \log_{10}\gamma_{\text{Mg}^{2+}} = -0.510\,\text{mol}^{-1/2}\,\text{dm}^{3/2} \times 4 \times \frac{0.548}{1+0.548}\,\text{mol}^{1/2}\,\text{dm}^{-3/2} = -0.722$$

$$\therefore \gamma_{\text{Mg}^{2+}} = 0.190$$

$$\log_{10}\gamma_{\text{Cl}^-} = -0.510\,\text{mol}^{1/2}\,\text{dm}^{-3/2} \times 1 \times \frac{0.548}{1+0.548}\,\text{mol}^{1/2}\,\text{dm}^{-3/2} = -0.181$$

$$\therefore \gamma_{\text{Cl}^-} = 0.660$$

(ii)   The non-ideal part of ΔG for dissolution of $MgCl_2$ is $RT\log_e\gamma_{\text{Mg}^{2+}}\gamma_{\text{Cl}^-}^2 = 3RT\log_e(\gamma_{\text{Mg}^{2+}}\gamma_{\text{Cl}^-}^2)^{1/3}$. The non-ideal part of ΔG can be written in terms of the mean activity coefficient of the electrolyte as a whole, $\gamma_\pm$, as $3RT\log_e\gamma_\pm$.

$$\therefore \gamma_\pm = (\gamma_{\text{Mg}^{2+}}\gamma_{\text{Cl}^-}^2)^{1/3} = \{0.190 \times (0.660)^2\}^{1/3} = (0.0828)^{1/3} = 0.436$$

(iii) The Debye-Hückel equation for the mean activity coefficient of an electrolyte, $\gamma_\pm$:

$$\log_{10}\gamma_\pm = -\frac{A|z_1 z_2|\sqrt{I}}{1+\sqrt{I}}$$

$$\log_{10}\gamma_\pm = -0.510\,\text{mol}^{-1/2}\,\text{dm}^{3/2} \times 2 \times 1 \times \frac{0.548}{1+0.548}\,\text{mol}^{1/2}\,\text{dm}^{-3/2} = -0.361$$

$$\therefore \gamma_\pm = 0.436$$

(iv)   As expected, calculation of $\gamma_\pm$ by the two methods gives the same value for $\gamma_\pm$, though that using the Debye-Hückel equation for $\gamma_\pm$ is the simpler.

8.23   A generalisation

Section 8.22 and Worked Problems 8.13 and 8.14 derive the expressions for the mean activity coefficient explicitly for specific electrolytes. This is preferable to simply fitting into a generalised expression which is often quoted for unsymmetrical electrolytes as it makes quite clear exactly what the thermodynamic argument actually is. The generalised expressions are:

- for symmetrical electrolytes: $AB(s) \rightarrow A^{x+}(aq) + B^{x-}(aq)$

$$\gamma_\pm = \{\gamma_{A^{x+}}\gamma_{B^{x-}}\}^{1/2} \tag{8.74}$$

- for unsymmetrical electrolytes: $A_xB_y(s) \rightarrow xA^{y+}(aq) + yB^{x-}(aq)$

$$\gamma_\pm = \{(\gamma_{A^{y+}})^x(\gamma_{B^{x-}})^y\}^{1/(x+y)} \tag{8.75}$$

Worked problem 8.15

Question

(i) $Hg_2Cl_2(s)$ is sparingly soluble in water at 25°C, solubility $= 6.5 \times 10^{-7} mol \, dm^{-3}$. Calculate the solubility product and explain why it is not necessary to correct for non-ideality.

(ii) Calculate the solubility in $0.100 \, mol \, dm^{-3}$ KNO_3 solution.

(iii) Compare the solubilities in the two solutions.

Answer

(i) $Hg_2Cl_2(s) \rightleftharpoons Hg_2^{2+}(aq) + 2\,Cl^-(aq)$

$$K_s = ([Hg_2^{2+}][Cl^-]^2)_{eq}$$

If the solubility in water is s, then $K_s = s \times (2s)^2 = 4s^3 = 4 \times (6.5 \times 10^{-7})^3 \, mol^3 \, dm^{-9} = 1.1 \times 10^{-18} \, mol^3 \, dm^{-9}$.

The solubility of $Hg_2Cl_2(s)$ is so low that the saturated solution will be so dilute as to be ideal. Hence corrections for non-ideality are not needed.

(ii) In $0.100 \, mol \, dm^{-3}$ $KNO_3(aq)$ correction for non-ideality will have to be made.

$$I = 0.100 \, mol \, dm^{-3} \qquad \therefore \sqrt{I} = 0.3162 \, mol^{1/2} \, dm^{-3/2}$$

$$K_s = a_{Hg_2^{2+}} a_{Cl^-}^2 = [Hg_2^{2+}]\gamma_{Hg_2^{2+}}[Cl^-]^2\gamma_{Cl^-}^2 = 4(s')^3 \gamma_{Hg_2^{2+}}\gamma_{Cl^-}^2 = 4(s')^3\gamma_\pm^3$$

since $\gamma_\pm = \left(\gamma_{Hg_2^{2+}}\gamma_{Cl^-}^2\right)^{1/3}$, and where s' is the solubility in the $KNO_3(aq)$ solution. The calculation can proceed in two ways:

- the individual activity coefficients can be calculated from the Debye-Hückel equation, and substituted direct into the expression for K_s in terms of individual activity coefficients;

- the mean activity coefficient for the electrolyte as a whole can be calculated from the Debye-Hückel expression for a mean activity coefficient.

In terms of the individual activity coefficient, γ_i: $\log_{10}\gamma_i = -\dfrac{Az_i^2\sqrt{I}}{1+\sqrt{I}}$

$$\log\gamma_{Hg_2^{2+}} = -\frac{0.510 \, mol^{-1/2} \, dm^{3/2} \times 4 \times 0.3162 \, mol^{1/2} \, dm^{-3/2}}{1.3162} = -0.490$$

$$\therefore \quad \gamma_{Hg_2^{2+}} = 0.323$$

$$\log\gamma_{Cl^-} = -\frac{0.510 \, mol^{-1/2} \, dm^{3/2} \times 1 \times 0.3162 \, mol^{1/2} \, dm^{-3/2}}{1.3162} = -0.123$$

$$\therefore \gamma_{Cl^-} = 0.754$$

But $K_s = a_{Hg_2^{2+}} a_{Cl^-}^2 = [Hg_2^{2+}]\gamma_{Hg_2^{2+}}[Cl^-]^2\gamma_{Cl^-}^2 = 4(s')^3\gamma_{Hg_2^{2+}}\gamma_{Cl^-}^2$
where s' is the solubility in $KNO_3(aq)$.

$$\therefore K_s = 4(s')^3 \times 0.323 \times (0.754)^2 = (s')^3 \times 0.736 = 1.1 \times 10^{-18}\,mol^3\,dm^{-9}$$
$$\therefore (s')^3 = 1.495 \times 10^{-18}\,mol^3\,dm^{-9}$$
$$\therefore s' = 1.14 \times 10^{-6}\,mol\,dm^{-3}$$

In terms of the mean activity coefficient, $\gamma_\pm$: $\log_{10}\gamma_\pm = -\dfrac{A|z_1 z_2|\sqrt{I}}{1 + \sqrt{I}}$

$$\log_{10}\gamma_\pm = -\frac{A \times z_{Hg_2^{2+}} z_{Cl^-}\sqrt{I}}{1 + \sqrt{I}} = -\frac{0.510\,mol^{-1/2}\,dm^{3/2} \times 2 \times 1 \times 0.3162\,mol^{1/2}\,dm^{-3/2}}{1.3162}$$
$$= -0.245$$
$$\therefore \gamma_\pm = 0.569$$
$$K_s = 4(s')^3\gamma_\pm^3 = 4(s')^3 \times (0.569)^3 = 0.736(s')^3 = 1.1 \times 10^{-18}\,mol^3\,dm^{-9}$$
$$(s')^3 = 1.495 \times 10^{-18}\,mol^3\,dm^{-9}$$
$$s' = 1.14 \times 10^{-6}\,mol\,dm^{-3}$$

As expected both calculations give the same result.

(iii) Non-ideality causes the solubility to increase by a factor of 1.75.

Worked problem 8.16

Question

The calculations given in Worked Problem 8.3 ignored non-ideality. Rework this problem taking account of non-ideality.

Answer

In this problem $[Ca^{2+}] = [CO_3^{2-}] = 1.00 \times 10^{-4}\,mol\,dm^{-3}$.

$$I = \frac{1}{2}\left(1.00 \times 10^{-4} \times 2^2 + 1.00 \times 10^{-4} \times 2^2\right)mol\,dm^{-3} = 4.00 \times 10^{-4}\,mol\,dm^{-3}$$
$$\sqrt{I} = 2.00 \times 10^{-2}\,mol^{1/2}\,dm^{-3/2}$$
$$\log_{10}\gamma_\pm = -\frac{A|z_1 z_2|\sqrt{I}}{1 + \sqrt{I}} = -\frac{0.510\,mol^{-1/2}\,dm^{3/2} \times 2 \times 2 \times 2.00 \times 10^{-2}\,mol^{1/2}\,dm^{-3/2}}{1.020}$$
$$= -0.0400$$
$$\therefore \gamma_\pm = 0.912$$

(i) The values of ΔG^θ are unaffected, but the equilibrium relation is now:

$$K = \left(a_{Ca^{2+}} a_{CO_3^{2-}}\right)_{eq} = \{\gamma_\pm^2[Ca^{2+}][CO_3^{2-}]\}_{eq}$$

The values of K are:

$$K_{calcite} = 8.9 \times 10^{-9} \, mol^2 \, dm^{-6} \quad and \quad K_{aragonite} = 12.2 \times 10^{-9} \, mol^2 \, dm^{-6}.$$

The value:

$$\gamma_\pm^2 [Ca^{2+}][CO_3^{2-}] = (0.912)^2 \times 1.00 \times 10^{-4} \times 1.00 \times 10^{-4} \, mol^2 \, dm^{-6} = 8.32 \times 10^{-9} \, mol^2 \, dm^{-6}$$

has now to be compared with the solubility products for calcite and aragonite. In both cases $\gamma_\pm^2 [Ca^{2+}][CO_3^{2-}]$ is smaller than K, and so both forms could dissolve in this solution.

(ii) $\Delta G = \Delta G^\theta + RT \log_e a_{Ca^{2+}} a_{CO_3^{2-}} = \Delta G^\theta + RT \log_e \{\gamma_\pm^2 [Ca^{2+}][CO_3^{2-}]\}$

- For calcite:

$$\Delta G = 45.95 \, kJ \, mol^{-1} + 8.3145 \times 10^{-3} \, kJ \, mol^{-1} \, K^{-1} \times 298.15 \, K \times \log_e \{(0.912)^2 \times 1 \times 10^{-8}\}$$
$$= (45.95 - 46.12) \, kJ \, mol^{-1} = -0.17 \, kJ \, mol^{-1}$$

- For aragonite:

$$\Delta G = 45.17 \, kJ \, mol^{-1} + 8.3145 \times 10^{-3} \, kJ \, mol^{-1} \, K^{-1} \times 298.15 \, K \times \log_e \{(0.912)^2 \times 1 \times 10^{-8}\}$$
$$= (45.17 - 46.12) \, kJ \, mol^{-1} = -0.95 \, kJ \, mol^{-1}$$

Both these values are negative and so both forms can dissolve. This is the same conclusion as in (i) which, of course it should be.

However, when this conclusion is compared with that reached in Worked Problem 8.3, it can be seen that when non-ideality is taken into consideration, calcite will now dissolve. This demonstrates how crucial it is to take non-ideality into account. This is particularly so when multiply charged species are involved, even when the concentration is low.

Worked problem 8.17

Question

The mean activity coefficient for an aqueous solution of KBr at 25°C has been found at the following ionic strengths:

$\dfrac{I}{mol \, dm^{-3}}$	0.001	0.002	0.005	0.010	0.020	0.050
$\gamma_\pm$	0.965	0.952	0.927	0.903	0.872	0.822

(i)   Plot $\log_{10} \gamma_\pm$ vs. $\sqrt{I}$ and $\sqrt{I}/1 + \sqrt{I}$ and comment on the shape of the graphs.

(ii) Plot $\log_{10} \gamma_\pm + A|z_1 z_2| \sqrt{I}/(1 + \sqrt{I})$ vs. I and comment on the results.

(iii) The following table gives data at higher ionic strengths. Include these points on the three graphs already drawn and comment on the results.

$\dfrac{I}{\text{mol dm}^{-3}}$	0.100	0.200	0.500
$\gamma_\pm$	0.777	0.728	0.665

Answer

$\dfrac{I}{\text{mol dm}^{-3}}$	0.001	0.002	0.005	0.010	0.020	0.050
$\dfrac{\sqrt{I}}{\text{mol}^{1/2}\,\text{dm}^{-3/2}}$	0.0316	0.0447	0.0707	0.1000	0.1414	0.2236
$\dfrac{\sqrt{I}}{1+\sqrt{I}}\Big/\text{mol}^{1/2}\,\text{dm}^{-3/2}$	0.0307	0.0428	0.0660	0.0909	0.1239	0.1827
$\dfrac{0.510\sqrt{I}}{1+\sqrt{I}}\Big/\text{mol}^{1/2}\,\text{dm}^{-3/2}$	0.0156	0.0218	0.0337	0.0464	0.0632	0.0932
$\gamma_\pm$	0.965	0.952	0.927	0.903	0.872	0.822
$-\log_{10}\gamma_\pm$	0.0155	0.0214	0.0329	0.0443	0.0595	0.0851
$\log_{10}\gamma_\pm +$ $\dfrac{0.510\sqrt{I}}{1+\sqrt{I}}\Big/\text{mol}^{1/2}\,\text{dm}^{-3/2}$	0.0002	0.0004	0.0008	0.0021	0.0037	0.0081

(i) and (ii) The graph of $\log_{10}\gamma_\pm$ vs. $\sqrt{I}$ is a quite decided curve (Figure 8.2(a)), with systematic deviations occurring at ionic strengths greater than 1×10^{-3} mol dm^{-3}. This is in line with statements that the Debye-Hückel limiting equation should not be used for ionic strengths greater than 1×10^{-3} mol dm^{-3}. The Debye-Hückel equation, $\log_{10}\gamma_\pm = -A|z_1 z_2|\sqrt{I}/(1+\sqrt{I})$, fits the data at the low end of the ionic strengths, but systematic deviations start to occur at ionic strengths greater than 0.02 mol dm^{-3} (Figure 8.2(b)). Use of the Debye-Hückel extended equation, $\log_{10}\gamma_\pm = -A|z_1 z_2|\sqrt{I}/(1+\sqrt{I}) + bI$, gives an approximately linear graph going through the origin and with gradient $= +0.171$ (Figure 8.2(c)).

(iii) Using the higher ionic strength data gives:

	0.100	0.200	0.500
$\dfrac{I}{\text{mol dm}^{-3}}$			
$\dfrac{\sqrt{I}}{\text{mol}^{1/2}\,\text{dm}^{-3/2}}$	0.316	0.4472	0.7071
$\dfrac{\sqrt{I}}{1+\sqrt{I}}\Big/\text{mol}^{1/2}\,\text{dm}^{-3/2}$	0.2402	0.3090	0.4142
$\dfrac{0.510\sqrt{I}}{1+\sqrt{I}}\Big/\text{mol}^{1/2}\,\text{dm}^{-3/2}$	0.1225	0.1576	0.2112
$\gamma_\pm$	0.777	0.728	0.665
$-\log_{10}\gamma_\pm$	0.1096	0.1379	0.1772
$\log_{10}\gamma_\pm +$ $\dfrac{0.510\sqrt{I}}{1+\sqrt{I}}\Big/\text{mol}^{1/2}\,\text{dm}^{-3/2}$	0.0129	0.0197	0.0340

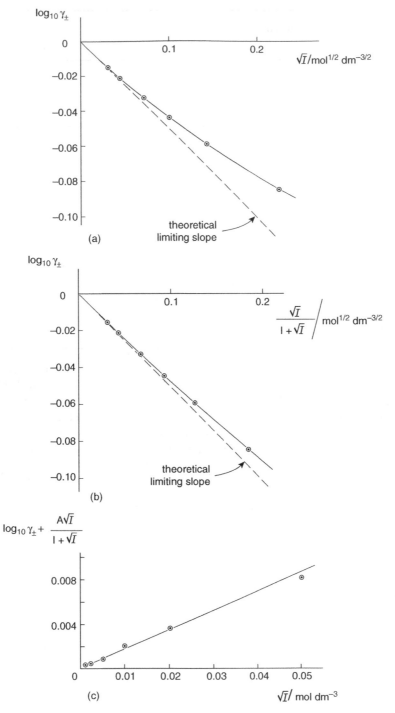

Figure 8.2   (a) Graph of $\log_{10}\gamma_{\pm}$ vs. $\sqrt{I}$; (b) Graph of $\log_{10}\gamma_{\pm}$ vs. $\sqrt{I}/(1+\sqrt{I})$; (c) Graph of $\log_{10}\gamma_{\pm} + A|z_1z_2|\sqrt{I}/(1+\sqrt{I})$ vs. I; (d) Graph of $\log_{10}\gamma_{\pm}$ vs. $\sqrt{I}$ for high ionic strength data; (e) Graph of $\log_{10}\gamma_{\pm}$ vs. $\sqrt{I}/(1+\sqrt{I})$ for high ionic strength data; (f) Graph of $\log_{10}\gamma_{\pm} + A|z_1z_2|\sqrt{I}(1+\sqrt{I})$ vs. I for high ionic strength data.

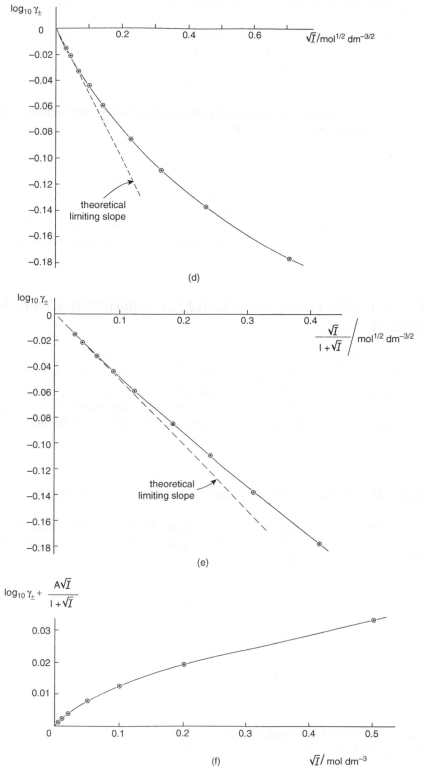

Figure 8.2 (*Continued*)

The graphs relating to the Debye-Hückel limiting law and the Debye-Hückel equation are now very decidedly curved and show considerable deviations from the theoretical behaviour (Figure 8.2(d) and 8.2(e)). The graph of $\log_{10}\gamma_\pm + A|z_1z_2|\sqrt{I}/(1+\sqrt{I})$ vs. I is interesting. Including the higher ionic strength data shows this graph also is a curve (Figure 8.2(f)).

Note: this result shows the importance of covering a wide enough range, here in ionic strength. A totally wrong interpretation results if only a limited range is considered.

Note also: Many electrolytes show similar behaviour. Graphs of $\log_{10}\gamma_\pm + A|z_1z_2|\sqrt{I}/(1+\sqrt{I})$ vs. I are only linear at the low end of ionic strengths and the deviations from linearity observed are specific to the particular electrolyte under study. Values of b found from graphs over a limited ionic strength range are specific to the electrolyte. It is observations such as these which have pushed theoreticians into deciding that further work should not involve the Debye-Hückel calculation, and other approaches should be used.

8.24   Corrections for non-ideality for experimental equilibrium constants

Experimental determinations made in terms of concentrations give concentration quotients which are **non-ideal** constants. Corrections for non-ideality are made in terms of the calculated ionic strength and the various Debye-Hückel expressions. However, emf experiments, including pH measurements, can **sometimes** furnish equilibrium constants directly in terms of activities, and as such these will be **ideal** equilibrium constants.

8.24.1   Dependence of equilibrium constants on ionic strength

The relation between K and the ionic strength should be worked out for each given reaction. The argument will be given for the reaction used previously to derive the algebraic form of the equilibrium constant (see Section 8.14).

$$Pb(s) + 2CH_3COOH(aq) \rightleftharpoons 2CH_3COO^-(aq) + Pb^{2+}(aq) + H_2(g)$$

For this reaction it was shown that $K = \left(\dfrac{a_{CH_3COO^-(aq)}^2 \, a_{Pb^{2+}} \, p_{H_2(g)}}{a_{CH_3COOH(aq)}^2} \right)_{eq}$ (8.49)

where K depends on T and slightly on p.
This is a quotient of activities and thus is an **ideal** equilibrium constant.

Equation (8.49) can also be written in terms of concentrations:

$$K_{\text{ideal}} = \left(\frac{[CH_3COO^-]^2 [Pb^{2+}] p_{H_2}}{[CH_3COOH]^2} \right) \left(\frac{\gamma_{CH_3COO^-}^2 \, \gamma_{Pb^{2+}}}{\gamma_{CH_3COOH}^2} \right) \tag{8.76}$$

$$= K_{\text{non-ideal}} \left(\frac{\gamma_{CH_3COO^-}^2 \, \gamma_{Pb^{2+}}}{\gamma_{CH_3COOH}^2} \right) \tag{8.77}$$

where $K_{\text{non-ideal}}$ is the quotient of concentrations at equilibrium.

The Debye-Hückel theory can be used to express activity coefficients in terms of the ionic strength of the solution.

- For an uncharged solute γ can be approximated to unity, $\therefore \gamma_{CH_3COOH} = 1$.

- For a charged solute the Debye-Hückel expressions are used: $\log_{10} \gamma_i = -A z_i^2 \sqrt{I}$ for dilute solutions, and for more concentrated solutions $\log_{10} \gamma_i = -A z_i^2 \sqrt{I}/(1 + \sqrt{I})$, where, for 25°C, $A = 0.510 \, \text{mol}^{1/2} \, \text{dm}^{-3/2}$.

Writing Equation (8.77) in the logarithmic form gives:

$$\log_{10} K_{\text{ideal}} = \log_{10} K_{\text{non-ideal}} + 2 \log_{10} \gamma_{CH_3COO^-} + \log_{10} \gamma_{Pb^{2+}} - 2 \log_{10} \gamma_{CH_3COOH} \tag{8.78}$$

where the logarithm is to base 10 because the Debye-Hückel theory is given to base 10.

$$\log_{10} K_{\text{ideal}} = \log_{10} K_{\text{non-ideal}} - \frac{2 \times (-1)^2 A \sqrt{I}}{1 + \sqrt{I}} - \frac{(2)^2 A \sqrt{I}}{1 + \sqrt{I}} \tag{8.79}$$

$$= \log_{10} K_{\text{non-ideal}} - \frac{6A \sqrt{I}}{1 + \sqrt{I}} \tag{8.80}$$

$$\therefore \log_{10} K_{\text{non-ideal}} = \log_{10} K_{\text{ideal}} + \frac{6A \sqrt{I}}{1 + \sqrt{I}} \tag{8.81}$$

From this it is clear that the equilibrium constant expressed in concentrations, i.e. $K_{\text{non-ideal}}$, is not a true constant but depends on the particular ionic strength at which the measurements are made. This applies, in particular, to any equilibrium which involves charged species or charge-separated species, since these are more likely to cause non-ideality than molecular species. All concentration equilibrium constants should thus be corrected for non-ideality.

There are two main ways of doing this, and both require use of the Debye-Hückel theory.

- Activity coefficients for each species can be calculated from the Debye-Hückel theory. The actual ionic strength must be calculated from the experimental conditions, and this generally involves successive approximations (see Worked Problem 8.22).

- Corrections can be made graphically. For the reaction discussed above a plot of $\log_{10} K_{\text{non-ideal}}$ vs. $\sqrt{I}/(1 + \sqrt{I})$ should be linear with slope equal to $6A$ and intercept equal to $\log_{10}K_{ideal}$. This, however, will hold only at moderate ionic strengths, and more accurate results will be found using the extended Debye-Hückel expression:

$$\log_{10}K_{non\text{-}ideal} - \frac{6A\sqrt{I}}{1+\sqrt{I}} = \log_{10}K_{ideal} + bI \tag{8.82}$$

where $\log_{10}K_{non\text{-}ideal} - \dfrac{6A\sqrt{I}}{1+\sqrt{I}}$ is plotted against I.

This method of correcting for non-ideality is the more straightforward and accurate, and should be used in preference to that using calculated activity coefficients found at one given ionic strength. Since what is required is the ideal K, a procedure giving an accurate extrapolation to zero ionic strength will suffice; with detailed consideration of the validity or otherwise of the Debye-Hückel theory being of less importance here than in calculations of the actual activity coefficients at any specific ionic strength.

Worked problem 8.18

Question

Predict the effect of ionic strength on the following equilibria:

1. $Cu^{2+}(aq) + SO_4^{2-}(aq) \rightleftharpoons Cu^{2+}SO_4^{2-}(aq)$

2. $CaF_2(s) \rightleftharpoons Ca^{2+}(aq) + 2F^-(aq)$

3. $Co^{2+}(aq) + 4NH_3(aq) \rightleftharpoons Co(NH_3)_4^{2+}(aq)$

Answer

1. $Cu^{2+}(aq) + SO_4^{2-}(aq) \rightleftharpoons Cu^{2+}SO_4^{2-}(aq)$

$$K_{ideal} = \left(\frac{a_{Cu^{2+}SO_4^{2-}}}{a_{Cu^{2+}}a_{SO_4^{2-}}}\right)_{eq} = \left(\frac{[Cu^{2+}SO_4^{2-}]}{[Cu^{2+}][SO_4^{2-}]}\right)_{eq}\frac{\gamma_{Cu^{2+}SO_4^{2-}}}{\gamma_{Cu^{2+}}\gamma_{SO_4^{2-}}} = K_{conc}\frac{\gamma_{Cu^{2+}SO_4^{2-}}}{\gamma_{Cu^{2+}}\gamma_{SO_4^{2-}}}$$

$$\therefore K_{conc} = K_{ideal}\frac{\gamma_{Cu^{2+}}\gamma_{SO_4^{2-}}}{\gamma_{Cu^{2+}SO_4^{2-}}}$$

Taking logarithms to base 10:

$$\log_{10}K_{conc} = \log_{10}K_{ideal} + \log_{10}\gamma_{Cu^{2+}} + \log_{10}\gamma_{SO_4^{2-}} - \log_{10}\gamma_{Cu^{2+}SO_4^{2-}}$$

- $\log_{10}\gamma_i = -Az_i^2\sqrt{I}/(1+\sqrt{I}) = -0.510\,\text{mol}^{-1/2}\,\text{dm}^{3/2} \times 4\sqrt{I}/(1+\sqrt{I})$ for the two divalent ions,

- since the overall charge is zero, γ for the ion pair is taken to be unity and $\log_{10}\gamma_{\text{ion-pair}} = 0$. However, this can only be an approximation since the ion pair is likely to be a charge-separated species.

$$\therefore \log_{10}K_{\text{conc}} = \log_{10}K_{\text{ideal}} - \frac{2 \times 4A\sqrt{I}}{1+\sqrt{I}} = \log_{10}K_{\text{ideal}} - \frac{8A\sqrt{I}}{1+\sqrt{I}}$$

$\therefore$ a plot of $\log_{10}K_{\text{conc}}$ vs. $\sqrt{I}/(1+\sqrt{I})$ should be linear with slope $= -8A = -4.08\,\text{mol}^{-1/2}\,\text{dm}^{3/2}$, and intercept $\log_{10}K_{\text{ideal}}$ leading to a value for K_{ideal}. The observed experimental equilibrium constant decreases as the ionic strength increases.

A more accurate extrapolation taking account of deviations at higher ionic strengths and the likely dipolar nature of the ion-pair involves a plot of:

$$\log_{10}K_{\text{conc}} + \frac{8A\sqrt{I}}{1+\sqrt{I}} \quad \text{vs.} \quad I.$$

Even if this graph shows deviations from non-linearity at the high ionic strength end, the extrapolation to zero ionic strength will be accurate and give an accurate value of K_{ideal}.

(2) $CaF_2(s) \rightleftharpoons Ca^{2+}(aq) + 2F^-(aq)$

Using the same arguments:

$$\log_{10}K_{\text{ideal}} = \log_{10}K_{\text{conc}} + \log_{10}\gamma_{Ca^{2+}} + 2\log_{10}\gamma_{F^-}$$

$$\therefore \log_{10}K_{\text{conc}} = \log_{10}K_{\text{ideal}} - \log_{10}\gamma_{Ca^{2+}} - 2\log_{10}\gamma_{F^-}$$

$$= \log_{10}K_{\text{ideal}} + \frac{(2)^2A\sqrt{I}}{1+\sqrt{I}} + \frac{2 \times (1)^2\sqrt{I}}{1+\sqrt{I}} = \log_{10}K_{\text{ideal}} + \frac{6A\sqrt{I}}{1+\sqrt{I}}$$

$\therefore$ a plot of $\log_{10}K_{\text{conc}}$ vs. $\sqrt{I}/(1+\sqrt{I})$ should be linear with slope $= +6A$ and intercept equal to $\log_{10}K_{\text{ideal}}$. The observed experimental equilibrium constant will therefore increase as the ionic strength increases.

$\log_{10}K_{\text{conc}} - 6A\sqrt{I}/(1+\sqrt{I})$ vs. I will give a more accurate extrapolation to zero ionic strength.

(3) $Co^{2+}(aq) + 4NH_3(aq) \rightleftharpoons Co(NH_3)_4^{2+}(aq)$

$$\log_{10}K_{\text{ideal}} = \log_{10}K_{\text{conc}} + \log_{10}\gamma_{Co(NH_3)_4^{2+}} - \log_{10}\gamma_{Co^{2+}} - 4\log_{10}\gamma_{NH_3}$$

$$= \log K_{\text{conc}} - \frac{(2)^2A\sqrt{I}}{1+\sqrt{I}} + \frac{(2)^2A\sqrt{I}}{1+\sqrt{I}} - 0 = \log_{10}K_{\text{conc}}$$

$$\therefore \log_{10}K_{\text{conc}} = \log_{10}K_{\text{ideal}}$$

In this case the experimental concentration equilibrium constant should be independent of ionic strength. However, deviations from non-ideality not accounted for by the Debye-Hückel expression could be taken care of by plotting $\log_{10}K_{\text{conc}}$ vs. I. This is likely to be necessary at high ionic strengths.

Worked problem 8.19

Question

The following data at 25°C give the dependence of the equilibrium constant for protonation of methylamine, $CH_3NH_2(aq)$, on ionic strength. Use a graphical procedure to find the ideal constant.

$\dfrac{10^4 K_b}{\text{mol dm}^{-3}}$	4.48	4.90	5.11	6.76	7.41
$\dfrac{I}{\text{mol dm}^{-3}}$	0.0010	0.0050	0.0100	0.0500	0.1000

Answer

$$CH_3NH_2(aq) + H_2O(aq) \rightleftharpoons CH_3NH_3^+(aq) + OH^-(aq)$$

$$K_{\text{ideal}} = \frac{a_{CH_3NH_3^+} a_{OH^-}}{a_{CH_3NH_2}} = \left(\frac{[CH_3NH_3^+][OH^-]}{[CH_3NH_2]}\right)_{eq} \frac{\gamma_{CH_3NH_3^+}\gamma_{OH^-}}{\gamma_{CH_3NH_2}} = K_{\text{conc}} \frac{\gamma_{CH_3NH_3^+}\gamma_{OH^-}}{\gamma_{CH_3NH_2}}$$

$$\log_{10}K_{\text{ideal}} = \log_{10}K_{\text{conc}} - \frac{(+1)^2 A\sqrt{I}}{1+\sqrt{I}} - \frac{(-1)^2\sqrt{I}}{1+\sqrt{I}} + 0 = \log_{10}K_{\text{conc}} - \frac{2A\sqrt{I}}{1+\sqrt{I}}$$

$$\therefore \log_{10}K_{\text{conc}} = \log_{10}K_{\text{ideal}} + \frac{2A\sqrt{I}}{1+\sqrt{I}}$$

$\log_{10}K_{\text{conc}}$ vs. $\sqrt{I}/(1+\sqrt{I})$ should be linear with slope $= +2A$ and intercept $= \log_{10}K_{\text{ideal}}$.

$\dfrac{I}{\text{mol dm}^{-3}}$	0.0010	0.0050	0.0100	0.0500	0.1000
$\dfrac{\sqrt{I}}{\text{mol}^{1/2}\,\text{dm}^{-3/2}}$	0.0316	0.0707	0.100	0.2236	0.3162
$\dfrac{\sqrt{I}}{1+\sqrt{I}}\Big/\text{mol}^{1/2}\,\text{dm}^{-3/2}$	0.0307	0.0660	0.0909	0.1827	0.2402
$\dfrac{10^4 K_b(\text{conc})}{\text{mol dm}^{-3}}$	4.48	4.90	5.11	6.76	7.41
$-\log_{10}K_b(\text{conc})$	3.349	3.310	3.292	3.170	3.130

The graph using the experimental data is linear and no further corrections for non-ideality are needed. The intercept is -3.379, and $K_b = 4.18 \times 10^{-4}\,\text{mol dm}^{-3}$.

8.25   Some specific examples of the dependence of the equilibrium constant on ionic strength

For most equilibrium constant measurements the amount of a reactant or product present is found as a concentration, e.g. spectrophotometric or conductance analyses. However, pH and some emf methods determine the activity of a species directly rather than a concentration, and so corrections for non-ideality for these species will not be necessary. But, there are also some situations where, although the basic experimental measurement is an activity, subsequent calculations involve stoichiometric relations given in concentrations. Unless care is taken, the final equilibrium constant could end up involving terms in activities and concentrations, i.e. is mixed. Here corrections for non-ideality will still have to be made. Specific cases will make this clearer.

8.25.1   The case of acid/base equilibria

The standard method of determination of K for acid/base equilibria is from emf or pH measurements. These strictly lead direct to the activity of $H_3O^+(aq)$ rather than $[H_3O^+]$.

$$pH = -\log_{10} a_{H_3O^+} = -\log_{10}\left([H_3O^+]\gamma_{H_3O^+}\right) = -\log_{10}\left([H_3O^+]\gamma_{\pm}\right) \qquad (8.83)$$

where $\gamma_{H_3O^+}$ has been replaced by the measurable quantity $\gamma_{\pm}$.

Note: in previous chapters where concentration quotients have been discussed the standard approximation that $pH = -\log_{10}[H_3O^+]$ has been consistently used. This, however, strictly applies only in very dilute solutions which can be taken to be ideal.

Whether or not the resultant K based on the approximation, $pH = -\log_{10}[H_3O^+]$, depends strongly on ionic strength is decided by the strength of the acid or base.

8.25.2   The weak acid where both approximations are valid

If the acid is weak then it is possible to ignore the self ionisation of $H_2O(l)$ and assume only slight ionisation of the weak acid.

The ionisation of ethanoic acid is a typical example:

$$CH_3COOH(aq) + H_2O(l) \rightleftharpoons CH_3COO^-(aq) + H_3O^+(aq)$$

$$K_a = \left(\frac{[CH_3COO^-]_{actual}[H_3O^+]_{actual}}{[CH_3COOH]_{actual}}\right)_{eq} \approx \left(\frac{[H_3O^+]^2_{actual}}{[CH_3COOH]_{total}}\right)_{eq} \qquad (8.84)$$

Since this equilibrium constant involves concentrations it is, by definition, a **non-ideal** constant, and in principle may show an ionic strength dependence. Generally the experimental measurement is the pH, found either directly from the pH of the given solution, or, more accurately, from a pH titration. The rigorous definition of pH is:

$$pH = -\log_{10} a_{H_3O^+} = -\log_{10}\left([H_3O^+]\gamma_{\pm}\right) \qquad (8.83)$$

and this means that it is strictly the activity of $H_3O^+(aq)$ which **actually** appears in the experimentally determined equilibrium relation which, therefore, is a mixture of concentration and activity:

K_a could be given as the quotient $\left(\dfrac{[CH_3COO^-]_{actual} a_{H_3O^+}}{[CH_3COOH]_{total}} \right)$:

but

$$[CH_3COO^-]_{actual} = \frac{a_{CH_3COO^-}}{\gamma_{\pm}} \tag{8.85}$$

and

$$[H_3O^+]_{actual} = \frac{a_{H_3O^+}}{\gamma_{\pm}} \tag{8.86}$$

and assuming both approximations hold,

$$[CH_3COO^-]_{actual} = [H_3O^+]_{actual} \tag{8.87}$$

then

$$a_{CH_3COO^-} = a_{H_3O^+} \tag{8.88}$$

since $\gamma_{\pm}$ relates to both ions

$$(K_a)_{experimental} = \left(\frac{a_{H_3O^+}^2}{[CH_3COOH]_{total}} \right) \tag{8.89}$$

Since CH_3COOH is an uncharged species, γ_{CH_3COOH} is expected to be approximately unity so that $a_{CH_3COOH} \approx [CH_3COOH]$ and:

$$(K_a)_{experimental} \approx \left(\frac{a_{H_3O^+}^2}{(a_{CH_3COOH})_{total}} \right) \tag{8.90}$$

Thus, for a weak acid where both approximations hold, the experimentally determined K_a found from the fundamental experimental measurement of pH is approximately a K_{ideal}.

However, if the experimental method had measured a concentration rather than an activity, then the experimental K_a:

$$K_a = \left(\frac{[CH_3COO^-]_{actual}[H_3O^+]_{actual}}{[CH_3COOH]_{actual}} \right)_{eq} \approx \left(\frac{[H_3O^+]_{actual}^2}{[CH_3COOH]_{total}} \right)_{eq} \tag{8.91}$$

would have been a non-ideal equilibrium constant which would have had to be corrected for non-ideality via the ionic strength.

$$(K_a)_{ideal} = \left(\frac{a_{H_3O^+} a_{CH_3COO^-}}{a_{CH_3COOH}} \right)_{eq} = \left(\frac{[H_3O^+]_{actual}\gamma_{H_3O^+}[CH_3COO^-]_{actual}\gamma_{CH_3COO^-}}{[CH_3COOH]_{total}\gamma_{CH_3COOH}} \right)_{eq} \tag{8.92}$$

$$\log_{10}(K_a)_{ideal} = \log_{10}(K_a)_{non\text{-}ideal} + \log_{10}\gamma_{H_3O^+} + \log_{10}\gamma_{CH_3COO^-} - \log_{10}\gamma_{CH_3COOH} \tag{8.93}$$

$$= \log_{10}(K_a)_{\text{non-ideal}} - A(1)^2\sqrt{I} - A(-1)^2\sqrt{I} + A(0)^2\sqrt{I} \qquad (8.94)$$

$$= \log_{10}(K_a)_{\text{non-ideal}} - 2A\sqrt{I} \qquad (8.95)$$

$$\therefore \log_{10}(K_a)_{\text{non-ideal}} = \log_{10}(K_a)_{\text{ideal}} + 2A\sqrt{I} \qquad (8.96)$$

where the ionic strength can be found from the known concentrations of the ions present.
  Similar arguments apply to the case of a weak base where both approximations are valid.

Worked problem 8.20

Question

The pH of a 0.050 mol dm^{-3}solution of a weak acid is 3.53 at 25°C.

1. Calculate K_a, making the standard approximation that pH $= -\log_{10}[H_3O^+]$.

2. A conductance measurement on this solution gives $[H_3O^+]_{\text{actual}} = 3.04 \times 10^{-4}$ mol dm^{-3} directly. Comment on the difference between this value and that found using the approximation in **1**. What implications does this have?

Answer

$$HA(aq) + H_2O(l) \rightleftharpoons H_3O^+(aq) + A^-(aq)$$

$$K_a = \left(\frac{[H_3O^+]_{\text{actual}}[A^-]_{\text{actual}}}{[HA]_{\text{actual}}} \right)_{\text{eq}}$$

1. If the usual approximation that pH can be taken to be a measure of the concentration, rather than the activity, is used:

$$\text{pH} = -\log_{10}[H_3O^+] = 3.53, \text{giving} [H_3O^+]_{\text{actual}} = 2.95 \times 10^{-4} \text{ mol dm}^{-3}.$$

This value suggests that the acid is sufficiently weak for there to be little ionisation, and for the self ionisation of water to be ignored.

$$\therefore K_a = \left(\frac{[H_3O^+]^2_{\text{actual}}}{[HA]_{\text{total}}} \right)_{\text{eq}}$$

$$= \frac{(2.95 \times 10^{-4})^2 \text{ mol}^2 \text{ dm}^{-6}}{0.05 \text{ mol dm}^{-3}} = 1.74 \times 10^{-6} \text{ mol dm}^{-3}$$

$$pK_a = 5.76$$

2. The measured $[H_3O^+]_{\text{actual}} = 3.04 \times 10^{-4}$ mol dm^{-3} is to be compared with that found from the **approximation** pH $= -\log_{10}[H_3O^+] = 3.53$ which gives

$[H_3O^+]_{actual} = 2.95 \times 10^{-4}$ mol dm^{-3}. These two values are very similar suggesting that the solution is close to ideality.

If the calculation is carried out using the formal rigorous definition of pH, $pH = -\log_{10} a_{H_3O^+}$, then it is $a_{H_3O^+}$ which is equal to 2.95×10^{-4} mol dm^{-3}, rather than $[H_3O^+]$. Since $a_{H_3O^+} = [H_3O^+]\gamma_\pm$ then:

$$\gamma_\pm = \frac{a_{H_3O^+}}{[H_3O^+]} = \frac{2.95 \times 10^{-4} \text{ mol dm}^{-3}}{3.04 \times 10^{-4} \text{ mol dm}^{-3}} = 0.970.$$

This value is close to unity, the value of $\gamma_\pm$ for an ideal solution, confirming the previous conclusion that the solution is close to ideality.

8.25.3   The weak acid where there is extensive ionisation

In this case there is only one valid approximation, i.e. the self ionisation of water can be ignored.

The ionisation of chloro-ethanoic acid is a typical example:

$$CH_2ClCOOH(aq) + H_2O(l) \rightleftharpoons CH_2ClCOO^-(aq) + H_3O^+(aq)$$

$$K_a = \left(\frac{[CH_2ClCOO^-]_{actual}[H_3O^+]_{actual}}{[CH_2ClCOOH]_{actual}} \right)_{eq} \tag{8.97}$$

$$= \left(\frac{[CH_2ClCOO^-]_{actual}[H_3O^+]_{actual}}{[CH_3ClCOOH]_{total} - [CH_3ClCOO^-]_{actual}} \right)_{eq} \tag{8.98}$$

$$\approx \left(\frac{[H_3O^+]^2_{actual}}{[CH_2ClCOOH]_{total} - [H_3O^+]_{actual}} \right)_{eq} \tag{8.99}$$

Since this involves a quotient of concentrations it is a non-ideal equilibrium constant.

However, if the fundamental experimental measurement is a pH, then what is strictly measured is $a_{H_3O^+}$. **But**, the stoichiometric relation between $[CH_2ClCOOH]_{total}$, $[CH_2ClCOOH]_{actual}$ and $[CH_2ClCOO^-]_{actual}$ remains the same, i.e. is **concentration** based and not **activity** based:

$$[CH_2ClCOOH]_{total} = [CH_2ClCOOH]_{actual} + [CH_2ClCOO^-]_{actual} \tag{8.100}$$

which becomes:

$$[CH_2ClCOOH]_{total} = [CH_2ClCOOH]_{actual} + [H_3O^+]_{actual} \tag{8.101}$$

and this involves a concentration of $H_3O^+(aq)$ rather than an activity. To be able to use this relation the concentration of $H_3O^+(aq)$ must be found from the measured activity. The ideal expression is:

$$(K_a)_{ideal} = \left(\frac{a_{H_3O^+} a_{CH_2ClCOO^-}}{a_{CH_2ClCOOH}} \right)_{eq} = \left(\frac{a_{H_3O^+} a_{CH_2ClCOO^-}}{[CH_2ClCOOH]_{actual}\gamma_{CH_2ClCOOH}} \right)_{eq} \tag{8.102}$$

$$= \left(\frac{a^2_{H_3O^+}}{\{[CH_2ClCOOH]_{total} - [H_3O^+]_{actual}\}\gamma_{CH_2ClCOOH}} \right)_{eq} \tag{8.103}$$

This quotient involves the activity of $H_3O^+(aq)$ in the numerator and the concentration of $H_3O^+(aq)$ in the denominator, and is thus a mix of activity and concentration.

In this case non-ideality must be taken into account explicitly, in contrast to the situation which obtained when the approximation of only slight ionisation is considered where there is no term in $[H_3O^+]_{actual}$ in the denominator, and the quotient of concentrations is virtually the same as the quotient of activities.

Worked problem 8.21

Question

An aqueous solution of an acid, HX, at 25°C has a stoichiometric concentration of 6.24×10^{-2} mol dm^{-3}. A spectrophotometric determination of $[X^-]$ carried out at a frequency at which HX(aq) does not absorb gives $[X^-] = 2.08 \times 10^{-2}$ mol dm^{-3}.

1. What can be said about the strength of this acid?

2. Find pK_a if non-ideality is totally ignored.

3. Find pK_a when corrections are made for non-ideality.

4. Comment on the results.

Answer

1. $[X^-]_{actual} = 2.08 \times 10^{-2}$ mol dm^{-3}; $[HX]_{stoich} = 6.24 \times 10^{-2}$ mol dm^{-3}.

$$\text{Fraction ionised} = \frac{[X^-]_{actual}}{[HX]_{stoich}} = \frac{2.08 \times 10^{-2} \text{ mol dm}^{-3}}{6.24 \times 10^{-2} \text{ mol dm}^{-3}} = 0.333$$

This shows that the acid is considerably dissociated, and no approximations, other than ignoring the self ionisation of water, can be made.

2. Ignoring non-ideality:

$$[HX]_{total} = [HX]_{actual} + [X^-]_{actual}$$

Since there is only one source of $H_3O^+(aq)$ and $X^-(aq)$, viz. the ionisation of HX(aq):

$$[H_3O^+]_{actual} = [X^-]_{actual}$$

$$[HX]_{actual} = [HX]_{total} - [X^-]_{actual} = [HX]_{total} - [H_3O^+]_{actual}$$

$$= (6.24 \times 10^{-2} - 2.08 \times 10^{-2}) \text{mol dm}^{-3}$$

$$= 4.16 \times 10^{-2} \text{ mol dm}^{-3}$$

If non-ideality is ignored then the equilibrium constant is given as a quotient of concentrations at equilibrium:

$$K_a = \left(\frac{[H_3O^+][X^-]}{[HX]}\right)_{eq} = \frac{2.08 \times 10^{-2} \text{ mol dm}^{-3} \times 2.08 \times 10^{-2} \text{ mol dm}^{-3}}{4.16 \times 10^{-2} \text{ mol dm}^{-3}}$$

$$= 1.04 \times 10^{-2} \text{ mol dm}^{-3}$$

$$\therefore pK_a = 1.98$$

3. Making corrections for non-ideality:

The activity coefficients have to be calculated from the ionic strength, found from the actual concentrations of the ionic species.

$$I = 1/2 \sum_i c_i z_i^2 = 1/2(2.08 \times 10^{-2} \text{ mol dm}^{-3} + 2.08 \times 10^{-2} \text{ mol dm}^{-3})$$

$$= 2.08 \times 10^{-2} \text{ mol dm}^{-3}$$

Since this is a fairly concentrated solution, then the Debye-Hückel equation must be used rather than the limiting law:

$$\log_{10}\gamma_i = -Az_i^2 \frac{\sqrt{I}}{1+\sqrt{I}} = -0.510 \text{ mol}^{-1/2} \text{ dm}^{-3/2} \times 1^2 \frac{\sqrt{2.08 \times 10^{-2}}}{1+\sqrt{2.08 \times 10^{-2}}} \text{ mol}^{1/2} \text{ dm}^{-3/2}$$

$$= 0.510 \text{ mol}^{-1/2} \text{ dm}^{3/2} \times \frac{0.1442}{1.1442} \text{ mol}^{1/2} \text{ dm}^{-3/2} = -0.0643$$

$$\therefore \gamma_i = 0.862$$

Since HX is uncharged $\gamma_{HX} = 1$

$$K_{a(ideal)} = \left(\frac{a_{H_3O^+}a_{X^-}}{a_{HX}}\right)_{eq} = \left(\frac{[H_3O^+]_{actual}\gamma_{H_3O^+}[X^-]_{actual}\gamma_{X^-}}{[HX]_{actual}\gamma_{HX}}\right)_{eq}$$

$$= \left(\frac{2.08 \times 10^{-2} \text{ mol dm}^{-3} \times 0.862 \times 2.08 \times 10^{-2} \text{ mol dm}^{-3} \times 0.862}{4.16 \times 10^{-2} \text{ mol dm}^{-3}}\right)$$

$$= 7.74 \times 10^{-3} \text{ mol dm}^{-3}$$

$$\therefore pK_{a(ideal)} = 2.11$$

In this case, allowing for non-ideality changes the value found for K_a from 1.04×10^{-2} mol dm^{-3} to 7.74×10^{-3} mol dm^{-3}, with the change in pK_a from 1.98 to 2.11. This is a significant change, and shows that, for this acid where the **basic measurement is a concentration**, corrections **must** be made for non-ideality. This should be contrasted with the situation which obtains when the basic measurement is from an emf or from a pH, i.e. where the **basic measurement is an activity**. This is illustrated in the following worked problem where the pH of the same solution is the basic experimental measurement.

Worked problem 8.22

Question

The same aqueous solution of the acid, HX, of stoichiometric concentration equal to 6.24×10^{-2} mol dm^{-3} has a pH of 1.75 at 25°C.

1. Find pK_a if non-ideality is totally ignored.

2. Find pK_a when corrections are made for non-ideality.

3. Comment on the results.

Answer

1. Ignoring non-ideality:

This means that the relation, $pH = -\log_{10}[H_3O^+]$ can be used rather than $pH = -\log_{10}a_{H_3O^+}$.
$-\log_{10}[H_3O^+] = 1.75$, giving $[H_3O^+]_{actual} = 1.78 \times 10^{-2} \, mol \, dm^{-3}$.
Since there is only one source of $H_3O^+(aq)$ and $X^-(aq)$:

$$[H_3O^+]_{actual} = [X^-]_{actual} = 1.78 \times 10^{-2} \, mol \, dm^{-3}.$$

$$[HX]_{actual} = [HX]_{total} - [H_3O^+]_{actual} = (6.24 \times 10^{-2} - 1.78 \times 10^{-2}) \, mol \, dm^{-3}$$
$$= 4.46 \times 10^{-2} \, mol \, dm^{-3}$$

$$K_{a(non-ideal)} = \left(\frac{[H_3O^+][X^-]}{[HX]} \right)_{eq} = \frac{1.78 \times 10^{-2} \, mol \, dm^{-3} \times 1.78 \times 10^{-2} \, mol \, dm^{-3}}{4.46 \times 10^{-2} \, mol \, dm^{-3}}$$
$$= 7.1 \times 10^{-3} \, mol \, dm^{-3}$$

$$pK_{a(non-ideal)} = 2.15.$$

2. Correcting for non-ideality:

This requires calculation of I and γ_i which is carried out by a series of successive approximations. A simple calculation as in the previous problem is not possible since the fundamental quantity which is known is the activity of $H_3O^+(aq)$, whereas it is $[H_3O^+]$ which is required.

- **first approximation**:

This is based on taking the concentrations as worked out in part 1. i.e. approximating the activities to concentrations.

$$[H_3O^+] = [X^-] = 1.78 \times 10^{-2} \, mol \, dm^{-3} \text{ giving } I = 1.78 \times 10^{-2} \, mol \, dm^{-3}.$$

$$\log_{10}\gamma_i = -\frac{0.510 \, mol^{-1/2} \, dm^{3/2} \times 1^2 \sqrt{1.78 \times 10^{-2}} \, mol^{1/2} \, dm^{-3/2}}{1 + \sqrt{1.78 \times 10^{-2}}} = -0.060$$

$$\therefore \gamma_{H_3O^+} = \gamma_{X^-} = 0.871$$

But $\gamma_{H_3O^+}[H_3O^+] = 1.78 \times 10^{-2} \, mol \, dm^{-3}$

$$\therefore [H_3O^+] = \frac{1.78 \times 10^{-2}}{0.871} \, mol \, dm^{-3} = 2.04 \times 10^{-2} \, mol \, dm^{-3}$$

- This can be used to give a **second approximate** $I = 2.04 \times 10^{-2} \, mol \, dm^{-3}$ from which $\gamma_{H_3O^+}$ and $[H_3O^+]$ can be found.

Successive approximations give as final values:

$$[H_3O^+]_{actual} = [X^-]_{actual} = 2.06 \times 10^{-2} \, \text{mol dm}^{-3}$$

$$[HX]_{actual} = (6.24 \times 10^{-2} - 2.06 \times 10^{-2}) \, \text{mol dm}^{-3} = 4.18 \times 10^{-2} \, \text{mol dm}^{-3}$$

$$\gamma_{H_3O^+} = \gamma_{X^-} = 0.863 \text{ and } \gamma_{HX} = 1$$

The pH gives $a_{H_3O^+}$ direct: $pH = -\log_{10} a_{H_3O^+} = 1.75$
so that $a_{H_3O^+} = a_{HX^-} = 1.78 \times 10^{-2} \, \text{mol dm}^{-3}$.

$$K_{a(ideal)} = \left(\frac{a_{H_3O^+} a_{X^-}}{a_{HX}} \right)_{eq} = \left(\frac{1.78 \times 10^{-2} \, \text{mol dm}^{-3} \times 1.78 \times 10^{-2} \, \text{mol dm}^{-3/2}}{4.18 \times 10^{-2} \, \text{mol dm}^{-3} \times 1} \right)$$

$$= 7.6 \times 10^{-3} \, \text{mol dm}^{-3}$$

$$\therefore pK_{a(ideal)} = 2.12$$

The values found for pK_a are:

- ignoring non-ideality, $pK_a = 2.15$.

- correcting for non-ideality, $pK_a = 2.12$

Correcting for non-ideality gives a much smaller correction here, where pH measurements are the basis, than was found **for the same solution** when spectrophotometry was the basis.

8.26   Graphical corrections for non-ideality

The most straightforward way to take account of non-ideality is to make measurements over a range of ionic strengths, and then extrapolate to zero ionic strength. Solutions of differing ionic strengths can be prepared in two ways:

- by making up differing concentrations of the weak acid on its own in water. This has the disadvantage that even fairly high concentrations of a weak acid may not give much of a range of ionic strengths simply because there never are sufficient ions present. However, the ionic strengths obtained would be sufficiently low for the solutions to be close to ideality.

- by making up solutions containing the same stoichiometric concentration of acid but with added inert electrolyte of differing amounts so that the ionic strengths of the solutions vary over a wide range of ionic strengths. This method is by far the most systematic and accurate.

8.27   Comparison of non-graphical and graphical methods of correcting for non-ideality

Calculated values of K_a could be found from experimental data for each solution, with corrections for non-ideality made by calculating the activity coefficients $\gamma_{H_3O^+}$ and γ_{A^-} for the ionic species and assuming γ_{HA} equal to unity. This would include any uncertainty associated with the Debye-Hückel and Debye-Hückel extended equations, and more importantly the inherent uncertainties resulting from the assumption that γ_{HA} is always unity, no matter how high the ionic strength. The beauty of the graphical procedure is that it automatically corrects for these uncertainties. In addition, since the ideal K is that found at zero ionic strength what is needed is an accurate extrapolation. Deviations at high ionic strengths should not affect this, but would affect a calculation of the ideal equilibrium constant found by using the Debye-Hückel equation to calculate activity coefficients.

8.28 Dependence of fraction ionised and fraction protonated on ionic strength

Calculations show that the fraction ionised and fraction protonated increase as the ionic strength increases. These calculations show that there should never be an automatic assumption of only slight ionisation or slight protonation, especially at moderate or high ionic strengths.

In general there are three important conclusions which can be reached:

- Ionic strength can have a quite dramatic effect on the pK of an acid or base, and consequently on the fraction ionised or the fraction protonated, **even though** the stoichiometric concentration remains the same.

- There could be situations where the effect of ionic strength could alter the fraction ionised or fraction protonated to such an extent that the approximation of only slight ionisation or protonation is no longer valid.

- Non-ideality must never be ignored when dealing with electrolyte solutions.

8.29 Thermodynamic quantities and the effect of non-ideality

If the equilibrium constant depends on ionic strength then so do the other thermodynamic quantities relating to the equilibrium process such as ΔH^{θ}, ΔG^{θ}, ΔS^{θ}, ΔV^{θ} and ΔC_p^{θ}. Hence these must be calculated from the ideal K, or, if found independently of K, they must be extrapolated to zero ionic strength.

9
Electrochemical Cells and EMFs

When a current is passed through a solution two contributing effects have to be considered:

- what the chemical effect is at the electrodes: this is considered in **electrolysis**;

- what the chemical effect is in the solution: this is considered in **conductance** studies;

- while in **electrochemical cells** and **emf** studies both effects must be considered.

Electrolysis and electrochemical studies will be discussed in this chapter, while conductance will be discussed in Chapters 11 and 12.

In this chapter attention will be focused on **carrying out calculations** relating to emfs of cells as well as explaining the principles involved.

Aims

By the end of this chapter you should be able to:

- discuss electrolysis in terms of the cathode and anode reactions, and migration of the ions in solution;

- describe what is meant by an electrochemical cell, describe what is meant by a cell operating reversibly and irreversibly, and formulate the conditions for reversibility of cells;

- describe the reactions occurring at the electrodes of an electrochemical cell and the migration processes taking place in the solutions around the electrodes;

An Introduction to Aqueous Electrolyte Solutions. By Margaret Robson Wright
© 2007 John Wiley & Sons Ltd   ISBN 978-0-470-84293-5 (cloth) ISBN 978-0-470-84294-2 (paper)

- list the types of electrodes commonly used in electrochemical cells;

- state the conventions for writing down electrochemical cells;

- understand the processes occurring in a cell with transport and a cell without transport;

- appreciate the migration processes occurring in a cell where (i) a narrow tube, (ii) a porous pot, (iii) a salt bridge joins the two electrode compartments;

- formulate thermodynamic quantities derivable from the emf of a cell;

- derive the Nernst equation for a variety of cells and correct for non-ideality;

- discuss the uses of concentration cells and 'concealed' concentration cells;

- carry out calculations leading to thermodynamic quantities for a wide variety of reactions including acid/base equilibria, pH measurements, solubility products, ion pair and complex formation;

- describe the use of cells in studying non-ideality and determination of $\gamma_{\pm}$ values;

- explain the use of cells with and without transport to determine transport numbers.

9.1 Chemical aspects of the passage of an electric current through a conducting medium

An electric current is the result of movement of an electric charge through a medium. Quantitatively, the current is the charge passing per unit time.

In a metallic solid, the current is carried by movement of electrons in the solid. In a solution, however, the current is carried by movement of charged particles, ions, present in the solution.

A battery is often used as a source of current with the electron flow a result of the chemical reactions occurring at the electrodes of the battery. The direction of the flow of electrons is determined by the reactions at the electrodes of the battery. The negative terminal which is connected to the negative electrode acts as the source of the electrons and so the chemical reaction occurring at this electrode must be **oxidation**. The positive terminal which is connected to the positive electrode acts as a sink for the electrons and so the reaction occurring there must take up electrons, i.e. **reduction**.

If a metal is connected to a battery, then the current flowing is caused by movement of electrons from the negative terminal of the battery through the solid metal and then back to the positive terminal of the battery.

Study of the flow of a current in solution requires the solution to be connected to the terminals of a battery by metal connectors connected to electrodes dipping into the solution.

A different situation now results. The current is carried through the solution by means of any charged particles in the solution, i.e. ions. Both cations and anions can carry the current. The electrode which dips into the solution and is connected to the negative terminal of the battery will be negatively charged, and so the reaction occurring on that electrode will be one which **takes up electrons** and is therefore **reduction**. By convention, the electrode at which **reduction** occurs is called the **cathode**. The electrode which dips into the solution and is connected to the positive terminal is positively charged, because electrons must flow from here back to the positive terminal of the battery. The reaction occurring here will be one which **produces electrons,** i.e. **oxidation**. The electrode at which **oxidation** occurs is called the **anode**. The overall flow is from the negative terminal of the battery through the solution back to the positive terminal of the battery.

The terms cathode and anode are almost exclusively confined to a discussion of electrolysis and are rarely used when discussing batteries and electrochemical cells where the important feature is a statement of the **polarity** of the electrodes of the battery or cell. The terms cathode and anode do not imply anything about the polarity, i.e. the charge, or sign, of the electrode.

In the solution being electrolysed the current is carried by the cations moving to the negative electrode, the **cathode**, and by the anions moving to the positive electrode, the **anode**. The fraction of the current carried by one ionic species in a solution is called the **transport number**. The fraction of the current carried by the cations moving to the negative electrode is written as t_+, the fraction carried by the anions moving to the positive electrode is t_-, where $t_+ + t_- = 1$.

If there is more than one type of cation present in the solution then, to a first approximation, the cation which is in higher concentration will carry the greater fraction of the total current. The same will apply to the anions. The other factor which influences the fraction of the current carried by a given ion is the speed at which the ion moves in solution. However, with the exception of $H_3O^+(aq)$ and possibly $OH^-(aq)$, the speed at which the ion moves is of much lower importance in determining the fraction of the current carried by a particular ion than is the concentration of the ion. $H_3O^+(aq)$ and $OH^-(aq)$ are exceptions to this rule because they have very high mobilities in aqueous solution relative to other ions, and this may counterbalance the concentration effect.

9.2 Electrolysis

If two inert wires, such as platinum wires, acting as electrodes dip into an aqueous solution and are connected to a source of electricity, such as a battery, a current will flow around the circuit. After some time it becomes clear that something is happening at the electrodes dipping into the solution. Bubbles of gas appear at each electrode. $H_2(g)$ appears at the negative electrode, and $O_2(g)$ at the positive electrode. Likewise, a current will flow if two copper electrodes dip into a solution of $CuSO_4(aq)$. In this case $Cu(s)$ will be deposited at the negative electrode, and at the positive electrode copper will pass into solution as $Cu^{2+}(aq)$ ions.

Processes such as these are termed **electrolysis**, i.e. a chemical reaction produced by passage of an electric current through a solution. The type of reaction produced, e.g. evolution of a gas, dissolution of a metal, deposition of a metal depends critically on what substance is used as an electrode, and on the solution.

Worked problem 9.1

Question

Interpret the following results in terms of:

- the cathode reaction;

- the anode reaction;

- migration of the ions in solution;

- the ions which carry the current.

1. Ag electrodes dipping into $AgNO_3$(aq): Ag(s) goes into solution at the anode, and is deposited at the cathode.

2. Pt electrodes dipping into $AgNO_3$(aq): O_2 is evolved at the anode, and Ag(s) is deposited at the cathode.

3. Pt electrodes dipping into H_2SO_4(aq): O_2 is evolved at the anode, and H_2(g) is evolved at the cathode.

4. Pt electrodes dipping into NaCl(aq): Cl_2(g) is evolved at the anode and H_2(g) is evolved at the cathode.

Answer

1. Ag(s) dipping into $AgNO_3$(aq)

The solution contains: Ag^+(aq), H_3O^+(aq), NO_3^-(aq), and OH^-(aq). Reactions at the electrodes are seen to involve deposition or dissolution of Ag(s) only. Thus H_3O^+(aq), NO_3^-(aq), and OH^-(aq) do not contribute to the electrode processes.

Cathode reaction: involves uptake of electrons, Ag^+(aq) $+ e \rightarrow$ Ag(s)
Anode reaction: involves release of electrons, Ag(s) $\rightarrow Ag^+$(aq) $+ e$
Migration of ions: Ag^+(aq), H_3O^+(aq) migrate to the cathode
$\qquad\qquad\qquad\quad NO_3^-$(aq), and OH^-(aq) migrate to the anode.
Current carried by: mainly by Ag^+(aq) since $[Ag^+] \gg [H_3O^+]$
$\qquad\qquad\qquad\quad$ mainly by NO_3^-(aq) since $[NO_3^-] \gg [OH^-]$

2. Pt(s) dipping into $AgNO_3$(aq)

The solution contains Ag^+(aq), H_3O^+(aq), NO_3^-(aq), and OH^-(aq). Ag(s) is deposited at the cathode while O_2(g) is evolved at the anode.

Cathode reaction: involves uptake of electrons, Ag^+(aq) $+ e \rightarrow$ Ag(s)
Anode reaction: involves release of electrons. However, in this electrolysis there is no Ag(s) to go into solution. The electrode is platinum which is a very inert metal and does not readily

dissolve in aqueous solution to give ions and release electrons. Also some species in the solution must be the source of the $O_2(g)$ and must also be able to release electrons. Out of the two anions present only $OH^-(aq)$ can produce $O_2(g)$ and the chemical reaction must be:

$$2OH^-(aq) \rightarrow \tfrac{1}{2}O_2(g) + H_2O(l) + 2e$$

Migration of ions: $Ag^+(aq)$, $H_3O^+(aq)$ migrate to the cathode

$NO_3^-(aq)$, and $OH^-(aq)$ migrate to the anode.

Current carried by: mainly by $Ag^+(aq)$ since $[Ag^+] \gg [H_3O^+]$

mainly by $NO_3^-(aq)$ since $[NO_3^-] \gg [OH^-]$

3. Pt dipping into $H_2SO_4(aq)$

Here there is no metal involved in the electrode reactions and the ions in solution are $H_3O^+(aq)$, $SO_4^{2-}(aq)$ and $OH^-(aq)$.

Cathode reaction: involves uptake of electrons; the only species which could do this is $H_3O^+(aq)$, $2H_3O^+(aq) + 2e \rightarrow H_2(g) + 2H_2O(l)$

Anode reaction: involves release of electrons. Of the two anions in solution it is only $OH^-(aq)$ which could release electrons and give $O_2(g)$:

$$2OH^-(aq) \rightarrow \tfrac{1}{2}O_2(g) + H_2O(l) + 2e$$

Migration of ions: $H_3O^+(aq)$ migrates to the cathode

$SO_4^{2-}(aq)$, and $OH^-(aq)$ migrate to the anode.

Current carried by: by $H_3O^+(aq)$ only, there being no other cation present

mainly by $SO_4^{2-}(aq)$ since $[SO_4^{2-}] \gg [OH^-]$

4. Pt dipping into $NaCl(aq)$

The ions in solution are $Na^+(aq)$, $H_3O^+(aq)$, $Cl^-(aq)$ and $OH^-(aq)$.

Cathode reaction: involves uptake of electrons; the only species which could do this are $Na^+(aq)$ and $H_3O^+(aq)$. It is very difficult to deposit $Na(s)$ from solution – **remember**: $Na(s)$ reacts violently with $H_2O(l)$. The only way to deposit $Na(s)$ from solution would be to have it come immediately into contact with $Hg(l)$ to form a non-reactive amalgam, and this is not possible here. Hence discharge of $H_3O^+(aq)$ is the only possible cathode reaction:

$$2H_3O^+(aq) + 2e \rightarrow H_2(g) + 2H_2O(l)$$

Anode reaction: involves release of electrons. The only two species which could do this are $Cl^-(aq)$ and $OH^-(aq)$ giving $Cl_2(g)$ and $O_2(g)$. $Cl_2(g)$ is observed, and so the anode reaction is:

$$2Cl^-(aq) + 2e \rightarrow Cl_2(g)$$

Migration of ions: $Na^+(aq)$ and $H_3O^+(aq)$ migrate to the cathode

$Cl^-(aq)$ and $OH^-(aq)$ migrate to the anode.

Current carried by: mainly by $Na^+(aq)$: $[Na^+] \gg [H_3O^+]$

mainly by $Cl^-(aq)$ since $[Cl^-] \gg [OH^-]$

Note: Discharge of a particular ion at the electrode depends on **the nature of the electrolyte** in this case, whereas in (1) and (2) it depended on **the nature of the electrode**.

9.2.1   Quantitative aspects of electrolysis

Experiment shows that the number of mol of a substance evolved, deposited or going into solution at an electrode is proportional to the amount of electrical charge passed. The charge is given by the product of the current passed and the time for which it is passed. The units of current are ampere, A, and so the units of charge are ampere second, A s, or coulomb, C. If one mol of electrons is passed, the charge passing is called one faraday, and this is found to be 9.6485×10^4 C. The charge on one electron is 1.60218×10^{-19} C, so that the charge on one mol of electrons is 1.60218×10^{-19} C $\times 6.0221 \times 10^{23} = 9.6485 \times 10^4$ C.

The quantitative aspect of electrolysis is an exercise in stoichiometry, identical to those given in other chapters, but, **in addition**, it includes a consideration of the number of electrons involved. Since reaction at the positively charged anode involves release of electrons the chemical reaction occurring is one of oxidation, while at the negatively charged cathode the reaction involves uptake of electrons and is thus reduction. The amount of substance oxidised or reduced at the electrodes is related to the amount of electricity passed, i.e. the charge passed, through the stoichiometry of the redox processes occurring.

When Zn electrodes dip into $ZnSO_4$ solution, $Zn(s)$ goes into solution at the anode and $Zn(s)$ is deposited at the cathode and the reactions occurring are:

$$\text{at the anode} \quad Zn(s) \rightarrow Zn^{2+}(aq) + 2e$$
$$\text{at the cathode} \quad Zn^{2+}(aq) + 2e \rightarrow Zn(s)$$

and 2 mol of electrons, i.e. 2 faraday, would cause 1 mol of $Zn^{2+}(aq)$ to be formed around the anode and 1 mol of $Zn(s)$ to be deposited at the cathode,
or n faraday would cause $n/2$ mol of Zn^{2+} (aq) to be formed around the anode and $n/2$ mol of $Zn(s)$ to be deposited at the cathode.

Worked problem 9.2

Question

Carry out the same sort of analysis as given above in Section 9.2.1 for reaction 2 of Worked Problem 9.1.

Answer

$$\text{at the anode :} \quad 2OH^-(aq) \rightarrow \tfrac{1}{2}O_2(g) + H_2O(l) + 2e$$
$$\text{at the cathode :} \quad Ag^+(aq) + e \rightarrow Ag(s)$$

In this case the number of electrons have to be the same, and so the cathode reaction will have to be written as:

$$2Ag^+(aq) + 2e \rightarrow 2Ag(s)$$

and so 2 mol of electrons, i.e. 2 faraday, will cause $\tfrac{1}{2}$ mol of $O_2(g)$ to be formed around the anode and 2 mol of $Ag(s)$ to be deposited at the cathode. Likewise $2n$ faraday will cause $n/2$ mol of $O_2(g)$ to be formed around the anode and $2n$ mol of $Ag(s)$ to be deposited at the cathode.

An alternative procedure would be to halve the initial reaction occurring at the anode giving for the electrode reactions:

at the anode : $\qquad OH^-(aq) \longrightarrow \frac{1}{4}O_2(g) + \frac{1}{2}H_2O(l) + e$

at the cathode : $\qquad Ag^+(aq) + e \longrightarrow Ag(s)$

with the corresponding statement:

1 mol of electrons, i.e. 1 faraday, will cause ¼ mol of $O_2(g)$ to be formed around the anode and 1 mol of $Ag(s)$ to be deposited at the cathode. Likewise n faraday will cause $n/4$ mol of $O_2(g)$ to be formed around the anode and n mol of $Ag(s)$ to be deposited at the cathode.

It is immaterial which stoichiometry is used, **provided** everything is **internally self consistent**.

Worked problem 9.3

Question

A 0.1 mol dm^{-3} solution of H_2SO_4 is electrolysed using $Pt(s)$ electrodes. If a current of 0.500 A is passed for 1 hour 40 minutes, find the number of mol deposited or released at the electrodes.

Answer

$$\text{Charge passed} = 0.500 \text{ A} \times 100 \times 60 \text{ s} = 3000 \text{ C} = \frac{3000}{9.6485 \times 10^4}\text{faraday}$$
$$= 3.11 \times 10^{-2}\text{faraday}$$

Anode reaction: release of electrons and evolution of $O_2(g)$

$$2OH^-(aq) \rightarrow \frac{1}{2}O_2(g) + H_2O(l) + 2e$$

Cathode reaction: uptake of electrons and evolution of $H_2(g)$

$$2H_3O^+(aq) + 2e \rightarrow H_2(g) + 2H_2O(l)$$

Passage of 2 mol of electrons, i.e. 2 faraday will release ½ mol of $O_2(g)$ and 1 mol of $H_2(g)$. $\therefore$ Passage of 3.11×10^{-2} faraday will release $\frac{1}{4} \times 3.11 \times 10^{-2}$ mol of $O_2(g) = 7.77 \times 10^{-3}$ mol of O_2 (g) and $\frac{1}{2} \times 3.11 \times 10^{-2}$ mol of $H_2(g) = 1.555 \times 10^{-2}$ mol of $H_2(g)$.

9.2.2 A summary of electrolysis

• The **direction of flow** of electrons **determines** the **charge** on an electrode, and this in turn determines the **type of reaction** occurring at that electrode.

• If electrons **flow to** an electrode, the electrode becomes **negatively** charged. The reaction which occurs is one which can **take up** electrons, i.e. **reduction**, and the electrode is called the **cathode**.

• If electrons **flow from** an electrode, the electrode becomes **positively** charged. The reaction which occurs is one which **gives up** electrons, i.e. **oxidation**, and the electrode is called the **anode**.

- The current **flows in the solution** from the negatively charged electrode, the cathode, to the positively charged electrode, the anode.

- **In solution**, the current is carried by the ions in solution, with the cations moving to the cathode and the anions moving to the anode.

- The **fraction of the current** carried by each type of ion is called the **transport number**.

9.3   Electrochemical cells

What is being studied here is how a chemical effect, at the battery electrodes, leads to an electrical effect, a current, in contrast to the situation in electrolysis where an electrical effect, the current, leads to a chemical effect, electrolysis.

An electrochemical cell is a source of electricity. Such cells can operate irreversibly when being used as a source of a current, or reversibly, as in emf studies. The terms irreversible and reversible are being used in the thermodynamic sense of the terms (see Section 8.3).

9.3.1   The electrochemical cell operating irreversibly or reversibly

A battery is an electrochemical cell operating irreversibly, and is capable of generating a current and a potential difference. The potential difference is set up by virtue of the chemical reactions occurring at the **electrodes of the battery**. This potential difference can be measured by connecting a voltmeter across the terminals of the battery. When current is taken from the battery, e.g. in electrolysis, it is operating irreversibly.

If, however, the battery is connected to another cell and the potential difference is balanced by an equal and opposite potential difference the cell can operate reversibly (see Sections 9.3.4 and 9.3.5 below).

The chemical reactions occurring **externally** by virtue of the battery generating a current are studied in **electrolysis**. In contrast, the chemical processes occurring **at the electrodes of the cell itself** are the subject of **emf studies**.

9.3.2   Possible sources of confusion

Confusion can result from the terminology used. When a cell is being used to generate an electric current, it is termed a battery and it is then operating irreversibly. When a cell is being used reversibly to study emfs then it is called an electrochemical cell, or galvanic cell. However, the processes occurring **in the cell** are the same in both cases. What differs is the **use being made** of the cell.

Confusion can also arise in the use of the terms anode and cathode. These should be confined to electrolysis. The electrodes of an electrochemical cell are best described in terms of their polarity.

9.3.3   Cells used as batteries as a source of current, i.e. operating irreversibly

The Daniell cell is most commonly used for illustrative purposes. In this cell, a zinc electrode dips into an aqueous solution of $ZnSO_4$ and a copper electrode dips into an aqueous solution of

$CuSO_4$. These two solutions are separated by a porous partition to avoid bulk physical mixing of the two solutions, but which allows electrical contact between the two solutions.

If a Zn(s) wire dips into an aqueous solution of $ZnSO_4$ two things might happen. The Zn(s) might go into solution as Zn^{2+}(aq) or the Zn^{2+}(aq) might deposit out from the solution as Zn(s). Likewise for a Cu(s) wire dipping into an aqueous solution of $CuSO_4$, either Cu(s) might go into solution as Cu^{2+}(aq) or Cu(s) might deposit out. It will be shown later that both Zn(s) and Cu(s) have a tendency to form Zn^{2+}(aq) ions and Cu^{2+}(aq) ions in aqueous solutions of their ions, but that the tendency for Zn(s) to do this is much greater than that for Cu(s). Similarly, when Zn(s) and Cu(s) are placed in a solution containing both Zn^{2+}(aq) and Cu^{2+}(aq), the tendency for Cu(s) to deposit out of solution is greater than for Zn(s).

If the Zn(s) and Cu(s) are connected together via a voltmeter, a potential difference is set up and electrons flow in this external part of the circuit. The electron flow direction is found from the direction of deflection of the voltmeter needle, and for this cell electrons flow from the Zn(s) through the voltmeter to the Cu(s). Since electrons flow from the Zn(s), then electrons are being generated on the Zn(s) electrode by virtue of the chemical reaction occurring at this electrode. Putting electrons into the electrode causes it to be the **negative** electrode. Generation of electrons corresponds to an **oxidation** reaction. By definition, an electrode at which oxidation occurs is the anode, but as indicated earlier this term is best restricted to electrolysis. The important point is that the Zn electrode has a **negative polarity**. The reaction will be conversion of the Zn(s) electrode to Zn^{2+}(aq) in solution. The electrons generated by reaction then leave this electrode to flow, via the terminal of the cell, through the **external** circuit to the Cu electrode.

At the Cu electrode the reaction is, therefore, uptake of electrons, i.e. **reduction**, and at this electrode Cu^{2+}(aq) comes out of solution, takes up the electrons and is deposited as Cu(s). Since the electrons are removed from the electrode it becomes the **positive** electrode, and more importantly it has **positive polarity**. Again, this electrode could be termed the cathode.

The current in the external circuit corresponds to a flow of electrons from the negative Zn electrode to the positive Cu electrode. But, **in the cell itself** the current is carried by the migrating ions. Some of the Cu^{2+}(aq) ions migrate to the electrode at which they are **being used up**, i.e. they migrate **to around the positive electrode**. Electrical neutrality around the Cu(s) electrode is maintained partly by Cu^{2+}(aq) migrating in and partly by SO_4^{2-}(aq) moving out, i.e. some SO_4^{2-}(aq) migrates **to the negative electrode**. It is important to realise that the amount of Cu^{2+}(aq) migrating in only partially replaces the Cu^{2+}(aq) removed by reaction at the electrode (see Worked Problem 9.4).

Some of the Zn^{2+}(aq) ions migrate away from the electrode at which they are being produced, i.e. they migrate **away from around the negative electrode**. Electrical neutrality around the Zn electrode is maintained partly by Zn^{2+}(aq) migrating out and partly by SO_4^{2-}(aq) moving in, i.e. some SO_4^{2-}(aq) migrate **to the negative electrode**. It is also important to realise that not all of the Zn^{2+}(aq) produced at the negative electrode will migrate away.

Take note:

• The migration processes are contrary to what might be expected:

Zn^{2+}(aq) ions migrate **away from** the negative Zn(s) electrode;
Cu^{2+}(aq) ions migrate **to** the positive Cu(s) electrode;
SO_4^{2-}(aq) ions migrate **to** the negative Zn(s) electrode **from** the positive Cu(s) electrode.

- Migration of the ions is the physical process by which the current is carried **through the solution**. Removal or production of the ions at the electrodes is the chemical process by which electrons can **flow to or from the battery to the external circuit**.

- The magnitude of the potential difference set up between the two electrodes of the Daniell cell is a measure of the relative tendencies for the two following processes to occur:

$$Zn(s) \longrightarrow Zn^{2+}(aq) + 2e$$
$$Cu(s) \longrightarrow Cu^{2+} + 2e$$

and depends on the relative magnitudes of the activities of the two ions in solution, and on the temperature. When the cell is used as a source of current it continues to operate until either all of the $Zn(s)$ is used up, or all of the $Cu^{2+}(aq)$ is deposited out of solution. In this respect it is acting irreversibly.

- The **fundamental principle** governing all aspects of electrolyte solutions is that **electrical neutrality** must be maintained throughout the solution (see Section 2.12).

Worked problem 9.4

Question

Explain the statements that not all of the $Zn^{2+}(aq)$ going into solution migrates away from the negative electrode, and that the amount of $Cu^{2+}(aq)$ migrating into the solution around the positive electrode only partially replaces the $Cu^{2+}(aq)$ removed by reaction at that electrode. Use an argument based on transport numbers.

Answer

Processes occurring around the negative electrode

$$Zn(s) \longrightarrow Zn^{2+}(aq) + 2e$$

2 faraday of electricity cause 1 mol of $Zn(s)$ to go into solution as $Zn^{2+}(aq)$. The fraction of the current flowing through the solution carried by the migrating $Zn^{2+}(aq)$ will be t_+, and the **charge** carried by the migrating $Zn^{2+}(aq)$ will be $2t_+$ faraday. Consequently the **amount** of $Zn^{2+}(aq)$ migrating away from the electrode and through the solution will be t_+ mol and not 1 mol. Hence not all of the $Zn^{2+}(aq)$ which is produced at the negative electrode will migrate away, only t_+ mol do. The fraction of the current carried by the $SO_4^{2-}(aq)$ migrating in the opposite direction will be $(1 - t_+)$, i.e. t_-, and the **charge** carried by the migrating $SO_4^{2-}(aq)$ will be $2t_-$ faraday. The **amount** of $SO_4^{2-}(aq)$ migrating to the electrode will be t_- mol not 1 mol.

Remember

both t_+ and t_- are necessarily less than unity: by definition $t_+ + t_- = 1$.

Processes occurring around the positive electrode

$$Cu^{2+}(aq) + 2e \longrightarrow Cu(s)$$

2 faraday of electricity cause 1 mol of $Cu^{2+}(aq)$ to come out of solution as $Cu(s)$. The fraction of the current flowing through the cell carried by the $Cu^{2+}(aq)$ as it migrates towards the $Cu(s)$ electrode will be t_+ and the **charge** carried by the migrating $Cu^{2+}(aq)$ will be $2t_+$ faraday. Consequently the **amount** of $Cu^{2+}(aq)$ migrating to the electrode will be t_+ mol and not 1 mol. Hence not all of the $Cu^{2+}(aq)$ which is removed at the positive electrode will be compensated by the $Cu^{2+}(aq)$ migrating to the electrode, only t_+ mol. The fraction of the current carried by the $SO_4^{2-}(aq)$ migrating in the opposite direction will be $(1 - t_+)$, i.e. t_-, and the **charge** carried by the migrating $SO_4^{2-}(aq)$ will be $2t_-$ faraday. The **amount** of $SO_4^{2-}(aq)$ migrating away will be t_- mol not 1 mol.

These processes **taken together** will maintain **electrical neutrality** throughout the solution.

Worked problem 9.5

Question

Discuss the processes occurring in the following electrochemical set up. A Daniell cell is being used to generate an electric current which causes electrolysis in a solution of $AgNO_3(aq)$. The electrodes are $Ag(s)$ wires dipping into the $AgNO_3(aq)$ solution. It is observed that electrons flow from the $Zn(s)$ electrode of the Daniell cell to the electrolytic cell, and from the electrolytic cell back to the $Cu(s)$ electrode of the Daniell cell.

Answer

- Since electrons flow from the $Zn(s)$ electrode of the Daniell cell they must be generated at that electrode, and so the reaction occurring is release of electrons, i.e. oxidation: $Zn(s) \rightarrow Zn^{2+}(aq) + 2e$. The $Zn(s)$ electrode is the negative electrode of the Daniell cell.

- Electrons flow through the conducting wire to the $Ag(s)$ electrode of the electrolytic cell forcing this electrode to be negative. The reaction occurring is uptake of electrons i.e. reduction: $Ag^+(aq) + e \rightarrow Ag(s)$ and the electrode is the cathode. The $Ag^+(aq)$ in the solution **being electrolysed** will migrate to it.

- The other $Ag(s)$ electrode which is connected to the $Cu(s)$ of the Daniell cell must be a positive electrode since electrons must flow from this electrode to the $Cu(s)$ electrode to complete the circuit. The reaction occurring must generate these electrons, i.e. oxidation occurs: $Ag(s) \rightarrow Ag^+(aq) + e$, and the electrode is an anode. The $NO_3^-(aq)$ anions in solution **being electrolysed** will migrate towards it.

- Since electrons are flowing from the anode of the electrolytic cell to the $Cu(s)$ electrode of the Daniell cell, the reaction occurring at this electrode must take up these electrons i.e. be reduction:

$$Cu^{2+}(aq) + 2e \longrightarrow Cu(s).$$

- Since electrons are being removed at this electrode it is positively charged.

- The current is carried between the two electrodes of the Daniell cell by migration of the ions in the solution through the porous pot separating the two electrode compartments. $Zn^{2+}(aq)$ ions migrate away from the negative electrode of the Daniell cell and $Cu^{2+}(aq)$ ions move towards the positive electrode of the Daniell cell, with SO_4^{2-} (aq) ions moving to the negative electrode.

9.3.4 Cells operating reversibly and irreversibly

Provided the argument is restricted to what happens at the electrodes, the Daniell cell can also be used to demonstrate what is meant by a cell operating reversibly and operating irreversibly.

The Daniell cell, an opposing second cell and a galvanometer are joined up in series in a complete circuit. The galvanometer is present to demonstrate whether a current is flowing and, if so, the direction of the flow of electrons.

- If the second cell has an equal potential difference, then the two potential differences are **exactly balanced**, there will be no current flowing from the Zn(s) through the external circuit to the Cu(s) and the Daniell cell **will operate reversibly**. There will be no reactions occurring at the electrodes.

- If, however, the emf supplied by the cell in the external circuit is **less than** that of the Daniell cell, then electrons will flow from the Zn(s) electrode through the external circuit to the Cu(s) electrode, and the cell is operating **irreversibly** and **spontaneously**. Reaction will occur at the electrodes of the Daniell cell with the Zn(s) electrode being the negative electrode from which electrons will flow to the external circuit, and thence to the Cu(s) electrode which will act as a sink for these electrons and be positively charged. Current will thus flow through the external circuit. The reactions occurring are:

$$\text{at the negative electrode}: \ Zn(s) \rightarrow Zn^{2+}(aq) + 2e$$
$$\text{at the positive electrode}: \ Cu^{2+}(aq) + 2e \rightarrow Cu(s)$$

- If, however, the external emf is **greater than** that of the Daniell cell, then this will force electrons to flow from the cell in the external circuit to the Zn(s) where reaction will be uptake of electrons leaving the Zn(s) electrode positively charged. Electrons will then flow back to the external cell from the Cu(s) which is thus being forced to behave as a supplier of electrons, making it be the negative electrode of the Daniell cell. Current will thus flow through the external circuit. The Daniell cell is again operating **irreversibly** and **spontaneously**, but the electrode reactions occurring in the Daniell cell are now reversed, i.e.:

$$\text{at the negative electrode}: \ Cu(s) \rightarrow Cu^{2+}(aq) + 2e$$
$$\text{at the positive electrode}: \ Zn^{2+}(aq) + 2e \rightarrow Zn(s)$$

This type of experiment defines the conditions for reversibility.

9.3.5 Conditions for reversibility of cells

For an electrochemical cell to be reversible in the thermodynamic sense, it must satisfy **all** of the following conditions:

- When there is no current flowing through the cell, as when it is connected to a balancing external potential difference, then there must be no chemical change in the cell.

- Starting from a state in which no current flows and no chemical change occurs, it must be possible to produce a flow of current through the cell and chemical change in the cell by a **vanishingly small** change in the external circuit.

- It must be possible to exactly reverse the flow of current and the accompanying chemical change by a corresponding **vanishingly small** displacement in the opposite direction.

9.4 Some examples of electrodes used in electrochemical cells

9.4.1 Gas electrodes

The H_2 electrode, $Pt(s)|H_2(g)|HCl(aq)$, is the most common and most often used gas electrode. The vertical lines represent boundaries between different phases, and the $Pt(s)$ is the part of the electrode which connects to any external circuit. This electrode consists of a platinum wire covered by finely divided platinum and saturated with bubbles of $H_2(g)$ at a known pressure dipping into aqueous HCl of known concentration. The reactions occurring at the electrode depend on whether the electrode functions as a positively charged electrode or as a negatively charged electrode in an electrochemical cell.

- **As a positive electrode**: electrons are removed from the electrode leaving it positively charged, reaction is uptake of electrons and **reduction** occurs: $H_3O^+(aq) + e \rightarrow \frac{1}{2}H_2(g) + H_2O(l)$

- **As a negative electrode**: electrons are generated on the electrode making it negatively charged, reaction releases electrons and **oxidation** occurs: $\frac{1}{2} H_2(g) + H_2O(l) \rightarrow H_3O^+(aq) + e$

- Which occurs depends on the electrode to which the H_2 electrode is paired in the electrochemical cell.

- Both $H_2(g)$ and $H_3O^+(aq)$ must be simultaneously present and the electrode is termed **reversible with respect to the species in solution**, here $H_3O^+(aq)$. $H_3O^+(aq)$ is the species in solution which appears in the electrode reactions occurring when it functions as a positive or as a negative electrode.

Other typical gas electrodes are:

(i) chlorine gas bubbling over a platinum wire dipping into a chloride solution, $Pt(s)|Cl_2(g)|Cl^-(aq)$.

(ii) oxygen gas bubbling over a platinum wire dipping into a hydroxide solution, $Pt(s)|O_2(g)|\ OH^-(aq)$

Worked problem 9.6

Question

Write down the reactions occurring at the electrodes when each of these electrodes quoted above functions as (i) a positive electrode, (ii) a negative electrode. What is the species to which these are reversible?

Answer

(i) as a positive electrode: electrons are used up and the electrode is positively charged,

$$Cl_2(g) + 2e \longrightarrow 2Cl^-(aq)$$

$$\tfrac{1}{2}O_2(g) + H_2O(l) + 2e \longrightarrow 2OH^-(aq)$$

(ii) as a negative electrode: electrons are generated and the electrode is negatively charged,

$$2Cl^-(aq) \longrightarrow Cl_2(g) + 2e$$

$$2OH^-(aq) \longrightarrow \tfrac{1}{2}O_2(g) + H_2O(l) + 2e$$

The **species in solution which appears in the electrode reaction** is the species with respect to which the electrode is reversible. This is $Cl^-(aq)$ for the chlorine electrode and $OH^-(aq)$ for the oxygen electrode.

9.4.2 Metal electrode dipping into an aqueous solution of its ions

Typical examples are: $Cu(s)|Cu^{2+}(aq)$, $Zn(s)|Zn^{2+}(aq)$, $Sn(s)|Sn^{2+}(aq)$, $Pb(s)|Pb^{2+}(aq)$, $Ag(s)|Ag^+(aq)$. The reactions occurring again depend on the other electrode in the electrochemical cell. The **species in solution which appears in the electrode reaction** is the species with respect to which the electrode is reversible. In general when the electrode functions:

• as a **positive electrode**: reaction is uptake of electrons, i.e. **reduction**, $M^{z+}(aq) + ze \rightarrow M(s)$,

• as a **negative electrode**: reaction is release of electrons, i.e. **oxidation**, $M(s) \rightarrow M^{z+}(aq) + ze$,

• the metal cation in solution is the species to which these electrodes are reversible.

9.4.3 Metal coated with a sparingly soluble compound of the metal dipping into an aqueous solution containing the anion of the sparingly soluble compound

One of the most common examples is that of a Ag(s) wire coated with a silver halide and dipping into an aqueous solution of the halide. It is worthwhile remembering here that **a source of the halide must always be present in solution**.

1. $Ag(s)|AgCl(s)|HCl(aq)$

The reaction occurring at this electrode depends on the electrode with which it is coupled. It is more difficult to work out what happens at this electrode than with a $Ag(s)|Ag^+(aq)$ electrode. The $AgCl(s)$ is made up of a lattice of Ag^+ and Cl^- ions. The Ag^+ ions in the solid can take up electrons generating $Ag(s)$ which is part of the electrode. The excess Cl^- ions of the lattice end up as ions in the solution. The reverse process can occur. It is important to realise that for this electrode to work all of $Ag(s)$, $AgCl(s)$ and $Cl^-(aq)$ must be simultaneously present and part of the electrode assembly, and the electrode reaction involves the $AgCl(s)$ and the $Cl^-(aq)$ as well as the $Ag(s)$.

When it acts:

- as a **positive electrode**: reaction takes up of electrons, i.e. **reduction**, $AgCl(s) + e \rightarrow Ag(s) + Cl^-(aq)$,

- as a **negative electrode**: reaction releases electrons, i.e. **oxidation**, $Ag(s) + Cl^-(aq) \rightarrow AgCl(s) + e$.

This electrode is called an electrode which is reversible to $Cl^-(aq)$. The electrode is reversible with respect to the **species in solution** which is involved in the electrode reaction. In this case $Cl^-(aq)$ is formed in the solution around the electrode when it acts as a positive electrode, and is removed from the solution around the electrode when it is acting as a negative electrode. If there were no $Cl^-(aq)$ in solution, the electrode would not function as an electrode.

2. $Hg(l)|Hg_2Cl_2(s)|Cl^-(aq)$

This is the calomel electrode and is often used in electrochemical studies. $Hg(l)$, $Hg_2Cl_2(s)$ and $Cl^-(aq)$ must be part of the electrode assembly and **all must be present**, and the electrode reaction involves the $Hg_2Cl_2(s)$ and the $Cl^-(aq)$ as well as the $Hg(l)$.

The solid Hg_2Cl_2 acts as a source of Hg_2^{2+} and Cl^- ions. The Hg_2^{2+} ions in the solid can take up electrons generating $Hg(l)$ which is part of the electrode. The excess Cl^- ions generated end up as ions in the solution. The reverse process can occur.

When it acts:

- as a **positive electrode**: reaction takes up electrons, i.e. **reduction**, $Hg_2Cl_2(s) + 2e \rightarrow 2Hg(l) + 2Cl^-(aq)$;

- as a **negative electrode**: reaction releases electrons, i.e. **oxidation**, $2Hg(l) + 2Cl^-(aq) \rightarrow Hg_2Cl_2(s) + 2e$.

The species in solution is the $Cl^-(aq)$, and the electrode is termed reversible with respect to $Cl^-(aq)$.

Worked problem 9.7

Question

Write down the reactions occurring when the $Pb(s)|PbI_2(s)|KI(aq)$ electrode functions as a (i) positive electrode, (ii) negative electrode. What is the ion with respect to which this electrode is reversible?

Answer

(i) as a positive electrode: reaction must involve uptake of electrons

$$PbI_2(s) + 2e \longrightarrow Pb(s) + 2I^-(aq)$$

(ii) as a negative electrode: reaction must involve generation of electrons

$$Pb(s) + 2I^-(aq) \longrightarrow PbI_2(s) + 2e$$

The electrode is reversible with respect to $I^-(aq)$, and all three species $Pb(s)$, $PbI_2(s)$, $I^-(aq)$ must be present for this to act as an electrode.

9.4.4   The redox electrode

This electrode is made up of a platinum wire dipping into an aqueous solution of ions of a metal in two oxidation states. Alternatively the solution could be that of a complex of a metal in two oxidation states. The electrode is reversible to the oxidised and reduced forms of the species in solution which appear in the electrode reactions.

Typical examples are:

- $Pt(s)$ dipping into $Fe(NO_3)_2(aq)$ and $Fe(NO_3)_3(aq)$, i.e. a solution of $Fe^{2+}(aq)$ and $Fe^{3+}(aq)$.

- Pt dipping into $Fe(CN)_6^{4-}(aq)$ and $Fe(CN)_6^{3-}(aq)$, essentially a solution of $Fe^{2+}(aq)$ and $Fe^{3+}(aq)$.

- $Pt(s)$ dipping into a solution containing $Sn^{2+}(aq)$ and $Sn^{4+}(aq)$ ions.

- $Pt(s)$ dipping into a solution containing $Ce^{3+}(aq)$ and $Ce^{4+}(aq)$.

9.4.5   Reactions occurring at the electrodes in a redox cell

1. $Pt(s)|Fe^{2+}(aq), Fe^{3+}(aq)$

Note: **both** oxidised and reduced forms must be present in the **same** solution, and so are written as separated only by a comma and not by a line.
When this electrode functions:

- as a **positive electrode**: reaction takes up electrons, i.e. **reduction**. The $Fe^{3+}(aq)$ can be reduced, and the electrode reaction is: $Fe^{3+}(aq) + e \rightarrow Fe^{2+}(aq)$.

- as a **negative electrode**: reaction releases electrons, i.e. oxidation. The $Fe^{2+}(aq)$ can be oxidised, and reaction is: $Fe^{2+}(aq) \rightarrow Fe^{3+}(aq) + e$

The species in solution involved in the electrode reactions are $Fe^{2+}(aq)$ and $Fe^{3+}(aq)$. The electrode is reversible with respect to **both** $Fe^{2+}(aq)$ and $Fe^{3+}(aq)$.

2. $Pt(s)|Fe(CN)_6^{4-}(aq), Fe(CN)_6^{3-}(aq)$

This electrode consists of $Pt(s)$ dipping into an aqueous solution containing $Fe(CN)_6^{4-}(aq)$ and $Fe(CN)_6^{3-}(aq)$. But a solution of $Fe(CN)_6^{4-}$ is an equilibrium mixture of $Fe^{2+}(aq)$, $CN^-(aq)$, and the complex $Fe(CN)_6^{4-}(aq)$,

$$Fe(CN)_6^{4-}(aq) \rightleftharpoons Fe^{2+}(aq) + 6CN^-(aq)$$

and functions as a source of $Fe^{2+}(aq)$.

Likewise a solution of $Fe(CN)_6^{3-}(aq)$ is an equilibrium mixture of $Fe^{3+}(aq)$, $CN^-(aq)$, and the complex $Fe(CN)_6^{3-}(aq)$,

$$Fe(CN)_6^{3-}(aq) \rightleftharpoons Fe^{3+}(aq) + 6CN^-(aq)$$

and functions as a source of $Fe^{3+}(aq)$.

The cell is thus essentially a redox cell where $Fe^{2+}(aq)$ can be oxidised, and $Fe^{3+}(aq)$ can be reduced, and the reactions occurring at the electrodes can be written as for $Pt(s)|Fe^{2+}(aq)$, $Fe^{3+}(aq)$

When this electrode functions:

- as a **positive electrode**: **reduction** occurs, $Fe^{3+}(aq) + e \longrightarrow Fe^{2+}(aq)$

- as a **negative electrode**: **oxidation** occurs, $Fe^{2+}(aq) \longrightarrow Fe^{3+}(aq) + e$

The electrode is reversible with respect to **both** $Fe^{2+}(aq)$ and $Fe^{3+}(aq)$, but it could also be said to be reversible with respect to **both** $Fe(CN)_6^{4-}(aq)$ and $Fe(CN)_6^{3-}(aq)$ since these are the principal species present in solution and the complexes are the sources of $Fe^{2+}(aq)$ and $Fe^{3+}(aq)$. When the electrodes are thought of in this way, the reactions are:

- as a **positive electrode**: **reduction** occurs, $Fe(CN)_6^{3-}(aq) + e \rightarrow Fe(CN)_6^{4-}(aq)$

- as a **negative electrode**: **oxidation** occurs, $Fe(CN)_6^{4-}(aq) \rightarrow Fe(CN)_6^{3-}(aq) + e$

Worked problem 9.8

Question

- Giving your reasoning, write down the electrode reactions occurring when the following electrode, $Pt(s)|Co(edta)^{2-}(aq), Co(edta)^-(aq)$, acts as:

(i) a positive electrode, (ii) a negative electrode. What are the species to which this electrode is reversible?

Answer

Both oxidised and reduced forms must be present in the **same** solution, and these can be considered to be either $Co^{3+}(aq)$ or $Co(edta)^-(aq)$ as the oxidised species, and $Co^{2+}(aq)$ or $Co(edta)^{2-}(aq)$ as the reduced species.

A solution of $Co(edta)^{2-}$ is an equilibrium mixture of $Co^{2+}(aq)$, $edta^{4-}(aq)$, and the complex $Co(edta)^{2-}(aq)$, and so functions as a source of $Co^{2+}(aq)$.

Likewise a solution of $Co(edta)^-(aq)$ is an equilibrium mixture of $Co^{3+}(aq)$, $edta^{4-}(aq)$ and the complex $Co(edta)^-(aq)$, and so functions as a source of $Co^{3+}(aq)$.

The cell is thus essentially a redox cell where $Co^{2+}(aq)$ can be oxidised, and $Co^{3+}(aq)$ can be reduced.

When this electrode functions:

- as a **positive electrode**: reaction takes up electrons, i.e. **reduction**, and reaction is $Co^{3+}(aq) + e \rightarrow Co^{2+}(aq)$, or equivalently $Co(edta)^-(aq) + e \rightarrow Co(edta)^{2-}(aq)$.

- as a **negative electrode**: reaction releases electrons, i.e. **oxidation**, and reaction is $Co^{2+}(aq) \rightarrow Co^{3+}(aq) + e$, or equivalently $Co(edta)^{2-}(aq) \rightarrow Co(edta)^-(aq) + e$.

The electrode is reversible with respect to **both** $Co^{2+}(aq)$ and $Co^{3+}(aq)$, or equivalently to **both** $Co(edta)^{2-}(aq)$ and $Co(edta)^-(aq)$.

9.4.6 The amalgam electrode

These are mainly used in the study of reactions involving alkali metals. Because of the extreme reactivity of alkali metals it is impossible to have a reversible electrode of the type $M(s)|M^+(aq)$ or $M(s)|MX(s)|X^-(aq)$. This can be overcome by dissolving the alkali metal in $Hg(l)$ which gives an alkali metal amalgam which is attacked much less vigorously by water. The $Na(Hg)|Na^+(aq)$ electrode is a common example of this type of electrode which behaves reversibly.

When this electrode functions:

- as a **positive electrode**: reaction takes up electrons, i.e. **reduction**, $Na^+(aq) + e \rightarrow Na(Hg)$,

- as a **negative electrode**: reaction releases electrons, i.e. **oxidation**, $Na(Hg) \rightarrow Na^+(aq) + e$.

This electrode is reversible with respect to $Na^+(aq)$ since this is the species in solution.

A similar electrode is the cadmium amalgam electrode, $Cd(Hg)|Cd^{2+}(aq)$ where:

$$\text{reaction as a positive electrode} : Cd^{2+}(aq) + 2e \rightarrow Cd(Hg)$$
$$\text{reaction as a negative electrode} : Cd(Hg) \rightarrow Cd^{2+}(aq) + 2e$$

The electrode is reversible with respect to $Cd^{2+}(aq)$.

Worked problem 9.9

Question

Write down the reactions occurring when the following electrodes act as:

- a positively charged electrode;

- a negatively charged electrode;

- write down the species to which the electrodes are reversible.

1. $Pb(s|Pb^{2+}(aq)$

2. $Pt(s)|Sn^{2+}(aq), Sn^{4+}(aq)$ ions.

3. $Cu(s)|CuCl(s)|Cl^-(aq)$

4. $Pt(s)|Ce^{3+}(aq), Ce^{4+}(aq)$.

5. $Pb(s)|PbSO_4(s)|SO_4^{2-}(aq)$

Answer

Remember – when an electrode functions as:

- a positively charged electrode, reaction is uptake of electrons, i.e. reduction;

- a negatively charged electrode, reaction is release of electrons, i.e. oxidation.

- The number of electrons involved is determined by the charges on the oxidised and reduced forms.

1. $Pb(s)|Pb^{2+}(aq)$
   positive electrode: $Pb^{2+}(aq) +2e \rightarrow Pb(s)$
   negative electrode: $Pb(s) \rightarrow Pb^{2+}(aq)+2e$
   reversible with respect to: $Pb^{2+}(aq)$

2. $Pt(s)|Sn^{2+}(aq), Sn^{4+}(aq)$ ions.
   positive electrode: $Sn^{4+}(aq) +2e \rightarrow Sn^{2+}(aq)$
   negative electrode: $Sn^{2+}(aq) \rightarrow Sn^{4+}(aq)+2e$
   reversible with respect to: $Sn^{2+}(aq)$ and $Sn^{4+}(aq)$

3. $Cu(s)|CuCl(s)|Cl^-(aq)$
   positive electrode: $CuCl(s) + e \longrightarrow Cu(s) + Cl^-(aq)$

negative electrode: $Cu(s) + Cl^-(aq) \rightarrow CuCl(s) + e$

reversible with respect to: $Cl^-(aq)$.

4. $Pt(s)|Ce^{3+}(aq), Ce^{4+}(aq)$.

positive electrode: $Ce^{4+}(aq) + e \rightarrow Ce^{3+}(aq)$

negative electrode: $Ce^{3+}(aq) \rightarrow Ce^{4+}(aq) + e$

reversible with respect to: $Ce^{3+}(aq)$ and $Ce^{4+}(aq)$

5. $Pb(s)|PbSO_4(s)|SO_4^{2-}(aq)$

positive electrode: $PbSO_4(s) + 2e \rightarrow Pb(s) + SO_4^{2-}(aq)$

negative electrode: $Pb(s) + SO_4^{2-}(aq) \rightarrow PbSO_4(s) + 2e$

reversible with respect to: $SO_4^{2-}(aq)$

9.4.7   Glass electrodes

A glass electrode generally consists of a tube and bulb containing a known concentration of $HCl(aq)$ into which there is inserted a $Ag(s)|AgCl(s)$ electrode. The tube and bulb are made of special glass which is permeable to, and reversible with respect to $H_3O^+(aq)$. The glass electrode is coupled with a second electrode, often a calomel electrode, and both are placed in an external solution whose properties are being studied. The emf of the cell set up is governed by the pH of the external solution. This set-up is particularly suited to the measurement of pH, and is the standard method for determining the pH of unknown solutions. The theory of the glass electrode is complicated, and expressions for the emfs of such cells are not derived for routine work. Instead, the cell is calibrated against solutions of accurately known pH. Special glasses are used for solutions which are strongly acidic or strongly basic.

'Ion-selective' electrodes sensitive to ions other than $H_3O^+(aq)$ can also be used. Here the glass of the electrode is permeable and reversible with respect to the specific ion in question. Silver halide electrodes are used to determine $Cl^-(aq)$, $Br^-(aq)$, $I^-(aq)$ and $CN^-(aq)$; lanthanum fluoride to determine $F^-(aq)$; Ag_2S electrodes to determine $S^{2-}(aq)$, and $Ag^+(aq)$, and electrodes made from a mixture of divalent metal sulphides and Ag_2S are used to determine $Pb^{2+}(aq)$, $Cu^{2+}(aq)$ and $Cd^{2+}(aq)$. Other types determine $Na^+(aq)$, $K^+(aq)$, $Ca^{2+}(aq)$, organic cations, and anions such as $NO_3^-(aq)$, $ClO_4^-(aq)$ and organic anions. Ion-selective electrodes have proved very useful and particularly suitable for monitoring the concentration of ions other than $H_3O^+(aq)$ or $OH^-(aq)$ in biological systems.

9.5   Combination of electrodes to make an electrochemical cell

If a $Pt(s)|H_2(g)|HCl(aq)$ electrode is coupled with a $Hg(l)|Hg_2Cl_2(s)|Cl^-(aq)$ electrode the resulting cell will be a potential source of an electric current. If the two electrodes are connected through an external circuit containing a voltmeter, deflection of the needle will indicate that a potential difference has been set up between the two terminals and that an electric current is flowing. The direction of the flow of electrons along the wires of the external circuit is given by the direction of the deflection (see Section 9.6.1), and in this example the electrons flow from the $Pt(s)|H_2(g)|HCl(aq)$ electrode to the $Hg(l)|Hg_2Cl_2(s)|Cl^-(aq)$ electrode. This means that oxidation, i.e. $\frac{1}{2}H_2(g) + H_2O(l) \rightarrow H_3O^+(aq) + e$, must be occurring at the $Pt(s)|H_2(g)|HCl(aq)$ electrode, which is thus the negative electrode. Reduction must be occurring at the $Hg(l)|Hg_2Cl_2(s)|Cl^-(aq)$ electrode, i.e. $Hg_2Cl_2(s) + 2e \rightarrow 2Hg(l) + 2Cl^-(aq)$, which is thus the positive electrode.

If, however, the $Pt(s)|H_2(g)|HCl(aq)$ electrode is linked to a $Mg(s)|Mg^{2+}(aq)$ electrode, deflection occurs in the opposite direction indicating that in this cell the potential difference set up when the circuit is completed causes the electrons to flow from the $Mg(s)|Mg^{2+}(aq)$ electrode to the $Pt(s)|H_2(g)|HCl(aq)$ electrode. Reduction is now occurring at the $Pt(s)|H_2(g)|$ $HCl(aq)$ electrode, i.e. $H_3O^+(aq) + e \rightarrow \frac{1}{2}H_2(g) + H_2O(l)$, making it the positive electrode. Oxidation occurs at the $Mg(s)|Mg^{2+}(aq)$ electrode, i.e. $Mg(s) \rightarrow Mg^{2+}(aq) + 2e$, making this the negative electrode.

Experiment can thus determine the direction of flow of electrons in the external circuit and can determine which is the positive electrode and which is the negative electrode of the cell set up. It now remains to decide on a standardised manner of writing down the cell which would allow inferences to be made about the polarity of the electrodes, reaction at the two individual electrodes and the direction of flow of the electrons in the external circuit.

9.6 Conventions for writing down the electrochemical cell

The cell is written down from **left to right** so that the inference is that:

- the reaction occurring at the **left** hand electrode **produces** electrons, i.e. is **oxidation, and this electrode has negative polarity**;

- the reaction occurring at the **right** hand electrode **uses up** electrons, i.e. is **reduction, and this electrode has positive polarity**;

- the overall cell reaction is a balanced sum of the processes occurring at each electrode;

- solid lines represent phase boundaries. Double broken lines represent a salt bridge, which keeps the solutions physically apart, but allows the current to be carried between the two separated solutions by virtue of migration of the ions of the salt bridge (see Section 9.8.6). A single broken line between two solutions means that they are physically separated by a narrow tube or a porous partition which allows migration of the ions of the solution through the porous partition or tube, but inhibits physical mixing to give a homogeneous solution (see Sections 9.8.4 and 9.8.5).

It now remains to find out experimentally whether the reactions which **actually occur** are those inferred from the way in which the cell is written down. This requires finding the polarity of the electrodes, or put otherwise, finding the direction of the flow of electrons in the external circuit.

- If the left hand electrode has a negative polarity and electrons flow from the left hand electrode to the right hand electrode which will therefore be of positive polarity, then the emf is deemed to be positive.

- If the electron flow is from the right hand electrode, which has negative polarity, to the left hand electrode which will have a positive polarity, then the emf is deemed to be negative.

- This requires finding the polarity of the electrodes. This is found using a digital voltmeter which measures both the potential difference and the direction of the current.

9.6.1 Use of a voltmeter to determine the polarity of the electrodes

When a voltmeter is connected to a cell the polarity of the electrodes determines the polarity of the terminals of the voltmeter, and this, in turn, determines the direction of deflection of the needle of the voltmeter.

For some cells it is possible to determine the polarity of each electrode by direct observation of the reactions at the electrodes when the cell is operating irreversibly as a source of current. One such cell is $Zn(s)|ZnCl_2(aq) \vdots CuCl_2(aq)|Cu(s)$. When run over a period of time it becomes clear that $Zn(s)$ is going into solution making it the negative electrode and $Cu(s)$ is coming out of solution making it the positive electrode. The standard convention (Section 9.6) states the emf to be positive since the actual reactions occurring are the same as those inferred from the way the cell is written down. The direction of the deflection of the voltmeter needle will correspond to a positive emf, and in modern digital voltmeters this often is a deflection to the right. The polarities of the terminals of the voltmeter corresponding to a positive emf are now known.

If another cell is taken where reduction is occurring at the left hand electrode, then this electrode will have a positive polarity and a deflection in the opposite direction is found. This will correspond to a negative emf since the polarities of the terminals have been reversed.

From experiments such as these, the direction of the deflection will indicate the polarities of the terminals of the voltmeter which, in turn, allows the polarities of the electrodes of the cell to be inferred. If these turn out to be what is inferred from the way in which the cell is written down, i.e. the left hand electrode has negative polarity corresponding to oxidation, then the emf is positive. If, however, the left hand electrode turns out to have a positive polarity corresponding to reduction, then the emf is negative.

Worked problem 9.10

Question

The cell, $Zn(s)|ZnSO_4(aq)|Zn(Hg)$, is shown to have a positive emf. Using the convention given above, write down, with explanation, (i) what the electrode reactions are, (ii) what the overall reaction is, (iii) the polarity of the electrodes, (iv) the direction of flow of the current in the external circuit, and (v) explain how the current is carried in solution.
What differences would it make if the electrolyte were changed to $ZnCl_2(aq)$?

Answer

$Zn(s)|ZnSO_4(aq)|Zn(Hg)$:

Using the convention:

- left hand electrode: produces electrons, i.e. oxidation $\therefore$ the negative electrode reaction is:
  $Zn(s) \rightarrow Zn^{2+}(aq) + 2e$

- right hand electrode: uses up electrons, i.e. reduction $\therefore$ the positive electrode reaction is:
$Zn^{2+}(aq) + 2e \rightarrow Zn(Hg)$

- The overall reaction is the sum of these two electrode reactions.

But watch: check that these balance in terms of electrons; they do:

$$Zn(s) \rightarrow Zn(Hg)$$

Since the measured emf is positive this is the actual reaction occurring in the cell.

The current is carried through the **external circuit** by electrons flowing from the $Zn(s)|ZnSO_4(aq)$ electrode to the $ZnSO_4(aq)|Zn(Hg)$ electrode.

The current is carried **across the solution** to around the electrodes by migration of the ions in solution.

At the **negative** electrode $Zn(s)$ goes into solution and reaction is $Zn(s) \rightarrow Zn^{2+}(aq) + 2e$. Passage of 2 faraday of electricity forms 1 mol of $Zn^{2+}(aq)$ around the electrode, causes t_+ mol of $Zn^{2+}(aq)$ to migrate from around this electrode and across the solution giving a net gain of $(1 - t_+)$ mol of $Zn^{2+}(aq)$. To maintain electrical neutrality $(1 - t_+)$ mol, i.e. t_- mol of $SO_4^{2-}(aq)$ must migrate across the solution to the negative electrode.

At the **positive** amalgam electrode $Zn^{2+}(aq)$ is discharged and reaction is $Zn^{2+}(aq) + 2e \rightarrow Zn(Hg)$. Passage of 2 faraday of electricity will discharge, i.e. remove 1 mol of $Zn^{2+}(aq)$ from around the electrode, causes t_+ mol of $Zn^{2+}(aq)$ to migrate across the solution towards the positive electrode, giving a net loss of $(1 - t_+)$ mol of $Zn^{2+}(aq)$. To maintain electrical neutrality $(1 - t_+)$ mol, i.e. t_- mol of $SO_4^{2-}(aq)$ must migrate away from this electrode across the solution to the negative electrode.

Note: the direction of migration of the ions to the electrodes is the **opposite** of what is found in **electrolysis**. For the electrochemical cell the anions migrate to the negatively charged electrode and the cations migrate to the positively charged electrode (Section 9.3.3).

$Zn(s)|ZnCl_2(aq)|Zn(Hg)$

The electrode reactions and the overall reaction are the same as above. The difference lies in the **migration of the ions in the solution**. It is $Cl^-(aq)$, not $SO_4^{2-}(aq)$, which now migrates. And **more importantly** $2(1 - t_+)$ mol or $2t_-$ mol of $Cl^-(aq)$ migrates in contrast to $(1 - t_+)$ mol or t_- mol of $SO_4^{2-}(aq)$. The $Cl^-(aq)$ has only one negative charge compared with two for the $SO_4^{2-}(aq)$ and so twice as much must migrate to enable electrical neutrality to be maintained.

At the **negative** electrode $Zn(s)$ goes into solution and reaction is:

$$Zn(s) \rightarrow Zn^{2+}(aq) + 2e$$

Passage of 2 faraday of electricity:

- forms 1 mol of $Zn^{2+}(aq)$ around the electrode;

- causes t_+ mol of $Zn^{2+}(aq)$ to migrate from this electrode and across the solution;

- giving a net gain of $(1 - t_+)$ mol of $Zn^{2+}(aq)$;

- to maintain electrical neutrality $2(1 - t_+)$ mol, i.e. $2t_-$ mol of $Cl^-(aq)$ must migrate across the solution to the negative electrode.

At the **positive** amalgam electrode $Zn^{2+}(aq)$ is discharged and reaction is:

$$Zn^{2+}(aq) + 2e \rightarrow Zn(Hg)$$

Passage of 2 faraday of electricity:

- will remove 1 mol of $Zn^{2+}(aq)$ from around the electrode;

- causes t_+ mol of $Zn^{2+}(aq)$ to migrate across the solution towards the positive electrode;

- giving a net loss of $(1-t_+)$ mol of $Zn^{2+}(aq)$

to maintain electrical neutrality $2(1-t_+)$ mol i.e. $2t_-$ mol of $Cl^-(aq)$ must migrate across the solution to the negative electrode.

Note well: It is $Cl^-(aq)$ which now migrates. And **more importantly** $2(1 - t_+)$ mol or $2t_-$ mol of $Cl^-(aq)$ migrates in contrast to $(1 - t_+)$ mol or t_- mol of $SO_4^{2-}(aq)$ in the cell $Zn(s)|ZnSO_4(aq)|Zn(Hg)$. The factor of two is required since the $Cl^-(aq)$ has only one negative charge compared with two for the $SO_4^{2-}(aq)$ and so twice as much must migrate to enable electrical neutrality to be maintained.

Worked problem 9.11

Question

The cell, $Pt(s)|H_2(g)|HCl(aq)|AgCl(s)|Ag(s)$, is shown to have a positive emf, but the cell, $Pt(s)|H_2(g)|HCl(aq) \vdots\vdots CdCl_2(aq)|Cd(s)$, is shown to have a negative emf. Using the convention given above, write down, with explanation, (i) what the electrode reactions are, (ii) the polarity of the electrodes, (iii) what the overall reaction is, (iv) the direction of flow of the current in the external circuit, and (v) indicate how the current is carried in solution.

Answer

$Pt(s)|H_2(g)|HCl(aq)|AgCl(s)|Ag(s)$

The inference from the way the cell is written down is:

- left hand electrode: oxidation $\therefore$ negative electrode: $\frac{1}{2}H_2(g) + H_2O(l) \rightarrow H_3O^+(aq) + e$

- right hand electrode: reduction $\therefore$ positive electrode: $AgCl(s) + e \rightarrow Ag(s) + Cl^-(aq)$

- the overall reaction is the sum of these two electrode reactions, **but watch**: these must be balanced so that the same number of electrons appears in each:

$$\frac{1}{2}H_2(g) + H_2O(l) + AgCl(s) \longrightarrow H_3O^+(aq) + Ag(s) + Cl^-(aq)$$

- since the measured emf is positive this is the actual reaction occurring in the cell.

- electrons are formed at the left hand electrode, i.e. $Pt(s)|H_2(g)|HCl(aq)$ and flow through the **external circuit** to the right hand electrode, i.e. $Cl^-(aq)|AgCl(s)|Ag(s)$

- the current is carried **across the solution** by the ions in solution.

For passage of 1 faraday of electricity,

- 1 mol of $H_3O^+(aq)$ will be formed at the $Pt(s)|H_2(g)|HCl(aq)$ electrode and t_+ mol will migrate away from the electrode and across the solution giving a net gain of $(1 - t_+)$ mol of $H_3O^+(aq)$. $(1-t_+)$ mol, i.e. t_- mol of $Cl^-(aq)$ migrates across the solution to the electrode to maintain electrical neutrality.

- 1 mol $Cl^-(aq)$ is formed at the $Cl^-(aq)|AgCl(s)|Ag(s)$ electrode and t_- mol of $Cl^-(aq)$ migrates away from this electrode across the solution giving a net gain of $(1-t_-)$ mol of $Cl^-(aq)$. $(1-t_-)$ mol i.e. t_+ mol of $H_3O^+(aq)$ will migrate across the solution to this electrode to maintain electrical neutrality.

$Pt(s)|H_2(g)|HCl(aq) \vdots CdCl_2(aq)|Cd(s)$

The inference from the way the cell is written down is:

- left hand electrode: oxidation $\therefore$ negative electrode: $\frac{1}{2}H_2(g) + H_2O(l) \rightarrow H_3O^+(aq) + e$

- right hand electrode: reduction $\therefore$ positive electrode: $Cd^{2+}(aq) + 2e \rightarrow Cd(s)$

- since the measured emf is **negative** then the actual reactions occurring at the electrodes are the **opposite** from those inferred from the way the cell is written down, i.e.

- left hand electrode: reduction is occurring $\therefore$ positive electrode: $H_3O^+(aq) + e \rightarrow \frac{1}{2}H_2(g) + H_2O(l)$

- right hand electrode: oxidation $\therefore$ negative electrode: $Cd(s) \rightarrow Cd^{2+}(aq) + 2e$

- the overall reaction is the sum of these two electrode reactions.

   But watch: the same number of electrons must appear in both electrode reactions:

$$Cd(s) + 2H_3O^+(aq) \rightarrow Cd^{2+}(aq) + H_2(g) + 2H_2O(l)$$

- the electron flow in the **external circuit** is from the right hand negative electrode, i.e. $CdCl_2(aq)|Cd(s)$ to the left hand positive electrode i.e. $HCl(aq)|H_2(g)|Pt(s)$.

- the current is carried **across the solution** by the ions in solution.

Passage of 2 faraday of electricity will cause:

- 2 mol of $H_3O^+(aq)$ to be removed from around the $Pt(s)|H_2(g)|HCl(aq)$ electrode and $2t_+$ mol of $H_3O^+(aq)$ to migrate across the solution towards it, leaving a net loss of $2(1 - t_+)$ mol of $H_3O^+(aq)$. To maintain electrical neutrality $Cl^-(aq)$ will move away from the electrode across the solution to the salt bridge.

- 1 mol of $Cd^{2+}(aq)$ is formed at the $Cd^{2+}(aq)|Cd(s)$ electrode but only t_+ mol $Cd^{2+}(aq)$ will migrate from the electrode leaving a net gain of $(1 - t_+)$ mol of $Cd^{2+}(aq)$. To maintain electrical neutrality $2(1 - t_+)$ mol $= 2t_-$ mol of $Cl^-(aq)$ will migrate across the solution to the $Cd^{2+}(aq)|Cd(s)$ electrode.

But note: There will also be migration occurring at the junctions of the ends of the salt bridge with the solution. These will be described in Section 9.8.6.

Watch: The factor of two is required since the $Cl^-(aq)$ and $H_3O^+(aq)$ have only one negative charge compared with two for the $Cd^{2+}(aq)$ and so twice as much must migrate to enable electrical neutrality to be maintained.

Note well: the important point is the direction of the flow of electrons in the external circuit. This is found from the experimentally measured emf, and it determines the polarity of the electrodes which in turn determines the type of reaction occurring at the electrode, viz., whether it be uptake of electrons or production of electrons, and this in turn determines the migration process in solution.

9.7   One very important point: cells corresponding to a 'net chemical reaction'

In the above description of what happens in a cell, an overall reaction has been found by combination of the reactions occurring at the two electrodes. This overall cell reaction is a 'formal representation' in the sense that it **does not actually take place** in the cell. The only chemical reactions which actually occur are those at the electrodes, but their net effect corresponds in quantitative terms to what would be expected if the overall chemical reaction did actually occur. The observed potential difference or emf is **related** to the ΔG for the overall cell reaction. It is this property of electrochemical cells which makes them so useful as they allow determination of thermodynamic quantities which are impossible to study directly.

The cell, $Zn(s)|ZnSO_4(aq) \vdots CuSO_4(aq)|Cu(s)$, corresponds to the overall reaction $Zn(s) + Cu^{2+}(aq) \rightarrow Zn^{2+}(aq) + Cu(s)$. But this reaction does not actually happen in the cell. The cell is **totally** distinct from the situation when some solid Zn is placed in a solution of $CuSO_4(aq)$ where the chemical reaction does actually occur, viz. the lump of Zn(s) disappears gradually and Cu(s) appears instead and only one solution is involved.

9.8   Liquid junctions in electrochemical cells

An electrochemical cell must have two electrodes coupled together so that the current can flow around the whole electrical circuit. There are several ways in which this can be achieved.

9.8.1   Cells without liquid junction

Two electrodes can dip into one solution with each electrode being reversible with respect to one of the ions of the solution, e.g. a hydrogen electrode, $Pt(s)|H_2(g)|HCl(aq)$ could be coupled with a $Ag(s)|AgCl(s)|Cl^-(aq)$. The hydrogen electrode is reversible with respect to the $H_3O^+(aq)$ of the HCl(aq). The $Ag(s)|AgCl(s)|Cl^-(aq)$ is reversible with respect to $Cl^-(aq)$, and this can be supplied by the HCl(aq) or by NaCl(aq) added to the solution as a source of $Cl^-(aq)$. The current is carried by the migrating ions of the **common** solution. The two possible cells would be: $Pt(s)|H_2(g)|HCl(aq)|AgCl(s)|Ag(s)$, or $Pt(s)|H_2(g)|HCl(aq),NaCl(aq)|AgCl(s)|Ag(s)$.

Such a cell can operate rigorously as a reversible cell, and is termed a cell **without a liquid junction**. Such cells are ideal since, by operating reversibly, they furnish exact thermodynamic quantities.

9.8.2 Cells with liquid junction

These are cells where the two electrode solutions must be kept physically separate. Basically there are two types:

- where the two electrode solutions must be kept physically separate because chemical reaction would occur between the ions of the two solutions, e.g. the cell $Zn(s)|ZnCl_2(aq) \vdots AgNO_3(aq)|Ag(s)$. Unless these two solutions were kept separate, here by a salt bridge, AgCl(s) would precipitate out. Another example is the Daniell cell, $Zn(s)|ZnSO_4(aq) \vdots CuSO_4(aq)|Cu(s)$ where reaction of Zn(s) with $Cu^{2+}(aq)$ would occur if the electrode compartments were not separated by the porous pot.

- where there are two chemically identical electrodes with the solutions around the electrode at different concentrations (see Section 9.18). These solutions must be kept separate otherwise physical mixing would rapidly occur to produce a homogeneous solution at one concentration. One example would be copper electrodes dipping into $CuSO_4$ solutions of different concentrations. These are called **concentration** cells.

9.8.3 Types of liquid junctions

There are three ways in which this physical separation can be achieved:

- by means of a narrow tube separating the solutions;

- by means of a porous pot separating the two electrode solutions;

- or by means of a salt bridge.

Under these conditions the current is carried in solution by ions migrating in the solutions **and also** through the narrow tube, or the porous pot, or through the salt bridge. Such cells do not operate reversibly in a rigorous manner because the processes occurring at the junctions will contribute to the thermodynamic quantities. Such cells are called cells **with liquid junctions**.

9.8.4 Cells with a liquid junction consisting of a narrow tube: a cell with transference

In the $Zn(s)|ZnCl_2(aq) \vdots CuCl_2(aq)|Cu(s)$ cell the two solutions must not come into direct contact otherwise chemical reaction would occur. Direct physical mixing can be avoided by connecting the two electrode solutions by a long narrow tube which initially has only air present.

When the solutions are connected by the tube, some $ZnCl_2(aq)$ flows into one end and some $CuCl_2(aq)$ flows into the other end and the two solutions will meet somewhere along the tube. Initially, the solutions in the tube each have their bulk concentrations, but they will immediately start to mix to give a uniform gradient of concentration along the tube for both solutions. This concentration gradient is set up very rapidly, before any measurements will be made.

Thereafter, a very much slower diffusive process is set up along the tube by virtue of the composition gradients along the tube. This occurs irrespective of whether or not a current flows. $Zn^{2+}(aq)$ diffuses along the tube towards the $CuCl_2(aq)|Cu(s)$ electrode compartment, while $Cu^{2+}(aq)$ diffuses in the opposite direction. $Cl^-(aq)$ diffuses in a direction which will maintain electrical neutrality along the length of the tube. Eventually, if the cell were left indefinitely in the absence of any electrical activity, this very slow diffusive process would replace the concentration gradients with uniform concentrations of both electrolytes. In practice, only an infinitesimal amount of diffusion takes place when the cell is being used to gather electrochemical quantities, and so this diffusive process is generally ignored in setting up the theoretical expressions for the emf of the cell. This is fortunate since the diffusive process is thermodynamically irreversible, making any theoretical description extremely difficult. In general, diffusion is also minimised by having a very narrow tube.

When the cell is operating, e.g. when the emf is being measured, then the current is carried through the tube by migration of the ions of the two electrolytes, here $Zn^{2+}(aq)$, $Cu^{2+}(aq)$ and $Cl^-(aq)$. In the $Zn(s)|ZnCl_2(aq)$ electrode compartment, the current is carried by $Zn^{2+}(aq)$ migrating away from the $Zn(s)$ with electrical neutrality being maintained by $Cl^-(aq)$ migrating towards the $Zn(s)|ZnCl_2(aq)$ electrode. In the $CuCl_2(aq)|Cu(s)$ electrode compartment the current is carried by $Cu^{2+}(aq)$ migrating to the $Cu(s)$ with electrical neutrality being again maintained by $Cl^-(aq)$ migrating away from the $CuCl_2(aq)|Cu(s)$ electrode (see Section 9.3.3). Within the tube itself $Zn^{2+}(aq)$ will continue **migrating towards** the $CuCl_2(aq)|Cu(s)$ electrode compartment and $Cu^{2+}(aq)$ will migrate along the tube in **the same direction**, with electrical neutrality being maintained by the $Cl^-(aq)$ moving in **the opposite direction**. Derivation of a theoretical expression for the contribution to the overall emf of the cell from this migration of the ions through the tube can be made, but this is complicated by there being two different cations moving through the tube.

Because there are concentration gradients along the tube which may not necessarily balance out, there will be a small charge gradient along the tube. As a consequence a potential difference is set up along the tube. This is called the liquid junction potential. It can be positive, or negative, and provided it is reproducible can often be ignored. Calculations of the magnitude of this potential difference can be made for the simple case where one cation only migrates through the tube, but even this calculation is difficult. An indication of the magnitude of this potential difference can be made in certain cases.

Worked problem 9.12

Question

The cell $Zn(s)|ZnCl_2(aq) \vdots CuCl_2(aq)|Cu(s)$ has been used to illustrate the liquid junction using a narrow tube. It has a positive emf. Explain the direction of migration of the ions $Zn^{2+}(aq)$, $Cu^{2+}(aq)$ and $Cl^-(aq)$ in this cell and in the tube.

Answer

Using the convention:

- left hand electrode: produces electrons, i.e. oxidation ∴ the negative electrode:
  $Zn(s) \rightarrow Zn^{2+}(aq) + 2e$

- right hand electrode: uses up electrons, i.e. reduction ∴ the positive electrode:
  $Cu^{2+}(aq) + 2e \rightarrow Cu(s)$

- The overall reaction is the sum of these two electrode reactions.
  $Zn(s) + Cu^{2+}(aq) \rightarrow Cu(s) + Zn^{2+}(aq)$

Since the measured emf is positive this is the actual reaction occurring in the cell.

At the Zn(s)|ZnCl$_2$(aq) electrode, passage of 2 faraday of electricity forms 1 mol of Zn^{2+}(aq) at the electrode, causes t_+ mol of Zn^{2+}(aq) to migrate from around this electrode, giving a net gain of $(1 - t_+)$ mol of Zn^{2+}(aq). To maintain electrical neutrality $2(1 - t_+)$ mol, i.e. $2t_-$ mol of Cl^-(aq) must migrate to the solution around the negative electrode and this will be replaced by Cl^-(aq) migrating in **from the tube**.

Remember Cl^- has only one negative charge while Zn^{2+} has two positive charges.

At the CuCl$_2$(aq)|Cu(s) electrode, passage of 2 faraday of electricity discharges 1 mol of Cu^{2+}(aq), causes t_+ mol of Cu^{2+}(aq) to migrate towards the positive electrode, giving a net loss of $(1 - t_+)$ mol of Cu^{2+}(aq). To maintain electrical neutrality $2(1 - t_+)$ mol, i.e. $2t_-$ mol of Cl^-(aq) must migrate away from around the negative electrode and **into the tube**.

Migration along the tube
There are three ions in the tube, i.e. Zn^{2+}(aq), Cu^{2+}(aq) and Cl^-(aq). Since Zn^{2+}(aq) is being formed at the Zn(s)|ZnCl$_2$(aq) electrode and is migrating away from the electrode, it will continue migrating along the tube towards the CuCl$_2$(aq)|Cu(s) electrode. Since Cu^{2+}(aq) is being removed from the solution around the CuCl$_2$(aq)|Cu(s) electrode, Cu^{2+}(aq) will migrate to this electrode from the solution around the electrode and from the tube. The Cl^-(aq) will migrate along the tube from the CuCl$_2$(aq)|Cu(s) electrode to the Zn(s)|ZnCl$_2$(aq) electrode, i.e. in the opposite direction to the migrating Zn^{2+}(aq) and Cu^{2+}(aq).

The current is carried through the tube by these migratory processes. This allows the electrical circuit to be completed.

9.8.5 Cells with a porous pot separating two solutions: a cell with transference

The Daniell cell, Zn(s)|ZnSO$_4$(aq) $\vdots$ CuSO$_4$(aq)|Cu(s), is the standard example of this type of junction. Here a porous pot replaces the narrow tube to separate the two electrode compartments. The porous pot is a more complex situation since there are a large number of channels through the pot linking each compartment. Each individual channel is equivalent to a very narrow tube. The processes occurring in this arrangement are essentially the same as those described for the narrow tube. A liquid junction potential is set up across the porous pot, and, as with the tube, very slow diffusion takes place through the channels. As previously discussed this process is ignored in the theoretical discussion of what contributes to the overall emf of the cell. The migration processes are similar to those described for the cell where a narrow tube is used to provide the liquid junction.

9.8.6   Cells with a salt bridge: cells without transference

The main method of minimising the liquid junction potential and making it reproducible is to use a salt bridge instead of the tube. When a narrow tube is used it operates by virtue of the solutions in the individual electrode compartments flowing along the tube until they meet: there are no extraneous ions in the tube. With the salt bridge, however, the tube is wide and is filled with a saturated or highly concentrated solution, generally of KCl(aq), KNO_3(aq) or NH_4NO_3(aq), and it is principally the ions of the salt bridge electrolyte which do the migration into or out from the salt bridge. Under reversible conditions, there is only a limited amount of migration into and out of the ends of the salt bridge.

Cells using a salt bridge are called **cells without transference** since the migration process is largely independent of the ions of the electrode compartments.

The liquid junction potential set up by these salt bridges is often highly reproducible and very small, provided that the solution of the bridge is highly concentrated or saturated. Studies of the emf of a given cell using salt bridges with differing concentrations show that, as the concentration of the solution of the salt bridge increases while everything else remains the same, the observed emf of the cell approaches a constant value. This suggests that the liquid junction potential is dropping to a minimal, constant and reproducible value. Saturated solutions of the bridge electrolyte are generally used.

For comparison with the cell discussed above, i.e. Zn(s)|$ZnCl_2$(aq) $\vdots$ $CuCl_2$(aq)|Cu(s), the processes involved in the cell with a salt bridge, i.e. Zn(s)|$ZnCl_2$(aq) $\vdots\vdots$ $CuCl_2$(aq)|Cu(s) will be described. The electrode reactions and the overall reaction will be the same as for the cell Zn(s)|$ZnCl_2$(aq) $\vdots$ $CuCl_2$(aq)|Cu(s) cell. The migration processes are different.

At the Zn(s)|$ZnCl_2$(aq) electrode, passage of 2 faraday of electricity forms 1 mol of Zn^{2+}(aq) around the electrode, causes t_+ mol of Zn^{2+}(aq) to migrate away from around this electrode, giving a net gain of $(1 - t_+)$ mol of Zn^{2+}(aq) around this electrode. To maintain electrical neutrality in this compartment, some Cl^-(aq) will migrate across the solution towards the electrode and these will be replaced by **anions migrating out of the salt bridge** into the electrode compartment. These will normally be singly charged anions since the salt bridge electrolyte is generally KCl(aq), KNO_3(aq) or NH_4NO_3(aq). Hence $2(1-t_+)$ mol, i.e. $2t_-$ mol of anions migrate out from the salt bridge.

At the $CuCl_2$(aq)|Cu(s) electrode, for passage of 2 faraday of electricity 1 mol of Cu^{2+}(aq) is removed from around this electrode, t_+ mol of Cu^{2+}(aq) migrates to this electrode, giving a net loss of $(1-t_+)$ mol of Cu^{2+}(aq) around the electrode. To maintain electrical neutrality in this compartment Cl^-(aq) will migrate across the solution in the electrode compartment to the salt bridge and **cations will also move in from the salt bridge**. Since the concentration of the ions in the salt bridge is much higher than [Cl^-] in the electrode compartment the dominant effect will be migration of cations from the salt bridge.

Carrying the current through the salt bridge is effectively done by cations of the salt bridge migrating through the salt bridge and **into** the positive electrode compartment, i.e. $CuCl_2$(aq)| Cu(s) and anions of the salt bridge migrating in the opposite direction **into** the negative electrode compartment, i.e. Zn(s)|$ZnCl_2$(aq).

Worked problem 9.13

Question

The cell $Pt(s)|H_2(g)|HCl(aq) \vdots CuSO_4(aq)|Cu(s)$ has a positive emf. Discuss the processes occurring at the electrodes and in the salt bridge.

Answer

The inference from the way the cell is written down is:

- left hand electrode: oxidation $\therefore$ negative electrode: $\tfrac{1}{2}H_2(g) + H_2O(l) \rightarrow H_3O^+(aq) + e$

- right hand electrode: reduction $\therefore$ positive electrode: $Cu^{2+}(aq) + 2e \rightarrow Cu(s)$

- the overall reaction is the sum of these two electrode reactions, **but watch**: these must be balanced so that the same number of electrons appears in each.

$$H_2(g) + 2H_2O(l) + Cu^{2+}(aq) \rightarrow 2H_3O^+(aq) + Cu(s)$$

- since the measured emf is positive this is the actual reaction occurring in the cell.

- electrons are formed at the left hand electrode, i.e. $Pt(s)|H_2(g)|HCl(aq)$ and flow through the **external circuit** to the right hand electrode, i.e. $CuSO_4(aq)|Cu(s)$.

- the current is carried **across the solution** by the ions in solution and **across the salt bridge** by migration of the ions of the salt bridge.

At the $Pt(s)|H_2(g)|HCl(aq)$ electrode, for passage of 2 faraday of electricity 2 mol of $H_3O^+(aq)$ is formed, $2t_+$ mol of $H_3O^+(aq)$ migrates away from the electrode, giving a net gain of $2(1-t_+)$ mol of $H_3O^+(aq)$ around the electrode. To maintain electrical neutrality some $Cl^-(aq)$ will move across the solution to the electrode and will be replaced by anions of the salt bridge migrating into the solution in the electrode compartment. Since 2 faraday of electricity is passed, then either $2(1-t_+)$ mol, i.e. $2t_-$ mol of a singly charged anion, or $(1-t_+)$ mol, i.e. t_- mol of a doubly charged anion of the salt bridge migrates into the $Pt(s)|H_2(g)|HCl(aq)$ electrode compartment.

At the $CuSO_4(aq)|Cu(s)$ electrode, for passage of 2 faraday of electricity 1 mol of $Cu^{2+}(aq)$ is removed: $Cu^{2+}(aq) + 2e \rightarrow Cu(s)$, t_+ mol of $Cu^{2+}(aq)$ migrates to the $Cu^{2+}(aq)|Cu(s)$ electrode, giving a net loss of $(1-t_+)$ mol of $Cu^{2+}(aq)$. To maintain electrical neutrality $(1-t_+)$ mol, i.e. t_- mol of $SO_4^{2-}(aq)$ could leave the compartment area and migrate to the salt bridge. But the cations of the salt bridge could also migrate **into** the electrode compartment to maintain electrical neutrality. Since the ions in the salt bridge are at a much higher concentration than the $SO_4^{2-}(aq)$ in the electrode compartment, the current will be carried mainly by migration of the cations along the salt bridge and into the solution around the $CuSO_4(aq)|Cu(s)$ electrode.

Worked problem 9.14

Question

Suggest cells which correspond to the following reactions:

(i) $Cu(s) + 2Ag^+(aq) \rightarrow Cu^{2+}(aq) + 2Ag(s)$

(ii) $Sn(s) + 2Eu^{3+}(aq) \rightarrow Sn^{2+}(aq) + 2Eu^{2+}(aq)$

(iii) $Hg_2Cl_2(s) + 2Ag(s) \rightarrow 2Hg(l) + 2AgCl(s)$

(iv) $AgBr(s) + I^-(aq) \rightarrow AgI(s) + Br^-(aq)$

(v) $Cu(s) + 2H_3O^+(aq) \rightarrow Cu^{2+}(aq) + H_2(g) + 2H_2O(l)$

Answer

Here the overall chemical process **must be** split up into two parts, each of which is the chemical equation for one of the electrode reactions, i.e. each must involve electrons. The reaction which releases electrons would correspond to the left hand electrode, and the reaction which takes up electrons would correspond to the right hand electrode.

(i) $Cu(s) \rightarrow Cu^{2+}(aq) + 2e$ would correspond to Cu(s) in $Cu^{2+}(aq)$
      $2e + 2Ag^+(aq) \rightarrow 2Ag(s)$ would correspond to Ag(s) in $Ag^+(aq)$

      A possible cell is $Cu(s)|Cu^{2+}(aq) \overset{..}{:} Ag^+(aq)|Ag(s)$, with salt bridge.

(ii) $Sn(s) \rightarrow Sn^{2+}(aq) + 2e$ would correspond to Sn(s) in $Sn^{2+}(aq)$
      $2e + 2Eu^{3+}(aq) \rightarrow 2Eu^{2+}(aq)$ would correspond to a mixture of $Eu^{3+}(aq)$ and $Eu^{2+}(aq)$ with a Pt electrode

      A possible cell is $Sn(s)|Sn^{2+}(aq) \overset{..}{:} Eu^{2+}(aq), Eu^{3+}(aq)|Pt(s)$, with a salt bridge.

(iii) $2Ag(s) + 2Cl^-(aq) \rightarrow 2AgCl(s) + 2e$ corresponds to Ag(s) coated with AgCl(s) dipping into $Cl^-(aq)$.
      $2e + Hg_2Cl_2(s) \rightarrow 2Hg(l) + 2Cl^-(aq)$ corresponds to Hg(l) surrounded by $Hg_2Cl_2(s)$ dipping into $Cl^-(aq)$.

      A possible cell is $Ag(s)|AgCl(s)|Cl^-(aq)|Hg_2Cl_2(s)|Hg(l)$. A salt bridge is not necessary; both electrodes are reversible with respect to $Cl^-(aq)$ and both dip into $Cl^-(aq)$.

(iv) $Ag(s) + I^-(aq) \rightarrow AgI(s) + e$ corresponds to Ag(s) coated with AgI(s) dipping into $I^-(aq)$.
      $e + AgBr(s) \rightarrow Ag(s) + Br^-(aq)$ corresponds to Ag(s) coated with AgBr(s) dipping into $Br^-(aq)$.

      A possible cell is $Ag(s)|AgI(s)|I^-(aq) \overset{..}{:} Br^-(aq)|AgBr(s)|Ag(s)$. The salt bridge is needed to separate the solutions. The left hand electrode is reversible with respect to $I^-(aq)$ while the right hand electrode is reversible with respect to $Br^-(aq)$.

(v) $Cu(s) \rightarrow Cu^{2+}(aq) + 2e$ corresponds to $Cu(s)$ in $Cu^{2+}(aq)$.

$2H_3O^+(aq) + 2e \rightarrow H_2(g) + 2H_2O(l)$ corresponds to $H_2(g)$ bubbling through $H_3O^+(aq)$ with Pt as electrode

A possible cell is $Cu(s)|Cu^{2+}(aq) \overset{..}{:} H_3O^+(aq)|H_2(g)|Pt(s)$ with salt bridge

In all these cases the actual reactions occurring can be inferred from the observed emf. If the emf is positive, then reaction will be as given above. If it is negative, then both the electrode reactions and the overall reaction will be the reverse.

9.9 Experimental determination of the direction of flow of the electrons, and measurement of the potential difference

In a cell, the electron flow in the external circuit is from the electrode at which electrons are generated to the electrode at which they are removed. Sometimes the direction of the current can be seen directly by observation of the chemical reactions at the electrodes. But for most cells, however, the chemical reactions occurring are not obvious, and the polarity of the electrodes is determined by instrument. The direction of the deflection of the needle of a dc ammeter or galvanometer gives the direction of the current (see Section 9.6.1). Modern digital voltmeters give directly both the emf and the direction of the current.

9.10 Electrode potentials

The overall reaction in a cell is made up of the reactions occurring at the electrodes. It is reasonable to assume that the overall emf of the cell is also made up of contributions from each electrode. To put this on a quantitative and comparative basis it is necessary to couple each electrode to one standard electrode. The $Pt(s)|H_2(g)$ 1 atm|HCl(aq) (unit activity) electrode is such a standard and by definition is taken to have zero electrode potential. Every other electrode potential is relative to this standard hydrogen electrode **which is always written on the left of the cell**.

Since the emf of the standard hydrogen electrode is taken to be zero, the measured emf gives the electrode potential of the electrode coupled with the hydrogen electrode. If the emf is positive, the polarity of the hydrogen electrode will be negative and that of the electrode whose potential is being measured is positive. Reaction at this electrode will use up electrons, i.e. will be reduction. If the emf is negative then the electrode potential will be negative. The electrode in question will have a negative polarity and oxidation will take place on the electrode generating electrons.

9.10.1 Redox potentials

This is the emf of a cell made up from the standard hydrogen electrode at the left coupled with a redox electrode (see Section.9.4.4). For example the cell:

$Pt(s)|H_2(g)$ 1 atm|HCl(aq) unit activity $\overset{..}{:} Cr^{2+}(aq), Cr^{3+}(aq)|Pt(s)$

is a redox cell and its emf will give the redox potential for the solution containing the ions, $Cr^{2+}(aq), Cr^{3+}(aq)$.

9.10.2   Electrode potentials for standard and non-standard conditions

For any given cell the measured emf will vary with the concentration and/or the partial pressures of the species making up the electrode. Because of this, two aspects must be considered:

- electrode potentials must be found for standard conditions so that meaningful comparisons between electrodes can be made. This will furnish a list of standard electrode potentials.

- a theoretical expression must be derived relating the observed emf of the cell to the concentrations and/or partial pressures of the species making up the electrode. Such relations are called Nernst equations; see Section 9.13.

9.11   Standard electrode potentials

These relate to electrodes where all the species making up the electrode are in their standard states, i.e.:

- solids and liquids are pure;

- gases are at 1 atm pressure;

- solutions are at unit activity.

Such electrodes are coupled with a standard hydrogen electrode and the emf measured. This will be the standard electrode potential. If the emf is positive, then the electrode will have positive polarity and electrons will be used up at it. If the emf is negative, then the electrode will have negative polarity and electrons will be produced on it.
   Examples of cells which would give standard electrode potentials are:

- $Pt(s)|H_2(g)$ 1 atm$|HCl(aq)$ unit activity $\vdots$ $Br^-(aq)$ unit activity$|AgBr(s)|Ag(s)$ has an emf $= 0.0732$ V,

- $Pt(s)|H_2(g)$ 1 atm$|HCl(aq)$ unit activity $\vdots$ $Cu^{2+}(aq)$ unit activity$|Cu(s)$ has an emf $= 0.340$ V

- $Pt(s)|H_2(g)$ 1 atm$|HCl(aq)$ unit activity $\vdots$ Zn^{2+} (aq) unit activity$|Zn(s)$ has an emf $= -0.763$ V

- $Pt(s)|H_2(g)$ 1 atm$|HCl(aq)$ unit activity $\vdots$ $OH^-(aq)$ unit activity$|O_2(g)$ 1 atm$|Pt(s)$ has an emf $= 0.401$ V

- $Pt(s)|H_2(g)$ 1 atm$|HCl(aq)$ unit activity $\vdots$ $OH^-(aq)$ unit activity$|H_2(g)$ 1 atm$|Pt(s)$ has an emf $= -0.828$ V

where the values quoted are for 25°C.

9.11.1 Standard redox potentials

These relate to redox electrodes where the oxidised and reduced components are present in their standard states. For example, if the emf of the cell:

$$Pt(s)|H_2(g) \ 1 \ atm|HCl(aq) \ unit \ activity \ \vdots \ Fe^{2+}(aq), Fe^{3+}(aq)|Pt(s)$$

is studied over a range of concentrations of Fe^{2+}(aq) and Fe^{3+}(aq), extrapolation to zero ionic strength will give E^θ, the standard electrode potential for the redox electrode, Fe^{2+}(aq), Fe^{3+}(aq)|Pt(s) (see Section 9.13.1 and Worked Problems 9.17 and 9.18).

As will be shown later (Section 9.17 and Worked Problem 9.20), the standard electrode potential is related to the equilibrium constant for the cell reaction. This has proved extremely useful for determining equilibrium constants for the redox reactions of inorganic systems and for redox reactions occurring in biological systems.

Worked problem 9.15

Question

What are the polarities of the electrodes in the five cells given above?

Answer

The sign of the emf allows the polarities to be assigned:

- If the emf is positive, the left hand electrode has a negative polarity and the right hand electrode is positive.

- If the emf is negative, then the left hand electrode has a positive polarity and the right hand electrode is negative.

−ve   +ve
$Pt(s)|H_2(g) \ 1 \ atm|HCl(aq) \ unit \ activity \ \vdots \ Br^-(aq) \ unit \ activity|AgBr(s)|Ag(s)$

−ve   +ve
$Pt(s)|H_2(g) \ 1 \ atm|HCl(aq) \ unit \ activity \ \vdots \ Cu^{2+}(aq) \ unit \ activity|Cu(s)$

+ve   −ve
$Pt(s)|H_2(g) \ 1 \ atm|HCl(aq) \ unit \ activity \ \vdots \ Zn^{2+}(aq) \ unit \ activity|Zn(s)$

−ve   +ve
$Pt(s)|H_2(g) \ 1 \ atm|HCl(aq) \ unit \ activity \ \vdots \ OH^-(aq) \ unit \ activity|O_2(g) \ 1 \ atm|Pt(s)$

+ve   −ve
$Pt(s)|H_2(g) \ 1 \ atm|HCl(aq) \ unit \ activity \ \vdots \ OH^-(aq) \ unit \ activity|H_2(g) \ 1 \ atm|Pt(s)$

However, sometimes it is impossible and often inconvenient to set up an electrode in its standard state, e.g. solubility problems may make it impossible to set up an electrode in which the ion to which the electrode is reversible is at unit activity. Also it may just be more convenient to set up the electrode not at standard conditions, in which case it becomes necessary to be able to find an alternative route to the standard electrode potential. This will be dealt with in Worked Problems 9.17 and 9.18.

9.12   Potential difference, electrical work done and ΔG for the cell reaction

The cell, $Zn(s)|ZnCl_2(aq)|AgCl(s)|Ag(s)$, can be used to illustrate the meanings of the terms: charge, current, potential, potential difference, electrical work done and the relation between the potential difference and the free energy change resulting from the combined effect of the chemical reactions occurring at the electrodes.

This cell has a positive emf. Hence the reactions actually occurring are the same as those inferred from the way in which the cell is written.

Left hand electrode is negatively charged, electrons are produced and reaction is $Zn(s) \rightarrow Zn^{2+}(aq) + 2e$.

Right hand electrode is positively charged, electrons flow to it from the $Zn(s)|Zn^{2+}(aq)$ electrode and the reaction is $2AgCl(s) + 2e \rightarrow 2Ag(s) + 2Cl^-(aq)$.

For every mol of $Zn(s)$ oxidised and every 2 mol of $AgCl(s)$ reduced, 2 mol of electrons flow through the external circuit, and this flow of electrons constitutes an electric current. The total charge conveyed is the total charge on the electrons, and for every mol of $Zn(s)$ oxidised and 2 mol of $AgCl(s)$ reduced the charge conveyed is $2 \times 6.0221 \times 10^{23} \times 1.6022 \times 10^{19} = 2 \times 9.6485 \times 10^4$ C (coulomb) = 2 faraday. The current passed is the charge passed per unit time having units C s^{-1} = A (ampere).

If $Zn(s)$ is dipped into a solution of $Zn^{2+}(aq)$ ions nothing happens. But if it is joined by a wire which allows electrons to flow along it, and if, at the end of the wire, there is something which will take up these electrons then the Zn will start going into solution as $Zn^{2+}(aq)$ and will generate electrons. It will continue to do so until either the $Zn(s)$ is all used up, or until the system receiving the electrons can no longer take them up.

The tendency for the reaction $Zn(s) \rightarrow Zn^{2+}(aq) + 2e$ to occur is greater than the tendency for $2Ag(s) + 2Cl^-(aq) \rightarrow 2AgCl(s) + 2e$ to occur. It is this difference which allows electrons to move along the external circuit to the $Cl^-(aq)|AgCl(s)|Ag(s)|$electrode and forces the reaction $2AgCl(s) + 2e \rightarrow 2Ag(s) + 2Cl^-(aq)$ to occur. The combined chemical reactions thus cause the electrons to flow.

The chemical reactions can be thought of as 'something pushing or pulling the electrons along the wire', depending on which is being considered. Work is therefore done 'pushing' electrons along the wire, and this work is done **by** the cell. Once the argument is translated into thermodynamic language this work done **by** the cell will have to be translated into work done **on** the cell, since all the relations generated in Chapter 8 relate to work done **on** the system, here the cell.

The term potential is often referred to in electrostatics. It is a quantity such that the potential difference (see below) between two points is equal to the difference between the value of the potential at one point and its value at the other. The term turns up in the Debye-Hückel theory (see Chapter 10, especially Section 10.4.5).

The electrostatic potential difference is the electrical work done by the cell per unit charge conveyed between two points, here the terminals of the cell. In physical terms the potential difference is what causes the current to flow through the wire of the external circuit and, as shown, this is a consequence of the differing tendencies of the reactions $Zn(s) \rightarrow Zn^{2+}(aq) + 2e$ and $2Ag(s) + 2Cl^-(aq) \rightarrow 2AgCl(s) + 2e$ to occur.

$$\text{potential difference} = \frac{\text{electrical work done by the cell}}{\text{charge conveyed}} = \frac{w_{\text{done by the cell}}}{Q} \tag{9.1}$$

where Q is the charge conveyed.

In the context of electrochemical cells, this potential difference is generally referred to as the electromotive force, E, or emf. This term will be used in the following discussion of cells.

9.12.1 Thermodynamic quantities in electrochemistry: relation of ΔG to E

The electrical work done by the cell by virtue of a charge being passed can be expressed in two ways:

- In terms of the charge conveyed, Q, and the potential difference, or emf, E, of the cell:

$$\text{electrical work done } \textbf{by} \text{ the cell in conveying the charge} = E \times Q \tag{9.2}$$

But the electrical work which is relevant to a thermodynamic analysis is the work done **on** the cell by virtue of the charge conveyed.

$$\therefore \text{electrical work done } \textbf{on} \text{ the cell in conveying the charge} = -E \times Q \tag{9.3}$$

- In thermodynamic terms, for reversible processes at constant T and p, the electrical work is given as:

$$\text{the electrical work done } \textbf{on} \text{ the cell in conveying the charge} = \Delta G \tag{9.4}$$

see Section 8.8 and Equations (8.17) and (8.18).

- From Equations (9.3) and (9.4) it follows that:

$$\Delta G = -EQ \tag{9.5}$$

$$\text{But} \quad Q = zF \tag{9.6}$$

(see Section 8.8).

$$\therefore \Delta G = -zFE \tag{9.7}$$

- For standard conditions: $\Delta G^\theta = -zFE^\theta$ $\tag{9.8}$

For the cell, $Zn(s)|ZnCl_2(aq)|AgCl(s)|Ag(s)$, operating reversibly, the overall reaction is thus $Zn(s) + 2AgCl(s) \rightarrow 2Ag(s) + 2Cl^-(aq) + Zn^{2+}(aq)$ for every 2 electrons produced. If 1 mol of $Zn(s)$ is removed and 2 mol of $Ag(s)$ deposited, then 2 mol of electrons, i.e. a charge of 2 faraday, flows along the external circuit.

$$\therefore \Delta G = -2FE \tag{9.9}$$

If the electrode reactions had been halved then the overall reaction becomes
½ Zn(s) + AgCl(s) → Ag(s) + Cl$^-$(aq) + ½ Zn^{2+}(aq). If ½ mol of Zn(s) is removed and
one mol of Ag(s) is formed, then one mol of electrons, i.e. a charge of one faraday, flows along
the external circuit. In this case $\Delta G = -FE$, where ΔG refers to conversion of ½ mol of Zn(s)
and formation of one mol of Ag(s), and is consequently ½ of the value of ΔG for the cell
reaction written in terms of one mol of Zn(s) removed and 2 mol of Ag(s) formed.

In general, a charge of z faraday will pass if z is the stoichiometric coefficient for the
electrons appearing in the equations for the electrode processes.
Note well:

- It is vitally important always to write out the electrode reactions and to formulate the overall
  cell reaction as a balanced sum of the electrode reactions. This procedure will ensure that
  the correct value of z is used, and that ΔG, z, the electrode reactions and the overall cell
  reaction are all internally self-consistent.

- If the emf is **positive** for the cell as written, then ΔG for the cell reaction is **negative** and so
  reaction as inferred from the way the cell is written down occurs spontaneously.

- If the emf is **negative**, then ΔG for the cell reaction as written is **positive** and so reaction as
  inferred from the way the cell is written down does not occur spontaneously, and the reverse
  reaction must occur.

9.12.2   Thermodynamic quantities in electrochemistry: effect of temperature on emf

Study of the dependence of the emf on temperature will give further thermodynamic
quantities. For processes occurring reversibly in a closed system (Section 8.7, Equation
8.14):

$$dG = V dp - S dT \qquad (9.10)$$

At constant p: $dG = -S dT$ $\qquad (9.11)$

$$\therefore \ d\Delta G = -\Delta S dT \qquad (9.12)$$

and

$$\left(\frac{\partial \Delta G}{\partial T}\right)_p = -\Delta S \qquad (9.13)$$

But, $\Delta G = -zFE$ $\qquad (9.7)$

$$\therefore \left(\frac{\partial \Delta G}{\partial T}\right)_p = -zF\left(\frac{\partial E}{\partial T}\right)_p = -\Delta S \qquad (9.14)$$

Hence the change in entropy for the cell reaction of a reversible cell can be found from the
variation of the emf with temperature.

Since $\Delta G = \Delta H - T\Delta S$ $\qquad (9.15)$

if both ΔG and ΔS have been found from the emf of the cell and its variation with temperature, ΔH can then be inferred.

Since
$$\Delta C_p = \left(\frac{\partial \Delta H}{\partial T}\right)_p = T\left(\frac{\partial \Delta S}{\partial T}\right)_p \tag{9.16}$$

a value can easily be found for ΔC_p from accurate emf studies.

Worked problem 9.16

Question

The standard emf of the cell:

$$Zn(s)|Zn^{2+}(aq) \; \vdots \; Fe^{2+}(aq), Fe^{3+}(aq)|Pt(s)$$

is $+1.534$ V at $25°C$ and $+1.576$ V at $65°C$. What is the cell reaction? Find $\Delta G^\theta, \Delta S^\theta, \Delta H^\theta$ and K at $25°C$, and state any assumptions which are made. What would be the values if the electrode reactions were changed by a factor of two?

Answer

Left hand electrode: $Zn(s) \rightarrow Zn^{2+}(aq) + 2e$

Right hand electrode: $2Fe^{3+}(aq) + 2e \rightarrow 2Fe^{2+}(aq)$

Overall cell reaction: $Zn(s) + 2Fe^{3+}(aq) \rightarrow Zn^{2+}(aq) + 2Fe^{2+}(aq)$

$\Delta G^\theta = -zFE^\theta$ where $z = 2$ for the cell reaction as written.

$$\therefore \Delta G^\theta = -2 \times 96485 \, C \, mol^{-1} \times 1.534V = -296.0 \, kJ \, mol^{-1} \text{ at } 25°C$$

Remember: $C \, V = J$

$$\Delta S^\theta = -\left(\frac{\partial \Delta G^\theta}{\partial T}\right)_p = zF\left(\frac{\partial E^\theta}{\partial T}\right)_p$$

If ΔS^θ and ΔH^θ are assumed to be independent of temperature then so will $(\partial E^\theta/\partial T)_p$ also be independent of temperature and a simple difference of the E^θ values can be taken.

$$\therefore \Delta S^\theta = \frac{2 \times 96485 \, C \, mol^{-1} \times (1.576 - 1.534) \, V}{40 \, K} = \frac{2 \times 96485 \times 0.042 \, V \, C \, mol^{-1} \, K^{-1}}{40}$$
$$= 203 \, J \, mol^{-1} \, K^{-1}$$

Again remember: $V \, C = J$

$$\Delta H^\theta = \Delta G^\theta + T\Delta S^\theta = -296 kJ \, mol^{-1} + 298.15 \, K \times 203 \times 10^{-3} kJ \, mol^{-1} \, K^{-1} = -236 kJ \, mol^{-1}$$

$$\Delta G^\theta = -RT \log_e K$$

$$\log_e K = -\frac{\Delta G^\theta}{RT} = \frac{296.0 kJ \, mol^{-1}}{8.3145 \times 10^{-3} kJ \, mol^{-1} \, K^{-1} \times 298.15 K} = 119.4$$

$$K = 7 \times 10^{51} \, mol \, dm^{-3}.$$

The cell reaction as written corresponded to $z = 2$. If, instead, the electrode reactions were written as:

left hand electrode: ½ $Zn(s) \rightarrow$ ½ $Zn^{2+}(aq) + e$

right hand electrode: $Fe^{3+}(aq) + e \rightarrow Fe^{2+}(aq)$

overall cell reaction: ½ $Zn(s) + Fe^{3+}(aq) \rightarrow$ ½ $Zn^{2+}(aq) + Fe^{2+}(aq)$

then the values for ΔG^θ, ΔS^θ and ΔH^θ would be halved.

$$K = \left(\frac{a_{Zn^{2+}} a^2_{Fe^{2+}}}{a^2_{Fe^{3+}}} \right)_{eq} = 7 \times 10^{51} \text{ mol dm}^{-3}.$$

is the equilibrium constant when 2 electrons are involved in the electrode reactions, while $K = \left(\frac{a^{1/2}_{Zn^{2+}} a_{Fe^{2+}}}{a_{Fe^{3+}}} \right)_{eq}$ is the constant when 1 electron is involved.

$$K = \left(\frac{a^{1/2}_{Zn^{2+}} a_{Fe^{2+}}}{a_{Fe^{3+}}} \right)_{eq} = \left(\frac{a_{Zn^{2+}} a^2_{Fe^{2+}}}{a^2_{Fe^{3+}}} \right)^{1/2}_{eq} = 8 \times 10^{25} \text{ mol}^{1/2} \text{ dm}^{-3/2}.$$

9.13 ΔG for the cell process: the Nernst equation

As will become apparent from many of the following problems there are two aspects of solving problems which, if understood and assimilated, make formulation of the equation relating the emf of a cell to the activities/concentrations and/or partial pressures a straightforward procedure. They are:

- writing down the electrode processes and thereby formulating the overall cell reaction;

- formulating ΔG for the cell reaction in terms of sums and differences of chemical potentials for the species involved in the overall cell reaction.

The following cell will be used to illustrate the procedure.

$$Pt(s)|H_2(g)|HCl(aq)|AgCl(s)|Ag(s)$$

reaction at the left hand electrode: ½ $H_2(g) + H_2O(l) \rightarrow H_3O^+(aq) + e$

reaction at the right hand electrode: $AgCl(s) + e \rightarrow Ag(s) + Cl^-(aq)$

overall cell reaction: ½ $H_2(g) + H_2O(l) + AgCl(s) \rightarrow H_3O^+(aq) + Ag(s) + Cl^-(aq)$

In Sections 8.11 to 8.13, Equations (8.31) and (8.33) it was shown that:

$$dG = \sum_i \mu_i dn_i \tag{9.17}$$

and for the general reaction, $aA + bB \rightleftharpoons cC + dD$

$$\Delta G = +c\mu_C + d\mu_D - a\mu_A - b\mu_B \tag{9.18}$$

$$\therefore \Delta G = \mu_{H_3O^+(aq)} + \mu_{Ag(s)} + \mu_{Cl^-(aq)} - \tfrac{1}{2}\mu_{H_2(g)} - \mu_{AgCl(s)} - \mu_{H_2O(l)} \tag{9.19}$$

$$= \mu^\theta_{H_3O^+(aq)} + RT\log_e a_{H_3O^+(aq)} + \mu^\theta_{Ag(s)} + \mu^\theta_{Cl^-(aq)} + RT\log_e a_{Cl^-(aq)}$$
$$- \tfrac{1}{2}\mu^\theta_{H_2(g)} - \tfrac{1}{2}RT\log_e p_{H_2} - \mu^\theta_{AgCl(s)} - \mu^\theta_{H_2O(l)} \tag{9.20}$$

$$= \Delta G^\theta + RT\log_e \frac{a_{H_3O^+(aq)}a_{Cl^-(aq)}}{p_{H_2}^{1/2}} \tag{9.21}$$

where ΔG^θ is a constant depending on T and p, and is a sum and difference of standard chemical potentials

Remember:

- μ for a pure solid or pure liquid $= \mu^\theta$ where μ^θ is a constant which depends on T and slightly on p. The standard state is pure solid at constant T.

- μ for a gas $= \mu^\theta + RT\log_e p$ where μ^θ is a constant which depends on T only. The standard state is pure gas at unit pressure, generally listed in tabular data as 1 atm, but rigorously should be 1 bar.

- μ for a solute $= \mu^\theta + RT\log_e a$ where μ^θ is a constant which depends on T and p, and $a = c\gamma$ where $\gamma \to 1$ as $c \to 0$. In dilute solutions, which are approximately ideal, the activity, a, can be approximated to the concentration, c. The standard state is solute at unit activity, or, for ideal solutions, a concentration of 1 mol dm^{-3}.

The relation between ΔG for the cell reaction and the emf generated by virtue of the electrode processes can now be used:

Since $$\Delta G = -zFE \text{ and } \Delta G^\theta = -zFE^\theta \tag{9.7) and (9.8}$$

$$E = E^\theta - \frac{RT}{zF}\log_e \frac{a_{H_3O^+(aq)}a_{Cl^-(aq)}}{p_{H_2}^{1/2}} = E^\theta - \frac{RT}{zF}\log_e \frac{[H_3O^+][Cl^-]}{p_{H_2}^{1/2}} - \frac{RT}{zF}\log_e \gamma_{H_3O^+}\gamma_{Cl^-} \tag{9.22}$$

Here, from the chemical equations representing the electrode reactions, $z = 1$.

If the ideal case is being considered, then this equation reduces to:

$$E = E^\theta - \frac{RT}{zF}\log_e \frac{[H_3O^+][Cl^-]}{p_{H_2}^{1/2}} \tag{9.23}$$

Note: E^θ is the emf generated by the cell when all the species involved are in their standard states.

9.13.1   Corrections for non-ideality

If non-ideality is being considered the full expression involving activity coefficients has to be used, i.e.

$$E = E^\theta - \frac{2.303RT}{zF}\log_{10}\frac{[H_3O^+][Cl^-]}{p_{H_2}^{1/2}} - \frac{2.303RT}{zF}\log_{10}\gamma_{H_3O^+}\gamma_{Cl^-} \tag{9.24}$$

where the logarithms have been converted to base 10, see note below.

This equation can be given in terms of the ionic strength, from which the effect of ionic strength on the emf can be predicted.

- If the Debye-Hückel limiting law is used, $\log_{10} \gamma_i = -z_i^2 A \sqrt{I}$ (9.25)

$$\therefore \log_{10}\gamma_{H_3O^+}\gamma_{Cl^-} = \log_{10}\gamma_{H_3O^+} + \log_{10}\gamma_{Cl^-} = -A \times 1^2 \times \sqrt{I} - A \times 1^2 \times \sqrt{I}$$
$$= -2A\sqrt{I} \tag{9.26}$$

$$\therefore E = E^\theta - \frac{2.303RT}{zF} \log_{10} \frac{a_{H_3O^+(aq)} a_{Cl^-(aq)}}{p_{H_2}^{1/2}}$$

$$= E^\theta - \frac{2.303RT}{zF} \log_{10} \frac{[H_3O^+][Cl^-]}{p_{H_2}^{1/2}} + 2.303 \frac{RT}{zF} \times 2A\sqrt{I} \tag{9.27}$$

Note: it is important to remember that in the basic expression for the emf the logarithmic terms are to base e. When Debye-Hückel equations are being used to relate the activity coefficients to the ionic strength, logarithms to base 10 are involved. Conversion between the two bases is essential. Hence the 2.303 in the above equation: $\log_e x = 2.303\log_{10} x$.

- If the Debye-Hückel equation is used, $\log_{10} \gamma_i = -z_i^2 A \dfrac{\sqrt{I}}{1 + \sqrt{I}}$ (9.28)

$$\therefore E = E^\theta - \frac{2.303RT}{zF} \log_{10} \frac{[H_3O^+][Cl^-]}{p_{H_2}^{1/2}} + 2.303 \frac{RT}{zF} \times 2A \frac{\sqrt{I}}{1 + \sqrt{I}} \tag{9.29}$$

This is a standard method for deducing the expression for the emf of a cell, and should be used under all circumstances.

It also furnishes a method for extrapolation to give the standard emf of a cell.

$$E + \frac{2.303RT}{zF} \log_{10} \frac{[H_3O^+][Cl^-]}{p_{H_2}^{1/2}} = E^\theta + 2.303 \frac{RT}{zF} \times 2A \frac{\sqrt{I}}{1 + \sqrt{I}} \tag{9.30}$$

and a plot of $E + \dfrac{2.303RT}{zF} \log_{10} \dfrac{[H_3O^+][Cl^-]}{p_{H_2}^{1/2}}$ vs $\dfrac{\sqrt{I}}{1 + \sqrt{I}}$ (9.31)

will give E^θ as intercept.

9.13.2   A further example deducing the Nernst equation and the dependence of emf on ionic strength

The cell:

$$Cu(s)|CuSO_4(aq)c_1 \;\vdots\; CuSO_4(aq)c_2|Cu(s)$$

is a concentration cell. The expression for the dependence of the emf on ionic strength can be deduced in the same manner as in Section 9.13.1, but it has one feature specific to concentration cells: it has no term in E^θ in the Nernst equation.

reaction at the left hand electrode, (lhs): $Cu(s) \rightarrow Cu^{2+}(aq)(lhs) + 2e$

reaction at the right hand electrode (rhs): $Cu^{2+}(aq)(rhs) + 2e \rightarrow Cu(s)$

overall cell reaction: $Cu(s) + Cu^{2+}(aq)(rhs) \rightarrow Cu^{2+}(aq)(lhs) + Cu(s)$

$$\Delta G = \mu_{Cu^{+2}(aq)}(lhs) + \mu_{Cu(s)} - \mu_{Cu^{2+}(aq)}(rhs) - \mu_{Cu(s)} \tag{9.32}$$

$$= \mu^\theta_{Cu^{2+}(aq)}(lhs) + RT \log_e a_{Cu^{2+}(aq)}(lhs) + \mu^\theta_{Cu(s)} - \mu^\theta_{Cu^{2+}(aq)}(rhs)$$

$$- RT \log_e a_{Cu^{2+}(aq)}(rhs) - \mu^\theta_{Cu(s)} \tag{9.33}$$

$$= RT \log_e \frac{a_{Cu^{2+}(aq)}(lhs)}{a_{Cu^{2+}(aq)}(rhs)} \tag{9.34}$$

Note very carefully: there is no ΔG^θ in this expression. This is because the standard values for $Cu(s)$ formed and removed at the electrodes cancel, and the standard values for the $Cu^{2+}(aq)$ formed and removed at the electrodes cancel. This always happens with concentration cells.

$$\Delta G = -zFE \tag{9.7}$$

$$E = -\frac{RT}{zF} \log_e \frac{a_{Cu^{2+}(aq)}(lhs)}{a_{Cu^{2+}(aq)}(rhs)} = -\frac{RT}{zF} \log_e \frac{[Cu^{2+}]_{lhs}}{[Cu^{2+}]_{rhs}} - \frac{RT}{zF} \log_e \frac{\gamma_{Cu^{2+}}(lhs)}{\gamma_{Cu^{2+}}(rhs)} \tag{9.35}$$

$$E = -\frac{2.303RT}{zF} \log_{10} \frac{[Cu^{2+}]_{lhs}}{[Cu^{2+}]_{rhs}} - \frac{2.303RT}{zF} \log_{10} \frac{\gamma_{Cu^{2+}}(lhs)}{\gamma_{Cu^{2+}}(rhs)} \tag{9.36}$$

$$\log_{10} \frac{\gamma_{Cu^{2+}}(lhs)}{\gamma_{Cu^{2+}}(rhs)} = \log_{10} \gamma_{Cu^{2+}}(lhs) - \log_{10} \gamma_{Cu^{2+}}(rhs)$$

$$= -A \times 2^2 \times \frac{\sqrt{I}}{1 + \sqrt{I}}(lhs) + A \times 2^2 \times \frac{\sqrt{I}}{1 + \sqrt{I}}(rhs) \tag{9.37}$$

$$E = -\frac{2.303RT}{zF} \log_{10} \frac{[Cu^{2+}]_{lhs}}{[Cu^{2+}]_{rhs}} + 2.303 \times 4A \times \frac{RT}{zF} \left(\frac{\sqrt{I}}{1 + \sqrt{I}}(lhs) - \frac{\sqrt{I}}{1 + \sqrt{I}}(rhs) \right) \tag{9.38}$$

where $z = 2$

Note: how important it is to distinguish between the solutions of $CuSO_4(aq)$ in the left hand and right hand electrode compartments.

9.14 Methods of expressing concentration

There are three ways of expressing concentration:

- As a **molar** concentration where the 'concentration' is given as: number of mol in a 1 dm^3 volume of solution. This is the usual definition, and is used in many aspects of quantitative chemistry. It is the quantity used throughout this book. Here ionic strengths are also molar quantities and mean ionic activity coefficients are thus based on molar quantities.

- As a **molal** concentration where the 'concentration' is given as: number of mol per 1 kg of solvent. This is used in highly accurate electrochemical studies, and quantities derived from experiment are then given in terms of molal concentrations. Most tabulated electrochemical

data is based on molal concentrations. In particular, compilations of mean activity coefficients are always given as molal $\gamma_{\pm}$ and awareness of this is crucial. For simplicity, in this
book all quantities are given in terms of molar concentrations, but care must be taken with
tabulated data. Molal concentrations do not vary with temperature since both quantities
involved, i.e. the number of mol and the amount of the solvent are based on masses, and
these do not vary with temperature. Molar concentrations, however, do depend on temperature since the volume of a given solution will vary from temperature to temperature.
This is of importance when the dependence of thermodynamic quantities such as ΔG^{θ}, ΔH^{θ},
ΔS^{θ} on temperature is being sought. The same considerations will apply for experiments at
high pressure where ΔV^{θ} are being determined. Molal concentrations do not vary with
pressure, but molar concentrations do.

- As a mol fraction where the 'concentration' is given as: the number of mol of the substance
per total number of mol of all substances present. Mol fractions are rarely used for
electrolyte systems, but again it is well to be aware of the possibility of results being
expressed in this manner.

- In electrolyte studies both molar and molal concentrations are extensively used, so again
attention must be paid to the manner in which data is presented.

Worked problem 9.17

Question

Deduce the expression for the dependence of the emf on ionic strength for the following
cell,

$$Pt(s)|Ag(s)|Ag^{+}(aq) \;\vdots\; Fe^{2+}(aq), Fe^{3+}(aq)|Pt(s)$$

and show how E^{θ} for the cell can be determined.

Answer

Reaction at the left hand electrode: $Ag(s) \rightarrow Ag^{+}(aq) + e$

Reaction at the right hand electrode: $Fe^{3+}(aq) + e \rightarrow Fe^{2+}(aq)$

Overall cell reaction: $Ag(s) + Fe^{3+}(aq) \rightarrow Ag^{+}(aq) + Fe^{2+}(aq)$

$$\Delta G = \mu_{Ag^{+}(aq)} + \mu_{Fe^{2+}(aq)} - \mu_{Fe^{3+}(aq)} - \mu_{Ag(s)}$$

$$= \mu^{\theta}_{Ag^{+}(aq)} + RT \log_e a_{Ag^{+}} + \mu^{\theta}_{Fe^{2+}(aq)} + RT\log_e a_{Fe^{2+}(aq)} - \mu^{\theta}_{Fe^{3+}(aq)} - RT\log_e a_{Fe^{3+}(aq)} - \mu^{\theta}_{Ag(s)}$$

$$= \Delta G^{\theta} + RT\log_e \frac{a_{Ag^{+}(aq)}a_{Fe^{2+}(aq)}}{a_{Fe^{3+}(aq)}} = \Delta G^{\theta} + RT\log_e \frac{[Ag^{+}][Fe^{2+}]}{[Fe^{3+}]} + RT\log_e \frac{\gamma_{Ag^{+}(aq)}\gamma_{Fe^{2+}(aq)}}{\gamma_{Fe^{3+}(aq)}}$$

where ΔG^{θ} is a constant depending on T and p.

Since $\Delta G = -zFE$ and $\Delta G^{\theta} = -zFE^{\theta}$

$$E = E^{\theta} - \frac{2.303RT}{zF} \log_{10} \frac{[Ag^+][Fe^{2+}]}{[Fe^{3+}]} - \frac{2.303RT}{zF} \log_{10} \frac{\gamma_{Ag^+(aq)}\gamma_{Fe^{2+}(aq)}}{\gamma_{Fe^{3+}(aq)}}$$

Dealing with the term in the activity coefficients:

$$\frac{2.303RT}{zF} \log_{10} \frac{\gamma_{Ag^+(aq)}\gamma_{Fe^{2+}(aq)}}{\gamma_{Fe^{3+}(aq)}} = \frac{2.303RT}{zF} \left\{ \log_{10}\gamma_{Ag^+(aq)} + \log_{10}\gamma_{Fe^{2+}(aq)} - \log_{10}\gamma_{Fe^{3+}(aq)} \right\}$$

$$= \frac{2.303RT}{zF} \left\{ -\frac{A \times 1^2\sqrt{I}}{1+\sqrt{I}} - \frac{A \times 2^2\sqrt{I}}{1+\sqrt{I}} + \frac{A \times 3^2\sqrt{I}}{1+\sqrt{I}} \right\} = \frac{2.303RT}{zF} \times \frac{4A\sqrt{I}}{1+\sqrt{I}}$$

$$\therefore E = E^{\theta} - \frac{2.303RT}{zF} \log_{10} \frac{[Ag^+][Fe^{2+}]}{[Fe^{3+}]} - \frac{2.303RT}{zF} \times \frac{4A\sqrt{I}}{1+\sqrt{I}}$$

$$\therefore E + \frac{2.303RT}{zF} \log_{10} \frac{[Ag^+][Fe^{2+}]}{[Fe^{3+}]} = E^{\theta} - \frac{2.303RT}{zF} \times \frac{4A\sqrt{I}}{1+\sqrt{I}}$$

$$\therefore \text{a plot of } E + \frac{2.303RT}{zF} \log_{10} \frac{[Ag^+][Fe^{2+}]}{[Fe^{3+}]} \text{ vs } \frac{\sqrt{I}}{1+\sqrt{I}} \text{ should give } E^{\theta} \text{ as intercept}$$

and $-\dfrac{2.303RT \times 4A}{zF}$ as gradient.

Be careful here: the equation relating ΔG and the activity is given in terms of logarithms to base e. The Debye-Hückel expression, however, is one in terms of logarithms to base 10. Hence the expression for E must be given in terms of logarithms to base 10 where $\log_e x = 2.303 \log_{10} x$.

9.15   Calculation of standard emf values for cells and ΔG^{θ} values for reactions

Cells can give an excellent route to equilibrium constants and thermodynamic quantities for reactions which cannot readily be carried out in the laboratory. They can also give insights into the practical aspects of non-ideality. Such studies can be best explained by worked examples.
  In all problems:

$$F = 9.6485 \times 10^4 \, C \, mol^{-1}$$

$$\frac{RT}{F} = 0.02569 \, V \quad \text{at } 25^{\circ}C \quad \text{and} \quad \frac{2.303RT}{F} = 0.05916 \, V \quad \text{at} \quad 25^{\circ}C$$

$$\text{Units of } \frac{RT}{F} \quad \text{are} \quad \frac{J \, mol^{-1} \, K^{-1} K}{C \, mol^{-1}} = \frac{J}{C} = \frac{C \, V}{C} = V$$

Worked problem 9.18

Question

The cell, $Pt(s)|H_2(g)$ 1 atm$|HCl$ unit activity $\vdots$ $AgNO_3(aq)|Ag(s)$ has the following dependence of emf on concentration at 25°C. Use a graphical procedure to find E^θ. Why is it necessary to have highly accurate emf data? What is the value of ΔG^θ for this cell? Comment on the fact that emf studies can furnish such a quantity. How could other thermodynamic quantities be found for this cell reaction?

$\dfrac{[AgNO_3]}{mol\ dm^{-3}}$	0.0004	0.0030	0.0100	0.0600	0.1000
$\dfrac{E}{V}$	1.000	0.9470	0.9148	0.8657	0.8514

Answer

Left hand electrode: $\frac{1}{2}H_2(g) + H_2O(l) \rightarrow H_3O^+(aq) + e$

Right hand electrode: $Ag^+(aq) + e \rightarrow Ag(s)$

Overall reaction: $\frac{1}{2}H_2(g) + H_2O(l) + Ag^+(aq) \rightarrow H_3O^+(aq) + Ag(s)$

Using the standard procedure:

$$\Delta G = \mu_{H_3O^+(aq)} + \mu_{Ag(s)} - \frac{1}{2}\mu_{H_2(g)} - \mu_{H_2O(l)} - \mu_{Ag^+(aq)}$$

$$= \mu^\theta_{H_3O^+(aq)} + RT\log_e a_{H_3O^+(aq)} + \mu^\theta_{Ag(s)} - \frac{1}{2}\mu^\theta_{H_2(g)}$$

$$- RT\log_e p^{1/2}_{H_2} - \mu^\theta_{H_2O(l)} - \mu^\theta_{Ag^+(aq)} - RT\log_e a_{Ag^+(aq)}$$

$$= \Delta G^\theta + RT\log_e \frac{a_{H_3O^+(aq)}}{p^{1/2}_{H_2} a_{Ag^+(aq)}}$$

Since HCl(aq) is present at unit activity, i.e. $a_{H_3O^+(aq)} = 1$ mol dm^{-3} and $p_{H_2(g)} = 1$ atm

$$\Delta G = \Delta G^\theta - RT\log_e a_{Ag^+(aq)}$$

Since $\Delta G = -zFE$ and $\Delta G^\theta = -zFE^\theta$ and for the cell reaction as written, $z = 1$:

$$E = E^\theta + \frac{2.303RT\log_{10}a_{Ag^+(aq)}}{F} = E^\theta + \frac{2.303RT\log_{10}[Ag^+]}{F} + \frac{2.303RT\log_{10}\gamma_{\pm AgNO_3(aq)}}{F}$$

where $\gamma_{Ag^+(aq)}$ is replaced by the measurable quantity $\gamma_{\pm AgNO_3(aq)}$.

$$E - \frac{2.303RT\log_{10}[Ag^+]}{F} = E^\theta + \frac{2.303RT\log_{10}\gamma_{\pm AgNO_3(aq)}}{F}$$

Using the Debye-Hückel equation to take account of non-ideality gives:

$$E - 0.05916\ V\log_{10}[Ag^+] = E^\theta - 0.05916\ V\frac{0.510\ mol^{-1/2}\ dm^{3/2}\sqrt{I}}{1 + \sqrt{I}}$$

$\therefore$ a plot of $E - 0.05916\,\text{V}\log_{10}[\text{Ag}^+]$ vs. $\sqrt{I}/(1 + \sqrt{I})$ should be linear with intercept equal to E^θ and slope equal to -0.0302.

$\dfrac{[\text{Ag}^+]}{\text{mol dm}^{-3}}$	0.0004	0.0030	0.0100	0.0600	0.1000
$\dfrac{I}{\text{mol dm}^{-3}}$	0.0004	0.0030	0.0100	0.0600	0.1000
$\dfrac{\sqrt{I}}{\text{mol}^{1/2}\,\text{dm}^{-3/2}}$	0.0200	0.0548	0.1000	0.2449	0.3162
$\dfrac{\sqrt{I}}{1+\sqrt{I}}/\text{mol}^{1/2}\,\text{dm}^{-3/2}$	0.0196	0.0519	0.9091	0.1968	0.2403
$\log_{10}[\text{Ag}^+]$	-3.3979	-2.5229	-2.0000	-1.2218	-1.0000
$0.05916\log_{10}[\text{Ag}^+]$	-0.2010	-0.1493	-0.1183	-0.0723	-0.0592
$\dfrac{E}{V}$	1.0000	0.9470	0.9148	0.8657	0.8514
$\dfrac{E}{V} - 0.05916\log_{10}[\text{Ag}^+]$	0.7990	0.7977	0.7965	0.7934	0.7922

Plot of $E - 0.05916\log_{10}[\text{Ag}^+]$ vs. $\sqrt{I}/(1 + \sqrt{I})$ is linear with intercept equal to $E^\theta = 0.7994\text{V}$ and slope equal to -0.029.

A look at the values for $E - 0.05916\log_{10}[\text{Ag}^+]$ shows why very accurate emf data are required. The changes in $E - 0.05916\log_{10}[\text{Ag}^+]$ as the ionic strength is varied are occurring in the third and fourth significant figures, hence data accurate to four significant figures is required. Changes in $E - 0.05916\log_{10}[\text{Ag}^+]$ are small compared with values for the quantity itself.

The cell reaction is $\tfrac{1}{2}\text{H}_2(g) + \text{Ag}^+(aq) + \text{H}_2\text{O}(l) \rightarrow \text{H}_3\text{O}^+(aq) + \text{Ag}(s)$

$$K_{\text{ideal}} = \left(\frac{a_{\text{H}_3\text{O}^+(aq)}}{p_{\text{H}_2}^{1/2}\,a_{\text{Ag}^+}}\right)_{eq} \quad \text{and} \quad \Delta G^\theta = -RT\log_e K$$

$$\Delta G^\theta = -zFE^\theta = -1 \times 96485\,\text{C mol}^{-1} \times 0.7994\,\text{V} = -77.13\,\text{kJ mol}^{-1}$$

Remember: C V = J

Since E^θ has been found by extrapolation of emfs to zero ionic strength it will generate an ideal K for the reaction.

$$\log_e K = -\frac{\Delta G^\theta}{RT} = -\frac{-77.13\,\text{kJ mol}^{-1}}{8.3145 \times 10^{-3}\,\text{kJ mol}^{-1}\,\text{K}^{-1} \times 298.15\text{K}} = 31.11$$

$$K = 3.2 \times 10^{13}\,\text{atm}^{-1/2}$$

This calculation illustrates one of the most useful functions of emf studies for chemists, viz., they can give information about reactions which do not take place in the laboratory. In the present case, the reaction of $\text{H}_2(g)$ with $\text{AgNO}_3(aq)$ solution to generate $\text{Ag}(s)$ and $\text{H}_3\text{O}^+(aq)$ just does not occur. However, by utilising the reactions occurring at the electrodes of this cell, information about the reaction can be found.

If the emf is studied over a range of temperatures ΔH^θ and ΔS^θ can be found, and if the cell could be set up at a variety of pressures ΔV^θ can also be found.

9.16 Determination of pH

Fundamental studies of the pH of a solution rest on the rigorous definition of pH:

$$pH = -\log_{10} a_{H_3O^+} = -\log_{10}[H_3O^+]\gamma_{H_3O^+}. \tag{9.39}$$

Highly accurate measurements of the emf of appropriate cells can lead to a highly accurate determination of the pH of one of the electrode compartments. Typical cells used are:

Pt(s)|H$_2$(g) 1 atm|HCl(aq) unknown concentration|Hg$_2$Cl$_2$(s) |Hg(l)

Pt(s)|H$_2$(g) 1 atm|HCl(aq) unknown concentration|AgCl(s)|Ag(s)

where the deduction of the relation between emf and the activity of H$_3$O$^+$(aq) is straightforward (see Worked Problems 9.19 and 9.23). 'Concealed' concentration cells are often used. In practice for routine measurements a glass electrode is used (see Section 9.4.7).

Worked problem 9.19

Question

The emf of the cell:
$\qquad$ Pt(s)|H$_2$(g) 1 atm|HCl(aq) unknown concentration|AgCl(s)|Ag(s)

is 0.4425V at 25°C, and E^θ for the cell is 0.2224 V at 25°C

(i) Find the pH of the electrode solution.

(ii) From this calculate [HCl].

Answer

(i) Left hand electrode: ½H$_2$(g) + H$_2$O(l) → H$_3$O$^+$(aq) + e

Right hand electrode: AgCl(s) + e → Ag(s) + Cl$^-$(aq)

Overall reaction: ½H$_2$(g) + H$_2$O(l) + AgCl(s) → H$_3$O$^+$(aq) + Ag(s) + Cl$^-$(aq)

$$\Delta G = \mu_{H_3O^+(aq)} + \mu_{Cl^-(aq)} + \mu_{Ag(s)} - \tfrac{1}{2}\mu_{H_2(g)} - \mu_{H_2O(l)} - \mu_{AgCl(s)}$$

$$= \mu^\theta_{H_3O^+_{(aq)}} + RT\log_e a_{H_3O^+} + \mu^\theta_{Cl^-(aq)} + RT\log_e a_{Cl^-(aq)} + \mu^\theta_{Ag(s)}$$

$$- 1/2\mu^\theta_{H_2(g)} - 1/2RT\log_e p_{H_2} - \mu^\theta_{H_2O(l)} - \mu^\theta_{AgCl(s)}$$

$$= \Delta G^\theta + RT\log_e \frac{a_{H_3O^+}a_{Cl^-}}{p_{H_2}^{1/2}}$$

$$\therefore E = E^\theta - \frac{RT}{zF}\log_e \frac{a_{H_3O^+}a_{Cl^-}}{p_{H_2}^{1/2}}$$

$z = 1$ for the cell reaction as written above, and $p_{H_2} = 1$ atm. Since $[H_3O^+] = [Cl^-]$ and $a_{H_3O^+} = [H_3O^+]\gamma_{\pm(HCl)} = [Cl^-]\gamma_{\pm(HCl)} = a_{Cl^-}$, then:

$$E - E^\theta = -\frac{RT}{F}\log_e a_{H_3O^+}^2 = -\frac{2 \times RT}{F}\log_e a_{H_3O^+}$$

$$0.4425\,\text{V} - 0.2224\,\text{V} = -(2 \times 0.02569 \times 2.303)\,\text{V}\log_{10} a_{H_3O^+}$$

$$= -(2 \times 0.05916)\,\text{V}\log_{10} a_{H_3O^+}$$

$$\therefore \log_{10} a_{H_3O^+} = -1.860$$

$$\therefore a_{H_3O^+} = 0.0138\,\text{mol dm}^{-3}$$

Using the rigorous definition of pH: $pH = -\log_{10} a_{H_3O^+} = 1.860$

(ii) What has been found is $a_{H_3O^+}$ where $a_{H_3O^+} = [H_3O^+]\gamma_{\pm(HCl)}$

But $[H_3O^+] = [Cl^-] = [HCl] = \dfrac{a_{H_3O^+}}{\gamma_{\pm(HCl)}}$

The calculation requires calculating $\gamma_{\pm(HCl)}$ for the unknown [HCl] and so has to find $[H_3O^+]$.

The Debye-Hückel equation is required: $\log_{10}\gamma_\pm = -A|z_{H_3O^+}z_{Cl^-}|\dfrac{\sqrt{I}}{1 + \sqrt{I}}$

But watch: the ionic strength is based on concentrations:

$$I = 1/2\sum_i c_i z_i^2 = 1/2[[H_3O^+] \times 1^2 + [Cl^-] \times 1^2] = [H_3O^+]$$

What is known, however, is $a_{H_3O^+}$. Calculating $[H_3O^+]$ requires a series of successive approximations.

First approximation:

Assume [HCl] = 0.0138 mol dm^{-3}

$$I = 0.0138\,\text{mol dm}^{-3} \quad \text{and} \quad \sqrt{I} = 0.1175\,\text{mol}^{1/2}\,\text{dm}^{-3/2}$$

$$\log_{10}\gamma_{\pm(HCl)} = -\frac{0.510\,\text{mol}^{-1/2}\,\text{dm}^{3/2} \times 1^2 \times 0.1175\,\text{mol}^{1/2}\,\text{dm}^{-3/2}}{1.1175} = -0.0536$$

$$\therefore \gamma_{\pm(HCl)} = 0.884$$

$$\therefore [HCl] = \frac{a_{H_3O^+}}{\gamma_{\pm(HCl)}} = \frac{0.0138\,\text{mol dm}^{-3}}{0.884} = 0.0156\,\text{mol dm}^{-3}$$

Second approximation:

$$I = 0.0156\,\text{mol dm}^{-3} \quad \text{and} \quad \sqrt{I} = 0.1249\,\text{mol}^{1/2}\,\text{dm}^{-3/2}$$

$$\log_{10}\gamma_{\pm(HCl)} = -\frac{0.510\,\text{mol}^{-1/2}\,\text{dm}^{3/2} \times 1^2 \times 0.1249\,\text{mol}^{1/2}\,\text{dm}^{-3/2}}{1.1249} = -0.0566$$

$$\therefore \gamma_{\pm(HCl)} = 0.878$$

$$\therefore [HCl] = \frac{a_{H_3O^+}}{\gamma_{\pm(HCl)}} = \frac{0.0138\,\text{mol dm}^{-3}}{0.878} = 0.0157\,\text{mol dm}^{-3}$$

Third approximation:

$$I = 0.0157 \, \text{mol dm}^{-3} \quad \text{and} \quad \sqrt{I} = 0.1254 \, \text{mol}^{1/2} \, \text{dm}^{-3/2}$$

$$\log_{10}\gamma_{\pm(\text{HCl})} = -\frac{0.510 \, \text{mol}^{-1/2} \, \text{dm}^{3/2} \times 1^2 \times 0.1254 \, \text{mol}^{1/2} \, \text{dm}^{-3/2}}{1.1254} = -0.0568$$

$$\therefore \gamma_{\pm(\text{HCl})} = 0.877$$

$$\therefore [\text{HCl}] = \frac{a_{\text{H}_3\text{O}^+}}{\gamma_{\pm(\text{HCl})}} = \frac{0.0138 \, \text{mol dm}^{-3}}{0.877} = 0.0157 \, \text{mol dm}^{-3}$$

There is no need to make any further approximations.

The pH based on the simple definition $-\log_{10}[\text{H}_3\text{O}^+] = 1.804$

This is to be contrasted with the rigorous calculation which takes non-ideality into account where pH = 1.860.

[HCl] is sufficiently high for non-ideality to be significant. Such considerations have to borne in mind when calculations of the pK of a fairly strong weak acid or a fairly strong weak base are made from pH titrations (see Section 8.25.3 and Worked Problems 8.21 and 8.22), or from emf measurements (see Worked Problem 9.23).

9.17 Determination of equilibrium constants for reactions where K is either very large or very small

Emf studies can also be used to determine the equilibrium constants for reactions where the equilibrium constant is so large or so small that experiment cannot determine it accurately. The reaction:

$$2\text{Cr}^{2+}(\text{aq}) + \text{Sn}^{4+}(\text{aq}) \rightarrow 2\text{Cr}^{3+}(\text{aq}) + \text{Sn}^{2+}(\text{aq})$$

can be carried out in the laboratory. However, it goes to virtual completion and the equilibrium constant cannot be measured. It is frequently found that the equilibrium constants for redox reactions are either much too large or much too small for the equilibrium constant to be found using standard measurements of concentrations at equilibrium. Emf measurements are particularly important in the determination of such equilibrium constants.

Another example is the redox reaction:

$$\text{Fe}^{2+}(\text{aq}) + \text{Ce}^{4+}(\text{aq}) \rightarrow \text{Fe}^{3+}(\text{aq}) + \text{Ce}^{3+}(\text{aq})$$

The cell which would be used to calculate ΔG^θ and K is:

$$\text{Pt(s)}|\text{Fe}^{2+}(\text{aq}), \text{Fe}^{3+}(\text{aq}) \vdots \text{Ce}^{3+}(\text{aq}), \text{Ce}^{4+}(\text{aq})|\text{Pt(s)}$$

The equilibrium constant for this reaction is 1.6×10^{14}.

Examples of reactions which have Ks which are too small to measure directly are:

$$\text{AgI(s)} + \text{Cl}^-(\text{aq}) \rightleftharpoons \text{AgCl(s)} + \text{I}^-(\text{aq})$$

$$\text{AgI(s)} + \text{Br}^-(\text{aq}) \rightleftharpoons \text{AgBr(s)} + \text{I}^-(\text{aq})$$

$$\text{AgBr(s)} + \text{Cl}^-(\text{aq}) \rightleftharpoons \text{AgCl(s)} + \text{Br}^-(\text{aq})$$

with Ks approximately equal to 5×10^{-7}, 3×10^{-4} and 2×10^{-3} respectively.

The cells which would correspond to these reactions are:

$Ag(s)|AgCl(s)|Cl^-(aq) \vdots\vdots I^-(aq)|AgI(s)|Ag(s)$

$Ag(s)|AgBr(s)|Br^-(aq) \vdots\vdots I^-(aq)|AgI(s)|Ag(s)$

$Ag(s)|AgCl(s)|Cl^-(aq) \vdots\vdots Br^-(aq)|AgBr(s)|Ag(s)$

The following problem illustrates the ease with which emf studies can furnish such equilibrium constants.

Worked problem 9.20

Question

The cell:

$$Pt(s)|Cr^{2+}(aq),\ Cr^{3+}(aq) \vdots\vdots Sn^{2+}(aq),\ Sn^{4+}(aq)|Pt(s)$$

has the following concentrations.

$[Cr^{2+}] = 2.80 \times 10^{-4}$ mol dm^{-3}; $[Cr^{3+}] = 9.5 \times 10^{-4}$ mol dm^{-3}; $[Sn^{2+}] = 1.55 \times 10^{-2}$ mol dm^{-3}; $[Sn^{4+}] = 6.50 \times 10^{-3}$ mol dm^{-3}.

The emf is 0.133 V at 25°C. Find E^θ and the equilibrium constant for the cell reaction.

Answer

Left hand electrode: $Cr^{2+}(aq) \rightarrow Cr^{3+}(aq) + e$

Right hand electrode: $Sn^{4+}(aq) + 2e \rightarrow Sn^{2+}(aq)$

Overall cell reaction: $2Cr^{2+}(aq) + Sn^{4+}(aq) \rightarrow 2Cr^{3+}(aq) + Sn^{2+}(aq)$

$$\Delta G = 2\mu_{Cr^{3+}(aq)} + \mu_{Sn^{2+}(aq)} - 2\mu_{Cr^{2+}(aq)} - \mu_{Sn^{4+}(aq)}$$

$$= 2\mu^\theta_{Cr^{3+}(aq)} + 2RT \log_e a_{Cr^{3+}(aq)} + \mu^\theta_{Sn^{2+}(aq)} + RT \log_e a_{Sn^{2+}(aq)}$$

$$- 2\mu^\theta_{Cr^{2+}(aq)} - 2RT \log_e a_{Cr^{2+}(aq)} - \mu^\theta_{Sn^{4+}(aq)} - RT \log_e a_{Sn^{4+}(aq)}$$

$$= \Delta G^\theta + RT \log_e \frac{a^2_{Cr^{3+}(aq)} a_{Sn^{2+}(aq)}}{a^2_{Cr^{2+}(aq)} a_{Sn^{4+}(aq)}}$$

In this problem this equation will have to be approximated by a logarithmic term in concentrations, i.e. ideality will have to be assumed. This is because there is insufficient information given to calculate the ionic strength for the solutions in each electrode compartment. Hence mean ionic activity coefficients cannot be calculated.

$$\therefore \Delta G = \Delta G^\theta + RT \log_e \frac{[Cr^{3+}]^2[Sn^{2+}]}{[Cr^{2+}]^2[Sn^{4+}]}$$

$z = 2$ for the cell reaction as written above.

$$\therefore E = E^\theta - \frac{2.303RT}{2F} \log_{10} \frac{[Cr^{3+}]^2 [Sn^{2+}]}{[Cr^{2+}]^2 [Sn^{4+}]}$$

$$\therefore 0.133\,V = E^\theta - \frac{0.05916\,V}{2} \log_{10} \frac{(9.50 \times 10^{-4})^2 \times 1.55 \times 10^{-2}}{(2.80 \times 10^{-4})^2 \times 6.50 \times 10^{-3}}$$

$$\therefore E^\theta = 0.133\,V + 0.02958\,V \log_{10} 27.45 = 0.133\,V + 0.0426\,V = 0.1756\,V$$

$$\therefore \Delta G^\theta = -zFE^\theta = -2 \times 96485\,C\,mol^{-1} \times 0.1756\,V = -33.9\,kJ\,mol^{-1}$$

Remember: V C = J

$$\Delta G^\theta = -RT \log_e K$$

$$\therefore \log_e K = -\frac{\Delta G^\theta}{RT} = \frac{33.9\,kJ\,mol^{-1}}{8.3145 \times 10^{-3}\,kJ\,mol^{-1}\,K^{-1} \times 298.15\,K} = 13.68$$

$$\therefore K = 8.7 \times 10^5$$

9.18 Use of concentration cells

These are cells where there are two chemically identical electrodes with the solutions around the electrodes at different concentrations. The electrode compartments must be kept physically separate generally by means of a salt bridge. The net reaction is the equivalent of transferring the solute from one electrode compartment to the other. This is effected by means of the electrode reactions and **not** by any actual transfer. The following problem illustrates this.

Worked problem 9.21

Question

What is the overall 'reaction' and the expression for the emf for each of the following cells:

(i) $Ag(s)|AgNO_3(aq)\ 0.01\ mol\ dm^{-3} \vdots\vdots AgNO_3(aq)\ 0.005\ mol\ dm^{-3}|Ag(s)$

(ii) $Cu(s)|CuSO_4(aq)\ 0.0020\ mol\ dm^{-3} \vdots\vdots CuSO_4(aq)\ 0.0075\ mol\ dm^{-3}|Cu(s)$

Calculate the emf of the second cell, ignoring non-ideality. Indicate how non-ideality can be taken into account.

Answer

(i) $Ag(s)|AgNO_3(aq)\ 0.01\ mol\ dm^{-3} \vdots\vdots AgNO_3(aq)\ 0.005\ mol\ dm^{-3}|Ag(s)$

Left hand electrode: $Ag(s) \rightarrow Ag^+(aq)(lhs) + e$

Right hand electrode: $Ag^+(aq)(rhs) + e \rightarrow Ag(s)$

Overall cell reaction: $Ag^+(aq)(rhs) \rightarrow Ag^+(aq)(lhs)$

$$\Delta G = \mu_{Ag^+(aq)}(lhs) - \mu_{Ag^+(aq)}(rhs)$$

$$= \mu^\theta_{Ag^+(aq)} + RT \log_e a_{Ag^+(aq)}(lhs) - \mu^\theta_{Ag^+(aq)} - RT \log_e a_{Ag^+(aq)}(rhs)$$

$$= RT \log_e \frac{[Ag^+]_{lhs} \gamma_{\pm(AgNO_3)}(lhs)}{[Ag^+]_{rhs} \gamma_{\pm(AgNO_3)}(rhs)}$$

Pay particular attention: there is no ΔG^θ for a concentration cell because the μ^θ's refer to the **same** species in solution and cancel out.

$$\therefore E = -\frac{2.303RT}{zF} \log_{10} \frac{[Ag^+]_{lhs} \gamma_{\pm(AgNO_3)}(lhs)}{[Ag^+]_{rhs} \gamma_{\pm(AgNO_3)}(rhs)}$$

Likewise take note: there is also no E^θ.

(ii) $Cu(s)|CuSO_4(aq)$ 0.0020 mol dm^{-3} $\vdots$ $CuSO_4(aq)$ 0.0075 mol dm$^{-3}|Cu(s)$

Left hand electrode: $Cu(s) \rightarrow Cu^{2+}(aq)(lhs) + 2e$

Right hand electrode: $Cu^{2+}(aq)(rhs) + 2e \rightarrow Cu(s)$

Overall cell reaction: $Cu^{2+}(aq)(rhs) \rightarrow Cu^{2+}(aq)(lhs)$

$$\Delta G = \mu_{Cu^{2+}(aq)}(lhs) - \mu_{Cu^{2+}(aq)}(rhs)$$

$$= \mu^\theta_{Cu^{2+}(aq)} + RT \log_e a_{Cu^{2+}(aq)}(lhs) - \mu^\theta_{Cu^{2+}(aq)} - RT \log a_{Cu^{2+}}(rhs)$$

$$= RT \log_e \frac{a_{Cu^{2+}(aq)}(lhs)}{a_{Cu^{2+}(aq)}(rhs)}$$

$$\therefore E = -\frac{2.303RT}{zF} \log_{10} \frac{a_{Cu^{2+}(aq)}(lhs)}{a_{Cu^{2+}(aq)}(rhs)}$$

For the cell reaction as written $z = 2$.

Ignoring non-ideality gives:

$$E = -\frac{2.303RT}{zF} \log_{10} \frac{[Cu^{2+}]_{(lhs)}}{[Cu^{2+}]_{(rhs)}}$$

$$= -\frac{0.05916\,V}{2} \log_{10} \frac{0.0020}{0.0075} = +0.02958\,V \times 0.5740 = +0.0170\,V$$

Taking account of non-ideality:

$$E = -\frac{2.303RT}{zF} \log_{10} \frac{[Cu^{2+}]_{(lhs)}\,\gamma_{\pm(CuSO_4)}(lhs)}{[Cu^{2+}]_{(rhs)}\,\gamma_{\pm(CuSO_4)}(rhs)}$$

Since there are no equilibria involved in either electrode compartment the actual ionic strength is found directly from the concentrations, and the mean activity coefficients for each of the electrode compartment solutions can be calculated from the Debye-Hückel equation. $I = 1/2 \sum_i c_i z_i^2$ where $z = 2$ for each of the two ions involved and

$$\log_{10}\gamma_\pm = -\frac{Az^2\sqrt{I}}{1 + \sqrt{I}} = -\frac{0.510\,mol^{-1/2}\,dm^{3/2} \times 2^2 \times \sqrt{I}}{1 + \sqrt{I}}$$

(a) for the left hand electrode compartment $I = 1/2[0.0020 \times 2^2 + 0.0020 \times 2^2] = 0.008 \, \text{mol dm}^{-3}$

$$\log_{10}\gamma_{\pm} = -\frac{0.510 \, \text{mol}^{-1/2} \, \text{dm}^{3/2} \times 2^2 \times \sqrt{0.0080} \, \text{mol}^{1/2} \, \text{dm}^{-3/2}}{1 + \sqrt{0.0080}} = 0.1675 \therefore \gamma_{\pm} = 0.680$$

(b) for the right hand electrode $I = 1/2[0.0075 \times 2^2 + 0.0075 \times 2^2] = 0.0300 \, \text{mol dm}^{-3}$

$$\log_{10}\gamma_{\pm} = -\frac{0.510 \, \text{mol}^{-1/2} \, \text{dm}^{3/2} \times 2^2 \times \sqrt{0.0300} \, \text{mol}^{1/2} \, \text{dm}^{-3/2}}{1 + \sqrt{0.0300}} = 0.3012 \therefore \gamma_{\pm} = 0.500$$

$$E = -\frac{2.303RT}{zF} \log_{10} \frac{[Cu^{2+}]_{(\text{lhs})} \, \gamma_{\pm(CuSO_4)}(\text{lhs})}{[Cu^{2+}]_{(\text{rhs})} \, \gamma_{\pm(CuSO_4)}(\text{rhs})}$$

$$= -\frac{0.05916 \, \text{V}}{2} \log_{10} \frac{0.0020 \times 0.6800}{0.0075 \times 0.4998} = +0.02958 \, \text{V} \times 0.4403 = +0.0130 \, \text{V}$$

It can be seen that accounting for non-ideality makes a considerable difference to the calculated emf, illustrating how important it is not to ignore non-ideality.

9.19 'Concealed' concentration cells and similar cells

These are used extensively in emf work to find, in particular, equilibrium constants, solubility products and equilibrium constants for complexing and ion-pair formation. Practice is necessary in recognising such situations and in handling them, and this is given in Worked Problems 9.23 to 9.27.

Worked problem 9.22

Question

Show how the following two cells can be regarded as equivalent.

$$Ag(s)|AgNO_3(aq) \, c_1 \,\vdots\, AgNO_3(aq) \, c_2|Ag(s)$$
$$Ag(s)|AgCl(s)|KCl(aq) \,\vdots\, AgNO_3(aq)|Ag(s)$$

Answer

- $Ag(s)|AgNO_3(aq) \, c_1 \,\vdots\, AgNO_3(aq) \, c_2|Ag(s)$

This is a typical concentration cell with electrode reactions as follows:

Left hand electrode: $Ag(s) \longrightarrow Ag^+(aq)(\text{lhs}) + e$

Right hand electrode: $Ag^+(aq)(\text{rhs}) + e \longrightarrow Ag(s)$

Overall cell reaction: $Ag^+(aq)(rhs) \rightarrow Ag^+(aq)(lhs)$

$$\Delta G = \mu_{Ag^+(aq)}(lhs) - \mu_{Ag^+(aq)}(rhs)$$
$$= \mu^\theta_{Ag^+(aq)} + RT \log_e a_{Ag^+(aq)}(lhs) - \mu^\theta_{Ag^+(aq)} - RT \log_e a_{Ag^+(aq)}(rhs)$$
$$= RT \log_e \frac{[Ag^+]_{lhs}\gamma_{\pm(AgNO_3)}(lhs)}{[Ag^+]_{rhs}\gamma_{\pm(AgNO_3)}(rhs)}$$

- $Ag(s)|AgCl(s)|KCl(aq) \vdots AgNO_3(aq)|Ag(s)$

The left hand electrode is generally considered to be described as:

$$Ag(s) + Cl^-(aq)(lhs) \rightarrow AgCl(s) + e$$

The right hand electrode reaction is:

$$Ag^+(aq)(rhs) + e \rightarrow Ag(s)$$

Overall reaction: $Ag^+(aq) (rhs) + Cl^-(aq) (lhs) \rightarrow AgCl(s)$

Pay particular attention: the two ions are in **different** solutions.

$$\Delta G = \mu_{AgCl(s)} - \mu_{Ag^+(aq)}(rhs) - \mu_{Cl^-(aq)}(lhs)$$
$$= \mu^\theta_{AgCl(s)} - \mu^\theta_{Ag^+(aq)} - RT \log_e a_{Ag^+(aq)}(rhs) - \mu^\theta_{Cl^-(aq)} - RT \log_e a_{Cl^-(aq)}(lhs)$$
$$\Delta G = \Delta G^\theta - RT \log_e a_{Ag^+(aq)}(rhs)a_{Cl^-(aq)}(lhs)$$

But $a_{Cl^-(aq)}(lhs)$ is related to $a_{Ag^+(aq)}(lhs)$ by the solubility product for AgCl(s), i.e.
$K_s = a_{Ag^+(aq)}(lhs)a_{Cl^-(aq)}(lhs)$

$$\therefore \Delta G = \Delta G^\theta - RT \log_e \left\{ a_{Ag^+(aq)}(rhs)\frac{K_s}{a_{Ag^+(aq)}(lhs)} \right\} = \Delta G^\theta - RT \log_e K_s - RT \log_e \frac{a_{Ag^+(aq)}(rhs)}{a_{Ag^+(aq)}(lhs)}$$

But $RT \log_e K_s = -\Delta G^\theta$ is a relation for the **dissolution** of AgCl(s)

$$AgCl(s) \rightleftharpoons Ag^+(aq) + Cl^-(aq)$$

where $\Delta G^\theta_{dissolution} = \mu^\theta_{Ag^+(aq)} + \mu^\theta_{Cl^-(aq)} - \mu^\theta_{AgCl(s)}$

and this is the **negative** of $\Delta G^\theta = \mu^\theta_{AgCl(s)} - \mu^\theta_{Ag^+(aq)} - \mu^\theta_{Cl^-(aq)}$ appearing earlier in the equation for ΔG for the cell reaction.

$$\therefore \Delta G = \Delta G^\theta - \Delta G^\theta - RT \log_e \frac{a_{Ag^+(aq)}(rhs)}{a_{Ag^+(aq)}(lhs)} = RT \log_e \frac{a_{Ag^+(aq)}(lhs)}{a_{Ag^+(aq)}(rhs)}$$

$$= RT \log_e \frac{[Ag^+]_{lhs}\gamma_{\pm(AgNO_3)}(lhs)}{[Ag^+]_{rhs}\gamma_{\pm(AgNO_3)}(rhs)}$$

and this is the same as the expression derived assuming that the cell can be treated as a concentration cell with electrode reactions as follows:

Left hand electrode: $Ag(s) \rightarrow Ag^+(aq)(lhs) + e$

Right hand electrode: $Ag^+(aq)(rhs) + e \rightarrow Ag(s)$

Overall cell reaction: $Ag^+(aq)(rhs) \rightarrow Ag^+(aq)(lhs)$

$$\Delta G = RT \log_e \frac{a_{Ag^+_{(aq)}}(lhs)}{a_{Ag^+_{(aq)}}(rhs)} = RT \log_e \frac{[Ag^+]_{lhs}\gamma_{\pm(AgNO_3)}(lhs)}{[Ag^+]_{rhs}\gamma_{\pm(AgNO_3)}(rhs)}$$

9.20   Determination of equilibrium constants and pK values for reactions which are not directly that for the cell reaction

These can be illustrated by the use of emf measurements to find the equilibrium constants for weak acids and bases, for the self ionisation of water, for the formation of a complex or ion pair and for the solubility of sparingly soluble salts. This, taken with the situations described in the previous worked problems, illustrates the extreme versatility of emf studies.

Modern digital voltmeters allow the measurement of highly accurate emfs. If the emfs are measured over a range of ionic strengths this will allow the effect of ionic strength on the equilibrium constants to be studied in detail. Furthermore if the dependence of emf on temperature is measured, the thermodynamic quantities, ΔH^θ, ΔS^θ and ΔC_p^θ can be found and interpretations of the magnitudes made. If the experimental set-up can be adapted to a determination of the emfs at various pressures to be made, then ΔV^θ values can also be found.

9.20.1   Determination of pK values for the ionisation of weak acids and weak bases, and for the self ionisation of $H_2O(l)$

When emf studies are used to determine these quantities the overall cell reaction is **not** that for the equilibrium being studied. The emf will give, however, the concentration of $H_3O^+(aq)$ which then allows pK_a, pK_b or pK_w to be found by standard equilibrium calculations. This is illustrated in Worked Problems 9.23 and 9.24. Determinations of pK_a, pK_b or pK_w from emf studies often make use of the 'concealed' concentration cell.

Worked problem 9.23

Question

The cell:

$$Pt(s), H_2(g)\,1\,atm|C_6H_5COOH(aq) \vdots\vdots KCl(aq)|Hg_2Cl_2(s)|Hg(l)$$

$$0.0100\,mol\,dm^{-3} \qquad 0.0200\,mol\,dm^{-3}$$

has an emf $= 0.5564$ V at 25°C. If E^θ is 0.2675 V at 25°C, find

(i) The pH of the left hand electrode compartment solution and K_a for the acid ionisation of $C_6H_5COOH(aq)$ at 25°C, ignoring non-ideality,

(ii) The pH and K_a correcting for non-ideality.

Answer

Left hand electrode: $\frac{1}{2}H_2(g) + H_2O(l) \rightarrow H_3O^+(aq) + e$

Right hand electrode: $\frac{1}{2}Hg_2Cl_2(s) + e \rightarrow Hg(l) + Cl^-(aq)$

Overall cell reaction: $\frac{1}{2}H_2(g) + \frac{1}{2}Hg_2Cl_2(s) + H_2O(l) \rightarrow Hg(l) + H_3O^+(aq) + Cl^-(aq)$

Take note: $H_3O^+(aq)$ and $Cl^-(aq)$ are in different electrode compartments.

$$\Delta G = \mu_{Hg(l)} + \mu_{H_3O^+(aq)} + \mu_{Cl^-(aq)} - \frac{1}{2}\mu_{H_2(g)} - \frac{1}{2}\mu_{Hg_2Cl_2(s)} - \mu_{H_2O(l)}$$

$$\therefore \Delta G = \mu^\theta_{Hg(l)} + \mu^\theta_{H_3O^+(aq)} + RT\log_e a_{H_3O^+(aq)} + \mu^\theta_{Cl^-(aq)} + RT\log_e a_{Cl^-}$$

$$- \frac{1}{2}\mu^\theta_{H_2(g)} - \frac{1}{2}RT\log_e p_{H_2} - \frac{1}{2}\mu^\theta_{Hg_2Cl_2(s)} - \mu^\theta_{H_2O(l)}$$

$$= \Delta G^\theta + 2.303RT\log_{10}\frac{a_{H_3O^+(aq)}a_{Cl^-(aq)}}{p_{H_2}^{1/2}} = \Delta G^\theta + 2.303RT\log_{10}a_{H_3O^+(aq)}a_{Cl^-(aq)}$$

Note: the term in $p_{H_2}^{1/2}$ drops out because $p_{H_2} = 1$ atm.

(i) Ignoring non-ideality enables the activities to be replaced by concentrations giving:

$$\Delta G = \Delta G^\theta + 2.303RT\log_{10}[H_3O^+][Cl^-]$$

Remember: $H_3O^+(aq)$ and $Cl^-(aq)$ are in different electrode compartments.

$$E = E^\theta - \frac{2.303RT}{zF}\log_{10}[H_3O^+] - \frac{2.303RT}{zF}\log_{10}[Cl^-]$$

where $z = 1$ for the cell reaction as written.

$$\frac{2.303RT}{zF}\log_{10}[H_3O^+] = E^\theta - E - \frac{2.303RT\log_{10}[Cl^-]}{zF}$$

$$= 0.2675V - 0.5564V - 0.05916V \times (-1.699)$$

$$= (-0.2889 + 0.1005)V$$

$$= -0.1884V$$

$$\therefore \log_{10}[H_3O^+] = \frac{-0.1884}{0.05916} = -3.185$$

$$\therefore [H_3O^+] = 6.54 \times 10^{-4}\ mol\ dm^{-3}$$

Assuming the approximate definition of pH, $pH = -\log_{10}[H_3O^+] = 3.185$

$$[C_6H_5COOH]_{stoich} = [C_6H_5COOH]_{actual} + [C_6H_5COO^-]_{actual}$$

Assuming that the self ionisation of $H_2O(l)$ can be ignored, then:

$$[H_3O^+]_{actual} = [C_6H_5COO^-]_{actual}$$

$$\therefore [C_6H_5COOH]_{actual} = [C_6H_5COOH]_{stoich} - [H_3O^+]_{actual}$$

$$= 0.0100 \, mol \, dm^{-3} - 6.54 \times 10^{-4} \, mol \, dm^{-3} = 9.35 \times 10^{-3} \, mol \, dm^{-3}$$

$$K_a = \left(\frac{[H_3O^+]_{actual}[C_6H_5COO^-]_{actual}}{[C_6H_5COOH]_{actual}} \right)_{eq} = \left(\frac{[H_3O^+]^2_{actual}}{[C_6H_5COOH]_{actual}} \right)_{eq}$$

$$= \frac{(6.54 \times 10^{-4})^2 \, mol^2 \, dm^{-6}}{9.35 \times 10^{-3} \, mol \, dm^{-3}} = 4.57 \times 10^{-5} \, mol \, dm^{-3}$$

$$pK_a = 4.34$$

(ii) Correcting for non-ideality:

$$\Delta G = \Delta G^\theta + 2.303RT \log_{10} a_{H_3O^+(aq)} a_{Cl^-(aq)}$$

Be careful here:

- Since the ionic strength is not known for the solution in the left hand electrode compartment, $a_{H_3O^+(aq)}$ cannot be replaced by $[H_3O^+]\gamma_\pm$.

- Since the ionic strength for the solution in the right hand electrode compartment can be found from the quoted concentration – there are no equilibria involved here and the stoichiometric ionic strength is the same as the actual ionic strength – $a_{Cl^-(aq)}$ can be replaced by $[Cl^-]\gamma_{\pm(KCl)}$.

$$\therefore E = E^\theta - \frac{2.303RT}{zF} \log_{10} a_{H_3O^+} - \frac{2.303RT}{zF} \log_{10}[Cl^-]\gamma_{\pm(KCl)} \quad \text{where} \quad z = 1$$

For the right hand electrode compartment:

$$I = 0.0200 \, mol \, dm^{-3} \therefore \sqrt{I} = 0.1414 \, mol^{1/2} \, dm^{-3/2}$$

$$\log_{10}\gamma_\pm = -\frac{A|z_{K^+} z_{Cl^-}|\sqrt{I}}{1 + \sqrt{I}} = -\frac{0.510 \, mol^{-1/2} \, dm^{3/2} \times 0.1414 \, mol^{1/2} \, dm^{-3/2}}{1.1414} = -0.0632$$

$$\therefore \gamma_{\pm KCl(aq)} = 0.865$$

$$\therefore 0.05916 V \log_{10} a_{H_3O^+(aq)} = E^\theta - E - 0.05916 \, V \log_{10}(0.0200 \times 0.865)$$

$$= 0.2675 \, V - 0.5564 \, V + 0.1042 \, V = -0.1847 \, V$$

$$\therefore \log_{10} a_{H_3O^+(aq)} = -3.121 \quad \therefore a_{H_3O^+(aq)} = 7.57 \times 10^{-4} \, mol \, dm^{-3}$$

If the formal definition of pH is used, i.e. $pH = -\log_{10} a_{H_3O^+(aq)}$, then the pH of the left hand solution is 3.12.

Taking account of non-ideality is now an exercise in successive approximations as illustrated in Worked Problem 9.19.

- **first approximation**

The activity found above is taken as a first approximate $[H_3O^+]$, i.e. $7.57 \times 10^{-4} \, mol \, dm^{-3}$, and since the self ionisation of $H_2O(l)$ can be ignored:

first approximate $[C_6H_5COO^-] = [H_3O^+] = 7.57 \times 10^{-4} \, mol \, dm^{-3}$

first approximate $[C_6H_5COOH] = [C_6H_5COOH]_{stoich} - [H_3O^+]$

$$= 0.0100 \, mol \, dm^{-3} - 7.57 \times 10^{-4} \, mol \, dm^{-3}$$

$$= 9.24 \times 10^{-3} \, mol \, dm^{-3}$$

first approximate ionic strength and activity coefficient:

$$I = 1/2 \sum_i c_i z_i^2 = 1/2\{7.57 \times 10^{-4} \times 1^2 + 7.57 \times 10^{-4} \times 1^2\} = 7.57 \times 10^{-4} \, mol \, dm^{-3}$$

$$\log_{10}\gamma_{H_3O^+} = -\frac{0.510 \, mol^{-1/2} \, dm^{3/2} \times 1^2\sqrt{I}\,mol^{1/2}\,dm^{-3/2}}{1+\sqrt{I}} = -\frac{0.510 \times 0.0275}{1.0275} = -0.0137$$

$$\therefore \gamma_{H_3O^+} = 0.969$$

$$[H_3O^+] = \frac{7.57 \times 10^{-4} \, mol \, dm^{-3}}{0.969} = 7.81 \times 10^{-4} \, mol \, dm^{-3}$$

continuing with successive approximations, see Worked Problem 9.19 gives:

$$[H_3O^+] = \frac{7.57 \times 10^{-4}}{0.969} = 7.82 \times 10^{-4} \, mol \, dm^{-3} \quad and \quad \gamma_{H_3O^+} = 0.969$$

$$[C_6H_5COO^-]_{actual} = [H_3O^+]_{actual} = 7.82 \times 10^{-4} \, mol \, dm^{-3}$$

$$[C_6H_5COOH]_{actual} = [C_6H_5COOH]_{stoich} - [H_3O^+]_{actual}$$

$$= 0.0100 \, mol \, dm^{-3} - 7.82 \times 10^{-4} \, mol \, dm^{-3}$$

$$= 9.22 \times 10^{-3} \, mol \, dm^{-3}$$

$$K_a = \frac{[H_3O^+]_{actual}[C_6H_5COO^-]_{actual}\gamma_{H_3O^+}\gamma_{C_6H_5COO^-}}{[C_6H_5COOH]_{actual}\gamma_{C_6H_5COOH}} = \frac{[H_3O^+]_{actual}^2\gamma_{H_3O^+}^2}{[C_6H_5COOH]_{actual}\gamma_{C_6H_5COOH}}$$

$$= \frac{(7.82 \times 10^{-4})^2 \, mol^2 \, dm^{-6} \times 0.969^2}{9.22 \times 10^{-3} \, mol \, dm^{-3}} = 6.23 \times 10^{-5} \, mol \, dm^{-3}$$

assuming that $\gamma_{C_6H_5COOH} = 1$.

$$pK_a = 4.21$$

The rigorous definition of $pH = -\log_{10}a_{H_3O^+} = 3.12$

Note: the approximate and rigorous pH values do not differ much for this acid. That this is so is explained in Section 8.25.2 and Worked Problem 8.20.

Worked problem 9.24

Question

How could the following cell be used to determine K_w for $H_2O(l)$ at 25°C?

$$Pt(s)|H_2(g) \, 1 \, atm|NaOH(aq) \, 0.015 \, mol \, dm^{-3}|AgCl(s)|Ag(s)$$
$$NaCl(aq) \, 0.015 \, mol \, dm^{-3}$$

If the observed emf is 1.0499 V and the standard emf for the $Cl^-(aq)|AgCl(s)|Ag(s)$ electrode is 0.2224 V at 25°C, find K_w.

Answer

Left hand electrode: $\tfrac{1}{2}H_2(g) + H_2O(l) \rightarrow H_3O^+(aq) + e$

Right hand electrode: $AgCl(s) + e \rightarrow Ag(s) + Cl^-(aq)$

Overall cell reaction: $\tfrac{1}{2}H_2(g) + H_2O(l) + AgCl(s) \rightarrow Ag(s) + H_3O^+(aq) + Cl^-(aq)$

$$\Delta G = \mu_{Ag(s)} + \mu_{H_3O^+(aq)} + \mu_{Cl^-(aq)} - \tfrac{1}{2}\mu_{H_2(g)} - \mu_{H_2O(l)} - \mu_{AgCl(s)}$$

$$\therefore \Delta G = \mu_{Ag(s)}^{\theta} + \mu_{H_3O^+(aq)}^{\theta} + RT\log_e a_{H_3O^+} + \mu_{Cl^-(aq)}^{\theta} + RT\log_e a_{Cl^-(aq)}$$

$$- 1/2\mu_{H_2(g)}^{\theta} - RT\log p_{H_2}^{1/2} - \mu_{H_2O(l)}^{\theta} - \mu_{AgCl(s)}^{\theta}$$

$$= \Delta G^{\theta} + RT\log_e \frac{a_{H_3O^+(aq)} a_{Cl^-(aq)}}{p_{H_2}^{1/2}}$$

The factor $p_{H_2}^{1/2}$ drops out since $p_{H_2} = 1$ atm.

$$\therefore E = E^{\theta} - 2.303\frac{RT}{zF}\log_{10} a_{H_3O^+(aq)} a_{Cl^-(aq)}$$

where $z = 1$ for the reaction as written.

$E^{\theta} = 0.2224\,V$ is the standard emf of the $Cl^-(aq)|AgCl(s)|Ag(s)$ electrode, and it is also by definition the standard E^{θ} of the cell:

$$Pt(s)|H_2(g)\ 1\ atm|HCl(aq)\ unit\ activity|AgCl(s)|Ag(s)$$

It is also the standard emf of the cell:

$$Pt(s)|H_2(g)\ 1\ atm|NaOH(aq)\ 0.015\ mol\ dm^{-3}|AgCl(s)|Ag(s),\ E^{\theta} = 0.2224\,V$$
$$NaCl(aq)\ 0.015\ mol\ dm^{-3}$$

because this is a 'concealed' type of cell analogous to the 'concealed' concentration cells. The mixed solution of NaOH(aq) and NaCl(aq) can be regarded as an extremely dilute solution of HCl(aq). Likewise the left hand electrode reaction can be written in two equivalent ways:

$$\tfrac{1}{2}H_2(g) + H_2O(l) \rightarrow H_3O^+(aq) + e \tag{1}$$

$$\tfrac{1}{2}H_2(g) + OH^-(aq) \rightarrow H_2O(l) + e \tag{2}$$

Subtracting the second equation from the first gives the equation for the self ionisation of water, where the concentrations of $H_3O^+(aq)$ and $OH^-(aq)$ are related to each other by K_w.

$$2H_2O(l) \rightarrow H_3O^+(aq) + OH^-(aq)$$

$$\therefore \frac{2.303RT}{zF}\log_{10} a_{H_3O^+(aq)} a_{Cl^-(aq)} = E^{\theta} - E = 0.2224\,V - 1.0499\,V = -0.8275\,V$$

$$\therefore \log_{10} a_{H_3O^+(aq)} a_{Cl^-(aq)} = -\frac{0.8275\,V}{0.05916\,V} = -13.987$$

$$a_{H_3O^+(aq)} a_{Cl^-(aq)} = 1.03 \times 10^{-14}\ mol^2\ dm^{-6}$$

Strictly, $K_w = a_{H_3O^+} a_{OH^-(aq)}$, although in the elementary chapters on equilibria K_w was taken to be equal to $[H_3O^+][OH^-]$.

$$a_{H_3O^+(aq)} = \frac{K_w}{a_{OH^-(aq)}}$$

$$\therefore a_{H_3O^+(aq)} a_{Cl^-} = \frac{K_w a_{Cl^-(aq)}}{a_{OH^-(aq)}} = \frac{K_w[Cl^-]\gamma_{Cl^-(aq)}}{[OH^-]\gamma_{OH^-(aq)}}$$

Since $OH^-(aq)$ and $Cl^-(aq)$ are in the same solution and thus relate to the same ionic strength and since both are univalent anions, $\gamma_{OH^-(aq)}$ and $\gamma_{Cl^-(aq)}$ can, to a first approximation, be taken to be equal and thus cancel out.

$$\therefore \frac{K_w[Cl^-]}{[OH^-]} = 1.03 \times 10^{-14}\ mol^2\ dm^{-6}$$

But $[OH^-] = [Cl^-] = 0.015\ mol\ dm^{-3}$

$$\therefore K_w = 1.03 \times 10^{-14}\ mol^2\ dm^{-6}$$

However, if the Debye–Hückel equation does not fully cope with non-ideality, then the extended Debye–Hückel equation is required to account for $\gamma_{OH^-(aq)}$ and $\gamma_{Cl^-(aq)}$ not quite cancelling each other out. Also it is possible that $\gamma_\pm$ for $NaOH(aq)$ and $NaCl(aq)$ are not quite equal because the value of b in the bI term of the extended Debye–Hückel equation $\log_{10}\gamma_\pm = -A|z_1 z_2|\sqrt{I}/(1 + \sqrt{I}) + bI$ may not be the same for the two electrolytes.

9.20.2 Solubility products

Determination of solubility products can be difficult using standard analytical procedures. These require that equilibration between the solid and the solution is complete and require constant shaking of the solution for a considerable length of time, followed by analysis of the saturated solution. Emf measurements allow a much faster and highly accurate determination. This is because emf values can generally be read accurately to 4 significant figures. 'Concealed' types of concentration cells with salt bridge are generally used.

Worked problem 9.25

Question

The cell:
Hg(l)|Hg$_2$Cl$_2$(s)|KCl(aq)   0.025   mol dm^{-3} $\vdots$ Hg$_2$(NO$_3$)$_2$(aq)   0.005   mol dm^{-3}|Hg(l)
has emf $= 0.3679$ V at 25°C. Find the solubility product of $Hg_2Cl_2(s)$.

Answer

This cell can be treated as a 'concealed' concentration cell.

Left hand electrode: $2Hg(l) \rightarrow Hg_2^{2+}(aq)(lhs) + 2e$

Right hand electrode: $2e + Hg_2^{2+}(aq)(rhs) \rightarrow 2Hg(l)$

Overall cell reaction: $Hg_2^{2+}(aq)(rhs) \rightarrow Hg_2^{2+}(aq)(lhs)$

$z = 2$ for the cell reaction as written.
If non-ideality is ignored, then:

$$\Delta G = \mu_{Hg_2^{2+}(aq)}(lhs) - \mu_{Hg_2^{2+}(aq)}(rhs)$$

$$= \mu_{Hg_2^{2+}(aq)}^{\theta}(lhs) + RT\log_e[Hg_2^{2+}]_{(lhs)} - \mu_{Hg_2^{2+}(aq)}^{\theta} - RT\log_e[Hg_2^{2+}]_{(rhs)}$$

$$= RT\log_e\frac{[Hg_2^{2+}]_{lhs}}{[Hg_2^{2+}]_{rhs}}$$

Note: there is no ΔG^{θ} term because the μ^{θ}s are the same for each electrode. As mentioned earlier, this always happens with concentration cells and with cells treated as 'concealed' concentration cells.

$$E = -\frac{2.303RT}{2F}\log_{10}\frac{[Hg_2^{2+}]_{lhs}}{[Hg_2^{2+}]_{rhs}} = 0.3679\,V$$

$$\therefore \log_{10}\frac{[Hg_2^{2+}]_{lhs}}{[Hg_2^{2+}]_{rhs}} = -0.3679\,V \times \frac{2F}{2.303RT} = -\frac{0.3679 \times 2\,V}{0.05916\,V} = -12.44$$

$$\therefore \frac{[Hg_2^{2+}]_{lhs}}{[Hg_2^{2+}]_{rhs}} = 3.65 \times 10^{-13}$$

$$\therefore [Hg_2^{2+}]_{lhs} = 3.65 \times 10^{-13} \times [Hg_2^{2+}]_{rhs} = 3.65 \times 10^{-13} \times 5 \times 10^{-3}\,mol\,dm^{-3}$$

$$= 1.8 \times 10^{-15}\,mol\,dm^{-3}$$

$$Hg_2Cl_2(s) \rightleftharpoons Hg_2^{2+}(aq)(lhs) + 2Cl^-(aq)(lhs)$$

$$K_s = \left([Hg_2^{2+}][Cl^-]^2\right)_{lhs} = 1.8 \times 10^{-15}\,mol\,dm^{-3} \times (0.025)^2\,mol^2\,dm^{-6} = 1.1 \times 10^{-18}\,mol^3\,dm^{-9}$$

Note: this calculation assumes ideality

9.20.3 A further use of cells to gain insight into what is occurring in an electrode compartment – ion pair formation

Emf and conductance studies have been used extensively in compilations of data on ion pair formation.

Worked problem 9.26

Question

The cell: $Cd(s)|CdCl_2(aq)\ 3.25 \times 10^{-4}\,mol\,dm^{-3} \,\vdots\vdots\, Cd(NO_3)_2\ 3.25 \times 10^{-4}\,mol\,dm^{-3}|Cd(s)$ has an emf $= 7.6 \times 10^{-4}$ V at 25°C. Estimate the concentration of $Cd^{2+}(aq)$ in the left hand electrode solution, stating any assumptions which are made, and from this find the equilibrium constant for

$$Cd^{2+}(aq) + Cl^-(aq) \rightleftharpoons CdCl^+(aq)$$

Assume that there are no $CdNO_3^+(aq)$ ion pairs present.

Answer

Left hand electrode: $Cd(s) \longrightarrow Cd^{2+}(aq)(lhs) + 2e$

Right hand electrode: $Cd^{2+}(aq)(rhs) + 2e \rightarrow Cd(s)$

Overall cell reaction: $Cd^{2+}(aq)(rhs) \rightarrow Cd^{2+}(aq)(lhs)$

$$\Delta G = \mu_{Cd^{2+}(aq)}(lhs) - \mu_{Cd^{2+}(aq)}(rhs)$$

$$= \mu_{Cd^{2+}(aq)}^\theta + RT \log_e a_{Cd^{2+}(aq)}(lhs) - \mu_{Cd^{2+}(aq)}^\theta - RT \log_e a_{Cd^{2+}(aq)}(rhs)$$

$$\therefore \Delta G = RT \log_e \frac{a_{Cd^{2+}(aq)}(lhs)}{a_{Cd^{2+}(aq)}(rhs)}$$

$$\therefore E = -\frac{2.303RT}{2F} \log_{10} \frac{a_{Cd^{2+}(aq)}(lhs)}{a_{Cd^{2+}(aq)}(rhs)}$$

Ignoring non-ideality, this approximates to:

$$E = -\frac{2.303RT}{2F} \log_{10} \frac{[Cd^{2+}]_{(lhs)}}{[Cd^{2+}]_{(rhs)}}$$

where $z = 2$ for the reaction as written.

$$\therefore \log_{10} \frac{[Cd^{2+}]_{(lhs)}}{[Cd^{2+}]_{(rhs)}} = -\frac{2 \times 7.6 \times 10^{-4}\,V}{0.05916\,V} = -0.0257$$

$$\therefore \frac{[Cd^{2+}]_{(lhs)}}{[Cd^{2+}]_{(rhs)}} = 0.9425$$

$$[Cd^{2+}]_{(lhs)} = 0.9425 \times 3.25 \times 10^{-4}\,mol\,dm^{-3} = 3.06 \times 10^{-4}\,mol\,dm^{-3}$$

Assuming that there are no $CdNO_3^+(aq)$ ion-pairs present:

$$[Cd^{2+}]_{total} = [Cd^{2+}]_{actual} + [CdCl^+]_{actual} = 3.25 \times 10^{-4}\,mol\,dm^{-3}$$

$$\therefore [CdCl^+]_{actual} = (3.25 \times 10^{-4} - 3.06 \times 10^{-4})\,mol\,dm^{-3} = 1.9 \times 10^{-5}\,mol\,dm^{-3}$$

$$[Cl^-]_{total} = [Cl^-]_{actual} + [CdCl^+]_{actual}$$

$$\therefore [Cl^-]_{actual} = (2 \times 3.25 \times 10^{-4} - 1.9 \times 10^{-5})\,mol\,dm^{-3} = 6.31 \times 10^{-4}\,mol\,dm^{-3}$$

$$K = \frac{[CdCl^+]_{actual}}{[Cd^{2+}]_{actual}[Cl^-]_{actual}} = \frac{1.9 \times 10^{-5}\,mol\,dm^{-3}}{3.06 \times 10^{-4}\,mol\,dm^{-3} \times 6.31 \times 10^{-4}\,mol\,dm^{-3}} = 98\,mol^{-1}\,dm^3$$

To take account of non-ideality it would be necessary to calculate a first approximate ionic strength assuming the concentrations as calculated above. From this, first approximate activity coefficients and first approximate activities could be found. Using the ideal expression for the emf the calculation could be repeated until successive approximations show no change in the concentrations.

9.20.4 Complex formation

Worked problem 9.27

Question

The cell:

$$Ag(s)|AgNO_3(aq)\ 0.10\ mol\ dm^{-3} \vdots AgNO_3(aq)\ 0.05\ mol\ dm^{-3}|Ag(s)$$
$$NH_3(aq)\ 0.50\ mol\ dm^{-3} \qquad NaNO_3(aq)\ 0.05\ mol\ dm^{-3}$$

has an emf $= 0.409$ V at $25°C$. Calculate the equilibrium constant for the complex formed between $Ag^+(aq)$ and $NH_3(aq)$.

Answer

The complex is formed as:

$$Ag^+(aq) + 2NH_3(aq) \rightleftharpoons Ag(NH_3)_2^+(aq)$$

$$K = \left(\frac{[Ag(NH_3)_2^+]}{[Ag^+][NH_3]^2} \frac{\gamma_{Ag(NH_3)_2^+(aq)}}{\gamma_{Ag^+(aq)}} \right)_{eq} \quad \text{assuming } \gamma_{NH_3} = 1$$

This can be treated as a 'concealed' concentration cell:

left hand electrode: $Ag(s) \rightarrow Ag^+(aq)$ (lhs) $+ e$

right hand electrode: $Ag^+(aq)$ (rhs) $+ e \rightarrow Ag(s)$

overall cell reaction: $Ag^+(aq)$ (rhs) $\rightarrow Ag^+(aq)$ (lhs)

$$\Delta G = \mu_{Ag^+(aq)}(lhs) - \mu_{Ag^+(aq)}(rhs)$$

$$= \mu^\theta_{Ag^+(aq)}(lhs) + RT \log_e a_{Ag^+(aq)}(lhs) - \mu^\theta_{Ag^+(aq)}(rhs) - RT \log_e a_{Ag^+(aq)}(rhs)$$

$$= RT \log_e \frac{[Ag^+]_{(lhs)}\gamma_{Ag^+(aq)}(lhs)}{[Ag^+]_{(rhs)}\gamma_{Ag^+(aq)}(rhs)}$$

$$\therefore E = -\frac{2.303RT}{F} \log_{10} \frac{[Ag^+]_{(lhs)}\gamma_{Ag^+(aq)}(lhs)}{[Ag^+]_{(rhs)}\gamma_{Ag^+(aq)}(rhs)}$$

where $z = 1$ for the reaction as written.

- In the left hand compartment the singly charged $Ag^+(aq)$ is being replaced by the singly charged complex, and to a first approximation the activity coefficients can be taken as equal and will cancel out in the equilibrium constant expression.

$$K = \left(\frac{[Ag(NH_3)_2^+]}{[Ag^+][NH_3]^2} \right)_{eq}$$

- The ionic strengths in the two electrode compartments are the same, i.e. 0.10 mol dm^{-3} and so the activity coefficients for the $Ag^+(aq)$ in each compartment will be the same and will cancel out in the emf expression.

$$\therefore E = -\frac{2.303RT}{F} \log_{10} \frac{[Ag^+]_{(lhs)}}{[Ag^+]_{(rhs)}}$$

$$\therefore \log_{10} \frac{[Ag^+]_{(lhs)}}{[Ag^+]_{(rhs)}} = -\frac{E}{0.05916 \text{ V}} = \frac{-0.409 \text{ V}}{0.05916 \text{ V}} = -6.913$$

$$\therefore \frac{[Ag^+]_{(lhs)}}{[Ag^+]_{(rhs)}} = 1.22 \times 10^{-7}$$

$$\therefore [Ag^+]_{(lhs)} = 5 \times 10^{-2} \times 1.22 \times 10^{-7} \text{ mol dm}^{-3} = 6.1 \times 10^{-9} \text{ mol dm}^{-3}$$

In the left hand compartment:

$$[Ag^+]_{total} = [Ag^+]_{actual} + [Ag(NH_3)_2^+]_{actual}$$

$$\therefore [Ag(NH_3)_2^+]_{actual} = [Ag^+]_{total} - [Ag^+]_{actual} = (0.10 - 6.1 \times 10^{-9}) \text{ mol dm}^{-3} = 0.10 \text{ mol dm}^{-3}$$

$$[NH_3]_{total} = [NH_3]_{actual} + 2[Ag(NH_3)_2]_{actual}$$

$$[NH_3]_{actual} = [NH_3]_{total} - 2[Ag(NH_3)_2]_{actual} = (0.50 - 2 \times 0.10) \text{ mol dm}^{-3} = 0.30 \text{ mol dm}^{-3}$$

Remember: $2NH_3$ are removed for every $Ag(NH_3)_2$ formed.

$$K = \left(\frac{[Ag(NH_3)_2^+]}{[Ag^+][NH_3]^2} \right)_{eq} = \frac{0.10 \text{ mol dm}^{-3}}{6.1 \times 10^{-9} \text{ mol dm}^{-3} \times (0.30)^2 \text{ mol}^2 \text{ dm}^{-6}} = 1.8 \times 10^8 \text{mol}^{-2} \text{ dm}^6$$

This is a further example of how useful emf measurements are in finding the equilibrium constant for a reaction which appears to have gone to virtual completion, but where a finite K can be found.

9.20.5   Use of cells to determine mean activity coefficients and their dependence on ionic strength

Emf studies have been the major route to the determination of highly accurate $\gamma_\pm$ values for electrolytes. They have also been used extensively in testing the Debye–Hückel limiting law at low concentrations, and the Debye–Hückel equation and extended equation over a large range of concentrations. Determinations at high concentrations have led to extensive compilations of data for a large number of electrolytes, and these are highly pertinent to the assessment of more advanced theories of non-ideality.

Worked problem 9.28

Question

At 25°C, the following cell:

$$Pt(s)|H_2(g) \ 1 \ atm|HCl(aq) \ unit \ activity \ \vdots \ CuSO_4(aq)|Cu(s)$$

has a standard emf $= 0.3394$ V and the following dependence on concentration of $CuSO_4(aq)$.

$\dfrac{[CuSO_4]}{mol \ dm^{-3}}$	0.0010	0.0030	0.0050	0.0100
$\dfrac{E}{V}$	0.2461	0.2571	0.2618	0.2679

- Find $\gamma_\pm$ for each $CuSO_4(aq)$ solution.

- Using the Debye–Hückel equation, calculate $\gamma_\pm$ for each $CuSO_4(aq)$ solution and comment on the two sets of values.

- Experiments at high concentrations give the following results:

$\dfrac{[Cu^{2+}]}{mol \ dm^{-3}}$	0.1000	0.2000	0.3000	0.4000
$\dfrac{I}{mol \ dm^{-3}}$	0.4000	0.8000	1.200	1.600
$\gamma_\pm(exptl)$	0.150	0.104	0.083	0.0705

- Compare these values with the corresponding values calculated from the Debye–Hückel equation, and comment.

Answer

Left hand electrode: $\tfrac{1}{2} H_2(g) + H_2O(l) \rightarrow H_3O^+(aq) + e$

Right hand electrode: $Cu^{2+}(aq) + 2e \rightarrow Cu(s)$

Overall cell reaction: $H_2(g) + 2H_2O(l) + Cu^{2+}(aq) \rightarrow 2H_3O^+(aq) + Cu(s)$

$$\Delta G = 2\mu_{H_3O^+(aq)} + \mu_{Cu(s)} - \mu_{H_2(g)} - 2\mu_{H_2O(l)} - \mu_{Cu^{2+}(aq)}$$

$$\Delta G = 2\mu_{H_3O^+(aq)}^\theta + 2RT \log_e a_{H_3O^+(aq)} + \mu_{Cu(s)}^\theta - \mu_{H_2(g)}^\theta - RT \log_e p_{H_2}$$

$$- 2\mu_{H_2O(l)}^\theta - \mu_{Cu^{2+}(aq)}^\theta - RT \log_e a_{Cu^{2+}(aq)}$$

$$\therefore \Delta G = \Delta G^\theta - RT \log_e \frac{a_{Cu^{2+}(aq)} p_{H_2(g)}}{a_{H_3O^+(aq)}^2}$$

Since the electrode on the left hand side is a standard hydrogen electrode, $a_{H_3O^+(aq)} = 1$ mol dm^{-3} and $p_{H_2} = 1$ atm, and the above equation reduces to:

$$\Delta G = \Delta G^\theta - RT \log_e [Cu^{2+}] \gamma_{\pm(CuSO_4)}$$

$$\therefore E = E^\theta + \frac{2.303 \times RT}{2F} \log_{10}[Cu^{2+}] + \frac{2.303 \times RT}{2F} \log_{10} \gamma_{\pm(CuSO_4)}$$

where $z = 2$ for the cell reaction as written.

$$\therefore (0.02958)\, V \log_{10} \gamma_{\pm(CuSO_4)} = E - E^\theta - (0.02958)\, V \log_{10}[Cu^{2+}]$$

$\dfrac{E}{V}$	0.2461	0.2571	0.2618	0.2679
$\dfrac{(E - E^\theta)}{V}$	−0.0933	−0.0823	−0.0776	−0.0715
$\dfrac{[Cu^{2+}]}{\text{mol dm}^{-3}}$	0.0010	0.0030	0.0050	0.0100
$0.02958 \log_{10}[Cu^{2+}]$	−0.0887	−0.0746	−0.0681	−0.0592
$0.02958 \log_{10}\gamma_\pm = \dfrac{E - E^\theta}{V}$ $-0.02958 \log_{10}[Cu^{2+}]$	−0.0046	−0.0077	−0.0095	−0.0123
$\log_{10} \gamma_{\pm(CuSO_4)}$	−0.154	−0.259	−0.322	−0.417
$\gamma_{\pm(CuSO_4)}(\text{exptl})$	0.701	0.550	0.476	0.383

- The Debye–Hückel equation can be used to calculate the mean activity coefficients for the CuSO$_4$(aq) solutions.

$$\log \gamma_\pm = -\frac{A \times 2 \times 2 \times \sqrt{I}}{1 + \sqrt{I}} \quad \text{and} \quad I = 1/2 \sum_i c_i z_i^2 = 1/2\{2^2[Cu^{2+}] + 2^2[SO_4^{2-}]\}$$

$$= 4[CuSO_4]$$

$\dfrac{[CuSO_4]}{\text{mol dm}^{-3}}$	0.0010	0.0030	0.0050	0.0100
$\dfrac{I}{\text{mol dm}^{-3}}$	0.0040	0.0120	0.0200	0.0400
$\dfrac{\sqrt{I}}{\text{mol}^{1/2}\,\text{dm}^{-3/2}}$	0.0632	0.1095	0.1414	0.2000
$\dfrac{\sqrt{I}}{1 + \sqrt{I}} \Big/ \text{mol}^{1/2}\,\text{dm}^{-3/2}$	0.0595	0.0987	0.1239	0.1667
$-\dfrac{0.510\,\text{mol}^{-1/2}\,\text{dm}^{3/2} \times 4 \times \sqrt{I}}{1 + \sqrt{I}}$	−0.1213	−0.2014	−0.2528	−0.3400
$\gamma_{\pm(CuSO_4)}(\text{calc})$	0.756	0.629	0.559	0.457

Comparison of observed and calculated values:

$\gamma_{\pm(CuSO_4)}$ from emf data	0.701	0.550	0.476	0.383
$\gamma_{\pm(CuSO_4)}$ calc from Debye–Hückel	0.756	0.629	0.559	0.457

It can be seen that the observed $\gamma_{\pm(CuSO_4)}$ are considerably lower than the $\gamma_{\pm(CuSO_4)}$ calculated from the Debye–Hückel equation. This is most probably due to formation of the ion pair $Cu^{2+}SO_4^{2-}(aq)$. Since both ions are doubly charged, ionic interactions are high and formation of the ion-pair is favoured.

- The data at high concentrations give the following calculated $\gamma_{\pm}$.

$\dfrac{[CuSO_4]}{\text{mol dm}^{-3}}$	0.1000	0.2000	0.3000	0.4000
$\dfrac{I}{\text{mol dm}^{-3}}$	0.4000	0.8000	1.200	1.600
$\dfrac{\sqrt{I}}{\text{mol}^{1/2}\,\text{dm}^{-3/2}}$	0.6325	0.8944	1.095	1.265
$\dfrac{\sqrt{I}}{1+\sqrt{I}} \Big/ \text{mol}^{1/2}\,\text{dm}^{-3/2}$	0.3874	0.4721	0.5228	0.5585
$-\dfrac{0.510\,\text{mol}^{-1/2}\,\text{dm}^{3/2} \times 4 \times \sqrt{I}}{1+\sqrt{I}}$	−0.7903	−0.9632	−1.066	−1.139
$\gamma_{\pm(CuSO_4)}$ (calc) from Debye–Hückel theory)	0.1620	0.1089	0.0858	0.0726
$\gamma_{\pm(CuSO_4)}$ from emf data	0.150	0.104	0.083	0.0705

The difference between the observed and calculated values of $\gamma_{\pm}$ is now much smaller. Ion association would be expected to be greater at these high ionic strengths and the difference between the $\gamma_{\pm}$ values would be expected to be even greater than at the low ionic strengths. The fact that this is not so can only be attributed to the inadequacy of the Debye–Hückel equation at these high ionic strengths and as evidence for the necessity of a bI term as is postulated in the extended Debye–Hückel equation:

$$\log_{10}\gamma_{\pm} = -\frac{A\left|z_{Cu^{2+}}z_{SO_4^{2-}}\right|\sqrt{I}}{1+\sqrt{I}} + bI$$

Worked problem 9.29

Question

At 25°C, the cell:

$$Pt(s)|H_2(g)\ 1\ atm|HCl(aq)|AgCl(s)|Ag(s)$$

has the following dependence on [HCl].

$\dfrac{[\text{HCl}]}{\text{mol dm}^{-3}}$	7.84×10^{-4}	2.50×10^{-3}	4.76×10^{-3}	8.10×10^{-3}	1.44×10^{-2}
$\dfrac{E}{V}$	0.5915	0.5332	0.5010	0.4749	0.4466

- Find E^{θ} and calculate values for $\gamma_{\pm(\text{HCl})}$ for each solution.

- When $[\text{HCl}] = 2.15 \text{ mol dm}^{-3}$ the observed emf is 0.1782 V. Calculate $\gamma_{\pm(\text{HCl})}$ and comment on its value.

Answer

Left hand electrode: $\frac{1}{2} \, \text{H}_2(\text{g}) + \text{H}_2\text{O}(\text{l}) \rightarrow \text{H}_3\text{O}^+(\text{aq}) + \text{e}$

Right hand electrode: $\text{AgCl}(\text{s}) + \text{e} \rightarrow \text{Ag}(\text{s}) + \text{Cl}^-(\text{aq})$

Overall cell reaction: $\frac{1}{2} \, \text{H}_2(\text{g}) + \text{H}_2\text{O}(\text{l}) + \text{AgCl}(\text{s}) \rightarrow \text{Ag}(\text{s}) + \text{Cl}^-(\text{aq}) + \text{H}_3\text{O}^+(\text{aq})$

Note: the $\text{H}_3\text{O}^+(\text{aq})$ and $\text{Cl}^-(\text{aq})$ refer to the **same** solution.

$$\Delta G = \mu_{\text{Ag(s)}} + \mu_{\text{Cl}^-(\text{aq})} + \mu_{\text{H}_3\text{O}^+(\text{aq})} - \tfrac{1}{2}\mu_{\text{H}_2(\text{g})} - \mu_{\text{H}_2\text{O(l)}} - \mu_{\text{AgCl(s)}}$$

$$\Delta G = \mu^{\theta}_{\text{Ag(s)}} + \mu^{\theta}_{\text{Cl}^-(\text{aq})} + RT\log_e a_{\text{Cl}^-(\text{aq})} + \mu^{\theta}_{\text{H}_3\text{O}^+(\text{aq})} + RT\log_e a_{\text{H}_3\text{O}^+(\text{aq})}$$

$$- \tfrac{1}{2}\mu^{\theta}_{\text{H}_2(\text{g})} - RT\log_e p^{1/2}_{\text{H}_2} - \mu^{\theta}_{\text{H}_2\text{O(l)}} - \mu^{\theta}_{\text{AgCl(s)}}$$

$$\therefore \Delta G = \Delta G^{\theta} + RT\log_e \frac{a_{\text{H}_3\text{O}^+(\text{aq})} a_{\text{Cl}^-(\text{aq})}}{p^{1/2}_{\text{H}_2(\text{g})}}$$

Since $\text{H}_2(\text{g})$ is present at 1 atm:

$$\Delta G = \Delta G^{\theta} + RT\log_e [\text{H}_3\text{O}^+]\gamma_{\pm(\text{HCl})}[\text{Cl}^-]\gamma_{\pm(\text{HCl})}$$

$$\therefore E = E^{\theta} - \frac{2.303RT}{F}\log_{10}[\text{H}_3\text{O}^+][\text{Cl}^-]\gamma^2_{\pm(\text{HCl})}$$

where $z = 1$ for the cell reaction as written.

$$\therefore E + \frac{2.303RT}{F}\log_{10}[\text{H}_3\text{O}^+][\text{Cl}^-] = E^{\theta} - \frac{2 \times 2.303RT}{F}\log_{10}\gamma_{\pm(\text{HCl})}$$

$$\therefore E + \frac{2.303RT}{F}\log_{10}[\text{H}_3\text{O}^+][\text{Cl}^-] = E^{\theta} + \frac{2 \times 2.303RT}{F} \times \frac{A|z_1 z_2|\sqrt{I}}{1 + \sqrt{I}}$$

A plot of:

$E + (2.303RT/F)\log_{10}[\text{H}_3\text{O}^+][\text{Cl}^-]$ vs. $\sqrt{I}/(1 + \sqrt{I})$ should be linear with intercept E^{θ} and slope $0.0603 \text{ V mol}^{-1/2} \text{ dm}^{3/2}$

Since:

$$\frac{2 \times 2.303RT}{F} \log_{10}\gamma_{\pm(HCl)} = E^{\theta} - E - 2.303\frac{RT}{F} \log_{10}[H_3O^+][Cl^-]$$

$$= E^{\theta} - (E + 2.303\frac{RT}{F} \log_{10}[H_3O^+][Cl^-])$$

and E^{θ} has been found, then this equation will allow calculation of $\gamma_{\pm(HCl)}$ for any [HCl].

- The table below gives the data required for the graph of $E + (2.303RT/F) \log_{10}[H_3O^+][Cl^-]$ vs. $\sqrt{I}/(1 + \sqrt{I})$

$\dfrac{c}{\text{mol dm}^{-3}}$	7.84×10^{-4}	2.50×10^{-3}	4.76×10^{-3}	8.10×10^{-3}	1.44×10^{-2}
$\dfrac{E}{V}$	0.5915	0.5332	0.5010	0.4749	0.4466
$\dfrac{(2.303RT/F)\log_{10}[H_3O^+][Cl^-]}{V}$	−0.3675	−0.3079	−0.2748	−0.2475	−0.2179
$\dfrac{E + (2.303RT/F)\log_{10}[H_3O^+][Cl^-]}{V}$	0.2240	0.2253	0.2262	0.2274	0.2287
$\dfrac{\sqrt{I}}{\text{mol}^{1/2}\,\text{dm}^{-3/2}}$	0.0280	0.0500	0.0690	0.0900	0.1200
$\dfrac{\sqrt{I}/(1 + \sqrt{I})}{\text{mol}^{1/2}\,\text{dm}^{-3/2}}$	0.0272	0.0476	0.0645	0.0826	0.1071

The graph of $E + (2.303RT/F) \log_{10}[H_3O^+][Cl^-]$ vs. $\sqrt{I}/(1 + \sqrt{I})$ is linear with intercept $= E^{\theta} = 0.2225V$ and slope $= 0.059$ V $\text{mol}^{-1/2}$ $\text{dm}^{3/2}$.

- The table below gives the data required for the calculation of $(2 \times 2.303RT/F) \log_{10}\gamma_{\pm(HCl)}$ $= E^{\theta} - (E + (2.303RT/F) \log_{10}[H_3O^+][Cl^-])$

$\dfrac{[H_3O^+]}{\text{mol dm}^{-3}}$	7.84×10^{-4}	2.50×10^{-3}	4.76×10^{-3}	8.10×10^{-3}	1.44×10^{-2}
$\dfrac{E}{V}$	0.5195	0.5332	0.5010	0.4749	0.4466
$\dfrac{E + (2.303RT/F)\log_{10}[H_3O^+][Cl^-]}{V}$	0.2240	0.2253	0.2262	0.2274	0.2287
$\dfrac{2 \times (2.303RT/F)\log_{10}\gamma_{\pm}}{V}$ where $E^{\theta} = 0.2225$ V	−0.0015	−0.0028	−0.0037	−0.0049	−0.0062
$\log_{10}\gamma_{\pm}$ where $2 \times 2.303RT/F$ $= 2 \times 0.05916$ V	−0.013	−0.024	−0.031	−0.041	−0.052
$\gamma_{\pm}$	0.971	0.947	0.931	0.909	0.886

This calculation shows that the mean activity coefficient for HCl(aq) decreases with increasing concentration, and this reflects the increasing non-ideality as the concentration increases. When $[HCl] = 2.15$ mol dm^{-3}:

$$2 \times \frac{2.303RT}{F} \log_{10}\gamma_\pm = E^\theta - E - \frac{2.303RT}{F} \log_{10}[H_3O^+][Cl^-]$$
$$= \{(0.2225 - 0.1782 - (2 \times 0.05916\log_{10} 2.15)\} \, V$$
$$= (0.2225 - 0.1782 - 0.0393)V$$
$$= +0.0050 \, V$$
$$\therefore \log_{10}\gamma_\pm = +\frac{0.0050 \, V}{2 \times 0.05916 \, V} = +0.042$$
$$\therefore \gamma_\pm = 1.10$$

Not only is this $\gamma_\pm$ greater than the values calculated above, but it is also greater than unity. This is a very clear indication that a term opposing the $-A|z_1 z_2|\sqrt{I}/(1 + \sqrt{I})$ term is needed. This is precisely what the b term in the extended Debye–Hückel equation does: $\log_{10}\gamma_\pm = -A|z_1 z_2|\sqrt{I}/(1 + \sqrt{I}) + bI$.

Studies of the emf of this cell at high concentrations furnish very accurate data for the determination of b for HCl(aq).

9.21 Use of concentration cells with and without liquid junctions in the determination of transport numbers

There are three main techniques for measuring transport numbers: emf methods and the Hittorf and moving boundary methods described in Sections 11.19.1 and 11.19.2 to 11.9.4.

The transport number has been defined in Section 9.1 as the fraction of the total current carried by a given ion. This is the definition most useful to the determination of transport numbers from emfs. In Chapter 11 the transport number is defined in terms of ionic mobilities, and/or individual molar ionic conductances (see Section 11.17), which are more directly linked to the methods described in that chapter.

Transport numbers are a very important property of ions both in terms of their intrinsic interest, and also in giving a bridge between consideration of emf's and conductances. They are also very important when electrolytes where one of the ions is large are being studied, such as polyelectrolyte solutions.

The transport number comes directly from a comparison of a concentration cell with a narrow tube as the liquid junction and the same cell but with a salt bridge as the liquid junction (see Sections 9.8.4 and 9.8.6).

The following cells illustrate the basis of the measurements.

- Ag(s)|AgNO$_3$(aq) 0.0100 mol dm^{-3} $\vdots$ AgNO$_3$(aq) 0.100 mol dm^{-3}|Ag(s)

$E = +0.0541 \, V$

is a cell **without** transference since the current is carried between the electrode compartments by migration of the ions of the salt bridge. The electrode reactions are:

Left hand electrode: Ag(s) $\rightarrow$ Ag$^+$(aq)(lhs) + e

Right hand electrode: $Ag^+(aq)(rhs) + e \rightarrow Ag(s)$

Overall reaction: $Ag^+(aq)(rhs) \rightarrow Ag^+(aq)(lhs)$

$$\Delta G = \mu_{Ag^+(aq)}(lhs) - \mu_{Ag^+(aq)}(rhs) = RT \log_e \frac{a_{Ag^+(aq)}(lhs)}{a_{Ag^+(aq)}(rhs)} \qquad (9.40)$$

$$E = -2.303 \frac{RT}{zF} \log_{10} \frac{a_{Ag^+(aq)}(lhs)}{a_{Ag^+(aq)}(rhs)} = -2.303 \frac{RT}{zF} \log_{10} \frac{\{[Ag^+]\gamma_\pm\}(lhs)}{\{[Ag^+]\gamma_\pm\}(rhs)} \qquad (9.41)$$

where $z = 1$ for the cell as written.

- $Ag(s)|AgNO_3(aq)\ 0.0100\ mol\ dm^{-3} \vdots AgNO_3(aq)\ 0.100\ mol\ dm^{-3}|Ag(s)$

$E = 0.0548\,V$

has the electrode compartments joined by a narrow tube filled with a solution of $AgNO_3(aq)$ which will have a concentration gradient set up along the tube as a result of the tube being filled from each end by the two $AgNO_3(aq)$ solutions (see Section 9.8.4). This is a cell **with transference** since the current is carried between the electrode compartments by migration of the ions of the electrode solutions, here as a result of the $Ag^+(aq)$ and $NO_3^-(aq)$ migrating into and out of the narrow tube. These migration processes must be included in the terms which contribute to the change in free energy. The electrode reactions in themselves are the same as previously, viz.:

Left hand electrode: $Ag(s) \rightarrow Ag^+(aq)(lhs) + e$

Right hand electrode: $Ag^+(aq)(rhs) + e \rightarrow Ag(s)$

But the derivation now deviates from that for the cell without transference.

- **At the left hand electrode**, for one faraday of electricity passed:
  1 mol of $Ag^+(aq)$ is formed and t_+ mol of $Ag^+(aq)$ moves away from around the electrode across the solution and into the tube, so that there is a net gain of $(1 - t_+)$ mol of $Ag^+(aq)$. To maintain electrical neutrality $(1 - t_+)$ mol, i.e. t_- mol of $NO_3^-(aq)$ will move across the solution towards the left hand electrode and will be replaced by t_- mol of $NO_3^-(aq)$ migrating into the electrode solution from the narrow tube.

- **At the right hand electrode**, for one faraday of electricity passed:
  1 mol of $Ag^+(aq)$ is removed and t_+ mol of $Ag^+(aq)$ moves in to around the electrode from the solution around the electrode and this is replaced by $Ag^+(aq)$ moving out of the tube, so that there is a net loss of $(1 - t_+)$ mol of $Ag^+(aq)$. To maintain electrical neutrality $(1 - t_+)$ mol, i.e. t_- mol of $NO_3^-(aq)$ will move across the solution away from the right hand electrode and into the narrow tube.

- As a consequence of these processes the net result in the cell is:
  The **left hand electrode** solution gains $(1 - t_+)$ mol of $Ag^+(aq)$, which is equivalent to t_- mol of $Ag^+(aq)$. At the same time it gains t_- mol of $NO_3^-(aq)$.

The **right hand electrode** solution loses $(1 - t_+)$ mol of $Ag^+(aq)$, which is equivalent to t_- mol of $Ag^+(aq)$. At the same time it loses t_- mol of $NO_3^-(aq)$,

$$\Delta G = t_-(\mu_{Ag^+(aq)})(lhs) + t_-(\mu_{NO_3^-(aq)})(lhs) - t_-(\mu_{Ag^+(aq)}(rhs) - t_-(\mu_{NO_3^-(aq)}(rhs) \quad (9.42)$$

$$= t_-\mu^{\theta}_{Ag^+(aq)} + t_-RT\log_e a_{Ag^+(aq)}(lhs) + t_-\mu^{\theta}_{NO_3^-(aq)} + t_-RT\log_e a_{NO_3^-(aq)}(lhs)$$

$$- t_-\mu^{\theta}_{Ag^+(aq)} - t_-RT\log_e a_{Ag^+(aq)}(rhs) - t_-\mu^{\theta}_{NO_3^-(aq)} - t_-RT\log_e a_{NO_3^-(aq)}(rhs)$$

$$(9.43)$$

$$\therefore \Delta G = t_-RT\log_e \frac{a_{Ag^+(aq)}(lhs) \times a_{NO_3^-(aq)}(lhs)}{a_{Ag^+(aq)}(rhs) \times a_{NO_3^-(aq)}(rhs)} \quad (9.44)$$

$$= t_-RT\log_e \frac{[Ag^+]_{(lhs)}\gamma_\pm(lhs)[NO_3^-]_{(lhs)}\gamma_\pm(lhs)}{[Ag^+]_{(rhs)}\gamma_\pm(rhs)[NO_3^-]_{(rhs)}\gamma_\pm(rhs)} \quad (9.45)$$

Since in each solution $[Ag^+] = [NO_3^-]$ \quad (9.46)

$$\Delta G = t_-RT\log_e \frac{[Ag^+]^2_{(lhs)}\gamma^2_\pm(lhs)}{[Ag^+]^2_{(rhs)}\gamma^2_\pm(rhs)} \quad (9.47)$$

$$\Delta G = 2t_-RT\log_e \frac{\{[Ag^+]\gamma_\pm\}(lhs)}{\{[Ag^+]\gamma_\pm\}(rhs)} \quad (9.48)$$

$$\therefore E = -2t_-\frac{2.303RT}{zF}\log_{10}\frac{\{[Ag^+]\gamma_\pm\}(lhs)}{\{[Ag^+]\gamma_\pm\}(rhs)} \quad (9.49)$$

where $z = 1$ for the cell reaction as written.

By comparing this expression for the emf of the cell with transport with the expression for the cell without transport, the transport number t_- for the anion can be found from the ratio of the emfs. From this t_+, the transport number for the cation, can be found since $t_+ + t_- = 1$.

$$\therefore \frac{E_{\text{with transport}}}{E_{\text{without transport}}} = -2t_-\frac{2.303RT}{zF}\log_{10}\frac{\{[Ag^+]\gamma_\pm\}(lhs)}{\{[Ag^+]\gamma_\pm\}(rhs)}$$

$$\div\left\{-2.303\frac{RT}{zF}\log_{10}\frac{\{[Ag^+]\gamma_\pm\}(lhs)}{\{[Ag^+]\gamma_\pm\}(rhs)}\right\} \quad (9.50)$$

$$= 2t_-$$

$$\therefore t_- = t_{NO_3^-} = \frac{0.0584}{2 \times 0.0541} = 0.540 \quad \text{and} \quad t_{Ag^+} = 0.460$$

Worked problem 9.30

Question

Deduce the expression for the emf of the following cell:
$$Cu(s)|CuCl_2(aq)\ c_1 \vdots CuCl_2(aq)\ c_2|Cu(s)$$

Answer

This is a cell with transport and so the migration processes must be included in the argument leading to the expression for the emf of the cell. The current is carried between the electrode compartments by migration of the ions of the electrode solutions, here as a result of the

Cu^{2+}(aq) and Cl^-(aq) migrating into and out of the narrow tube. The reactions occurring at the electrodes are:

Left hand electrode: $Cu(s) \rightarrow Cu^{2+}$(aq)(lhs) $+ 2e$

Right hand electrode: Cu^{2+}(aq)(rhs) $+ 2e \rightarrow Cu(s)$

Overall reaction: Cu^{2+}(aq)(rhs) $\rightarrow Cu^{2+}$(aq)(lhs)

- At the left hand electrode, for two faraday of electricity passed 1 mol of Cu^{2+}(aq) is formed and t_+ mol of Cu^2(aq) moves away from around the electrode across the solution and into the tube, so that there is a net gain of $(1 - t_+)$ mol of Cu^{2+}(aq). To maintain electrical neutrality $2(1 - t_+)$ mol, i.e. $2t_-$ mol of Cl^-(aq) will move across the solution towards the left hand electrode and will be replaced by $2t_-$ mol of Cl^-(aq) migrating into the electrode solution from the narrow tube.

- At the right hand electrode, for two faraday of electricity passed 1 mol of Cu^{2+}(aq) is removed and t_+ mol of Cu^{2+}(aq) moves in to around the electrode from the solution around the electrode, and this is replaced by Cu^{2+}(aq) moving out of the tube. There is a net loss of $(1 - t_+)$ mol of Cu^{2+}(aq). To maintain electrical neutrality $2(1 - t_+)$ mol, i.e. $2t_-$ mol of Cl^-(aq) will move across the solution away from the right hand electrode and into the narrow tube.

- As a consequence of these processes the net result in the cell is:

The left hand electrode gains $(1 - t_+)$ mol of Cu^{2+}(aq), which is equivalent to t_- mol of Cu^{2+}(aq). At the same time it gains $2t_-$ mol of Cl^-(aq),

The right hand electrode loses $(1 - t_+)$ mol of Cu^{2+}(aq), which is equivalent to t_- mol of Cu^{2+}(aq). At the same time it loses $2t_-$ mol of Cl^-(aq),

$$\Delta G = t_-(\mu_{Cu^{2+}(aq)})(lhs) + 2t_-(\mu_{Cl^-(aq)})(lhs) - t_-(\mu_{Cu^{2+}(aq)})(rhs) - 2t_-(\mu_{Cl^-(aq)})(rhs)$$

$$= t_-\mu^\theta_{Cu^{2+}(aq)} + t_-RT \log_e a_{Cu^{2+}(aq)}(lhs) + 2t_-\mu^\theta_{Cl^-(aq)} + 2t_-RT \log_e a_{Cl^-(aq)}(lhs)$$

$$- t_-\mu^\theta_{Cu^{2+}(aq)} - t_-RT \log_e a_{Cu^{2+}(aq)}(rhs) - 2t_-\mu^\theta_{Cl^-(aq)} - 2t_-RT \log_e a_{Cl^-(aq)}(rhs)$$

$$\therefore \Delta G = t_-RT \log_e \frac{a_{Cu^{2+}(aq)}(lhs) \times a^2_{Cl^-}(aq)(lhs)}{a_{Cu^{2+}(aq)}(rhs) \times a^2_{Cl^-}(aq)(rhs)}$$

$$= t_-RT \log_e \frac{[Cu^{2+}]_{(lhs)} \gamma_\pm(lhs)[Cl^-]^2_{(lhs)} \gamma^2_\pm(lhs)}{[Cu^{2+}]_{(rhs)} \gamma_\pm(rhs)[Cl^-]^2_{(rhs)} \gamma^2_\pm(rhs)}$$

$$\therefore E = -t_- \frac{RT}{zF} \log_e \frac{[Cu^{2+}]_{(lhs)} \gamma_\pm(lhs)[Cl^-]^2_{(lhs)} \gamma^2_\pm(lhs)}{[Cu^{2+}]_{(rhs)} \gamma_\pm(rhs)[Cl^-]^2_{(rhs)} \gamma^2_\pm(rhs)}$$

$$\therefore E = -t_- \frac{2.303RT}{zF} \log_{10} \frac{[Cu^{2+}]_{(lhs)} \gamma_\pm(lhs)[Cl^-]^2_{(lhs)} \gamma^2_\pm(lhs)}{[Cu^{2+}]_{(rhs)} \gamma_\pm(rhs)[Cl^-]^2_{(rhs)} \gamma^2_\pm(rhs)}$$

$$\text{or} \quad E = -t_- \frac{2.303RT}{2F} \log_{10} \frac{[Cu^{2+}]_{(lhs)} [Cl^-]^2_{(lhs)} \gamma^3_\pm(lhs)}{[Cu^{2+}]_{(rhs)} [Cl^-]^2_{(rhs)} \gamma^3_\pm(rhs)}$$

where $z = 2$ for the cell reaction as written.

Remember: $\gamma_{\pm(CuCl_2(aq))}^3 = \gamma_{Cu^{2+}(aq)} \gamma_{Cl^-(aq)}^2$

By comparing this expression for the emf of the cell with transport with the expression for the cell without transport, the transport number t_- for the anion can be found from the ratio of the emfs. From this t_+, the transport number for the cation, can be found since $t_+ + t_- = 1$.

9.21.1   Use of cells with and without transference in determination of the transport numbers of large ions

Such cells are extremely useful for obtaining data on large ions and polyelectrolytes.
    For instance, the electrolyte Na^+(aq) (large anion) can be studied in the cells:

- with transport

$NaHg(l)|Na^+$(aq) (large anion)(aq) $c_1 \vdots Na^+$(aq) (large anion) $c_2|NaHg(l)$

- without transport

$NaHg(l)|Na^+$(aq) (large anion)(aq) $c_1 \vdots\vdots Na^+$(aq) (large anion) $c_2|NaHg(l)$

Comparison of the emfs will lead to the transport number for the large anion, from which inferences can be made about the mobility and possibly the degree of solvation of the large ion.
    Likewise cells can furnish the same sort of information for large cations. The two cells:

$Ag(s)|AgCl(s)|$large cation(aq) Cl^-(aq) $c_1 \vdots$ large cation(aq) Cl^-(aq) $c_2|AgCl(s)|Ag(s)$

$Ag(s)|AgCl(s)|$large cation(aq) Cl^-(aq) $c_1 \vdots\vdots$ large cation(aq) Cl^-(aq) $c_2|AgCl(s)|Ag(s)$

will give the value for the transport number of the large cation.

10

Concepts and Theory of Non-ideality

Electrolyte solutions are non-ideal, with non-ideality increasing with increase in concentration. When experimental results on aspects of electrolyte solution behaviour are analysed, this non-ideality has to be taken into consideration. The standard way of doing so is to extrapolate the data to zero ionic strength. However, it is also necessary to obtain a theoretical description of non-ideality, and to deduce theoretical expressions which describe non-ideality for electrolyte solutions. Non-ideality is taken to be a manifestation of the electrostatic interactions which occur as a result of the charges on the ions of an electrolyte, and these interactions depend on the concentration of the electrolyte solution. Theoretically this non-ideality is taken care of by an activity coefficient for each ion of the electrolyte.

The Debye-Hückel theory discusses **equilibrium** properties of electrolyte solutions and allows the calculation of an activity coefficient for an individual ion, or equivalently, the mean activity coefficient of the electrolyte. Fundamental concepts of the Debye-Hückel theory also form the basis of modern theories describing the **non-equilibrium** properties of electrolyte solutions such as diffusion and conductance. The Debye-Hückel theory is thus central to all theoretical approaches to electrolyte solutions.

Sections 10.16 to 10.22 give a brief description of modern developments in electrolyte theory. This is a much more difficult section conceptually and can be omitted until after the Debye-Hückel and Bjerrum theories have been assimilated.

Aims

By the end of the chapter you should be able to:

- list the features of the simple Debye-Hückel model;

- appreciate the relevance of aspects of electrostatics which are of importance for the Debye-Hückel theory;

- understand the significance of the ionic atmosphere;

An Introduction to Aqueous Electrolyte Solutions. By Margaret Robson Wright
© 2007 John Wiley & Sons Ltd   ISBN 978-0-470-84293-5 (cloth) ISBN 978-0-470-84294-2 (paper)

- follow through the arguments of the Debye-Hückel theory which result in an expression for the mean ionic activity coefficient;

- understand the significance of the approximations inherent in the derivation;

- compare the Debye-Hückel expression and the Debye-Hückel limiting law;

- list the shortcomings of the Debye-Hückel model;

- explain how the extended Debye-Hückel equation helps with fitting experimental data to theoretical equations;

- discuss the evidence for ion association from Debye-Hückel plots;

- follow through the reasoning behind the Bjerrum theory of ion association;

- gain an appreciation of modern advances in electrolyte theory.

10.1   Evidence for non-ideality in electrolyte solutions

The Debye-Hückel theory deals with departures from ideality in electrolyte solutions. The main experimental evidence for this non-ideality is that:

- Concentration equilibrium constants are not constant.

- Rate constants depend on concentration.

- Molar conductivities for strong electrolytes vary with concentration.

- EMF expressions are not satisfactory when concentrations appear in the log[ ] terms.

- Freezing points of electrolytes are different from what would be expected for ideal behaviour.

These departures from ideality become less as the concentration decreases, and behaviour approaches ideality. They are taken care of thermodynamically as the non-ideal part of the free energy, often called the 'excess free energy', G^E. This can alternatively be expressed in terms of activity coefficients.

Departures from ideality in electrolyte solutions have been shown (see Chapter 1) to be mainly due to:

interionic electrostatic interactions which obey Coulomb's law.

Although this is the major factor other smaller contributions must be considered:

- Modified ion–solvent interactions over and above those interactions present at infinite dilution.

- Modified solvent–solvent interactions over and above those interactions present at infinite dilution.

10.2 The problem theoretically

In the simple Debye-Hückel model, attention is focused on the coulombic interactions as the source of non-ideality. The aim of the Debye-Hückel theory is to calculate the mean activity coefficient for an electrolyte in terms of these electrostatic interactions between the ions.

All other interactions which lead to non-ideality are not considered, and will have to be superimposed on the simple model when consideration is given to the ways in which the simple treatment has to be modified to bring it more into line with experimental results.

The problem is to calculate from this simple model:

the electrostatic potential energy of all the interactions
of the ions in the non-ideal case.

One mode of attack is:

- (a) Choose a given ion and work out the electrostatic interaction energy between it and all the other ions. In turn, this requires a knowledge of the distribution of all the other ions around the central ion, but this is dependent on the electrostatic interaction energy. This mutual dependence is one of the major problems of this treatment.

  An assumed electrostatic interaction energy can be used to calculate a first approximation to the distribution of all the other ions around this central ion. This, in turn, will give a second approximate interaction energy, and successive approximations can be made to obtain the best values for the interaction energy and the distribution.

- (b) Take each of the other ions in turn and make it a central ion, and proceed as in (a).

- (c) Sum all the interaction energies, counting each pair once only.

This is very similar in essence to Milner's approach in 1912. It corresponds to a very simple physical model but results in some very complex and well-nigh intractable mathematics. Consequently this approach was dropped, though as will be shown later (Section 10.18), it has been revived in the modern Monte Carlo computer simulation methods of solving the problems of electrolyte solutions.

10.3 Features of the simple Debye-Hückel model

- Strong electrolytes are completely dissociated into ions.

- Random motion of these ions is not attained: electrostatic interactions impose some degree of order over random thermal motions.

- Non-ideality is due to these electrostatic interactions between the ions. Only electrostatic interactions obeying the Coulomb inverse square law are considered.

- Ions are considered to be spherically symmetrical, unpolarisable charges. They are assumed to have a definite ion-size, and this represents a distance of closest approach within which no other ion or solvent molecule can approach. Unpolarisable means that the ion is a simple charge with no possibility for displacement of the charge in the presence of an electric field imposed externally, or imposed by the presence of other ions.

- The solvent is considered to be a structureless, continuous medium whose sole purpose is to allow the ions to exist as ions, and whose sole property is manifested in the bulk macroscopic value of the relative permittivity. No microscopic structure is allowed for the solvent which means that is not necessary to consider any:

    (a)   specific ion–solvent interactions;

    (b)   specific solvent–solvent interactions;

    (c)   possibility of alignment of the dipoles of the solvent;

    (d)   polarisability of the solvent to give induced molecular dipoles;

    (e)   possibility of dielectric saturation.

- No electrostriction is allowed.

- The most important feature of the Debye-Hückel model is that each ion is taken to have an ionic atmosphere associated with it. This is made up from all the other ions in the electrolyte solution. Because the solution is overall electrically neutral, the charge on the central ion is balanced by the charge on the ionic atmosphere. Although the ions of the ionic atmosphere are **discrete charges**, the ionic atmosphere is **treated as though it were a smeared-out cloud of charge** whose charge density varies continuously throughout the solution.

In all aspects of the Debye-Hückel theory the electrolyte is considered to be made up of:

- a chosen central reference ion, called the j-ion;

- with all the other ions (cations and anions) which are present making up the ionic atmosphere.

- The ions of the ionic atmosphere are distributed around the central j-ion. Because of the electrostatic interactions which result from the charges on the ions of the electrolyte, the **distribution** of the ions in the ionic atmosphere is **not random**. If it were random, the chance of finding an ion of opposite charge to the central j-ion in a given region would be equal to the chance of finding an ion of the same charge. In the ionic atmosphere, however, the chance of finding an ion of opposite charge to the central ion in any given region is greater than the chance of finding an ion of the same charge. This is because the central ion will attract ions of opposite charge and repel ions of like charge, again a consequence of electrostatic interactions.

An alternative way of looking at this distribution states that the chance of finding an ion of opposite charge around the central j-ion in any given region is greater than the random value, and the chance of finding an ion of the same charge is less than the random value, see Section 10.19.

The distribution of the ionic atmosphere about the central j-ion lies somewhere in between that of the regular arrangement of a lattice of ions and the random arrangement in a gas at very low pressures, i.e. an ideal gas.

- Statistical mechanics is a theory which discusses probabilities and distributions and so is relevant to a discussion of situations like that of the ionic atmosphere, and the Maxwell-Boltzmann distribution will feature heavily in the theoretical development.

These points are the main features of the simple Debye-Hückel model. Other aspects of the Debye-Hückel theory are illustrated by the mode of approach considered shortly.

10.3.1 Naivety of the Debye-Hückel theory

A re-read of Chapter 1 taken in conjunction with the seven statements describing the model (Section 10.3) makes it quite clear just how naïve the Debye-Hückel simple model actually is. But when the complexities of fitting this very simple model into the mathematical framework, and the even greater complexities of solving the mathematics are considered, it becomes clear just how badly stuck the theory is with a model which is so physically unrealistic. Even today, much though it is to be desired that there should be incorporation of modifications to the simple model into the theory, this has proved mathematically extremely difficult.

10.4 Aspects of electrostatics which are necessary for an understanding of the procedures used in the Debye-Hückel theory and conductance theory

10.4.1 The electric field, force of interaction and work done

The term **electric field** of a charge, here an ion, describes the effect of the charge, and it depends on both the magnitude of the charge and on the distance from the charge at which the electric field is being considered. If the charge is put into a material medium the field at any distance, r, from the charge will be decreased by a factor, the relative permittivity, ε_r, whose magnitude depends on the medium.

Any effect which an ion possesses by virtue of its charge must be susceptible to experimental measurement, and this cannot be done for an ion in isolation. For the present discussion, a second charge or ion is introduced so that the effect of the first charge on the second, or vice versa, can be studied. The effect of an ion on a solvent molecule becomes relevant when solvation is considered.

A second charge or ion placed at some distance, r, from the first charge, or ion, results in an electrostatic interaction where the **force** between the two charges depends on:

  i) the charges;

 ii) the distance between the charges;

iii) whether the charges are in a vacuum or in a material medium: gas, liquid or solid.

One of Coulomb's experiments involved bringing one charged sphere up to another charged sphere. This experiment can be used to discuss electrostatic forces and work done on systems where electrostatic interactions are involved.

The work done by a force = the magnitude of the force × the displacement in the direction of the force

- If the direction of the externally applied force and the displacement which occurs are in the same direction, then the work done on the system is positive;

- if the direction of the externally applied force and the displacement which occurs are in opposite directions, then the work done on the system is negative.

From this it follows that, if the spheres have like charges, they will be repelled from each other, and work would have to be done by an **external** force to negate this repulsion and bring them together. Four situations can be envisaged:

- If one of the charges is held in a fixed position and no external force acts on the system, the charges will repel each other.

- If there is an external force acting which keeps the two charges in their initial positions, then this force will be equal and opposite to the repulsive force. In this case no work will be done since there is no displacement.

- If the external force pushes the charges towards each other, then work will be done. The force will act along the line of centres and will act in a direction towards the fixed charge. The displacement will also be in this direction and so the work done by the external force will be positive.

- If, despite the external force, the two charges move further apart, then the displacement will be in the opposite direction to the external force and the work done by this force will be negative.

If the charges are of unlike charge they will be attracted to each other, and work would have to be done by an **external** force to negate this attraction and force them further apart. Again four situations can envisaged with conclusions similar to those above.

10.4.2   Coulomb's Law

This summarises the results of experiments on bringing two charged spheres together. It is found that the force of electrostatic interaction between two charges in a medium of relative permittivity, ε_r, at a distance, r, apart is given by:

$$\text{force} = \frac{\text{product of the two charges}}{4\pi\varepsilon_0\varepsilon_r(\text{distance between them})^2} \tag{10.1}$$

where $4\pi\varepsilon_0$ is a universal constant.
For the case of two ions of charge z_1e and z_2e this becomes:

$$\text{force} = \frac{z_1z_2e^2}{4\pi\varepsilon_0\varepsilon_r r^2} \tag{10.2}$$

- If z_1e and z_2e are of like sign, then the force is positive and, as shown above, the work done in forcing them together is positive.

- If z_1e and z_2e are of unlike sign, then the force is negative and, as shown above, the work done in forcing them apart is positive.

The force depends on the distance between the charges and can be given schematically as in Figure 10.1(a). The curve approaches both axes asymptotically.

10.4.3   Work done and potential energy of electrostatic interactions

In the electrostatic situation:
the work done on a system = the change in the total energy of the system
= the change in the kinetic energy of the system
+ the change in the potential energy of the system

If the external force pushing or pulling at the charges as they are repelled or attracted to each other is very nearly equal and opposite to the electrostatic force between the charges, then all the changes in position can take place very, very slowly, and so the kinetic energy will remain effectively constant.
Under these conditions:

the **work done** on the system in bringing the second charge z_2e from a distance, a, to a distance, b, from the first charge z_1e

= the change in **potential energy** of the electrostatic interactions
between the two charges as the distance between them varies
between a and b
= the potential energy of the two charges at point, b,
− the potential energy of the two charges at point, a.

Hence to obtain an expression for the potential energy of electrostatic interactions between the two charges requires that an expression for the work done in bringing the two charges from a

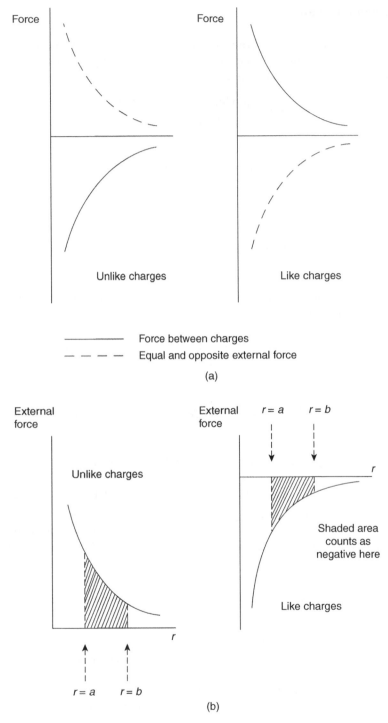

Figure 10.1 (**a**) Dependence of force on distance between two charges; (**b**) Work done on bringing two charges from $r = a$ to $r = b$.

distance, a, apart to a distance, b, apart is found. This can be done using the Coulomb's Law, Equation (10.2), and the definition of work done given above.

> work done bringing charge $z_2 e$ from a distance, a, apart
> to a distance, b, apart from the charge $z_1 e$
> $= $ force $\times$ displacement

where the displacement is given in terms of a change in r.

But, it has been shown that the force depends on the distance which the charges are apart (see Figure 10.1(a)), and so the work done is given by the area below the curve between a, and b, i.e. an integral must be taken (see Figure 10.1(b)).

$$w = \int_a^b f \, dr \qquad (10.3)$$

By Coulomb's law, the force of interaction, f, between the charges is:

$$\text{force} = \frac{z_1 z_2 e^2}{4\pi\varepsilon_0\varepsilon_r r^2} \qquad (10.2)$$

But to evaluate the work done, w, the force which is required is the external force which is either pushing the charges together or pulling them apart depending on whether there is repulsion or attraction between the charges. This force will be equal and opposite to the force of interaction.

$$\therefore w = -\int_a^b \frac{z_1 z_2 e^2}{4\pi\varepsilon_0\varepsilon_r} \frac{1}{r^2} \, dr = +\frac{z_1 z_2 e^2}{4\pi\varepsilon_0\varepsilon_r} \left[\frac{1}{r}\right]_a^b = \frac{z_1 z_2 e^2}{4\pi\varepsilon_0\varepsilon_r} \frac{1}{b} - \frac{z_1 z_2 e^2}{4\pi\varepsilon_0\varepsilon_r} \frac{1}{a} \qquad (10.4)$$

From this the change in potential energy can now be found since

> work done on the system in bringing   potential energy of the
> charge $z_2 e$ from a distance, a, from $=$ system at b $-$ potential
> $z_1 e$, to a distance, b, from $z_1 e$ energy of the system at a

From this, it follows that:

the potential energy of the system of two charges at point, $b = \dfrac{z_1 z_2 e^2}{4\pi\varepsilon_0\varepsilon_r} \dfrac{1}{b}$ $\qquad (10.5)$

the potential energy of the system of two charges at point, $a = \dfrac{z_1 z_2 e^2}{4\pi\varepsilon_0\varepsilon_r} \dfrac{1}{a}$ $\qquad (10.6)$

$\therefore$ the potential energy of the system of charges at any distance, $r = \dfrac{z_1 z_2 e^2}{4\pi\varepsilon_0\varepsilon_r} \dfrac{1}{r}$ $\qquad (10.7)$

and $\quad -\dfrac{d(\text{potential energy})}{dr} = \text{force at point } r$ $\qquad (10.8)$

i.e. the force at a point, r, is the gradient of the tangent to the graph of potential energy vs. r, at that point (see Figure 10.2).

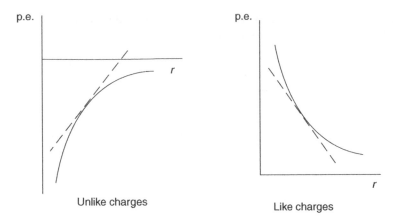

Figure 10.2   Dependence of potential energy on the distance between two charges.

10.4.4   The relation between the forces of interaction between two charges and the electric fields associated with each of them

The effect which charge $z_1 e$ has on charge $z_2 e$ and vice versa manifests itself as a force of interaction between the two charges. The force between charges $z_1 e$ and $z_2 e$ at a distance, r, apart in a medium of relative permittivity, ε_r has been given:

$$\text{force} = \frac{z_1 z_2 e^2}{4\pi\varepsilon_0\varepsilon_r r^2} \tag{10.2}$$

This can be written as:

$$\text{force} = \frac{z_1 e}{4\pi\varepsilon_0\varepsilon_r r^2} \times z_2 e \quad (10.9), \quad \text{or} \quad \text{force} = \frac{z_2 e}{4\pi\varepsilon_0\varepsilon_r r^2} \times z_1 e \tag{10.10}$$

- where the term, $z_1 e/4\pi\varepsilon_0\varepsilon_r r^2$, relates only to the charge $z_1 e$ and represents the effect which charge $z_1 e$ has on charge $z_2 e$ and defines the electric field, X_1, due to charge $z_1 e$ at the second charge $z_2 e$ at the distance, r, apart;

- and $z_2 e/4\pi\varepsilon_0\varepsilon_r r^2$ relates only to the charge $z_2 e$ and is the effect which charge $z_2 e$ has on charge $z_1 e$. It is thus the field, X_2, due to charge $z_2 e$ at the position of charge $z_1 e$.

These two terms, $z_1 e/4\pi\varepsilon_0\varepsilon_r r^2$ and $z_2 e/4\pi\varepsilon_0\varepsilon_r r^2$, give a quantitative measure of the effect of charge $z_1 e$ on charge $z_2 e$ which is at a distance r from $z_1 e$ in a medium of relative permittivity, ε_r, and vice versa, and so

- the force, f, between two charges at a distance r apart in a medium is given as:

$$f = \text{field} \times \text{charge on which this field acts}$$

Since the terms $z_1 e/4\pi\varepsilon_0\varepsilon_r r^2$ and $z_2 e/4\pi\varepsilon_0\varepsilon_r r^2$ depend only on the charge in question, then the **definition** of the field is independent of whether or not another charge is present, but the **experimental manifestation** of the field (via the measurement of a force) depends on a second charge being present.

The symbol, X, is used for the field.

10.4.5 The relation between the electrostatic potential energy and the electrostatic potential

The electrostatic potential plays a fundamental role in Debye-Hückel theory, and it is important to understand what it corresponds to physically.

The electrostatic potential energy of interaction between charges $z_1 e$ and $z_2 e$ at a distance, r, apart in a medium of relative permittivity, ε_r, has been given as:

$$\text{potential energy} = \frac{z_1 z_2 e^2}{4\pi\varepsilon_0\varepsilon_r} \frac{1}{r} \tag{10.7}$$

This can be written as:

$$\text{potential energy} = \frac{z_1 e}{4\pi\varepsilon_0\varepsilon_r r} \times z_2 e \quad (10.11) \quad \text{or} \quad \text{potential energy} = \frac{z_2 e}{4\pi\varepsilon_0\varepsilon_r r} \times z_1 e \quad (10.12)$$

- where the term, $z_1 e/4\pi\varepsilon_0\varepsilon_r r$, relates to the charge $z_1 e$ only. This quantity is the electrostatic potential, ψ_1, due to charge $z_1 e$ at the position of the charge $z_2 e$ at a distance, r, from charge $z_1 e$ and in a medium of relative permittivity, ε_r. Like the field it is another way of looking at the effect which one ion can have on another;

- and the term, $z_2 e/4\pi\varepsilon_0\varepsilon_r r$, relates to the charge $z_2 e$ only. This quantity is the electrostatic potential, ψ_2, due to charge $z_2 e$ at the position of charge $z_1 e$, distance, r, from charge $z_2 e$ and in a medium of relative permittivity, ε_r, and again this is another way of looking at the effect which charge $z_2 e$ has on charge $z_1 e$.

The electrostatic potential energy of interaction between two charges at a distance, r, apart in a medium of relative permittivity, ε_r, is given as:

$$\text{P.E.} = \text{potential} \times \text{charge on which the potential acts}$$

Since the terms, $z_1 e/4\pi\varepsilon_0\varepsilon_r r$ and $z_2 e/4\pi\varepsilon_0\varepsilon_r r$, depend only on the charge in question, then the definition of potential is independent of whether there is or is not another charge present.

The symbol, ψ, is often used for the electrostatic potential.

10.4.6 Relation between the electric field and the electrostatic potential

In general, the field and the potential due to a single charge can be written as:

$$\text{field} = \frac{ze}{4\pi\varepsilon_0\varepsilon_r r^2} \quad (10.13) \quad \text{and} \quad \text{potential} = \frac{ze}{4\pi\varepsilon_0\varepsilon_r r} \quad (10.14)$$

and so

$$\text{field} = -\frac{d(\text{potential})}{dr} \quad \text{or} \quad X = -\frac{d\psi}{dr} \quad (10.15)$$

i.e. the field at a given point is the gradient of the tangent to the curve of ψ vs. r at that point (see Figure 10.3). Thus, if the potential varies non-linearly as the distance alters, then the electric field also alters with distance. This is relevant to Debye-Hückel theory, see Sections 10.5 and 10.6.17.

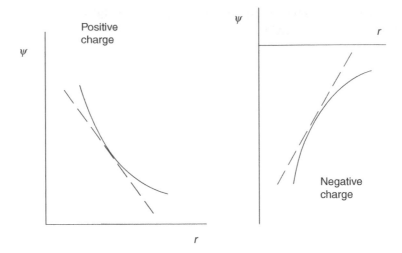

Figure 10.3   Dependence of potential due to a single charge on distance.

10.5   The ionic atmosphere in more detail

In the theory, all ions are considered to be spherically symmetrical with the charge on the ion taken to be at the **centre** of the ion. One ion is taken to be a central reference ion with all the other ions which make up the ionic atmosphere arranged around it. There is no preferred direction for this arrangement and this implies that the ionic atmosphere is **spherically symmetrical** around the central reference ion. Polar coordinates are therefore used and the centre of the central reference ion is taken as the origin of the coordinate system. It is also the position of the charge on the central reference ion.

Crystallographic measurements give sizes for unsolvated ions, and these vary considerably from ion to ion. But, in solution, most ions will be solvated and to differing extents. This is a problem which is not explicitly tackled by the simple Debye-Hückel theory.

In the theory, the distance of closest approach of one ion to another, $\mathring{a}$, is of fundamental importance. This is the distance between the centres of the two ions when they are in contact (see Figure 10.4). It is also the distance between the charges on the ions. For a given electrolyte it is assumed that the radius of the cation is the same as the radius of the anion. This is obviously a gross approximation, but it does allow the distance of closest approach to be simply twice the radius (see Figure 10.4).

From this diagram and Figure 10.5, it can be seen that the surface of any ion lies at a distance, $\mathring{a}/2$, from the centre of the ion, and that between $\mathring{a}/2$ and $\mathring{a}$ there is a region around any ion into which it is impossible for the **centre** of another ion to penetrate. This will have significance in the development of the theory.

There are two related potentials pertinent to these distances which are of fundamental importance in the Debye-Hückel theory:

- The potential due to the ionic atmosphere **at the surface** of the ion, i.e. at a distance $\mathring{a}/2$ from the centre of the central reference ion. Non-ideality in electrolyte solutions is a result of electrostatic interactions obeying Coulomb's Law. The potential energy of such interactions is given in terms of:

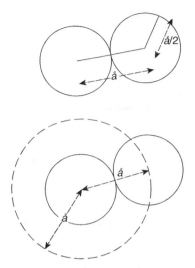

Figure 10.4   Diagram illustrating the meaning of the distance of closest approach for two ions.

(a) the charge on the central reference j-ion and;

(b) the potential at the j-ion, i.e. **at its surface**, due to the ionic atmosphere (see Sections 10.4.5 and 10.6.1).

- The potential due to the ionic atmosphere **at a distance**, $\mathring{a}$, from the central reference ion. This defines the distance around the reference ion into which no ion can penetrate, and is thus the distance from the centre of the reference ion at which the ionic atmosphere starts.

It is vital to realise that **these two potentials have the same magnitude**, as shown by the following standard argument of electrostatics.

Mathematically the ionic atmosphere can be treated **as though it were** a sphere of radius, $\mathring{a}$, with all the ions of the ionic atmosphere distributed on the **surface** of this sphere. Electrostatic theory states that inside such a sphere the potential due to the ionic atmosphere is constant at all points within the sphere and **is equal** to the value of this potential at the surface of the sphere. It follows that:

The potential due to the ionic atmosphere is the same at the distance $\mathring{a}/2$ as at the distance $\mathring{a}$, and is therefore constant in the region in between, i.e. in the region beyond the surface of the central ion into which the centre of any other ion cannot penetrate (see Figure 10.5).

A further potential of relevance is for distances $\geq \mathring{a}$ for which the potential decreases steadily, but non-linearly, with distance.

One other crucial feature of the model is that all the ions in the ionic atmosphere which would normally be thought of as a set of discrete charges are to be replaced by a 'smeared out' cloud of charge which varies continuously from point to point, with the charge density

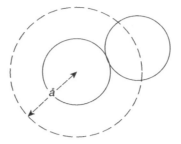

Figure 10.5 Diagram illustrating the region around an ion into which the centre of a second ion cannot penetrate.

being greater when the point considered is nearer to the central j-ion (see Sections 10.6.3 and 10.6.4).

This smeared out cloud of charge density is the ionic atmosphere of the central j-ion, and in it no discrete charges are allowed.

Furthermore, any ion can be chosen to be the central reference j-ion

- once a certain ion is chosen to be the central reference j-ion;

- all the other ions present constitute that central ion's ionic atmosphere.

But an ion from this ionic atmosphere could be picked out and be made to be a central reference ion, i.e.:

- choose one of these ions to be the new central reference ion;

- and all the other ions **including** the central ion in the first reference system will now be the ionic atmosphere of the second reference system.

10.5.1 A summary

- Each central ion which corresponds to one framework or reference system is always part of the ionic atmosphere of each of all the other ions which could instead have been chosen to be the central reference ion.

- Although the above is a description in terms of discrete charges for the central reference ion, the ionic atmosphere is treated **as though** there were no discrete charges in it, and it behaved as a 'smeared out' charge cloud.

- The reason and the absolute necessity for replacing the set of discrete ions which make up the ionic atmosphere by a 'smeared out' charge cloud **only becomes obvious** when these **verbal statements** describing the **model** are translated into the **mathematics** of the theory.

10.6 Derivation of the Debye-Hückel theory from the simple Debye-Hückel model

10.6.1 Step 1 Stating the problem

The aim is to calculate the mean ionic activity coefficient from the non-ideal part of the free energy. This is done in terms of the electrostatic potential energy of the coulombic interactions between the ion and its ionic atmosphere. These interactions give rise to non-ideality.

The electrostatic potential energy of interaction between a charge z_1e and a charge z_2e at a distance, r, apart in a medium of relative permittivity, ε_r, has been given (Section 10.4.5; Equations 10.11 and 10.12), and from these:

$$\text{potential energy} = \frac{z_1e}{4\pi\varepsilon_0\varepsilon_r r} \times z_2e = \psi_1 z_2e \tag{10.16}$$

or

$$\text{potential energy} = \frac{z_2e}{4\pi\varepsilon_0\varepsilon_r r} \times z_1e = \psi_2 z_1e \tag{10.17}$$

where ψ_1 is the potential due to charge z_1e at charge z_2e, and ψ_2 is the potential due to charge z_2e at charge z_1e.

In the present context, the charge is that of the central reference ion, z_je, and the potential is that due to the ionic atmosphere **at the surface** of the ion at a distance, $\mathring{a}/2$, from the centre of the j-ion.

Note well: the potential which appears in the equation for the potential energy of interaction between the ion and its ionic atmosphere is specified as that **at the surface** of the ion, i.e. at a distance $\mathring{a}/2$ from the central reference ion. It is **not specified** in terms of the potential at the distance from the central ion at which the ionic atmosphere begins, i.e. at a distance $\mathring{a}$, even though both these potentials take the same value (see Section 10.5). It is very important to be aware that the potential energy is defined in this way.

The electrostatic potential energy of interaction between the ion and its ionic atmosphere

$$= (z_je)\psi_j^* \tag{10.18}$$

charge on the central reference j-ion   potential **at the surface** of the central reference j-ion due to all the other ions which make up the ionic atmosphere

z_je is easy to calculate; therefore the problem is **to calculate ψ_j^***

Note well: ψ_j^* is the potential due to the ionic atmosphere at the surface of the central reference j-ion, i.e. at a distance $\mathring{a}/2$ from the centre of the j-ion. **But** (see Section 10.5), it is equal to the potential due to the ionic atmosphere at a distance $\mathring{a}$ from the centre of the j-ion.

Remember: $\mathring{a}$ is the distance of closest approach of two ions to each other and is the region around the central reference ion into which no ion can penetrate, i.e. it is **the distance from the j-ion beyond which the ionic atmosphere can lie.**

10.6.2   Step 2 The problem is to calculate ψ_j^* in terms of other calculable potentials, but what are these?

This focuses attention on potentials at a distance, r, from the centre of the central j-ion, where $r \geq \mathring{a}$. The restriction to distances where $r \geq \mathring{a}$ is a consequence of the assumption of an ion size $\mathring{a}$ where $\mathring{a}$ is both the diameter of an ion **and** the distance from the centre of the j-ion into which the centre of another ion cannot penetrate (see Figure 10.5). Hence the ionic atmosphere will start at a distance, $\mathring{a}$ from the centre of the j-ion. As a consequence of this assumption of a finite ion size, all of the potentials given in Equation (10.19) below apply only to distances r from the centre of the j-ion where $r \geq \mathring{a}$.

It is also important that the physical meaning of what is meant by each of the potentials which appear in the Debye-Hückel theory is explicitly stated.

$$\psi_j \qquad = \qquad \psi_j' \qquad + \qquad \psi_j'' \qquad (10.19)$$

$$\downarrow \qquad\qquad\qquad \downarrow \qquad\qquad\qquad \downarrow$$

| the **total potential** at any distance r from the central j-ion due to the **ion itself and its ionic atmosphere** | the potential at a distance r from the j-ion due to the **ion itself** | the potential at a distance r from the j-ion due to the **ionic atmosphere only** |

ψ_j is unknown but, as will be shown, is calculable from the Debye-Hückel theory,

ψ_j' is known and is equal to $z_j e / 4\pi\varepsilon_0\varepsilon_r r$ (see Section 10.4.5).

ψ_j'' is unknown but can be calculated once ψ_j is calculated, and this will then give ψ_j^* automatically; see Section 10.6.14, and below.

This equation shows an additive relation between the potentials and therefore fits a fundamental principle of electrostatics: the linear superposition of fields.

Once the equation relating ψ_j'' and r is found, substitution of $r = \mathring{a}$ for r in this equation gives the potential due to the ionic atmosphere at a distance from the centre of the j-ion due to the ionic atmosphere at a distance $r = \mathring{a}$.

But remember: This is the potential due to the ionic atmosphere at the distance from the central ion at which the ionic atmosphere starts.

The potential energy of interaction between an ion and its ionic atmosphere is, however, $(z_j e)\psi_j^*$ (Equation 10.18), where ψ_j^* is the potential at the surface of the central j-ion due to the ionic atmosphere. There is no problem here since ψ_j'' at $r = \mathring{a}$ is equal to ψ_j'' at $r = \mathring{a}/2$ and this is also equal to ψ_j^*, (see Section 10.5). Use of these arguments will be made in Sections 10.6.13 and 10.6.14.

The problem thus reduces to:

a calculation of ψ_j

10.6.3   Step 3 The question now is: is there anything in physics, that is, in electrostatic theory, which would enable this to be done?

Fortunately the answer is yes, and is given by Poisson's equation of electrostatics:

$$\nabla^2 \psi = -\frac{1}{\varepsilon_0 \varepsilon_r} \rho \tag{10.20}$$

and for **spherical symmetry** : $\nabla^2 \psi = \dfrac{1}{r^2} \dfrac{\partial}{\partial r} \left(r^2 \dfrac{\partial \psi}{\partial r} \right) = \dfrac{d^2 \psi}{dr^2} + \dfrac{2}{r} \dfrac{d\psi}{dr}$ (10.21)

ψ is the potential at a given point and ρ is the corresponding charge density, i.e. the charge per unit volume, at that point.

In the purely electrostatic context Poisson's equation is used to describe:

- a **continuous** distribution of charge in a medium of relative permittivity ε_r; or

- a set of **discrete charges** in **fixed** positions in a medium of relative permittivity ε_r.

However, the electrolyte solution conforms to neither of these two situations:

the electrolyte solution consists of discrete ions rather than a continuous distribution of charge, and these ions move around the solution in contrast to discrete charges at fixed positions in space.

The Poisson equation thus cannot be taken over **without modification** and **therein** lies a **major problem** for the theory. The two quantities, ψ_j and ρ_j, relate to a **continuous** distribution of charge, but there is no such thing as a continuous charge density for discrete charges, here ions.

And so, the next part of the problem **has to be**:

to think of a physical process which would allow the discrete charges of the ionic atmosphere to be replaced by a continuous charge density.

10.6.4   Translating the Poisson equation directly to the case of an electrolyte in solution

This step is a **crucial** aspect of the Debye-Hückel model. It is at this stage that it becomes **absolutely necessary** to introduce into the model the concept of 'smearing out' of **all** the discrete ions **in the ionic atmosphere** into a continuous distribution of charge, whose charge density is assumed to be a function of the distance from the central reference j-ion. The ionic atmosphere lies outside a sphere of radius $\mathring{a}$, and the continuous space charge is also spread over all distances greater than $\mathring{a}$ from the origin. **Only** the ions in the ionic atmosphere are replaced by the continuous charge. The central reference j-ion is a **discrete** charge lying at the origin and so the charge density between $r = 0$ and $r = \mathring{a}$ must be zero.

By use of some advanced statistical mechanical averaging procedures it is possible to translate a situation which corresponds to **discrete** ions moving around in the solution into one which corresponds to a **smeared out continuous cloud of charge** where the charge density varies **continuously** from the origin to any distance, r, from the origin. The charge density resulting from this averaging is often referred to as a '**time-averaged**' charge density'.

Likewise, similar averaging corresponding to 'smearing out' of the discrete charges can be carried out for the potential to give ψ as the '**time-averaged**' potential.

If the Poisson equation is **modified in this way** by using such time-averaged ψ and ρ, and then is applied to the electrolyte solution, there has been a '**smearing out**' of the discrete charges of the ionic atmosphere into a **continuous distribution** of charge. **Only then** is it legitimate to use the Poisson equation of electrostatics to discuss the theory of electrolyte solutions, i.e.

$$\nabla^2 \psi_j = -\frac{1}{\varepsilon_0 \varepsilon_r} \rho_j \qquad (10.20)$$

where ψ_j is **now** the total potential at a distance, r, from the central j-ion due to the j-ion and the ionic atmosphere, **with this being an average quantity**, and ρ_j is **now** the charge density **due to the ionic atmosphere** of the central j-ion at a distance, r, from the central j-ion, and again **this is an average quantity.**

Note: the whole of the charge of the central j-ion is at the origin, and so does not contribute to ρ_j.

10.6.5 Step 4 How can the distribution of the discrete ions in the ionic atmosphere of the j-ion be described?

Step 3 involved smearing out of the discrete ions of the ionic atmosphere into a continuous cloud of charge so that the Poisson equation could be used. The problem in this step, however, is to find the actual distribution of the **discrete** ions of the ionic atmosphere around the central j-ion. Statistical mechanics in the form of the Maxwell-Boltzmann distribution is used. The derivation of the Maxwell-Boltzmann distribution automatically involves an implicit averaging process, but this statistical mechanical averaging is different from that used for the Poisson equation when 'smearing out' has been done.

The ions of the ionic atmosphere are made up of at least two types of ions – the anions and the cations of the electrolyte. If there were another electrolyte present, as is often the case in experimental situations, then there will also be present the anions and cations of that electrolyte, e.g.

NaCl(aq) is made up of **two** types of ions, Na^+(aq) and Cl^-(aq);

NaCl(aq) $+$ Ca(NO$_3$)$_2$(aq) is made up of **four** types of ions, Na^+(aq), Cl^-(aq), Ca^{2+}(aq), and NO_3^-(aq).

Each type, i, of ion in the ionic atmosphere is treated separately and then a summation is taken over all types, i, present.

The distribution of the ions in the ionic atmosphere of the j-ion is governed by the total potential energy of electrostatic interactions at all positions, r, from the central reference j-ion. For any ion of type, i, in the ionic atmosphere of the j-ion, this potential energy is given by $z_i e \psi_j$. Here $z_i e$ is the charge on the ion of type i, and ψ_j is the potential at this ion due to the central j-ion and all the **other ions** of the ionic atmosphere, but it does **not include** a contribution from the ion $z_i e$ which is at position, r.

However, the ψ_j which appears in the equation $\psi_j = \psi'_j + \psi''_j$ (Equation 10.19), and in the Poisson equation $\nabla^2 \psi_j = -(1/\varepsilon_0 \varepsilon_r)\rho_j$ (Equation 10.20), includes contributions from **all** of the ions of the ionic atmosphere. The approximation is made of treating the two ψ_js as the same, although some modern work queries the validity of this.

The Maxwell-Boltzmann then gives separate expressions for the **average** number of ions for **each type** of ion, i, as a function of distance, r, from the central j-ion at the origin.

The form of the Maxwell-Boltzmann distribution for spherical symmetry usually quoted is:

$$n'_i \qquad = \qquad n_i \qquad \exp\left(-\frac{z_i e \psi_j}{kT}\right) \qquad (10.22)$$

$$\downarrow \qquad\qquad \downarrow \qquad\qquad \downarrow$$

the number of ions of one type, i, per unit volume at a given position, i.e. the **average** local concentration of ions of one type, i, at a given position	the total number of ions of one type, i, per unit volume in the solution as a whole i.e. the **bulk** concentration of ions of one type i, in the solution as a whole	ψ_j is taken to be the same as the quantity ψ_j appearing in the equation: $\psi_j = \psi'_j + \psi''_j$ and is taken to be the same as the quantity which is applicable to the Poisson equation, i.e. it includes contributions from **all** the ions of the ionic atmosphere

Equations of this form can be written for each type of ion in the solution, and the result is then summed over all ions of type, i, in the ionic atmosphere to give:

$$\sum_i \{n'_i(z_i e)\} = \sum_i \left\{ n_i(z_i e) \exp\left(-\frac{z_i e \psi_j}{kT}\right) \right\} \qquad (10.23)$$

10.6.6 The two basic equations

So far two basic equations of the theory have been deduced one involving ψ_j and ρ_j, and the other involving n'_i and ψ_j:

- Poisson's equation:

$$\nabla^2 \psi_j = -\frac{1}{\varepsilon_0 \varepsilon_r} \rho_j \qquad (10.20)$$

where the ionic atmosphere is treated as a continuous distribution of charge. This is an equation in two unknowns, ψ_j and ρ_j, and cannot be solved.

- The Maxwell-Boltzmann equation:

$$n'_i = n_i \exp\left(-\frac{z_i e \psi_j}{kT}\right) \qquad (10.22)$$

$$\sum_i \{n'_i(z_i e)\} = \sum_i \left\{ n_i(z_i e) \exp\left(-\frac{z_i e \psi_j}{kT}\right) \right\} \qquad (10.23)$$

which can be written for each type, i, of ion (Equation 10.22), and the result is then summed over all types of ion present (Equation 10.23). Here the ionic atmosphere is treated as a collection of discrete ions. These are equations in two unknowns, n_i' and ψ_j, and cannot be solved.

But note: both the Poisson and the Maxwell-Boltzmann equations have ψ_j in them, with the other quantities being ρ_j and n_i', and so the problem now reduces to:

Finding a relation between the smeared out ρ_j of Poisson's equation and the n_i' corresponding to the discrete ions of the Maxwell-Boltzmann distribution.

10.6.7   Step 5 The problem is to combine the Poisson equation with the Maxwell-Boltzmann equation – how can this be done?

The problem is to get some device which would substitute the average distribution of the discrete ions in the ionic atmosphere around the central j-ion, given by n_i' in the Maxwell-Boltzmann expression, by a continuous charge density which could be taken to be equivalent to ρ_j in the Poisson equation. This would enable Poisson's equation to be combined with a Maxwell-Boltzmann distribution.

It is reasonable to assume that a spherically symmetrical **continuous** distribution of charge could reflect in some way the spherically symmetrical distribution of all the **discrete** ions around the central j-ion. This requires looking at what a 'charge density', i.e. the average charge per unit volume at a given point, for **discrete** ions means.

- n_i' is the number of ions of one type per unit volume, at any point beyond $r = \mathring{a}$, $\therefore (z_i e) n_i'$ is the total charge per unit volume of ions of **one type**, i, at that distance.

- But there has to be **at least** one other type of ion present and so $(z_i e) n_i'$ has to be summed over all types of ion present. Doing so gives:

$$\text{Total charge per unit volume} = \sum_i \{n_i'(z_i e)\} \tag{10.24}$$

This expression for the total charge per unit volume **is taken to be** the charge density, ρ_j, at that point, i.e. the charge density 'seen' at that point by the central j-ion due to the ionic atmosphere.

The argument is thus equating:

the total charge per unit volume at a given point for a distribution of discrete charges with the charge density which would result if it were possible to convert the discrete ions in the ionic atmosphere into a 'smeared out' continuous charge lying at distances greater than $r = \mathring{a}$.

It then follows, if this can be done, that:

$$\rho_j = \sum_i \{n_i'(z_i e)\} \tag{10.25}$$

The Maxwell-Boltzmann distribution summed over all types, i, is

$$\sum_i \{n_i'(z_ie)\} = \sum_i \left\{ n_i(z_ie) \exp\left(-\frac{z_ie\psi_j}{kT}\right) \right\}$$ (10.23)

and so by substitution of $\sum_i \{n_i'(z_ie)\} = \rho_j$ into Equation (10.23)

$$\rho_j = \sum_i \{n_i'(z_ie\} = \sum_i \left\{ n_i(z_ie) \exp\left[-\frac{(z_ie)\psi_j}{kT}\right] \right\}$$ (10.26)

where ρ_j is taken to be the same as a charge density of **discrete** ions.

Electrical neutrality requires that the charge density summed over all distances from $r = \overset{\circ}{a}$ to $r = \infty$, and assuming spherical symmetry, must be equal and opposite to the charge on the central reference j-ion at the origin, $(-z_je)$.

$$\int_{\overset{\circ}{a}}^{\infty} 4\pi r^2 \rho_j dr = -z_je$$ (10.27)

where $4\pi r^2$ is the surface area of a sphere of radius, r. This integral can be evaluated as the area under the graph of $4\pi r^2 \rho_j$ vs r from $r = \overset{\circ}{a}$ to $r = \infty$ (Figure 10.6).

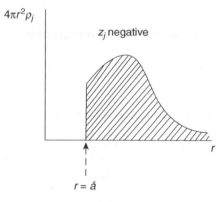

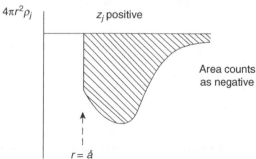

Figure 10.6   Graph of $4\pi r^2 \rho_j$ vs. r.

10.6.8   Step 6 Combining the Poisson and Maxwell-Boltzmann equations

The Maxwell-Boltzmann equation has been converted into the equation:

$$\rho_j = \sum_i \left\{ n_i(z_i e) \exp\left[-\frac{(z_i e)\psi_j}{kT} \right] \right\} \tag{10.26}$$

The Poisson equation is:

$$\nabla^2 \psi_j = -\frac{1}{\varepsilon_0 \varepsilon_r} \rho_j \tag{10.20}$$

Substituting the equation for ρ_j (Equation 10.26) into the Poisson equation gives:

$$\nabla^2 \psi_j = -\frac{1}{\varepsilon_0 \varepsilon_r} \sum_i \left\{ n_i(z_i e) \exp\left[-\frac{(z_i e)\psi_j}{kT} \right] \right\} \tag{10.28}$$

This equation is called the Poisson-Boltzmann equation and is in principle solvable to give an explicit expression for ψ_j in terms of r.
The problem has now reduced to:

**solving the Poisson-Boltzmann equation, calculating ψ_j^* and from this evaluating:
the electrostatic potential energy of interactions between the ion $= z_j e \psi_j^*$ and its
ionic atmosphere**

This then gives the contribution to the free energy due to non-ideality.

10.6.9   Step 7 Solving the Poisson-Boltzmann equation

This requires first that the exponential term is worked out and the 2nd order differential equation is then solved.

Handling the exponential term can be done in two ways

it can be done algebraically by expressing the exponential as a series and evaluating all the terms in the exponential, or the the expression could be integrated numerically. Both approaches were dropped because of mathematical complexity. **But,** with the advent of computing techniques, Guggenheim (1957) was able to integrate the expression numerically to get an exact solution of the Poisson-Boltzmann equation. This major advance in the theory was made possible, not by an advance in chemistry, but by an advance in another field, computer technology. Without computer techniques exact solution of the Poisson-Boltzmann equation is well-nigh impossible.

it can be done by **approximation,** by simple expansion of the exponential and dropping out all terms other than the first and second terms. It can be shown that, for **symmetrical** electrolytes, the third, fifth and higher order terms, drop out automatically by cancellation, but do not do so for **unsymmetrical** electrolytes. So the theory is more accurate for symmetrical electrolytes than for unsymmetrical electrolytes to the extent of including one more term, the third. For both types of electrolytes, the first term drops out because of electrical neutrality and the exponential is thus **approximated** to one term only.

The procedure used here is the expansion.

10.6.10   Expansion and approximation of the Poisson-Boltzmann equation to one non-zero term only

The Poisson-Boltzmann equation is:

$$\nabla^2 \psi_j = -\frac{1}{\varepsilon_0 \varepsilon_r} \sum_i \left\{ n_i(z_i e) \exp\left[-\frac{(z_i e)\psi_j}{kT} \right] \right\} \tag{10.28}$$

By expanding the exponential and restricting it to the first three terms:

$$\nabla^2 \psi_j = -\frac{1}{\varepsilon_0 \varepsilon_r} \sum_i \left\{ n_i z_i e - \frac{n_i z_i^2 e^2 \psi_j}{kT} + \frac{n_i z_i^3 e^3 \psi_j^2}{(kT)^2 2!} - \cdots \right\} \tag{10.29}$$

The term $\sum_i (n_i z_i e)$ is the total charge per unit volume of the ions of all types, i. Since this is summed over all the ions present and the solution has to be electrically neutral, then this term is zero.

The above equation then reduces to:

$$\nabla^2 \psi_j = -\frac{1}{\varepsilon_0 \varepsilon_r} \sum_i \left\{ 0 - \frac{n_i z_i^2 e^2 \psi_j}{kT} + \frac{n_i z_i^3 e^3 \psi_j^2}{(kT)^2 2!} - \cdots \right\} \tag{10.30}$$

When the equation is restricted to these three terms only, the final expression will be different for symmetrical electrolytes and unsymmetrical electrolytes.

- For symmetrical electrolytes:

$$\nabla^2 \psi_j = -\frac{1}{\varepsilon_0 \varepsilon_r} \sum_i \left\{ 0 - \frac{n_i z_i^2 e^2 \psi_j}{kT} + 0 \right\} \tag{10.31}$$

For unsymmetrical electrolytes:

$$\nabla^2 \psi_j = -\frac{1}{\varepsilon_0 \varepsilon_r} \sum_i \left\{ 0 - \frac{n_i z_i^2 e^2 \psi_j}{kT} + \text{non}-\text{zero term in } \psi_j^2 \right\} \tag{10.32}$$

The general form of the Poisson-Boltzmann equation **truncated** to **two terms only** is accurate to three terms for symmetrical electrolytes since the third term equals zero, but is only accurate to two terms for unsymmetrical electrolytes, the third term being non-zero. In the form truncated to two terms:

$$\nabla^2 \psi_j = -\frac{1}{\varepsilon_0 \varepsilon_r} \sum_i \left[-\frac{n_i z_i^2 e^2 \psi_j}{kT} \right] \tag{10.33}$$

$$= +\frac{e^2}{\varepsilon_0 \varepsilon_r kT} \sum_i n_i z_i^2 \psi_j \tag{10.34}$$

$$= \kappa^2 \psi_j \tag{10.35}$$

$$\text{where } \kappa^2 = \frac{e^2}{\varepsilon_0 \varepsilon_r kT} \sum_i n_i z_i^2 \tag{10.36}$$

Solving equation $\nabla^2 \psi_j = \kappa^2 \psi_j$ will enable ψ_j to be found.

κ is a function of:

- the number of ions of type i per unit volume, and this is summed over all ions;

- the ionic charge of each ion of type i;

- the absolute temperature;

- the relative permittivity.

κ has units, length^{-1}. As will be shown later (Section 10.6.17) this has considerable physical significance.

10.6.11 The ionic strength

The concept of the ionic strength was introduced into electrolyte solution studies some time prior to the development of the Debye-Hückel theory, but it rises to prominence in this theory. The ionic strength is a special type of summation over **all charged species** in solution (see Section 8.21.1) and is defined as:

$$I = \frac{1}{2} \sum_i c_i z_i^2 \tag{10.37}$$

where c_i is the concentration of ions of each type i of ions in solution, and z_i is the algebraic charge on each type i of ions in solution, i.e. it includes the sign of the charge. Worked Problems 8.12 and 8.14 give practice in calculating I for various solutions.

Converting I to molecular units gives:

$$I = \frac{1}{2} \sum_i c_i z_i^2 = \frac{\frac{1}{2} \sum_i n_i z_i^2}{N} \tag{10.38}$$

$$\therefore \kappa^2 = \frac{2e^2 N}{\varepsilon_0 \varepsilon_r kT} I \tag{10.39}$$

where κ^2 is now in molar units since Equation (10.39) contains the molar quantity, I.

The Poisson-Boltzmann equation truncated to two terms only is:

$$\nabla^2 \psi_j = -\frac{1}{\varepsilon_0 \varepsilon_r} \sum_i \left[-\frac{n_i z_i^2 e^2 \psi_j}{kT} \right] = +\frac{e^2}{\varepsilon_0 \varepsilon_r kT} \sum_i n_i z_i^2 \psi_j = \kappa^2 \psi_j \qquad \text{(10.33) to (10.35)}$$

Solving this equation would give ψ_j which is the total potential at distances $r \geq \mathring{a}$ from the central j-ion due to the ion itself and its ionic atmosphere, and $\psi_j = \psi_j' + \psi_j''$ (see Section 10.6.2).

10.6.12 The next step is to solve the truncated Poisson-Boltzmann equation

$$\nabla^2 \psi_j = \kappa^2 \psi_j \tag{10.35}$$

This equation involves ψ_j to power one and is a linear second-order differential equation because it involves second derivatives. It is also of a standard form.

Note: the non-truncated form of the Poisson-Boltzmann equation has terms in powers of ψ_j and would be a non-linear differential equation.

If $u = r\psi_j$ then Equation (10.35) becomes via a series of standard manipulations of derivatives:

$$\frac{d^2u}{dr^2} = \kappa^2 u \tag{10.40}$$

and has the general solution:

$$u = Ce^{-\kappa r} + De^{\kappa r} \tag{10.41}$$

and thus

$$\psi_j = C\frac{e^{-\kappa r}}{r} + D\frac{e^{\kappa r}}{r} \tag{10.42}$$

where C and D are constants of integration where:

$$C = \frac{z_j e}{4\pi\varepsilon_0\varepsilon_r} \frac{e^{\kappa \mathring{a}}}{1 + \kappa \mathring{a}} \tag{10.43}$$

and

$$D = 0 \tag{10.44}$$

$$\therefore \psi_j = \frac{z_j e}{4\pi\varepsilon_0\varepsilon_r} \frac{e^{\kappa \mathring{a}}}{1 + \kappa \mathring{a}} \frac{e^{-\kappa r}}{r} \tag{10.45}$$

And so, the next step in the development is

to calculate ψ_j^* and then evaluate $z_j e\psi_j^*$

10.6.13 Step 8 Calculation of the potential at the surface of the central *j*-ion due to the ionic atmosphere, and thence finding the electrostatic energy of interaction between an ion and its ionic atmosphere

ψ_j	$=$	ψ_j'	$+$	ψ_j'' (10.19)
↓		↓		↓
the **total** potential at any distance r from the central *j*-ion due to the **ion itself and its ionic atmosphere,** and has been found as above		the potential at distance r from the *j*-ion due to the **ion itself** and equals $z_j e/4\pi\varepsilon_0\varepsilon_r r$		the potential at distance r from the *j*-ion due to the **ionic atmosphere only**

Using the **approximate** solution to the Poisson-Boltzmann equation given in Equation (10.45) and substituting into Equation (10.19) gives:

$$\frac{z_j e}{4\pi\varepsilon_0\varepsilon_r} \frac{e^{\kappa \mathring{a}}}{1 + \kappa \mathring{a}} \frac{e^{-\kappa r}}{r} = \frac{z_j e}{4\pi\varepsilon_0\varepsilon_r r} + \psi_j'' \tag{10.46}$$

$$\psi_j'' = \frac{z_j e}{4\pi\varepsilon_0\varepsilon_r}\frac{e^{\kappa\mathring{a}}}{1+\kappa\mathring{a}}\frac{e^{-\kappa r}}{r} - \frac{z_j e}{4\pi\varepsilon_0\varepsilon_r r} \tag{10.47}$$

$$\therefore \psi_j'' = \frac{z_j e}{4\pi\varepsilon_0\varepsilon_r r}\left[\frac{e^{\kappa\mathring{a}}}{1+\kappa\mathring{a}}e^{-\kappa r} - 1\right] \tag{10.48}$$

and this holds **for all r** down to $r = \mathring{a}$, because between $r = 0$ and $\mathring{a}$ the potential ψ_j'' is constant, or, put otherwise, the ionic atmosphere starts at $r = \mathring{a}$ since the centre of any ion **cannot get closer** to the centre of the j-ion than the distance $\mathring{a}$ (see Sections 10.5).

Pay particular note to the following:

- the potential due to the ionic atmosphere, ψ_j'' at $r = \mathring{a}$ is equal to ψ_j'' at $r = \mathring{a}/2$, and ψ_j'' is constant over the region between $r = \mathring{a}$ and $r = \mathring{a}/2$;

- the value for ψ_j^* is the value of ψ_j'' at the surface of the central ion, i.e. at $r = \mathring{a}/2$, and is also the value of ψ_j'' at $r = \mathring{a}$;

- the charge density, ρ_j, is zero for all $r < \mathring{a}$, but varies with r for all distances beyond $r = \mathring{a}$;

- the solution of the truncated Poisson-Boltzmann equation (Equation 10.45) relates only to values of $r \geq \mathring{a}$, and **must not** be used for **any** $r < \mathring{a}$. This is because the ionic atmosphere **starts** at $r = \mathring{a}$. This is a point which is often not appreciated.

10.6.14 Calculation of ψ_j^* by substitution of $r = \mathring{a}$ in Equation (10.48)

Pay particular attention to the fact that $r = \mathring{a}$ is used for substitution here since Equation (10.48) only holds for $r \geq \mathring{a}$, despite the fact that the potential energy of interaction between the central j-ion and its ionic atmosphere is defined in terms of the potential due to the ionic atmosphere at the central j-ion, i.e. at its surface, ψ_j^*.

$$\psi_j'' = \frac{z_j e}{4\pi\varepsilon_0\varepsilon_r r}\left[\frac{e^{\kappa\mathring{a}}}{1+\kappa\mathring{a}}e^{-\kappa r} - 1\right] \tag{10.48}$$

$$= \frac{z_j e}{4\pi\varepsilon_0\varepsilon_r\mathring{a}}\left[\frac{e^{\kappa\mathring{a}}e^{-\kappa\mathring{a}}}{1+\kappa\mathring{a}} - 1\right] \tag{10.49}$$

$$= \frac{z_j e}{4\pi\varepsilon_0\varepsilon_r\mathring{a}}\left[\frac{1}{1+\kappa\mathring{a}} - 1\right] \tag{10.50}$$

$$= -\frac{z_j e}{4\pi\varepsilon_0\varepsilon_r\mathring{a}}\frac{\kappa\mathring{a}}{1+\kappa\mathring{a}} \tag{10.51}$$

$$= -\frac{z_j e}{4\pi\varepsilon_0\varepsilon_r}\frac{\kappa}{1+\kappa\mathring{a}} \tag{10.52}$$

This is the value ψ_j'' takes when $r = \mathring{a}$. But since the potential at $r = \mathring{a}$ is the same as the potential at $r = \mathring{a}/2$ (Sections 10.5 and 10.6.1), this is also the value of ψ_j^*, the potential **at the surface** of the j-ion due to its ionic atmosphere, and so

$$\psi_j^* = -\frac{z_j e}{4\pi\varepsilon_0\varepsilon_r}\frac{\kappa}{1+\kappa\mathring{a}} \tag{10.53}$$

It then follows that:

The electrostatic potential energy of the coulombic interactions between the ion and its ionic atmosphere

$$= z_j e \psi_j^*$$
(10.18)

$$= -\frac{z_j^2 e^2}{4\pi\varepsilon_0\varepsilon_r}\frac{\kappa}{1+\kappa\mathring{a}}$$
(10.54)

and so the next step in the problem is:

to relate this electrostatic potential energy to the non-ideal part of the free energy, and thence to the mean ionic activity coefficient.

10.6.15   Step 9 The problem is to calculate the mean ionic activity coefficient, $\gamma_\pm$

From the argument given above, the contribution to the total non-ideal free energy from the interaction of the central reference j-ion with its ionic atmosphere is:

$$G_{\text{non-ideal contribution from the } j\text{-ion and its ionic atmosphere}} = -\frac{z_j^2 e^2}{4\pi\varepsilon_0\varepsilon_r}\frac{\kappa}{1+\kappa\mathring{a}}$$
(10.55)

But this has to be summed over all ions where each ion in turn has been a central reference ion. This is done in two stages:

- sum over all ions of a single type, i,

$$G_{\text{non-ideal contribution from ions of a single type } i} = -\frac{z_i^2 e^2 N_i}{4\pi\varepsilon_0\varepsilon_r}\frac{\kappa}{1+\kappa\mathring{a}}$$
(10.56)

where N_i is the number of ions of type, i.

- sum over all types of ions, i, and divide by two to avoid counting each ion twice over, i.e. one ion can be a central reference ion and it can also be part of another ion's ionic atmosphere. This must not be counted as two ions.

$$G_{\text{non-ideal contribution from ions of all types } i, \text{ i.e. for the whole solution}} = -\frac{1}{2}\sum_i\left\{\frac{z_i^2 e^2 N_i}{4\pi\varepsilon_0\varepsilon_r}\frac{\kappa}{1+\kappa\mathring{a}}\right\}$$
(10.57)

$$= -\frac{e^2}{8\pi\varepsilon_0\varepsilon_r}\frac{\kappa}{1+\kappa\mathring{a}}\sum_i N_i z_i^2$$
(10.58)

The argument which eventually leads from this expression to an explicit expression for the mean ionic activity coefficient in terms of the ionic strength is complex, and only the final result will be quoted.

Symmetrical electrolytes

These are electrolytes such as NaCl(aq), CaSO$_4$(aq) and LaFe(CN)$_6$(aq) where the concentration of the cation equals the concentration of the anion.

Section 8.22 and Worked Problem 8.13 show that the non-ideal part of G can be related to the mean activity coefficient, $\gamma_\pm$, giving for NaCl(aq):

$$G_{\text{non-ideal contribution}} = nRT\ln\gamma_{Na^+}\gamma_{Cl^-} = 2nRT\log_e(\gamma_{Na^+}\gamma_{Cl^-})^{1/2} = 2nRT\log_e\gamma_\pm \quad (10.59)$$

But the non-ideal part of G has been shown by the Debye-Hückel arguments to be given by:

$$G_{\text{non-ideal contribution from ions of all types } i, \text{ i.e. for the whole solution}} = -\frac{e^2}{8\pi\varepsilon_0\varepsilon_r}\frac{\kappa}{1+\kappa\mathring{a}}\sum_i N_iz_i^2 \quad (10.58)$$

Using these equations and the expression already deduced for κ^2 (Equations 10.36 and 10.39) the Debye-Hückel equation for the mean activity coefficient of a symmetrical electrolytes such as NaCl is:

$$\log_{10}\gamma_\pm = -Az_i^2\frac{\sqrt{I}}{1+B\mathring{a}\sqrt{I}} \quad (10.60)$$

where

$$A = \frac{e^2B}{2.303 \times 8\pi\varepsilon_0\varepsilon_r kT} \quad (10.61)$$

and where

$$B = \left\{\frac{2e^2N}{\varepsilon_0\varepsilon_r kT}\right\}^{1/2} \quad (10.62)$$

The above derivation holds for all symmetrical electrolytes since $z_{\text{cation}} = z_{\text{anion}} = z_i$.

Unsymmetrical electrolytes

These are electrolytes such as CaCl$_2$(aq), K$_2$SO$_4$(aq), K$_3$Fe(CN)$_6$ (aq) and Al$_2$(SO$_4$)$_3$(aq) for which the concentrations of the cations and anions are not equal.

Consider Al$_2$(SO$_4$)$_3$(aq) (see Sections 8.22 and 8.23 and Worked Problem 8.13):

$$\begin{aligned} G_{\text{non-ideal contribution}} &= 2nRT\log_e\gamma_{Al^{3+}} + 3nRT\log_e\gamma_{SO_4^{2-}} = nRT\log_e\gamma_{Al^{3+}}^2\gamma_{SO_4^{2-}}^3 \\ &= 5nRT\log_e(\gamma_{Al^{3+}}^2\gamma_{SO_4^{2-}}^3)^{1/5} = 5nRT\log_e\gamma_\pm \end{aligned} \quad (10.63)$$

But the non-ideal part of G has been shown by the Debye-Hückel arguments to be given by:

$$G_{\text{non-ideal contribution from ions of all types i, i.e. for the whole solution}} = -\frac{e^2}{8\pi\varepsilon_0\varepsilon_r}\frac{\kappa}{1+\kappa\mathring{a}}\sum_i N_iz_i^2 \quad (10.58)$$

Again by a complex argument, the Debye-Hückel theory gives the following equation for the mean ionic activity coefficient:

$$\log_{10}\gamma_{\pm} = +Az_{+}z_{-}\frac{\sqrt{I}}{1+B\mathring{a}\sqrt{I}} \tag{10.64}$$

where the charges appearing in this form of the equation include both the magnitude and the sign of the charge, and A and B are defined in Equations (10.61) and (10.62) above.

However, the equation is usually written in the form:

$$\log_{10}\gamma_{\pm} = -A|z_{+}z_{-}|\frac{\sqrt{I}}{1+B\mathring{a}\sqrt{I}} \tag{10.65}$$

where $|z_{+}z_{-}|$ reads 'mod $z_{+}z_{-}$' and means that only the magnitudes of the charges appear. It does **not** include the signs.

The expression for a symmetrical electrolyte can be deduced from Equation (10.65) describing the unsymmetrical electrolyte.

For the symmetrical electrolyte:

$$z_{+} = -z_{-} \tag{10.66}$$

$$\therefore z_{+}z_{-} = -z^{2} \tag{10.67}$$

$$\text{and } |z_{+}z_{-}| = z^{2} \tag{10.68}$$

10.6.16 Constants appearing in the Debye-Hückel expression

The constants A and B involve the absolute temperature and the relative permittivity of the solvent as well as several universal constants.

$$B = \left\{\frac{2e^{2}N}{\varepsilon_{0}\varepsilon_{r}kT}\right\}^{1/2} = \left\{\frac{2e^{2}N}{\varepsilon_{0}k}\right\}^{1/2}\frac{1}{(\varepsilon_{r}T)^{1/2}} = \frac{50.29\times10^{8}\text{cm}^{-1}\text{mol}^{-1/2}\text{dm}^{3/2}\text{K}^{1/2}}{(\varepsilon_{r}T)^{1/2}}$$

For water at 25°C and taking $\varepsilon_{r} = 78.46$, $B = 3.288\,\text{nm}^{-1}\,\text{mol}^{-1/2}\,\text{dm}^{3/2}$.

$B\sqrt{I}$ is the fundamental quantity κ in the interionic attraction theory of Debye and Hückel.

$$A = \frac{e^{2}B}{2.303\times8\pi\varepsilon_{0}\varepsilon_{r}kT} = \frac{\sqrt{2}e^{3}N^{1/2}}{2.303\times8\pi(\varepsilon_{0}k)^{3/2}(\varepsilon_{r}T)^{3/2}} = \frac{1.8248\times10^{6}\text{mol}^{-1/2}\text{dm}^{3/2}\text{K}^{3/2}}{(\varepsilon_{r}T)^{3/2}}$$

For water at 25°C and taking $\varepsilon_{r} = 78.46$, $A = 0.510\,\text{mol}^{-1/2}\,\text{dm}^{3/2}$.

10.6.17 The physical significance of κ^{-1} and ρ_{j}

$$\kappa^{2} = \frac{e^{2}}{\varepsilon_{0}\varepsilon_{r}kT}\sum_{i}n_{i}z_{i}^{2} \tag{10.36}$$

κ has the units m^{-1} as shown below:

e^{2} has units: coulomb^{2} i.e. C^{2}

n_i is a number per unit volume and hence has units m^{-3}

z_i is a number and hence unitless

ε_0 has units: farad m^{-1} or coulomb^2 J^{-1} m^{-1}

ε_r is a number and has no units

k has units J K^{-1}

T has units K

$$\therefore \kappa^2 = \frac{e^2(\text{coulomb}^2)}{\varepsilon_0(\text{coulomb}^2\text{J}^{-1}\text{m}^{-1})\varepsilon_r k(\text{JK}^{-1})T(\text{K})} \sum_i n_i z_i^2 (\text{m}^{-3})$$

$$\therefore \kappa^2 \text{ has units m}^{-2} \text{ and } \kappa \text{ has units m}^{-1}.$$

The quantity, κ^{-1}, has units m, i.e. has the units of a length and as such it is often stated that κ^{-1} represents the extent of the ionic atmosphere around the central j cation. Such a description can only be given if the value of κ^{-1} is measured from $r = \mathring{a}$, since the ionic atmosphere must always lie at distances, $r > \mathring{a}$. Alternatively it has been stated that κ represents the reciprocal of the effective radius of the ionic atmosphere, again with the proviso that this is measured from $r = \mathring{a}$. However, these could be misleading statements, as is shown by the diagrams below.

What would be more correct would be to look at the equation:

$$\psi_j^* = -\frac{z_j e}{4\pi\varepsilon_0\varepsilon_r}\frac{\kappa}{1 + \kappa\mathring{a}} \tag{10.53}$$

where ψ_j^* is the potential at the surface of the j-ion due to its ionic atmosphere. It is also the potential at a distance, $\mathring{a}$, from the centre of the j-ion.

Mathematically, this is equivalent to stating that the **effect** on the potential of the central reference j-ion due to all the ions in the ionic atmosphere is what it would be if the whole of the ionic atmosphere were situated on a sphere lying at a distance $1/\kappa$ beyond $\mathring{a}$, i.e. on a sphere of radius $\mathring{a} + (1/\kappa)$ i.e. of radius $(1 + \kappa\mathring{a})/\kappa$.

Physically such a statement is not realistic. The atmosphere decreases smoothly from a distance, $\mathring{a}$, from the central ion to a situation where the distribution is almost random when $r \to \infty$ (see Figure 10.7). These diagrams show that there are regions of r values where the distribution is definitely not random and where there is significant charge per unit volume, and as shown on the diagrams these distances extend to values of r greater than $1/\kappa$ beyond $\mathring{a}$.

The dependence of the charge density on distance is given by:

$$\rho_j = \frac{z_j e\kappa^2}{4\pi}\frac{e^{\kappa\mathring{a}}}{1 + \kappa\mathring{a}}\frac{e^{-\kappa r}}{r} \tag{10.69}$$

For a given ionic strength, κ is known. The value of $z_j e$ is known for any charge type for the central j-ion and if a value is assumed for $\mathring{a}$, then

$$\rho_j = \text{constant}\frac{e^{-\kappa r}}{r} \tag{10.70}$$

from which values of ρ_j can be found as a function of r. Typical graphs are shown; see Figure 10.8.

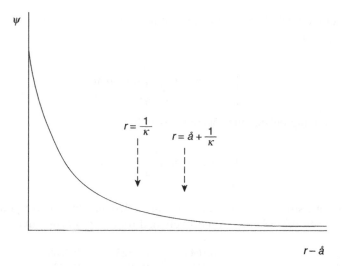

Figure 10.7 Graph of ψ vs. $r - \mathring{a}$ showing how the ionic atmosphere tails off to ∞.

These show the charge density to be non-zero for all values of $r > \mathring{a} + \kappa^{-1}$ and are similar to those for ψ_j as a function of r (Figure 10.7), confirming the conclusion that there is never any sharp cut-off point for the ionic atmosphere which actually extends to infinity:

$$\rho_j \to 0 \; as \; r \to \infty.$$

This can be demonstrated more effectively if the proportion of the charge inside a sphere of radius $r = \mathring{a} + (1/\kappa)$ is compared with the proportion of the charge found outside the sphere.

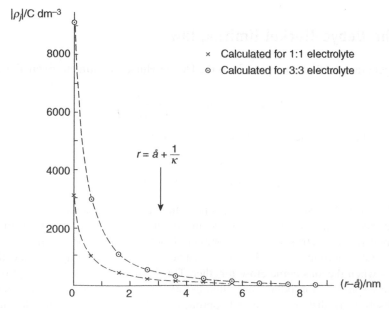

Figure 10.8 Graph of $|\rho_j|$ vs. $(r - \mathring{a})$ showing how the ionic atmosphere tails off to ∞.

The proportion of the total charge of the ionic atmosphere lying within a sphere of radius $r = \mathring{a} + (1/\kappa)$ is given by:

$$\int_{\mathring{a}}^{\mathring{a}+(1/\kappa)} 4\pi r^2 \rho_j dr \Bigg/ \int_{\mathring{a}}^{\infty} 4\pi r^2 \rho_j dr \tag{10.71}$$

and the proportion lying outside the sphere is given by:

$$\int_{\mathring{a}+(1/\kappa)}^{\infty} 4\pi r^2 \rho_j dr \Bigg/ \int_{\mathring{a}}^{\infty} 4\pi r^2 \rho_j dr \tag{10.72}$$

Calculations for ionic strengths of 0.0025, 0.005 and 0.010 mol dm^{-3} corresponding to values of 0.1644, 0.2325 and 0.3288 nm^{-1} respectively and for $\mathring{a} = 0.400$ nm give:

κ/nm^{-1}	0.1644	0.2325	0.3288
proportion outside	0.713	0.705	0.693
proportion inside	0.287	0.296	0.307

i.e. approximately 30% lies within a distance $1/\kappa$ from where the ionic atmosphere begins, or put otherwise within a distance $\mathring{a} + (1/\kappa)$ from the centre of the j-ion. This means that around 70% of the total charge lies outside the supposed radius of the ionic atmosphere.

Further calculations show that approximately 50% and 40% respectively of the total charge lies at distances greater than $\mathring{a} + (1.5/\kappa)$, and $\mathring{a} + (2/\kappa)$ respectively.

If κ^{-1} is taken to be the 'effective extent' of the ionic atmosphere outside the surface of the central j-ion, then the proportion outside should be effectively zero. The conclusion to be drawn is that the ionic atmosphere fades away towards zero as the distance from the surface of the ion tends to infinity.

10.7 The Debye-Hückel limiting law

The basic general equation emerging from the Debye-Hückel treatment given above is:

$$\log_{10}\gamma_{\pm} = -A|z_+z_-|\frac{\sqrt{I}}{1 + B\mathring{a}\sqrt{I}} \tag{10.65}$$

where the constants take the following values for water as solvent at 25°C:

$$A = 0.510 \, \text{mol}^{-1/2} \, \text{dm}^{3/2}$$
$$B = 3.288 \, \text{nm}^{-1} \, \text{mol}^{-1/2} \, \text{dm}^{3/2},$$

and the value of $\mathring{a}$, the distance of closest approach, is as yet unspecified.

The numerator in this equation takes account of the contribution to non-ideality arising from the long range electrostatic forces obeying Coulomb's inverse square law, while the denominator takes account, in a rather crude manner, of the non-ideality introduced by short range forces when the ions come close together. This corresponds to the assumption that the ions are hard spheres, non-polarisable, non-deformable and spherically symmetrical, and manifests itself as the distance of closest approach, $\mathring{a}$. There are, of course, other short range forces of interaction between the ions, between the ion and the solvent and between the solvent molecules which are not taken account of (see Sections 10.3, 10.8.3, 10.8.5 and 10.8.6 and the chapter on solvation).

At the surface of a large ion, the electric field will be considerably smaller than that at the surface of a small ion. Consequently, non-ideality is to be expected to be less pronounced for large ions than for small ions and the denominator in the Debye-Hückel expression may be regarded as a reflection of this.

The values of $\gamma_{\pm}$ calculated from Equation (10.65) can be compared with values of $\gamma_{\pm}$ calculated for the same ionic strengths from the Debye-Hückel **limiting law**:

$$\log_{10}\gamma_{\pm} = -A|z_1 z_2|\sqrt{I} \qquad (10.73)$$

where $A = 0.510 \, \text{mol}^{-1/2} \, \text{dm}^{+3/2}$.

This equation was proposed by Debye and Hückel as coping adequately with non-ideality in very dilute solutions when the correction for the short range forces arising from the finite size of the ions can be ignored.

The comparison shows that the limiting law can be taken to be reasonably accurate up to an ionic strength of $1.00 \times 10^{-2} \, \text{mol} \, \text{dm}^{-3}$ for a 1:1 electrolyte, but this decreases to $1.00 \times 10^{-3} \, \text{mol} \, \text{dm}^{-3}$ for a 2:2 electrolyte. This latter value corresponds to a concentration of $2.5 \times 10^{-4} \, \text{mol} \, \text{dm}^{-3}$. What should also be noted is that it will only be possible to ignore non-ideality in electrolyte solutions when $\gamma_{\pm} \to 1$, and the calculations show that this will not happen until ionic strengths $< 1 \times 10^{-5} \, \text{mol} \, \text{dm}^{-3}$ are reached.

A similar equation has been proposed by Guntelberg in which $B\mathring{a} = 1$, which corresponds to a value of $\mathring{a} = 0.304$ nm:

$$\log_{10}\gamma_{\pm} = -A|z_+ z_-|\frac{\sqrt{I}}{1 + \sqrt{I}} \qquad (10.74)$$

The same conclusions follow from a comparison of this equation and the Debye-Hückel limiting law.

However, it is recommended that one of the extended Debye-Hückel equations (see Sections 10.10.1 and 10.10.2) should always be used if the ionic strength range is extended, and especially so when extrapolation of experimental data is needed.

By considering under what conditions of ionic strength the denominator $1 + B\mathring{a}\sqrt{I}$ can be approximated to unity, the Debye-Hückel limiting law has been shown above to be a good approximation for dilute solutions. However, there was at that time no theoretical justification for this except for the arguments put forward by Kramers. However, the modern work of Kirkwood, Bogolyubov *et al.* (Section 10.18) has fully justified the limiting law.

At around the time that the Debye-Hückel theory was published, Kramers put forward an alternative theory which led to an equation of the same form as the limiting law. His calculations were for point charges in very dilute solutions and in a continuous medium, and did not involve an ion size. However, point charges imply the need to postulate some sort of repulsion to stop the point charges of opposite sign sticking together. Kramers showed that short range repulsions over very small distances on the microscopic scale would achieve this, **without the necessity of introducing a distance of closest approach.** This led to the Debye-Hückel limiting law, and is the only direct theoretical derivation of the limiting law.

10.8   Shortcomings of the Debye-Hückel model

When the Debye-Hückel equation is tested against experimental results it is very successful in accounting for behaviour at low concentrations, and it is believed that the theory is basically correct for low concentrations (see Section 10.10). Having to test the theory rigorously at very low concentrations proved a great stimulus in developing precision techniques for deriving experimental values of $\gamma_\pm$.

At moderate and higher concentrations deviations from theoretical behaviour become apparent, and ways of dealing with these problems are described later in the chapter.

Meanwhile, it is constructive to look again at the physical basis of the simple Debye-Hückel model and its mathematical development to see where both could be modified, and to consider whether this would be mathematically possible. What has been written in Chapter 1 on ions and solvent structure shows that the Debye-Hückel model is painfully naïve and cannot even approach physical reality. A brief reassessment of the features 1–7 of the simple Debye-Hückel model is given below, along with indications as to how these problems have been tackled.

10.8.1   Strong electrolytes are completely dissociated

As a result of the assumption that random motion is not attained, a consistent treatment should allow, in principle, the possibility that electrostatic interactions could be sufficiently large to result in some ion pairing. This is likely for ions of high charge type and for those ions which just happen to be close together, and the model must therefore be modified. This has been carried out by Bjerrum and others (see Section 10.12).

10.8.2   Random motion is not attained

As mentioned above, the model has to be extended to take account of the **short range** coulombic interactions of ion pairs over and above the **long range** coulombic interactions of the ionic atmosphere dealt with in the theory.

10.8.3   Non-ideality results from coulombic interactions between ions

This is not a sufficient description of what contributes to non-ideality. Many other types of interactions must be included, for example:

- short range coulombic interactions;

- short range quantum mechanical interactions;

- hard sphere repulsions;

- non-hard sphere repulsions;

- polarisability of ions – ion-induced dipole interactions;

- induced dipole-induced dipole interactions;

- modified ion–solvent interactions;

- modified solvent–solvent interactions.

Physically, all of these are simple ideas and are easy to discuss and explain, but unfortunately there are very considerable mathematical difficulties in incorporating them into the theory. Other than ion pairing and solvation, nothing fundamental in the form of a new 'working' equation representing non-ideality has been achieved for the others, but see Sections 10.18–10.21.

10.8.4   Ions are spherically symmetrical and are unpolarisable

Even some simple ions, such as NO_3^- (aq) and $PtCl_4^{2-}$ (aq), are not spherically symmetrical, and large ions such as $CH_3CH_2COO^-$ (aq) are decidedly non-spherically symmetrical. Other large ions, especially those encountered in solutions containing ions of biological importance certainly are not. Likewise, many ions are not unpolarisable; even some simple ones like I^- (aq) are highly polarisable, and the model needs to be grossly modified on both counts. But again this is so complex mathematically that not much has been done, but see Sections 10.17.3, 10.19 and 10.21.

10.8.5   The solvent is a structureless dielectric

This suggestion is manifestly untrue. The solvent consists of molecules which have structure. The only respect in which the theory takes account of this is the modification of the electrostatic interactions by the solvent through the macroscopic relative permittivity. This is a crude attempt to recognise the importance of the ion–dipole interactions which undoubtedly occur. To cover more physically realistic situations the model needs considerable modification to include:

- ion–solvent interactions;

- solvent–solvent interactions;

- the effect of dielectric saturation.

The complexity of the mathematics has meant that these features of the dielectric properties of the solvent have not been incorporated into a new development and suggests that a new model is required (see Sections 10.16, 10.19 and 10.21).

10.8.6   Electrostriction is ignored

Because of the intense fields near the ion, solvent molecules will get packed tighter together, but little consideration has been given to this, other than in the developments incorporating solvation effects.

10.8.7   Concept of a smeared out spherically symmetrical charge density

This is an absolutely crucial part of the model and has been dealt with by statistical mechanical averaging procedures. But only spherical symmetry has been assumed. And so for large non-spherical ions a modification to the smearing out procedure is needed, but see Sections 10.17.3 and 10.19.

Any one given distribution of ions around a spherically symmetrical central j-ion need not necessarily be spherically symmetrical, but **on average** all possible arrangements will correspond to spherical symmetry. A charge density necessarily corresponds to an **average distribution of ions**, so conversion of the Poisson-Boltzmann equation to spherical symmetry is purely formal.

But as has been hinted at above, there is one important limitation to this when considering large complex electrolytes such as are found in aqueous solutions of biological materials. Here the central ion is non-spherical. An ion which is **not spherically symmetrical** may impose a **non-spherically symmetrical distribution** of charge around it, and this ought to be taken care of, but is not, in the theory. The Debye-Hückel theory can thus only be approximate for non-spherical ions.

10.9   Shortcomings in the mathematical derivation of the theory

These have been discussed as the theory was developed, and will only be summarised here.

- **Use of the Poisson-Boltzmann equation in its spherically symmetrical form**

This is inadequate if the ions are non-spherical and a modified Poisson equation is needed.

- **The Poisson-Boltzmann equation in its complete form**

This may contradict one of the basic laws of electrostatics, that is the linear superposition of fields. This is a fundamental problem for the theory and to any other theory or development, such as that due to Guggenheim below (Section 10.13.1), which makes use of any such combination. In fact this is one of the big problems in the theory. The other big problem is that the ψ_js in the Poisson equation and in the Maxwell-Boltzmann distribution are different and have a different physical basis (see Section 10.6.5). This is believed by many to be yet another fundamental problem for the theory.

- **Truncation of the exponential**

This is serious even though it removes the problem of the linear superposition of fields by having both sides of the combined equation linear. But it imposes a **very severe restriction** to the theory because the approximation requires

$$z_i e \psi_j \ll kT \tag{10.75}$$

and this can **only** hold for **very dilute** solutions where the ions are, on the average, at **large distances** from each other, and so unhappily it must be accepted that, for normal working conditions such as an ionic strength of $0.010 \text{ mol dm}^{-3}$ or above, the Debye-Hückel theory can, at best, be only an approximate theory.

Extension to higher concentrations is required, and one of the main advances here has been Guggenheim's numerical integration by computer, but the problem of the superposition of fields and the different ψ_js still remain.

Furthermore, this approximation appears to contradict one of the assumptions of the theory that electrostatic interactions impose some degree of order on the random thermal motions of the ions.

- **Distance of closest approach, $\overset{\circ}{a}$**

The term $\overset{\circ}{a}$ is somewhat arbitrary, but has been defined as the distance of closest approach of two ions – they cannot get closer together than the sum of their respective radii. As shown earlier the theory has implicitly assumed $\overset{\circ}{a}$ to be the same for both cation and anion of all electrolytes. This is a quite considerable assumption, probably not valid for most electrolytes. This is especially so for large complex ions, such as found in biological situations, and for electrolytes which are not like the simple ions such as $Ca^{2+}(aq)$, $Na^+(aq)$, $OH^-(aq)$ or $Cl^-(aq)$.

There is thus quite a predicament here as to what $\overset{\circ}{a}$ actually means physically. There has been much argument, but no real major advance has been made theoretically.

Curve fitting of experimental determinations of $\gamma_\pm$ over a range of ionic strengths has attempted to overcome these difficulties (see Section 10.10.2), but on closer inspection this does not really advance matters very far, as the real status of $\overset{\circ}{a}$ in such procedures is as a disposable parameter.

10.10 Modifications and further developments of the theory

Precision experimental results show theory and experiment to deviate from each other considerably at moderate concentrations, and grossly at high concentrations. The theory under these conditions is obviously inadequate, and the following discusses attempts to overcome these problems.

10.10.1 Empirical methods

One way of coping with observed deviations from the behaviour predicted by the Debye-Hückel expression is not to do any further theoretical calculations, but to work at empirical extensions to the Debye-Hückel expression to take it to higher concentrations. These methods assume that the Debye-Hückel theory is valid at low concentrations.

As a rough and ready guide the following will be assumed:

- low ionic strengths are taken to be $< 1 \times 10^{-3}$ mol dm^{-3},

- moderate ionic strengths are in the range 1×10^{-3} mol dm^{-3} to 0.1 mol dm^{-3} and

- high ionic strengths are those > 0.1 mol dm^{-3}.

Testing the Debye-Hückel limiting law, the Debye-Hückel equation and the extended Debye-Hückel equation has demanded highly accurate experimental activity coefficient determinations.

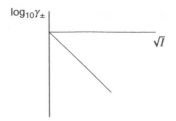

Figure 10.9   Graph of $\log_{10} \gamma_\pm$ vs. $\sqrt{I}$ illustrating the Debye-Hückel limiting law.

Calculations using the various equations of the Debye-Hückel theory are given in Worked Problems 8.12 to 8.17 and Sections 9.13.1 and 9.13.2, and Worked Problems 9.17 to 9.19, 9.21 to 9.23. Use of emf data to give experimental data to test the Debye-Hückel theory is given in Worked Problems 9.28 and 9.29.

Experimental activity coefficients are always quoted as stoichiometric values, $\gamma_\pm$, based on stoichiometric concentrations and ionic strengths. This is still the case for associated electrolytes. However, when association occurs the actual ionic strength will be less than the stoichiometric ionic strength, and it then becomes vital to distinguish between $\gamma_\pm^{actual}$ based on the **actual** concentration and the **actual** ionic strength, and $\gamma_\pm^{stoich}$ based on the **stoichiometric** concentration and the **stoichiometric** ionic strength.

Experiment shows that the **Debye-Hückel limiting law**:

$$\log_{10} \gamma_\pm = -A|z_1 z_2|\sqrt{I} \tag{10.73}$$

holds well for ionic strengths less than $1 \times 10^{-3} \,\mathrm{mol\,dm^{-3}}$ for 1–1 and 1–2 electrolytes but less well for 1–3, 2–2 and higher charge type electrolytes, and that as the ionic strength increases the full Debye-Hückel expression will be needed for all charge types.

Graphs of $\log_{10} \gamma_\pm$ vs. $\sqrt{I}$ should be linear with slope $= -A|z_1 z_2|$ (Figure 10.9). Accurate experimental $\gamma_\pm$ values show that, although linearity is found at very low ionic strengths, as the ionic strength is increased the graph of $\log_{10} \gamma_\pm$ vs. $\sqrt{I}$ lies **increasingly above** the Debye-Hückel limiting slope (see Figure 10.10). This is in keeping with the **Debye-Hückel expression**:

$$\log_{10} \gamma_\pm = -\frac{A|z_1 z_2|\sqrt{I}}{1 + B\mathring{a}\sqrt{I}} \tag{10.65}$$

The denominator, $1 + B\mathring{a}\sqrt{I}$, will cause $\log_{10} \gamma_\pm$ to be less negative than the limiting law value, and so values of $\log_{10} \gamma_\pm$ calculated from the Debye-Hückel equation will lie **above** values of $\log_{10} \gamma_\pm$ calculated from the limiting law. This is because:

$$\frac{A|z_1 z_2|\sqrt{I}}{1 + B\mathring{a}\sqrt{I}} < A|z_1 z_2|\sqrt{I}, \text{ and so } -\frac{A|z_1 z_2|\sqrt{I}}{1 + B\mathring{a}\sqrt{I}} > -A|z_1 z_2|\sqrt{I}$$

Worked Problems 9.28 and 9.29 illustrate this.

Fitting of very accurate experimental data at low concentrations to the Debye-Hückel equation, $\log_{10} \gamma_\pm = -A|z_1 z_2|\sqrt{I}/(1 + B\mathring{a}\sqrt{I})$, using electrolytes which are known, from conductance measurements, to show virtually no ion pairing has indicated that a value of $\mathring{a}$

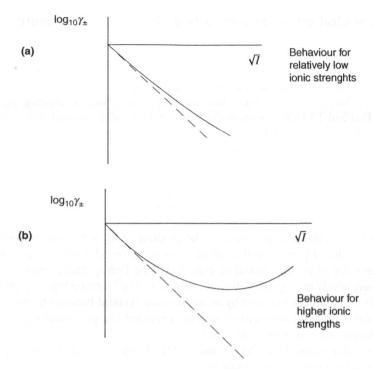

Figure 10.10 (a) and (b) Graphs of $\log_{10} \gamma_{\pm}$ vs. $\sqrt{I}$ showing deviations from the Debye-Hückel limiting law at high ionic strengths.

lying between 0.40 and 0.50 nm gives a good fit. But the higher the ionic strength, the more likely it is that $\mathring{a}$ is being forced to absorb other short range effects which are expected to give a term linear in ionic strength (see Section 10.10.2), making meaningful interpretations difficult.

The values of $\mathring{a}$ found using the Debye-Hückel equation are greater than the sum of the crystallographic radii. This is unsurprising as ions will be solvated in aqueous solution. Once solvation comes into the picture, it becomes even more difficult to assign a precise meaning to $\mathring{a}$. Does the closest approach mean:

- two bare ions;

- two ions with one H_2O between;

- two ions with two H_2O between;

- two ions with one 'squeezed' H_2O between?

There is a further problem. Is a unique value of $\mathring{a}$ appropriate for all electrolytes? Even for one given electrolyte there is no reason to assume that the distance of closest approach will remain the same under all circumstances.

10.10.2   Empirical extension using a term linear in ionic strength

Experiment shows that at higher ionic strengths deviations from the Debye-Hückel equation occur. Graphs of $\log_{10} \gamma_{\pm}$ vs. $\sqrt{I}/\left(1 + \sqrt{I}\right)$ are linear at the low end of ionic strength, but, with increasing ionic strength, the graphs become non-linear and eventually show a turn-up with values of $\log_{10} \gamma_{\pm}$ approaching zero and, in some cases, becoming positive (see Figures 10.11(a) and 10.11(b)). Such deviations can be handled by including a term linear in ionic strength (see Figure 10.11(c)).

The working equation which results is:

$$\log_{10}\gamma_{\pm} = -\frac{A|z_1 z_2|\sqrt{I}}{1 + B\mathring{a}\sqrt{I}} + bI \qquad (10.76)$$

where $\mathring{a}$ and b are adjustable parameters. The positive term, $+bI$, is of **opposite** sign to $-A|z_1 z_2|\sqrt{I}/\left(1 + B\mathring{a}\sqrt{I}\right)$ and so will result in a less negative value for $\log_{10}\gamma_{\pm}$, corresponding to a larger value of $\gamma_{\pm}$ than would be expected if the Debye-Hückel expression is used. This term progressively has a greater effect than the $1 + B\mathring{a}\sqrt{I}$ term on $\log_{10}\gamma_{\pm}$, and the graph typically shows a minimum followed by an upturn (see Worked Problem 9.29).

Experimental data shows conclusively that the extended Debye-Hückel equation must be used for moderate and higher ionic strengths.

There are two disposable parameters, $\mathring{a}$ and b, in the Debye-Hückel extended equation, and there are two ways to assign values to them.

- Assume a value of $\mathring{a}$ and plot a graph of $\log_{10} \gamma_{\pm} + A|z_1 z_2|\sqrt{I}/(1 + B\mathring{a}\sqrt{I})$ vs. I. Such graphs are often linear, and allow b to be found from the slope. However, if a constant $\mathring{a}$ is assumed the value of b then found could be absorbing errors, or changes in $\mathring{a}$ as the ionic strength increases, or vice versa. Furthermore, various combinations of $\mathring{a}$ and b can give good fits. Typical choices are:

- $\mathring{a} = 0.304$ nm corresponding to $B\mathring{a} = 1$,

- $\mathring{a} = 0.400$ nm corresponds to $B\mathring{a} = 1.315 \, \text{mol}^{-1/2} \, \text{dm}^{3/2}$,

- $\mathring{a} = 0.480$ nm corresponds to $B\mathring{a} = 1.578 \, \text{mol}^{-1/2} \, \text{dm}^{3/2}$.

Often it is impossible to discriminate between these choices. This makes it very difficult to make any meaningful interpretations of specific values.
- Curve fitting assuming $\mathring{a}$ and b to be disposable parameters can also be used. At first sight this appears to be a superior approach in so far as it allows $\mathring{a}$ and b to vary from electrolyte to electrolyte without any choice being made for either. However, it does carry all the ambiguities and problems of curve fitting, and again it must be realised that there is likely to be more than one combination of $\mathring{a}$ and b which would give a good fit, with similar limitations for specific interpretations.

Experimental values of b vary in magnitude and in sign from electrolyte to electrolyte, and as shown above depend on the value of $B\mathring{a}$. Furthermore, it is impossible to split up these values of (a)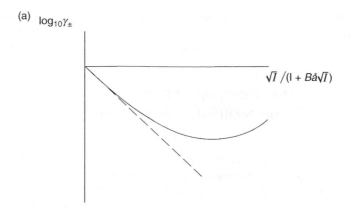

(b)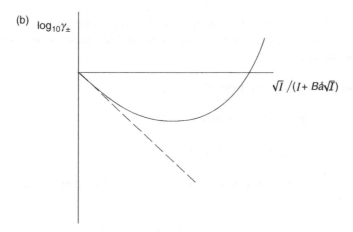

(c)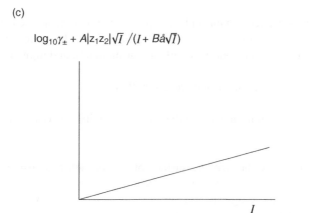

Figure 10.11 (a) and (b) Graphs of $\log_{10} \gamma_{\pm}$ vs. $\sqrt{I}/(1 + B\dot{a}\sqrt{I})$ showing deviations from the Debye-Hückel equation, and illustrating the need for a bI term. (c) Graph of $\log_{10} \gamma_{\pm} + A|z_1 z_2|\sqrt{I}/(1 + B\dot{a}\sqrt{I})$ vs. I showing how deviations from the Debye-Hückel equation can be handled by a term in ionic strength.

b into characteristic contributions from each ion. For instance, the results from a series of $Na^+(aq)$ and $K^+(aq)$ salts, e.g. the nitrates, chlorides, bromides and hydroxides can be considered for each cation. b values can be found for each set of electrolytes, and it is found that:

$$b \text{ for the NaCl(aq)} - b \text{ for KCl(aq)}$$
$$b \text{ for NaNO}_3(aq) - b \text{ for KNO}_3(aq)$$
$$b \text{ for NaOH(aq)} - b \text{ for KOH(aq)}$$

. . . .

are all different. Hence b values are not additive with respect to cation and anion, or put otherwise b values are **specific to the electrolyte** not to the individual ions. This is in keeping with modern work on electrolyte theory (see Section 10.16.3).

What perhaps will prove to be more significant will be the sign of b rather than its magnitude.

This is in stark contrast to the situation where limiting ionic conductivities are concerned (see Sections 11.11 to 11.13). Here limiting molar conductivities can be split up into individual limiting ionic conductivities for the cation and the anion, so that a table of these can be constructed, e.g.:

Λ^0 for NaCl(aq) can be made up from $\lambda^0_{Na^+} + \lambda^0_{Cl^-}$

Λ^0 for NaOH(aq) can be made up from $\lambda^0_{Na^+} + \lambda^0_{OH^-}$

where $\lambda^0_{Na^+}$ takes the same value in each case.

It has been shown theoretically that a term linear in the first power of the ionic strength could arise from:

• factors similar to those giving rise to non-ideality in solutions of uncharged solutes;

• short range interactions between an ion and the dipole of the solvent. These have an effect in an opposite direction to the long range forces between ions. The effect is likely to be negligible at low concentrations, but be significant at moderate and high concentrations.

• dispersion forces between the ions when close together;

• the effect of the presence of ions may cause the relative permittivity to be different from that for the pure solvent;

• ions will be solvated, and so the size and surface of the ion will be considerably different from that of the unsolvated ion. Ambiguities in the value assigned to $\mathring{a}$ can be reflected in a term linear in ionic strength. Assignment of the value used for the relative permittivity of the solution is also intimately tied up with how the theory will handle ion solvation.

• short range repulsions.

However, even this extended equation is not expected to be accurate at high ionic strengths where higher order terms, such as $c^{3/2}$, c^2 or even $c^{3/2}\log_{10}c$ are necessary.

10.11 Evidence for ion association from Debye-Hückel plots

Deviations from the Debye-Hückel equation:

$$\log_{10}\gamma_\pm = -\frac{A|z_1 z_2|\sqrt{I}}{1 + B\mathring{a}\sqrt{I}}$$

(10.65)

could be due to:

- a term linear in ionic strength;

- ion association;

- or both phenomena taken together.

Once the possibility of ion association is postulated it becomes absolutely vital that **stoichiometric** and **actual concentrations**, and **stoichiometric** and **actual ionic strengths** are clearly distinguished (see Section 3.11). It is also vital to make clear whether $\gamma_\pm^{\text{stoichiometric}}$ or $\gamma_\pm^{\text{actual}}$ is being referred to. If clear distinctions between all these quantities are not made, then clarity is lost and mistakes only too easily creep in. From now on these distinctions will be made explicitly. Experimental activity coefficients are always quoted as stoichiometric values, $\gamma_\pm^{\text{stoich}}$, and they generally refer to stoichiometric ionic strengths; but this again should always be made clear.

For symmetrical electrolytes, association will produce an ion pair of zero charge and to a first approximation $\gamma_\pm^{\text{actual}}$ for the ion pair is unity. For unsymmetrical electrolytes, association will produce an ion pair which is charged and so $\gamma_\pm^{\text{actual}}$ for the ion-pair is no longer unity. This complicates matters, and the arguments from now on will consequently be mainly **confined** to **symmetrical** electrolytes.

- If highly accurate experimental $\gamma_\pm$ are available, evidence for association can be found from typical plots of $\log_{10}\gamma_\pm$ vs. $\sqrt{I}/(1 + \sqrt{I})$ where $B\mathring{a} = 1$, or from plots of $\log_{10}\gamma_\pm$ vs. $\sqrt{I}/(1 + B\mathring{a}\sqrt{I})$ where a specific value other than 0.304 nm is assigned to $\mathring{a}$.

- If there is no association the experimental curves should approach the limiting slope **from above** and at higher ionic strengths go through a minimum (see Sections 10.10.1 and 10.10.2). However, where there is association the experimental plots approach the Debye-Hückel limiting slope **from below** and at higher ionic strengths cross the limiting slope and then lie above the limiting slope.

- Graphs of $\log_{10}\gamma_\pm$ vs. $\sqrt{I}/(1 + \sqrt{I})$ are given below (Figure 10.12). Figure 10.12(a) shows no definite evidence of association but definite evidence for a term linear in I_{stoich}. Figure 10.12(b) shows evidence of both association and a term linear in I_{stoich}. The Debye-Hückel limiting slope $-A|z_1 z_2|$ is drawn on both graphs.

10.11.1 An explanation of the statement that if an electrolyte is associated then the graph of $\log_{10}\gamma_\pm$ vs. $\sqrt{I}/(1 + B\mathring{a}\sqrt{I})$ will approach the Debye-Hückel slope from below

If association occurs then the actual concentration of free ions will be less than the stoichiometric concentration. Likewise, the actual ionic strength is less than the stoichiometric value.

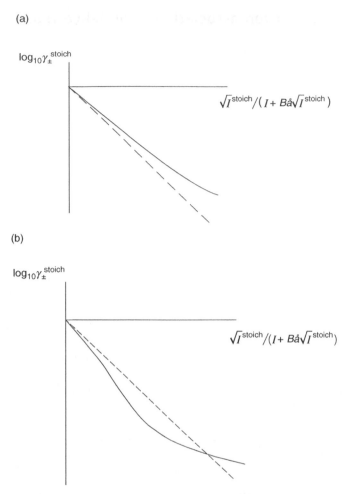

Figure 10.12 (a) Graph of $\log_{10} \gamma_\pm^{\text{stoich}}$ vs. $\sqrt{I^{\text{stoich}}}/\left(1 + B\dot{a}\sqrt{I^{\text{stoich}}}\right)$ showing deviations from the Debye-Hückel equation due to non-ideality. (b) Graph of $\log_{10} \gamma_\pm^{\text{stoich}}$ vs. $\sqrt{I^{\text{stoich}}}/\left(1 + B\dot{a}\sqrt{I^{\text{stoich}}}\right)$ showing deviations from the Debye-Hückel equation due to non-ideality and also to association.

Two situations can be considered:

Association in the absence of non-ideality	Non-ideality in the absence of association
$c_{\text{actual}} < c_{\text{stoich}}$, because of association, but because the solution is considered under ideal conditions:	If a solution is non-ideal, then according to the Debye-Hückel theory $\gamma_\pm < 1$, and so, because $a_{\text{observed}} = \gamma_\pm c$:
$a_{\text{observed}} = c_{\text{actual}}$ where $a =$ the activity, and so:	$a_{\text{observed}} < c$
$a_{\text{observed}} < c_{\text{stoich}}$	where $c_{\text{stoich}} = c_{\text{actual}} = c$, in the absence of association, and therefore:
and this is **purely a result** of removal of ions and **must not** be confused with a situation of	$a_{\text{observed}} < c_{\text{stoich}}$
$a_{\text{observed}} < c_{\text{stoich}}$ because $\gamma_\pm < 1$.	and this is **purely an effect** of non-ideality and **must not** be confused with the effect of removal of ions due to association

What this argument shows is that removal of ions due to association and non-ideality will both independently cause the activity to be less than the stoichiometric concentration. For an associated electrolyte, both association and non-ideality will be involved and both factors will contribute to bring $a_{observed}$ down more extensively than either of them would do in the absence of the other.

Since $a_{observed} < c_{stoich}$ this can be quantified as $a_{observed} = \left(\gamma_{\pm}^{stoich}\right)_{observed} c_{stoich}$, and will manifest itself in a $\left(\gamma_{\pm}^{stoich}\right)_{observed}$ being less than would be expected for a corresponding $\left(\gamma_{\pm}^{stoich}\right)_{observed}$ obeying the Debye-Hückel equation. For an **associated** electrolyte $\log_{10}\left(\gamma_{\pm}^{stoich}\right)_{observed}$ vs. $\sqrt{I^{stoich}}\big/\left(1 + B\mathring{a}\sqrt{I^{stoich}}\right)$ plots will therefore approach the Debye-Hückel slope from **below.**

10.11.2 Unknown parameters in the Debye-Hückel extended equation when association occurs

If an electrolyte is believed to be associated there are three quantities which are unknown:

- $\mathring{a}$, the distance of closest approach;

- b, the coefficient of the term linear in I; and

- β, the fraction of the ions associated into ion pairs. The latter can alternately be quantified in an association constant, K_{assoc}.

These three unknowns can be incorporated into the Debye-Hückel equation as three disposable parameters, the other quantities being dependent only on charge type and ionic strength. If $\mathring{a}$ is taken to be known, the problem reduces to two disposable parameters, b and β or K_{assoc}. There are several ways to tackle this, but what is often done is to assume a value for $\mathring{a}$, and take b and K_{assoc} as disposable parameters and curve fit to the observed graph of $\log_{10}\gamma_{\pm}^{stoich}$ vs $\sqrt{I^{stoich}}\big/\left(1 + B\mathring{a}\sqrt{I^{stoich}}\right)$. However, the standard cautionary warning must be given: a range of pairs of acceptable values of b and K_{assoc} will occur and chemical judgement will be needed if a distinction is to be made among them. It would also be possible to curve fit using the three disposable parameters $\mathring{a}$, b and K_{assoc}. The same cautionary note must again be given.

10.12 The Bjerrum theory of ion association

An essential feature of the Debye-Hückel model is that electrostatic interactions between ions impose some degree of order on the random thermal motion of the ions. It is thus possible that some interactions are so strong that two ions could move around together as an independent ion pair. This pair of ions would be counted as an ion pair if it survived long enough on a time scale of several collisions. Ion pair formation is more likely to happen in concentrated solutions where the ions are close together and the coulombic interaction is likely to be sufficiently large. It also becomes more likely as the charge type of the electrolyte increases. Absorption and Raman spectra observations and conductance studies show evidence for ion pairing in solutions of high charge types, for instance most 2–2 electrolytes are associated.

10.12.1   The graph at the heart of the Bjerrum theory

This looks at the interaction of one ion with another of opposite charge. An ion of charge z_1e is chosen to be the reference ion, and this is placed at the origin of the coordinate system where spherical symmetry is assumed. The distance of the second ion of opposite charge, z_2e, from the first ion is r, where r is the radius of a sphere around the central ion.

- (a) The chances of finding a second ion of opposite charge to the central ion become less as the distance from the origin increases. This is a direct consequence of the decrease, as r increases, in the potential energy of electrostatic interaction between the two ions; this interaction being a **short range** coulombic interaction (see Figure 10.13(a)).

- (b) However, as r increases the volume of the sphere around the central ion increases and so the number of ions present will increase. Consequently the chances of finding an ion of opposite charge will increase (see Figure 10.13(b)).

These two effects are in opposite directions with (a) dominant at short distances and (b) dominant at large distances. This results in a minimum in the chance of finding an ion of opposite charge to the central ion (Figure 10.13(c)), and theory shows that this occurs when:

$$\frac{|z_1 z_2| e^2}{4\pi \varepsilon_0 \varepsilon_r r} = 2kT \tag{10.77}$$

This value of r at which the minimum occurs is given the symbol, q, and the name, 'the Bjerrum distance', and takes a value:

$$q = \frac{|z_1 z_2| e^2}{8\pi \varepsilon_0 \varepsilon_r kT} \tag{10.78}$$

Bjerrum then defined ion pairs as:

- ions which are within a distance, q, of each other are paired or associated;

- ions which are at distances from each other greater than q are unpaired, or unassociated.

Bjerrum thus considered that the cut-off between unpaired and paired ions occurs when the electrostatic potential energy of interaction equals $2kT$. This is a somewhat arbitrary distance defined by the minimum in the graph (Figure 10.13(c)) and there has been much discussion as to the validity of choosing this distance to define the cut-off.

However, it is still useful to see how the Bjerrum critical distance, q, varies with charge type (see Table 10.1).

Since simple ions have mean ionic diameters of the order 0.3–0.6 nm, it is evident that deviations from the Debye-Hückel expression due to ion association are likely to occur for charge types of 2–1 or greater. These deviations are then handled by Bjerrum's ion-pair

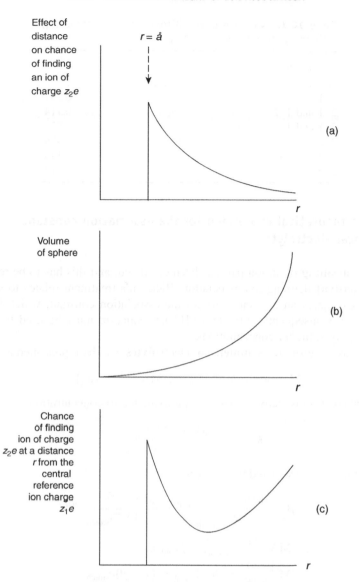

Figure 10.13   (a) Graph showing the effect of distance on the chance of finding an ion of charge z_2e of opposite sign to the central reference ion of charge z_1e. (b) Dependence of the volume of a sphere on the radius, r. (c) Graph showing the chance of finding an ion of charge z_2e of opposite sign to the central reference ion of charge z_1e at a distance r from the central reference ion of charge z_1e.

concept. For 1–1 electrolytes the Debye-Hückel approximations are generally valid, and for these electrolytes (provided $\mathring{a} > 0.357$ nm) ion pairing is not to be expected in water, and experimentally this is what is usually found. For large complex 1–1 electrolytes, ion pairing in water is not expected since $\mathring{a}$, the distance of closest approach, for such electrolytes will certainly not be 0.357 nm – it will be much larger. For higher charge types similar reasoning can hold. What will be important is the relative values of $\mathring{a}$ and q.

Table 10.1. Variation of the Bjerrum critical distance, q, with charge type for water at 25°C

Electrolyte type	$\dfrac{q}{\text{nm}}$
1–1	0.357
2–1 and 1–2	0.714
3–1 and 1–3	1.071
2–2	1.429
3–2 and 2–3	2.143
3–3	3.214

10.12.2 The theoretical expression for the association constant for symmetrical electrolytes

So far only the **meaning** of an ion pair has been discussed, and this has to be related to an **equilibrium constant** defining ion association. Bjerrum's treatment relates to **very dilute** solutions, and calculates an explicit value for the association constant, which is therefore an **ideal** constant. Consequently, the Debye-Hückel equation must be used to enable the calculation to apply at higher concentrations.

Formally ion association, for **symmetrical electrolytes**, can be represented as:

$$M^{z+}(aq) + A^{z-}(aq) \rightleftharpoons MA^{(z^+)+(z^-)}(aq)$$

where, for ideal conditions, activities can be approximated to concentrations,

$$K_{assoc} = \frac{[MA^{(z^+)+(z^-)}]_{actual}}{[M^{z+}]_{actual}[A^{z-}]_{actual}} \tag{10.79}$$

The fraction of the ions associated to ion-pairs

$$\beta = \frac{[MA^{(z^+)+(z^-)}]_{actual}}{[M^{z+}]_{stoich}} = \frac{[MA^{(z^+)+(z^-)}]_{actual}}{[A^{z-}]_{stoich}} \tag{10.80}$$

$$\therefore [MA^{(z^+)+(z^-)}]_{actual} = \beta c_{stoich} \tag{10.81}$$

$$[M^{z+}]_{actual} = [A^{z-}]_{actual} = (1 - \beta)c_{stoich} \tag{10.82}$$

where $c_{stoich} =$ the stoichiometric concentration of the electrolyte solution.

The Bjerrum calculation gives β, the fraction of the ions associated to ion pairs, and this is calculated for very dilute solutions. From this a theoretical K_{assoc} at infinite dilution, i.e. for an ideal solution, can be found.

10.12.3 Calculation of β from the Bjerrum theory

Bjerrum looks at the direct interaction between an ion of charge $z_1 e$ and another ion with charge $z_2 e$ leading to the formation of the ion pair. The calculation is based on the Debye-Hückel model (Section 10.3) and its development as given in Section 10.6.

If an ion of charge z_1e is regarded as being placed at the origin, then the potential energy of interaction between z_1e and z_2e can be written as:

$$\text{potential energy} = z_2e\psi_1 \qquad (10.16)$$

where ψ_1 is the potential due to ion z_1e at ion z_2e which is at a distance r from z_1e.

Arguing along the lines of Section 10.6.2

$$\psi_1 \qquad = \qquad \psi_1' \qquad + \qquad \psi_1'' \qquad (10.19)$$

$$\downarrow \qquad\qquad\qquad \downarrow \qquad\qquad\qquad \downarrow$$

total potential at any distance, r, from the central z_1e ion due to **that ion and to its ionic atmosphere and this includes the z_2e ion**	total potential at any distance, r, from the central z_1e ion due to **that ion itself**	total potential at any distance, r, from the central z_1e ion due to **the ionic atmosphere and this includes the z_2e ion**

If the ionic atmosphere is ignored, i.e. ideal conditions, then Equation (10.19) reduces to:

$$\psi_1 = \psi_1' \qquad (10.83)$$

$$= \frac{z_1e}{4\pi\varepsilon_0\varepsilon_r r} \qquad (10.84)$$

and is the total potential at a distance r from the central ion at the origin of charge z_1e, **if non-ideality is ignored.**

Be very careful here: The total potential at any distance, r, from the origin can be identified with the potential at any distance, r from the central z_1e ion due to that ion itself, **only if** non-ideality is ignored. This means that corrections for non-ideality must be superimposed onto the Bjerrum theory after the association constant has been derived (see Section 10.12.4). If the second charge, z_2e, is placed at distance, r, from the first charge, z_1e, then:

Potential energy of interaction between the two charges, ignoring non-ideality

$$= z_2e\psi_1 = \frac{z_1z_2e^2}{4\pi\varepsilon_0\varepsilon_r}\frac{1}{r} \qquad (10.85)$$

This is the potential energy which appears in the Maxwell-Boltzmann distribution for the ion pair situation, see below.

In the Bjerrum theory, the potential which appears in the potential energy of interaction between the two ions of the ion pair is thus an **approximate** potential in so far as the term due to the ionic atmosphere is ignored.

This requires that any further argument will relate to ideal conditions only. The treatment will then simply be looking at the interaction between two ions neither of which has an ionic atmosphere. **Then, and only then**, do Equations (10.83) to (10.85) hold.

It is for reasons of **mathematical simplicity** that Bjerrum limits his argument in this way, and consequently his derivation **can only apply** to very dilute solutions in which the effects of the ionic atmosphere become negligible, and the K_{assoc}, as calculated, is an **ideal** K.

The Maxwell-Boltzmann distribution can now be used to find the average number of ions, n_2', of charge z_2e per unit volume at a distance, r, form the ion, z_1e, at the origin as:

$$n_2' = n_2 \exp\left(-\frac{z_1 z_2 e^2}{4\pi\varepsilon_0\varepsilon_r r}\frac{1}{kT}\right)$$

(10.86)

From this expression, by considering all distances $< q$, the probability that an ion of charge z_1e has an ion of opposite charge, z_2e, associated with it can be found. This, in turn, gives β the fraction of ions associated to ion pairs as:

$$\beta = \int_{\mathring{a}}^{q} 4\pi n_2 \exp\left(-\frac{z_1 z_2 e^2}{4\pi\varepsilon_0\varepsilon_r r kT}\right) \times r^2 dr$$

(10.87)

where $\quad q = \dfrac{|z_1 z_2|e^2}{8\pi\varepsilon_0\varepsilon_r kT}$

(10.78)

The integral has limits $\mathring{a}$ and q, since ions lying within the distance $\mathring{a}$ to q are defined as paired. By carrying out this integration:

$$\beta = 4\pi n_2 \times \left(-\frac{z_1 z_2 e^2}{4\pi\varepsilon_0\varepsilon_r kT}\right)^3 Q(b)$$

(10.88)

where $\quad Q(b) = \displaystyle\int_{2}^{b} \frac{e^x}{x^4} dx$

(10.89)

$$b = -\frac{z_1 z_2 e^2}{4\pi\varepsilon_0\varepsilon_r \mathring{a} kT} \quad \text{and} \quad x = -\frac{z_1 z_2 e^2}{4\pi\varepsilon_0\varepsilon_r r kT}$$

(10.90) and (10.91)

Values of the integral $Q(b)$ have been tabulated for various values of b.

β as given by Equation (10.88), is in molecular quantities, but the K_{assoc} given in Equation (10.79) is in molar quantities, i.e. the β required is

$$\beta = 4\pi N c_2 \times \left(-\frac{z_1 z_2 e^2}{4\pi\varepsilon_0\varepsilon_r kT}\right)^3 Q(b)$$

(10.92)

where c_2 is the bulk actual concentration of ions of charge z_2e. For symmetrical electrolytes which is what this derivation pertains to, $c_2 = c_1$ which is the bulk actual concentration of the ions of charge z_1e and of the bulk solution.

This calculation relates to very dilute solutions. For these **very dilute** solutions the fraction of ions associated to ion pairs is very small, i.e. β approaches zero and $(1 - \beta) \to 1$. Under these conditions the activity coefficients approach unity, corresponding to ideality.

$$\therefore K_{assoc} = \frac{\beta}{(1-\beta)^2 c_2} \approx \frac{\beta}{c_2}$$

(10.93)

Substituting for β gives:

$$K_{\text{assoc}} = \frac{4\pi Nc_2}{c_2} \times \left(-\frac{z_1 z_2 e^2}{4\pi\varepsilon_0\varepsilon_r kT} \right)^3 Q(b) = 4\pi N \left(-\frac{z_1 z_2 e^2}{4\pi\varepsilon_0\varepsilon_r kT} \right)^3 Q(b) \tag{10.94}$$

This is the expression for an **ideal** solution.

10.12.4 Extension to account for non-ideality

For higher concentrations, K_{assoc} will have to be corrected for non-ideality by including activity coefficients calculated from the Debye-Hückel equation.

$$K_{\text{assoc}} = \frac{\beta}{(1-\beta)^2 c^{\text{stoich}}} \frac{(\gamma_{\pm\text{MA}})_{\text{actual}}}{(\gamma_{\pm}^2)_{\text{actual}}} \tag{10.95}$$

where:

- β is the quantity which is calculated from Bjerrum's theory;

- $\gamma_{\pm}$ is the mean activity coefficient for the electrolyte ions which are unassociated, and is calculated from the Debye-Hückel expression:

$$\log_{10}\gamma_{\pm}^{\text{actual}} = -\frac{A|z_1 z_2|\sqrt{I^{\text{actual}}}}{1 + Bq\sqrt{I^{\text{actual}}}} \tag{10.96}$$

But take note: The Debye-Hückel expression is **not** that which is used for unassociated electrolytes. There is one **very important difference**:

the distance of closest approach which appears in the denominator is not $\mathring{a}$, but q, the distance beyond which the ions are considered to be free and unassociated.

and it is to distances greater than q that the Debye-Hückel theory will apply.

- $\gamma_{\pm\text{MA}}$ is the mean activity coefficient for the ion pair, and takes a value dependent on the type of electrolyte involved:

  - if the electrolyte is symmetrical, then the ion pair will be overall uncharged, though it will be a dipole. Its mean activity coefficient is taken to be unity, though this assumption can be queried because of the dipolar nature of the ion pair.

  - if the electrolyte is unsymmetrical, the ion pair will be charged and the Debye-Hückel theory is used to calculate the activity coefficient but the stoichiometry will be different from that given above.

These activity coefficients must refer to activity coefficients calculated in terms of the **actual** ionic strength not the **stoichiometric** ionic strength, since they relate to the free ions only, and thus to a c^{actual}. Their calculation will thus involve a series of successive approximations.

If the electrolyte is unsymmetrical, the whole theoretical approach becomes much more complex.

10.12.5   Critique of Bjerrum's theory

Bjerrum's theory has been criticised because it involves the arbitrary cut-off at a critical distance q between ion pairs and free ions. It is felt that a more realistic situation would be one which would allow more of a 'fall-off' between paired and free ions as the distance between them alters.

For a symmetrical electrolyte, the ion pair is taken to behave as though it were a neutral species. It is, however, a dipole and as such will interact with the other ions. Consideration of this would lead to an activity coefficient for the ion pair which is not unity. There is also the question as to whether the ions of the ion pair are in contact, or are separated by one or more water molecules. Bjerrum himself recognised these limitations to his treatment: 'the distinction between free and associated ions is not a chemical one, but only a mathematical device'.

There have been attempts to modify Bjerrum's treatment to remove this arbitrariness, but none has been used universally to any great extent in the interpretation of experimental data. Nevertheless, despite this artificiality, the Bjerrum theory coupled with the Debye-Hückel theory has proved a very useful and relatively successful tool in discussing electrolyte solutions. This success is especially noteworthy when the Bjerrum-Debye-Hückel theory is compared with the alternative approach of Guggenheim's numerical integration which gives similar results (see Section 10.13.1). Table 10.2 gives values of K_{assoc} for various charge types and for various values of $\overset{\circ}{a}$ and q.

10.12.6   Fuoss ion pairs and others

Bjerrum's concept of an ion pair and his theoretical development is the most successful treatment used in conjunction with the Debye-Hückel theory for analysing experimental data, despite all the ambiguities in deciding at what stage to introduce the cut-off distance between ion pairs and free ions. His theory was basically a device to account for the short range electrostatic interactions not included in the Debye-Hückel theory (see Guggenheim's treatment, Section 10.13.1).

Fuoss has also attempted to define what is meant by an ion pair and to deduce a theoretical treatment to allow a calculation of the expected $K_{association}$ based on his model. He stated that an ion pair formed when the ions come in contact at the distance of closest approach, i.e. to a distance $\overset{\circ}{a}$ apart. There are, however, theoretical reasons for saying that this should more correctly read as 'ions at around the distances close to the distance of closest approach'.

Table 10.2.   Values of $K_{assoc}/mol^{-1}dm^3$ for various charge types at 25°C and $q = 0.3572$ nm

$\dfrac{\overset{\circ}{a}}{nm}$	0.238	0.286	0.357	0.476
1–1	0.9	0.5	0	0
1–2	23.0	17	12.1	7.2
1–3	220	124	77.5	49
2–2	2370	817	352	184
2–3	4.96×10^5	5.48×10^4	7.99×10^3	1.76×10^3
3–3	$>10^6$	$>10^6$	1.67×10^6	6.82×10^4

The theoretical basis of his calculation is less secure than Bjerrum's, but his work had the merit of inspiring high precision work on the behaviour of electrolyte solutions over a range of relative permittivities. Bjerrum's and Fuoss' theories predicted different dependencies of $K_{association}$ on ε_r (see Section 12.16).

Other workers have also discussed this concept of a 'contact ion pair', but detailed analysis indicates that there is a considerable degree of latitude in what physically is meant by a 'contact ion pair'. Fuoss revisited this dilemma in his 1978 conductance theory (see Section 12.17).

10.13 Extensions to higher concentrations

10.13.1 Guggenheim's numerical integration

Guggenheim used the **complete** Poisson-Boltzmann Equation (10.28), and solved it by numerical integration using a computer. He included all terms in his calculations, and his results allow for all electrostatic interactions **including those giving rise to ion pairs**. Hence with his work there is no need to refer explicitly to ion association since this is automatically included in his calculation. His work was of considerable significance, and the conclusions drawn were:

- The Debye-Hückel treatment is correct and rigorous for very low concentrations for all charge types.

- His treatment extended the range of the theory to ionic strengths of 0.1 molal, and showed quite unambiguously that $\gamma_{\pm}$ values predicted by the numerical integration for 1–1 electrolytes fit almost exactly those predicted by the Debye-Hückel theory. It follows that the Debye-Hückel expression is an **excellent base-line** for describing the properties of **1–1** electrolyte solutions.

- The same analysis showed that the predicted $\gamma_{\pm}$ values for **1–2 and 2–1** electrolytes fit the Debye-Hückel equation within 2%. So again the Debye-Hückel expression can be taken as a **reasonable base-line** for these electrolytes.

- However, for higher charge types matters become more difficult. When 2–2 electrolytes were considered, the predicted values and the Debye-Hückel values for $\gamma_{\pm}$ differed by up to 20%. Values for 1–3 and 3–1 electrolytes are expected to lie between these values and those for 1–2 electrolytes. The inevitable conclusion is that the Debye-Hückel expression **cannot be used as a base-line** for these **high charge** types.

- No conclusions could be reached for 3–3 electrolytes because the integrals used in the numerical integration diverged – a sure sign that there may be some hidden problems in the analysis, physical or mathematical, which have become acute for 3–3 electrolytes, but do not show up for lower charge types.

- Ion association was shown to be negligible for 1–1 electrolytes, possible, but low, for 1–2, 2–1 electrolytes, but becomes important for 2–2 electrolytes.

- Results for 2–2 electrolytes were of considerable importance since he showed that the numerical integration which uses the complete Poisson-Boltzmann equation fitted well with the Debye-Hückel equation coupled with Bjerrum Ks for association, **provided** that the Bjerrum q was used instead of $\mathring{a}$ in the denominator, i.e. $1 + Bq\sqrt{I}$ must be used.

The numerical integration can, therefore, give a **possible base-line** for the treatment of electrolyte solutions. Provided that association is taken care of by the Bjerrum treatment, the Debye-Hückel expression is also a good base-line for the behaviour of free unassociated ions in solution, **provided** that the appropriate value of q appears in the denominator rather than the conventional $\mathring{a}$.

Guggenheim did, however, point out that there are also inherent problems in this approach because of the distinct physical status of the ψs used in the Poisson and Maxwell-Boltzmann equations (see Section 10.6.5).

10.13.2   Extensions to higher concentrations: Davies' equation

Davies put forward the equation:

$$\log_{10}\gamma_{\pm}^{\text{actual}} = -\frac{A|z_1 z_2|\sqrt{I^{\text{actual}}}}{1 + \sqrt{I^{\text{actual}}}} + 0.15|z_1 z_2|I^{\text{actual}} \qquad (10.97)$$

to describe the behaviour of the **free unassociated** ions of any electrolyte. Taking data for electrolytes known to be associated, and using the known K_{assoc} for such electrolytes he was able to show that a large variety of electrolytes of all charge types, fitted his equation within acceptable margins for any deviations. He proposed that this equation could be used to calculate the activity coefficients for the free ions.

10.14   Modern developments in electrolyte theory

Because of the mathematical complexity of modern electrolyte theory only a qualitative introduction to the concepts will be given.

10.15   Computer simulations

Computer simulation techniques represent a very useful theoretical approach. Modern high speed, high memory computers give a realistic hope that a much more direct solution to the theory of electrolyte solutions will be forthcoming. A large amount of exciting work has been done in this field, some of a very complex nature.

Essentially what is done is to choose a model for the electrolyte solution specifying what interactions are inherent in the model. For each type of interaction, the total potential energy of interaction calculated pair-wise, i.e. as a sum of contributions from all pairs of particles, can be formulated.

It is crucial to realise that the choice of model is the most important part of the technique, and this requires a considerable degree of insight into, and understanding of, what goes on at the microscopic level in an electrolyte solution. This requires that a lot of thought must go into the setting up of the physical model. The computing part of computer simulation techniques basically makes use of an extremely useful and versatile tool which extends considerably the type of calculation able to be handled. However, the physical model should always remain the central issue.

The techniques are of two types:

- Monte Carlo calculations, and

- molecular dynamics.

10.15.1   Monte Carlo calculations

These are essentially an updated version of the Milner type calculation outlined earlier (Section 10.2). The intractable difficulties which faced Milner can now be handled by computer. The biggest problem is to suggest a credible potential on which to base the Milner type calculation.

The essence is to calculate all interactions between all the ions and average them. More advanced models would include interactions between ions and solvent molecules and between solvent molecules. The particles making up the solution are displaced from an initial random distribution by a small amount and the total energy calculated for this new distribution. The procedure is repeated taking only displacements which led to a lowering of the total energy, since the states of lower energy contribute the most to all statistically averaged properties of the electrolyte solution at equilibrium. Doing this on a computer opens up almost limitless possibilities for the model, and since computing techniques are involved it is possible to incorporate or discard whatever feature seems appropriate, and compare the results with each other and with experiment.

For instance, Monte Carlo procedures start from a set of conceivable distributions, and

- for each distribution chosen any number of possibilities of interaction could be included in the energy. For instance, all the interactions listed in Section 10.8.3 could be included as well as the coulombic long range electrostatic interactions.

- Attempts could be made to make the parameters for the ions as realistic as possible, for instance terms to account for size, shape, distribution of charge in the ions including charge separated ions could be included.

- Also terms to describe all sorts of ion–solvent, solvent–solvent, and modified ion–solvent, and modified solvent–solvent interactions could be included.

If all sorts of possible combinations of the features suggested above are to be incorporated in the calculations and each result is compared with the others and with experiment, a much deeper and more accurate picture of what is happening at the microscopic level emerges. Consequently, a better theory of electrolyte solutions has been forthcoming. Much highly promising work has been done and modifications both to the Debye-Hückel model and its mathematical development have resulted. Also new theoretical models have been proposed, some with considerable success. Unfortunately, many of these advances are complex conceptually and mathematically which renders an elementary account less feasible.

Some of the statistical mechanical developments which depend on the power of the computer to carry out the complex calculations are outlined in Sections 10.17 to 10.22 below. Studies on solvation of ions using, in particular, Monte Carlo calculations are given in Chapter 13 on solvation.

10.15.2   Molecular dynamics

Such studies are even more demanding of computing skill and time than Monte Carlo calculations.

For molecular dynamics, the computer is used to describe how an assembly of molecules and ions would behave over a period of time. A model is chosen and the resulting equations describing this are fed into the computer. Because of the interactions allowed in the model, rearrangements in the distribution of the particles in the solution follow inevitably. In molecular dynamics what can be watched are the rearrangements which follow inevitably from the first description of the model fed into the computer, and this is given as a function of time. The simulation is open to sampling for all rearrangements which lead to equilibrium, with sampling intervals of 10^{-15}s possible.

Molecular dynamics is open to all the possibilities listed for the Monte Carlo calculations. One very important feature of these techniques is that the limitations involved are not limitations forced onto the model or onto the theory; they are not physical but are computer limitations.

There has been a large amount of work published in this field, and there is a vast potential and excitement in computer simulation techniques, and they can equally well be applied to other areas of chemistry, physics and biology.

10.16   Further developments to the Debye-Hückel theory

Modern theoretical developments focus on two areas:

- modifications to the Debye-Hückel model;

- new approaches to the theoretical treatment.

However, because the mathematical aspects of these developments are so complex, only a qualitative description of each will be given.

The fundamental requirement for any computer simulation study is a knowledge of the interactions which are being included in the physical model of the electrolyte solution. The expression for the potential energy for each interaction and its dependence on distance, or distances, between the particles chosen is set up. The simplest calculations involve interactions where the potential energy is calculated pairwise, i.e. only interactions between two particles are considered. But the calculations are not limited to this simple model. In principle, simultaneous interactions between many particles can be handled with modern computers. Once these have been decided upon an expression for the total potential energy is set up using the expressions for the potential energy for each of the interactions postulated. Obviously, the more interactions chosen, the more complex is the total potential energy and the more complex are the calculations involved in the computer simulation.

In Sections 10.16.1 to 10.16.7 following, some of the ideas incorporated into a model of an electrolyte solution are developed. In this way, the alterations to the total potential energy can be illustrated qualitatively as a model is progressively refined. The model chosen is the Debye-Hückel model which has been described in detail earlier in this chapter.

The Debye-Hückel model considered the solvent to be a structureless medium whose only property is to reduce the interactions between ions in a vacuum by a factor given by the macroscopic relative permittivity, ε_r. No cognisance was taken of the possibility of ion–solvent interactions, and the size of the ion was assumed to be that of the bare ion. Gurney in the 1930s introduced the concept of the co-sphere and this has proved to be a useful concept in the theory of electrolyte solutions. Many recent theories of conductance are based on the Gurney co-sphere concept (see Section 12.17).

10.16.1 Developments from the Gurney concept of the co-sphere: a new model

Gurney introduced the idea of a co-sphere around each ion, which can loosely be identified with a region of hydration. Outside each co-sphere the water is treated as unmodified bulk water with all the properties of pure water. However, within the co-sphere the water is no longer treated as unmodified bulk water. Allowance is also made for the possibility that the individual co-spheres of ions could overlap with water being 'squeezed out' (see Figures 10.14(a) and 10.14(b)).

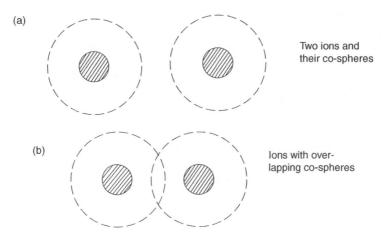

Figure 10.14   (a) A Gurney diagram showing two ions with their co-spheres. (b) A Gurney diagram showing ions with over-lapping co-spheres.

10.16.2   The unmodified Debye-Hückel model

The potential energy of interaction of two ions is considered to be of the form:

$$\frac{z_i z_j e^2}{4\pi\varepsilon_0\varepsilon_r r} + \{\text{a 'core' term equal to} +\infty \text{ for } r < \mathring{a}, \text{ zero for } r > \mathring{a}\}$$

'Core' in this sense means something at the centre, and physically it can be identified with the 'hard sphere repulsion'. Ions cannot get closer together than the 'distance of closest approach' in the hard sphere model. And so the 'core' term for the simple Debye-Hückel model contributes nothing except to stop the ions getting closer than $\mathring{a}$. Furthermore, in the model the ions are considered to be non-deformable in the mechanical sense, i.e. they cannot be 'squeezed'. This is reflected in the statement that the potential energy becomes $+\infty$ for $r < \mathring{a}$. The Debye-Hückel theory considers only situations where $r > \mathring{a}$, and for these the 'core' term is zero. Hence the potential energy of interaction of two ions can be given simply as $z_i z_j e^2 / 4\pi\varepsilon_0\varepsilon_r r$ in the simple Debye-Hückel treatment.

The following treatment due to workers such as Gurney, Rasaiah, Friedman and Levine and their co-workers can best be explained qualitatively by considering how each modification to the Debye-Hückel model will alter the potential energy of interaction of two ions as a function of distance apart.

10.16.3   A first modification to the simple Debye-Hückel model

This is where the Gurney co-sphere concept makes its first appearance in the theory. An extra term is added to the potential energy, and this discusses what happens between $r = \mathring{a}$ and an upper limit. This upper limit is $r_i + r_j + d$, where d is twice the thickness of a co-sphere, taken by some workers to be equal to the diameter of a molecule of solvent, here $H_2O(l)$, and r_i and r_j are the crystallographic radii of the ions.

This then modifies the equation for the potential energy, giving for the potential energy of interaction of two ions:

$$\frac{z_i z_j e^2}{4\pi\varepsilon_0\varepsilon_r r} + \{\text{the Debye-Hückel 'core' term}\} + \{\text{a 'Gurney' term for the range } r = \mathring{a} \text{ to}$$

$$r_i + r_j + d, \text{equal to some constant}\}$$

The constant in the Gurney term can be positive or negative, and corresponds to the following:

(i) When this term is positive, repulsion comes into play as the ions approach each other **before** $r = r_i + r_j$ is reached. This repulsion is attributable to 'pushing the waters of the co-sphere out of the way' as the ions approach each other, and occurs whether the ions are of like charge or of opposite charge (see Figure 10.14(b)).

(ii) When the term is negative, an additional attraction is coming into play as the ions approach each other, but again **before** $r = r_i + r_j$ is reached. This attraction is attributable to ε_r not having the bulk value when the ions lie in the region between $r = r_i + r_j$ and $r = r_i + r_j + d$. ε_r is less than the bulk value because of dielectric saturation near the ions, and this will increase the energy of the attractive interaction of the two ions. Since attraction is

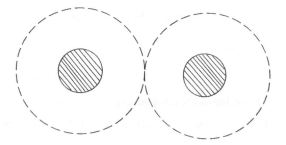

Figure 10.15   A Gurney diagram showing co-spheres just touching.

involved this is only applicable to opposite charges. In addition, dispersion forces (Section 1.7.3) could also give an attraction, and this can happen whether the charges are like or opposite.

(iii) When two ions of like charge approach each other and dielectric saturation occurs in the region between $r = r_i + r_j$ and $r = r_i + r_j + d$, this will again result in ε_r being less than the bulk value. However, in this case where the ions have the same charge increased repulsion will occur and this manifests itself in a positive term.

This modification is applicable to distances between $r = r_i + r_j$ and $r = r_i + r_j + d$ and the upper limit corresponds to when the Gurney co-spheres cease to overlap (see Figure 10.15). Any value of r between the lower limit $r_i + r_j$ and the upper limit $r_i + r_j + d$ corresponds to overlap of the co-spheres. Overlap of the co-spheres contributes a **constant** term, **no matter how extensive the overlap is**.

The approach here is totally different from the Debye-Hückel approach in so far as the contributions to the 'Gurney' terms are specific to the ions in question. For instance, if a solution containing NaCl(aq) and one containing NaOH(aq) are discussed the contribution from the Na^+(aq) will be the same for each, but the contributions to the 'Gurney' term will be different for Cl^-(aq) and OH^-(aq). In the Debye-Hückel case it is only the **charge** which is important, the argument is independent of the **nature** of the ions. The Gurney modification would imply that $\gamma_{\pm}^{NaCl} \neq \gamma_{\pm}^{NaOH}$ in contrast to the identity which is predicted by the Debye-Hückel theory. It is interesting that when experimental $\gamma_{\pm}$ values are analysed the coefficient b for the term linear in I is found to be specific to the electrolyte in question, and 'non-additive'.

10.16.4   A less simple Gurney model: a second modification to the Debye-Hückel model

In this model the hard 'core' term is as before, but the 'Gurney' term is different. Instead of adding or subtracting a **constant** term for the region between $r = r_i + r_j$ and $r = r_i + r_j + d$, this term **is allowed to vary** and to depend on the extent to which the co-spheres of the two ions overlap. The new term which is used is taken to be proportional to the volume of the overlapping part of the co-spheres, and again the magnitude will depend on the identities of the ions.

In this modification, the potential energy of interaction of the two ions is:

$$\frac{z_i z_j e^2}{4\pi\varepsilon_0\varepsilon_r r} + \{\text{the Debye-Hückel 'core' term}\}$$

$$+ \{\text{a new 'Gurney' term, which is proportional to the volume}$$

$$\text{of overlap of the co-spheres}\}$$

The constant of proportionality in the new 'Gurney' term is taken as a disposable parameter.

10.16.5   A less simple Gurney model: a third modification to the Debye-Hückel model

This involves a modified core term where the hard 'core' or hard sphere repulsion is replaced by a non-hard sphere short range repulsion which is proportional to r^{-9}. This replaces the Debye-Hückel 'core' term. The 'Gurney' term remains proportional to the volume of overlap of the co-spheres.

The potential energy of interaction of the two ions now becomes:

$$\frac{z_i z_j e^2}{4\pi\varepsilon_0\varepsilon_r r} + \{\text{a new 'core' term proportional to } r^{-9} \text{ for all } r\}$$

$$+ \{\text{the 'Gurney' term which is proportional to the volume}$$

$$\text{of overlap of the co-spheres}\}$$

The constant of proportionality in the 'core' term is taken as known, being that which will give the correct value for the sum of the crystallographic radii.

Again, the final expression will depend on the identity of the ions.

10.16.6   A further modification involving a 'cavity' term

This adds another term, the 'cavity' term, which takes account of the departure of ε_r from its bulk value. This term is in addition to the Gurney term when it is negative. The cavity term is proportional to r^{-4}.

The potential energy of interaction of the two ions now becomes:

$$\frac{z_i z_j e^2}{4\pi\varepsilon_0\varepsilon_r r} + \{\text{a 'core' term proportional to } r^{-9} \text{ for all } r\}$$

$$+ \{\text{the 'Gurney' term which is proportional to the}$$

$$\text{volume of overlap of the co-spheres}\}$$

$$+ \{\text{a 'cavity' term, proportional to } r^{-4} \text{ for all } r\}$$

In this modification the constant of proportionality for the cavity term is taken to be completely determined by the crystallographic radii, r_i and r_j, the bulk ε_r and the 'cavity' ε_r. The 'cavity' ε_r could be treated as a second adjustable parameter, but some workers indicate that it is equal to the square of the refractive index of the solvent. The constants of proportionality for the 'core' and 'Gurney' terms are as given previously.

10.16.7 Use of these ideas in producing a new treatment

The potential energy expressions for each of these terms, the core terms, all the Gurney terms and the cavity terms, can all be written down, and each will have a specific dependence on the distance between the pair of ions considered. The final total potential energy is equal to the sum of each term which is taken to be incorporated in the new model, and this total potential energy will have an explicit dependence on distance.

These modifications to the simple Debye-Hückel model have now to be incorporated into the theoretical treatment. The potential energy term appearing in the exponential term in the Maxwell-Boltzmann Equation (10.23) is one deriving from the modified potential energy, e.g.:

$$\frac{z_i z_j e^2}{4\pi\varepsilon_0\varepsilon_r r} + \{\text{a 'core' term proportional to } r^{-9} \text{ for all } r\}$$

$$+ \{\text{the 'Gurney' term which is proportional to the volume of}$$
$$\text{overlap of the co-spheres}\}$$

$$+ \{\text{a 'cavity' term proportional to } r^{-4} \text{ for all } r\}$$

could be considered.

The Poisson equation (10.20), $\nabla^2\psi_j = -(1/\varepsilon_0\varepsilon_r)\rho_j$, has also to be modified to take into account the varying value for ε_r. All the relevant equations can now be set up, but the mathematical treatment here is complex. However, the computational details can be handled easily by computer. What finally emerges are quantities for which there are explicit expressions which enable μ, the chemical potential, for the solute to be found. Ultimately this enables $\gamma_\pm$ to be determined for various values of $\kappa(r_i + r_j)$. These can then be compared with experimental quantities over the same range of ionic strengths. Good agreement is found.

The modifications involved in this treatment are expected to give linear first order and higher order terms in ionic strength. Experimentally it has been found that values of the coefficient b of the linear term in ionic strength are specific to the electrolyte and are not just dependent on the charges. This treatment has an in-built dependency on the specific nature of the electrolyte. The limiting law, however, is in no way modified.

This brief qualitative account is given to indicate how theoreticians can go about modifying the Debye-Hückel model. It also shows how a similar treatment could be given for other models, viz. the important points are:

- defining the interactions involved;

- setting up the expressions for the potential energy of each interaction as a function of distance, giving the dependence on distance;

- formulating the total potential energy as a function of distance.

10.17 Statistical mechanics and distribution functions

Most of the modern developments in electrolyte theory rest on the use of statistical mechanical arguments, with the computational details carried out by computer simulation methods,

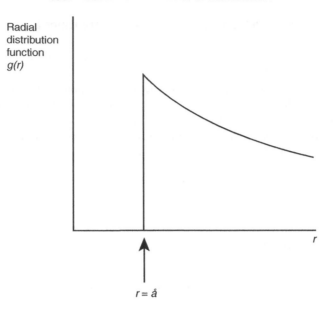

Figure 10.16   A radial distribution function $g(r)$ corresponding to a Maxwell-Boltzmann distribution as used in the Debye-Hückel theory.

particularly Monte Carlo calculations. The central part of these treatments rests on a set of quantities called 'distribution functions'.

10.17.1   The simplest situation: the radial distribution function of the Debye-Hückel theory

This distribution function has already appeared in the Debye-Hückel theory (Equation 10.22) when the Maxwell-Boltzmann distribution was used to describe the distribution of the ions of the ionic atmosphere around the central j ion (see Figure 10.16). Because the distribution function is given in terms of the distance between an ion of the ionic atmosphere and the central reference j ion it is termed a 'radial' distribution function. This reflects the spherical symmetry assumed in the Debye-Hückel theory.

$$n'_i = n_i \exp\left(-\frac{z_i e \psi_j}{kT}\right) \tag{10.22}$$

where ψ_j is the total potential at any distance, r, from the central j ion due to the ion itself and its ionic atmosphere (see Section 10.6.2).

The physical meaning of this equation can alternately be expressed in terms of probabilities, such that:

$$\frac{n'_i}{n_i} = \exp\left(-\frac{z_i e \psi_j}{kT}\right) \tag{10.98}$$

can be transformed into a probability relation.

$$g(r) = \frac{n_i'}{n_i} = \exp\left\{ -\frac{z_i e \psi_j}{kT} \right\} \tag{10.99}$$

$$= \exp\left(-\frac{\phi}{kT}\right) \tag{10.100}$$

where ϕ is the potential energy of the interaction between an ion of the ionic atmosphere and all other ions, and is a contribution to the total potential energy, Φ, and $g(r)$ is the radial distribution function.

This gives the probability of finding an ion of the ionic atmosphere in a given small region lying at a distance r from the central reference j ion at the origin. It is often written in the more explicit form $g^{(2)}(r_{12})$. Here the superscript represents the number of species involved, here the central reference ion, ion 1, and the ion in the ionic atmosphere, ion 2, and the bracket lists the distances on which the probability depends. Since only two ions are considered explicitly, then there can only be one distance to specify, i.e. the distance between the two ions, r_{12}. These probabilities are calculated for all conceivable r_{12} values of a central reference ion and an ion of the ionic atmosphere, and this allows the $g^{(2)}$ to be found for each value of r_{12}.

Physically, $g^{(2)}(r_{12})$ determines the probability of finding ion 2 in the ionic atmosphere in a given small region lying at a distance r_{12} from the central reference ion, ion 1.

Because there are cations and anions present in a solution of an electrolyte there are three distinct sorts of $g^{(2)}$. A radial distribution function can be defined in terms of:

- a cation in the ionic atmosphere and a cation as the central reference ion, written as $g^{(2)}{}_{++}(r_{12})$;

- an anion in the ionic atmosphere and an anion as the central reference ion, written as $g^{(2)}{}_{--}(r_{12})$;

- an anion in the ionic atmosphere and a cation as the central reference ion, written as $g^{(2)}{}_{+-}(r_{12})$. This is the same as the situation of a cation in the ionic atmosphere and an anion as the central reference ion. This is because a central reference ion can be part of the ionic atmosphere of any other ion.

In the Debye-Hückel model the interactions contributing to the potential energy, Φ, are long range coulombic interactions between the ions. However, because of the versatility of the computer simulation calculations, the statistical mechanical description of an electrolyte solution could include all conceivable electrostatic interactions such as the ion–ion, ion–dipole and dipole–dipole, dipole–quadrupole and quadrupole–quadrupole as well as induced dipole interactions between the ions, and between the ions and the solvent and between solvent molecules. The total potential energy, Φ, fed into the calculations which ultimately lead to $g^{(2)}(r_{12})$ could be made up of contributions such as these and those given in Sections 10.16.3–10.16.7 dealing with Gurney co-spheres. The computer simulation technique can therefore modify the Debye-Hückel model at will.

10.17.2 More complex distribution functions

In the Debye-Hückel theory the probability of finding one ion of the ionic atmosphere lying in a given small region at a distance, r, from the central reference ion was calculated. However, a more informative situation would be to consider the central reference ion and two ions in the ionic atmosphere. This would lead to a $g^{(3)}(r_{12}, r_{13}, r_{23})$ and would give the probability of finding ion 2 lying in a given small region distance r_{12} from the central reference ion, ion 1, while simultaneously finding ion 3 lying in a given small region at a distance r_{13} from the central ion. The geometry will fix r_{23} automatically, with these small regions being at a distance r_{23} from each other. As this is specifying three ions simultaneously, it is a more refined description than that given by $g^{(2)}(r_{12})$ where only one distance is specified.

If there are three ions being considered, i.e. a central reference ion and two ions in the ionic atmosphere, there are three distances to specify, r_{12}, r_{13}, r_{23}, and the distribution function is written as $g^{(3)}(r_{12}, r_{13}, r_{23})$.

However, this does not cover all possibilities. The possible charges on the three ions must also be considered. For three ions there are four distribution functions which can be found; $g_{+++}^{(3)}, g_{++-}^{(3)}, g_{+--}^{(3)}, g_{---}^{(3)}$.

- $g_{+++}^{(3)}$ corresponds to a cation as the central reference ion and two cations in the ionic atmosphere;

- $g_{++-}^{(3)}$ corresponds to a cation as the central reference ion and a cation and an anion in the ionic atmosphere;

- $g_{+--}^{(3)}$ corresponds to a cation as the central reference ion and two anions in the ionic atmosphere; and

- $g_{---}^{(3)}$ corresponds to an anion as the central reference ion and two anions in the ionic atmosphere.

Because the reference ion is part of the ionic atmosphere of another ion chosen to be the reference ion, there are only these four possibilities.

Distribution functions can be written for 4, 5, 6..... ions $g^{(4)}(r_{12}, r_{13}, r_{14}, r_{23}, r_{24}, r_{34})$ shortened to $g^{(4)}(r)$, and corresponding $g^{(5)}, g^{(6)}$.... Each of these distribution functions is calculated in a similar manner to the simple one of two ions, and can be described similarly. The more ions which can be selected, the more the situation will approximate to reality. If, say, 50 ions were selected and the probabilities found for each set of relative positions describing these ions, then this should be a more complete description of the behaviour of the ions under the effect of all the interactions between all the ions, than, for example, the Debye-Hückel theory where only two ions are selected. However, as the number of ions considered in the ionic atmosphere grows so does the complexity of the verbal description. Each of these distribution functions is given by an exponential energy term, $\exp(-\Phi/kT)$.

The power of this type of calculation is, in part, due to the ability of including all possible contributions to the potential energy from all conceivable interactions between the ions, and between the ions and the solvent and between solvent molecules. It is also open to the theoretician see how the distribution function alters as contributions to the total potential energy of the system are varied.

Since statistical mechanical calculations can, in principle, be extended to take account of the interactions between one ion and very many others there is no need to think in terms of the ionic atmospheres of the ions present. There also is no need to invoke the Poisson equation, since what is being calculated is a quantity which is defined by the total potential energy which is a sum of the contributing potential energies as a function of distance. These potential energies are specified by the model proposed. Since the Poisson equation is not used, the argument does not require the 'smearing out' of the ions of the ionic atmosphere into a continuous distribution of charge.

In principle, with Monte Carlo computer simulation techniques, statistical mechanical calculations can give a time shot of the average positions of all the ions in a solution, and if molecular dynamics techniques are used it can show how these average positions change with time.

10.17.3 Contributions to the total potential energy of the electrolyte solution

The expressions for any $g^{()}$ will derive from the quantity Φ, the total potential energy of all the interactions for the electrolyte solution. Contributions to Φ can be as simple or as complex as is wished, e.g. hard core terms or other short range repulsions, ion–dipole interactions, Gurney terms, cavity terms, ion–solvent interactions, solvent–solvent interactions and others where the molecular structure of the solvent is considered. All these can be included, or selections can be made, and all the resultant $g^{()}$ compared. In these cases, the total potential energy, Φ, is then taken to be a sum of pair-wise terms for interactions between the ions concerned:

$\Phi = \sum_{\text{all pairs}} \phi_{ab}$ where a and b are any two ions in the whole assembly of ions, or an ion and a solvent molecule, or two solvent molecules.

The fact that the interactions are summed pair-wise is an approximation which is inherent in this particular treatment of contributions to the total potential energy. The pair-wise approximation treats the interaction between two species as though it were independent of all the other species present. However crude though this may be, it is a standard approximation appearing in many aspects of theoretical chemistry.

The inclusion of many types of interactions is one of the respects in which this treatment is superior to the Debye-Hückel treatment. In the Debye-Hückel treatment the only interactions considered are the hard sphere repulsion and the long range coulombic interactions between the ions which are taken to be the only factor giving rise to non-ideality.

In the statistical mechanical treatments it has been shown that all conceivable interactions could, in principle, be included. This has become possible because of the considerable advances made in modern computing techniques and computer memory.

Furthermore, this treatment can be extended to cover situations other than the spherical symmetry of the Debye-Hückel approach, and of the simple treatments outlined above. To do so requires consideration of the distance, r, and the two angles θ and ϕ. Formally, this is achieved by presenting the argument in terms of vectors. If, instead of the simpler description in terms of spherical symmetry in which only the distances $r_{12}, r_{13}, r_{14}, \ldots$, are needed, the vector formulation were used then, in principle, all shapes of ions could be considered, e.g. typical charge separated ions such as $(CH_3)_3N^+(CH_2)_nCOO^-(aq)$ can be handled assuming cylindrical symmetry. This represents a vast improvement on the Debye-Hückel model.

Each $g^{0}(r)$ can also be more succinctly written as:

$$g^{()}(r) = \frac{\text{the actual probability}}{\text{the random probability}}$$

for all values of the various r. The random probability takes the same value irrespective of the position of the ions in the ionic atmosphere, since all positions are equally likely if there are no interactions between the ions. For the random probability $g^{()}(r)$ is, therefore, equal to unity.

The actual probability is for a situation where there are interactions between the particles involved. Since deviations from the random distribution of an assembly of ions are a result of electrostatic interactions between ions and all other interactions between ions, and interactions between ions and solvent and between solvent and solvent, it follows that each $g^{()}$ can be taken as giving some kind of measure of the non-ideality introduced by virtue of these interactions. The more ions which can be included explicitly in the calculations the more nearly the results will approximate to a quantitative evaluation of the non-ideality resulting from these interactions. However, there is a considerable increase in complexity as the number of ions is increased. If, for example, ten ions were taken to be in specified small regions then the expression for $g^{(10)}$(various r) becomes exceedingly complex. Some way of simplifying the procedure is necessary. Although modern computers should be able to handle large values of x in $g^{(x)}$, most of the theory and calculations have been worked out in terms of approximations which amount to taking, e.g.:

$$g^{(3)}(r_{12}, r_{13}, r_{23}) = g^{(2)}(r_{12}) g^{(2)}(r_{13}) g^{(2)}(r_{23})$$ (10.101)

which thus reduces the more complex $g^{(3)}$ to a product of three of the less complex $g^{(2)}$.

10.18 Application of distribution functions to the determination of activity coefficients due to Kirkwood; Yvon; Born and Green; and Bogolyubov

Since the distribution function, $g(r)$, is the statistical mechanical means of taking account of the interactions which lead to non-ideality it can eventually be linked to the non-ideal free energy and to the mean ionic activity coefficient. This is analogous to the Debye-Hückel procedure of equating the potential energy of interaction of an ion with its ionic atmosphere to the non-ideal free energy and to the mean ionic activity coefficient.

But remember: many types of interactions can be included in the potential energy which appears in the $g^{(2)}(r)$, but only one type – the long range potential energy of interaction between an ion and its ionic atmosphere – appears in the Debye-Hückel treatment. Also many ions can be included in the treatment since the approximation given in Section 10.17.3 means that a $g^{(35)}$ could be approximated to a product of many terms which involve $g^{(2)}$'s only, corresponding to the number of distinct pairs which can be counted for 35 ions. In this way the ionic atmosphere can be thought of as individual ions in this treatment, whereas in the Debye-Hückel treatment it **must** be thought of as a smeared out cloud of charge density.

The Debye-Hückel theory focuses on ψ_j and by using Poisson's equation of electrostatics finds an explicit equation for ψ_j, from which the potential, ψ_j^*, at the surface of the central j ion due to the ionic atmosphere is found. Thereafter the potential energy of interaction of the central reference j ion with its ionic atmosphere, $z_j e \psi_j^*$, leads directly to the non-ideal part of the free energy from which the Debye-Hückel expression for the mean ionic activity coefficient can be found. The treatment of Kirkwood *et al.* focuses attention directly onto the probability function from which statistical mechanical arguments, rather than the Debye-Hückel route, relate the probability function to the dependence of the mean ionic activity coefficient on concentration. This is done by obtaining relations between thermodynamic quantities and $g(r)$ using a new quantity G which describes the deviations of the actual probabilities from the random values.

10.18.1 Using distribution functions to formulate a new quantity G

The starting point is a $g^{(2)}$ describing the distribution function for two interacting species, A and B, i.e. $g_{AB}^{(2)}$, where $g_{AB}^{(2)}$ gives the probability of finding species B near species A in a given small region lying at a distance r from the species A which is at the origin.

If there were no interactions present: $g_{AB} = 1$. If there are interactions present, then $g_{AB} - 1$ will represent the deviation from the random value of unity.

A further quantity can now be defined. The integral G_{AB} is related to the quantity $g_{AB} - 1$ by the expression:

$$G_{AB} = \int_0^\infty 4\pi r^2 \{g_{AB}(r) - 1\}dr \qquad (10.102)$$

G_{AB} is related to the mean activity coefficient for the species AB.

In the solution there will be positive ions, negative ions and the solvent and there will be six radial distribution functions:

$$g_{++} \quad g_{+-}, \quad g_{--}, \quad g_{+solv}, \quad g_{-solv}, \quad g_{solv\,solv}$$

The probabilities involved are in turn:

probability of finding a +ve ion in a given small region at a given distance from a +ve ion;

probability of finding a −ve ion in a given small region at a given distance from a +ve ion;

probability of finding a −ve ion in a given small region at a given distance from a −ve ion;

probability of finding a solvent molecule in a given small region at a given distance from a +ve ion;

probability of finding a solvent molecule in a given small region at a given distance from a −ve ion;

probability of finding a solvent molecule in a given small region at a given distance from a solvent molecule.

Each of these gs will contain contributions to the total potential energy from all the interactions which are likely to be present in such an ionic solution, viz., all possible ion–ion interactions, all possible ion–solvent interactions and all possible solvent–solvent interactions.

Each of these radial distribution functions is related to one of the corresponding integrals:

$$G_{++} \quad G_{+-}, \quad G_{--}, \quad G_{+\text{solv}}, \quad G_{-\text{solv}}, \quad G_{\text{solv solv}}$$

Using these quantities and carrying through a very complex statistical mechanical argument results in the following equation for **symmetrical** electrolytes:

$$\left(\frac{\partial \log_e \gamma_\pm}{\partial c_+}\right)_{T,p} = \frac{\{1 - 2c_+ N[G_{+-} - G_{+\text{solv}}]\}}{2c_+^2 N[G_{+-} - G_{+\text{solv}}]} \tag{10.103}$$

or in terms of the electrolyte concentration, c:

$$\left(\frac{\partial \log_e \gamma_\pm}{\partial c}\right)_{T,p} = \frac{\{1 - 2cN[G_{+-} - G_{+\text{solv}}]\}}{2c^2 N[G_{+-} - G_{+\text{solv}}]} \tag{10.104}$$

or as:

$$\left(\frac{\partial \log_e \gamma_\pm}{\partial c}\right)_{T,p} = \frac{1}{2c^2 N[G_{+-} - G_{+\text{solv}}]} - \frac{1}{c} \tag{10.105}$$

where N is Avogadro's number.

What this result shows is that the mean activity coefficient can be calculated directly from the **two** radial distribution functions, g_{+-} and $g_{+\text{solv}}$, without any reference to $g_{++}, g_{--}, g_{-\text{solv}}$ and $g_{\text{solv solv}}$. This considerably simplifies the final calculations.

Important points to grasp:

- This treatment is a very powerful development in the theory of electrolyte solutions, and holds out the prospect that a full theory is forthcoming.

- The contributions to the total potential energy can be made as simple or as complex as wanted, and the treatment will enable all the shortcomings of the Debye-Hückel model to be removed.

- This treatment automatically includes all the interactions with the solvent. In this respect the treatment is vastly superior to any treatment based on the Debye-Hückel theory where the effect of the solvent has to be superimposed on the theory itself.

- The $g^{(2)}(r)$ depends on the concentration of the solution, as do any higher g.

- The statistical mechanical procedures do not need to invoke the concept of an ion pair as an aspect of non-ideality. The interactions which give rise to ion pairing are automatically included in the total potential energy. Consequently the treatment gives rise to a stoichiometric mean activity coefficient.

- What is perhaps important to emphasise is that the statistical mechanical derivation does not involve the use of the Poisson equation and hence the treatment does not involve the ambiguities involved in the combination of the Maxwell-Boltzmann distribution with the Poisson equation.

This statistical mechanical approach has been used in various forms by other workers, and further advances are still being made. It represents a very powerful theoretical tool, with a realistic expectation of providing considerable insight into the behaviour of electrolyte solutions at the microscopic level. A similar approach using distribution functions has also been used in modern developments of conductance theory (see Section 12.17).

10.19   A few examples of results from distribution functions

(A) The **value** of a $g(r)$ and its dependence on distance depends on whether it refers to a cation or anion. This dependence can be given qualitatively as:

- cation–cation: the $g_{++}^{(2)}$ values at given distances are expected to be less than unity because of repulsions between like charges, i.e. these probabilities will be less than the corresponding random values. This is equivalent to saying that the chance of finding a cation near another cation is less than the random value.

- anion–anion: the $g_{--}^{(2)}$ is also expected to be less than unity; this again is due to repulsions. Again this is equivalent to saying that the chance of finding an anion near another anion is less than the random value.

- cation–anion: in this case the $g_{+-}^{(2)}$ values at various distances are now expected to be greater than the random values. This is a consequence of the attractive interactions between the unlike charges. This is equivalent to saying that the chance of finding an ion of opposite charge to a given ion is greater than the random value.

Note well: these results can be expressed as the number of anions in a given volume around a central cation is greater than the number of cations in that volume. This is highly reminiscent of the Debye-Hückel description of the ionic atmosphere (Section 10.3): 'the chance of finding an ion of opposite charge to the central ion in any given region is greater than the chance of finding an ion of the same charge'. It is in this way that the two approaches have a common identity, even though the theoretical treatments are different, with the statistical mechanical approach independent of the concept of an ionic atmosphere. Also the contributions to Φ involved in the $g^{(2)}$s are vastly different in each approach.

(B) Recent work using this approach has applied the treatment to concentrated solutions. Here the above conclusions no longer hold. For moderate concentrations $g^{(2)}(r)$ decreases or increases smoothly (monotonically) with distance (see Figure 10.17). For more concentrated and highly concentrated solutions $g^{(2)}(r)$ can pass through one or more maxima and minima (see Figure 10.18). This is behaviour more typical of the fused crystal than of a solution, and fits in with the approach given in Section 10.22.

(C) Other recent work applies this treatment to ions of unequal size. In the Debye-Hückel theory it was assumed that the cations and anions are equal in size, and this is unlikely to be the case. In one study the anions were taken to have a given fixed size and the change in $g^{(2)}(r)$ values found as the size of the cations was altered:

- it was found that $g_{--}^{(2)}$ did not vary much as the cation size altered;

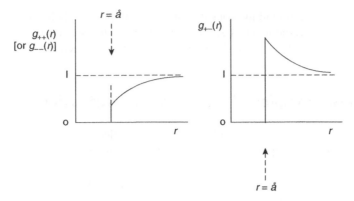

Figure 10.17   The radial distribution function $g(r)$ for moderate concentrations showing a monotonic increase, or monotonic decrease.

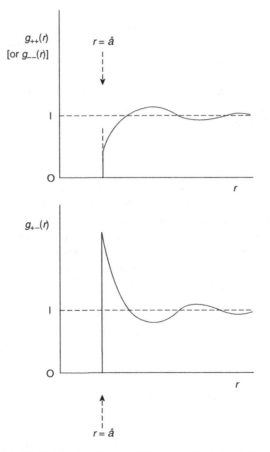

Figure 10.18   The radial distribution function $g(r)$ for concentrated solutions showing maxima and minima.

- but $g_{++}^{(2)}$ decreased if the size of the cation were decreased;

- $g_{+-}^{(2)}$ is increased if the size of the cation is decreased, and this is especially so for short distances.

These results are highly significant and mark a considerable improvement on the Debye-Hückel approach.

(D) As indicated in Section 10.17.3, the statistical mechanical approach can be used to describe the behaviour of a solution containing an electrolyte made up of at least one grossly non-spherically symmetrical ion. All shapes of ions can be considered and this represents a vast improvement on the Debye-Hückel theory.

10.20 'Born-Oppenheimer level' models

These are models which attempt to take into consideration the molecular structure of the solvent molecules. Previous treatments based on the Debye-Hückel model have approximated the solvent to a continuous medium characterised simply by the macroscopic relative permittivity. Qualitatively, the Born-Oppenheimer models could be taken as replacing the 'Gurney' terms and the 'cavity' terms of the previous model (Sections 10.16.3 to 10.16.6).

The model requires potential energies describing interactions between:

- ion and ion;

- ion and molecule of solvent;

- molecule of solvent and molecule of solvent.

These potential energies can be made as simple or as complex as desired.

The simplest description for the solvent–solvent interactions would be as hard spheres with a dipole at the centre, e.g. the charges could be large and close together. More complex models could be given in terms of quadrupoles or higher order multipoles, or as discrete charges specified as at given positions in the molecule, or with increasingly complex specifications.

Analysis of this model of the electrolyte solution takes the above quantities and by carrying out statistical mechanical calculations or Monte Carlo calculations leads eventually to a calculation of activity coefficients for a range of ionic strengths. The details of the procedures are complex. Results can be compared with experiment, or with the results of other theoretical treatments and indicate that this is a very fruitful line of enquiry. The extension to allow for the molecular nature of the solvent is a very necessary step in leading to an eventual full description of the behaviour of electrolyte solutions.

10.21 Lattice calculations for concentrated solutions

Some theoreticians believe that the best approach to electrolyte theory would be to work downwards from the fused electrolyte to the highly concentrated solution, to the concentrated solution. The fused electrolyte and the electrolyte solution have one feature in common, they both consist of ions, and so coulombic interactions are the dominant feature. In the crystal, a given cation is surrounded by nearest neighbour anions, and vice versa. In the solution a cation is surrounded by its ionic atmosphere and the charge on the cation is balanced by the negative charge of the atmosphere. Hopefully, this treatment will join up with the extension to the Debye-Hückel theory for dilute solutions given by the Guggenheim-Bjerrum approach and other more advanced approaches such as given in Sections 10.16 to 10.21.

This requires starting with the **known** short-range order of the crystal and the distribution function describing it. The distribution function for the fused salt is also known and is taken to be a good approximation to the distribution in solutions of very high concentration. The distribution function for concentrated solutions is also known from the statistical mechanical calculations outlined above , (see Section 10.19). As the concentration decreases this function should approach and should approximate to the Debye-Hückel, Guggenheim, Bjerrum approach. A considerable amount of work has been done here and this is a very promising and fruitful line of enquiry.

11
Conductance: The Ideal Case

The Debye-Hückel theory is a study of the **equilibrium** properties of electrolyte solutions, where departures from ideal behaviour are considered to be a result of coulombic interactions between ions in an **equilibrium** situation. It is for this reason that **equilibrium** statistical mechanics can be used to calculate an equilibrium Maxwell-Boltzmann distribution of ions.

When a current is passed through the solution, the current is carried by virtue of the ions moving under the influence of an externally applied field. This is now a **non-equilibrium** situation, and is an inherently more difficult situation to handle theoretically.

This chapter describes the behaviour of electrolyte solutions under **ideal** conditions. Chapter 12 describes the theoretical attempts made to cope with the problems introduced when non-ideality is considered.

This chapter contains a large number of worked examples. These are important as they illustrate many of the fundamental points made throughout the chapter and are an integral part of the development of the chapter. Carrying through these calculations is probably the best way to attain an understanding of what is in the chapter.

Aims

By the end of this chapter you should be able to:

- define conductance, G, resistance, R, conductivity, κ, resistivity, ρ, and molar conductivity, Λ;

- describe how the conductivity of a solution can be measured;

- explain the dependence of molar conductivity, Λ, on concentration for a strong electrolyte and for a weak electrolyte;

- demonstrate how Λ^0 and K_{assoc} can be found for a weak electrolyte;

An Introduction to Aqueous Electrolyte Solutions. By Margaret Robson Wright
© 2007 John Wiley & Sons Ltd   ISBN 978-0-470-84293-5 (cloth) ISBN 978-0-470-84294-2 (paper)

- show how conductance measurements can lead to solubility products;

- state what is meant by an ionic molar conductivity, λ_+, λ_-, and deduce their relation to Λ for symmetrical electrolytes and for unsymmetrical electrolytes;

- know what is meant by Kohlrausch's law, and be able to use it;

- explain what is meant by the transport number and the mobility of an ion;

- formulate the inter-relations between transport number, t, mobility, u, ionic conductivity, λ, and molar conductivity, Λ, for symmetrical and unsymmetrical electrolytes;

- describe the Hittorf and moving boundary methods for determining transport numbers;

- carry out calculations on aspects of the experimental study of conductance.

11.1 Aspects of physics relevant to the experimental study of conductance in solution

The fundamental concepts of the Debye-Hückel theory are also important in the description of the theoretical study of the passage of a current through a solution. Section 10.4 is therefore relevant here. There are other aspects of physics which are pertinent to the experimental study of conductance in solution. These will be discussed below.

11.1.1 Ohm's Law

Experiment shows that the current flowing in a solution is proportional to the potential difference acting across the solution, i.e.:

$$I \propto E \tag{11.1}$$

$$\therefore I = GE \tag{11.2}$$

where G is a constant of proportionality, the conductance of the solution, and is the quantity most relevant to this chapter. This is the most physically natural way of expressing the observations. However, historically, Ohm expressed the experimental results in terms of the potential difference set up across a solution being proportional to the current flowing, i.e.:

$$E \propto I \tag{11.3}$$

$$\therefore E = RI \tag{11.4}$$

where R is now the constant of proportionality, and is the resistance of the solution.

As will be seen later, this latter form has been that most commonly used by experimental workers, and it is best to be aware of the two forms in which Ohm's Law can be expressed.

Ohm's law usually holds for both solids and liquids, though it can break down for very high potential differences.

The resistance or conductance of a specimen depends on the temperature, on the chemical nature, the homogeneity, and on the size and shape of the specimen. For solutions, the resistance or conductance also depends on the number of ions present.

For a specimen uniform over its whole length, experiment shows that:

$$\text{The conductance, } G \propto \frac{A}{l} \tag{11.5}$$

$$\text{and so } \quad G = \kappa \frac{A}{l} \tag{11.6}$$

where A is the cross sectional area of the specimen and l is its length (see Figure 11.1), and κ is a constant of proportionality, called the conductivity.

The same physical facts can be stated in terms of the resistance of the solution.

$$\text{The resistance, } R \propto \frac{l}{A} \tag{11.7}$$

$$\text{and } \quad R = \rho \frac{l}{A} \tag{11.8}$$

where ρ is a constant of proportionality, called the resistivity.

$$\text{Hence} \quad \kappa = \frac{1}{\rho} = \rho^{-1} \quad \text{and } \quad \rho = \frac{1}{\kappa} = \kappa^{-1} \tag{11.9) and (11.10}$$

Ohm's law can thus be given in two ways:

$$\frac{I}{E} = G = \kappa \frac{A}{l} \quad \text{and } \quad \frac{E}{I} = R = \rho \frac{l}{A} \tag{11.11) and (11.12}$$

and:

$$\frac{I}{A} = \kappa \frac{E}{l} \tag{11.13}$$

where I/A is the current per unit area and E/l is the gradient of potential along the conductor, i.e. the electric field, X (see Section 10.4.1, and below).

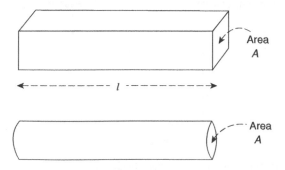

Figure 11.1 Geometry of a specimen showing length and cross section.

11.1.2   The electric field

In Section 10.4.6, the relation between the field and the potential was shown to be:

$$X = -\frac{\mathrm{d}\psi}{\mathrm{d}r} \qquad (11.14)$$

If the potential varies linearly with distance, then the field is uniform over that distance, but if the potential varies non-linearly with distance then the field over that distance is non-uniform.

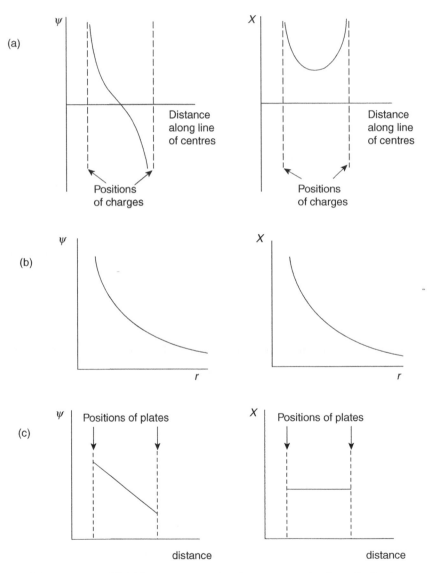

Figure 11.2 (a) Potential and field due to two opposite charges at a point along the line between them. **(b)** Potential and field in the ionic atmosphere. **(c)** Potential and field between two charged plates.

There are three situations which are relevant to electrolyte solutions:

- The field due to two charges at any point along the line of centres. In this case the potential depends non-linearly on distance and so the field between the two charges is non-uniform (see Figure 11.2(a)).

- The field due to an ion and its ionic atmosphere. The ionic atmosphere is spherically symmetrical and the potential drops off non-linearly with distance, and so the field is non-uniform (see Figure 11.2(b)).

- The field between the plates of a battery or between the electrodes in a conductance cell. In this case the electrodes have many charges on them, and these balance out to give an overall linear dependence of the potential on distance. In this case the field between the electrodes is uniform (see Figure 11.2 (c)).

In the case of conductance studies, the ions move under the influence of the electric field. Because the field is uniform this means that the ions migrate at a constant velocity. This will become relevant in the section on ionic mobilities (Section 11.17), and when discussing the relaxation and electrophoretic effects in the theories of conductance (see Sections 12.1, 12.2 and 12.4).

11.2 Experimental measurement of the conductivity of a solution

The discussion of the effect of passage of an electric current through a solution is usually carried through in terms of the conductivity, κ, where:

$$\kappa = G\left(\frac{l}{A}\right) = \frac{1}{R}\left(\frac{l}{A}\right)$$

(11.15)

The experimental measurement which determines κ is either the resistance of the solution or the conductance. Older textbooks will generally quote the raw data in terms of the resistance, but modern work uses high accuracy conductance bridges and will quote the conductance directly. The basic experimental measurement is in both cases the resistance of the solution, but the modern bridges are calibrated to read the conductance direct. It is well to be aware of this. The experimental set-up is based on a Wheatstone bridge type of apparatus.

When studying the conductivity of a solution it is essential that the concentration remains constant throughout the measurement. In Section 9.2 it was shown that passage of an electric current through a solution results in chemical reactions occurring at the electrodes, and these can result in changes in the concentration of the solution. The effect of these reactions must be nullified, and the procedure normally used is to keep reversing the direction of the current so that the reactions at the electrodes are reversed continually, giving no **net** chemical effect. Alternating current, such as 1000 Hz, is used and so in one half-cycle an electrode could be an anode and in the next half-cycle the electrode will be a cathode.

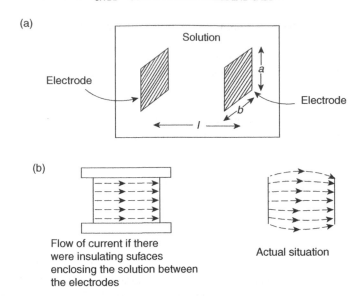

Figure 11.3 **(a)** Schematic diagram for an apparatus for determining the conductance of a solution. **(b)** Flow of current in a solution: idealised and actual.

The solution in question is contained in a conductance cell placed in a thermostat bath to maintain a constant temperature, and when a current is passed the experiment measures the resistance, or conductance, of the solution contained between the electrodes (see Figure 11.3(a)). The length, l, is the distance apart of the electrodes, and the cross section is the area, A, of the electrodes, i.e. $a \times b$ in the diagram. The quotient l/A is called the **cell constant**, and has units length^{-1}, generally cm^{-1}.

Knowing the resistance, R, or the conductance, G, the length, l, and the cross-sectional area, A, the conductivity, κ, can be found from either $\kappa = (1/R)\,(l/A)$, or $\kappa = G(l/A)$.

In practice, the situation is not as simple as this. The manner and direction of flow of the electric current (see Figure 11.3(b)) is such that the resistance measured is for a volume of the solution which is not the perfect cuboid given in Figure 11.3(a). This results in the length, l, which is the distance apart of the electrodes, being not exactly the length which should be used in Equation (11.15).

What is required is thus an effective $(l/A)_{\text{effective}}$, and there are two ways in which this can be found:

- **Experimentally**, by having a cell with very large electrodes which are very close together. This will result in the current flowing between the electrodes in an approximately horizontal manner so that the errors in assuming this are much less than the experimental error in measuring the resistance. In this case the dimensions will give the cell constant direct.

- **Theoretically**, by setting up the differential equations required to describe the flow of the current between the electrodes and from this obtaining a calculated cell constant.

Either of these ways would enable highly accurate values of the cell constant to be found. This cell constant can then be used to obtain highly accurate values of the conductivities of solutions of very accurately known concentrations, generally KCl(aq), and this data has been tabulated. This standard data can then be used to calibrate any other conductance cell.

The conductance cell to be used in experiments on determination of the conductivities of solutions of electrolytes is calibrated by measuring the resistance, or conductance, of solutions of known concentrations of the standard KCl(aq) solutions of known conductivity.

$$\kappa \quad = \quad \left(\frac{1}{R}\right) \quad \times \quad \left(\frac{l}{A}\right)_{\text{effective}} \tag{11.16}$$

is known for the	can be found for the	is thus directly
standard solution	standard solution	determined by experiment

By taking the average of the values found for $(l/A)_{\text{effective}}$, the effective cell constant can be found.

Once the cell constant has been found for a given conductance cell, this cell can be used, generally for some time, without further calibration being necessary. For highly accurate work the cell constant should be checked regularly in case distortion or alteration of the electrodes has occurred.

The conductivity of any solution of known concentration can now be found from the measured resistance, or conductance.

$$\kappa \quad = \quad \left(\frac{1}{R}\right) \quad \times \quad \left(\frac{l}{A}\right)_{\text{effective}} \tag{11.16}$$

can now be found	can be measured for	value found as above
for an unknown solution	the unknown solution	can be used

From this tabulated values of κ for various known concentrations of the electrolyte can be drawn up.

11.3 Corrections to the observed conductivity to account for the self ionisation of water

The self ionisation of water results in H_3O^+(aq) and OH^-(aq) always being present in any aqueous electrolyte solution. The measured resistance, or conductance, is therefore a composite quantity made up from the contribution made by the ions of the electrolyte and the H_3O^+(aq) and OH^-(aq). The contribution from the H_3O^+(aq) and OH^-(aq) can be obtained by measuring the resistance, or conductance, of the H_2O(l) used in preparing the electrolyte solutions under study. The conductivity of the water can then be found from this measured conductance and the cell constant. This conductivity must then be subtracted from the conductivities found for each electrolyte solution. Values below $5 \times 10^{-6}\,\text{S cm}^{-1}$ are acceptable. This may seem to be a small contribution, but with accurate work it could represent a significant contribution especially if low concentration solutions are being studied, e.g. solubilities of sparingly soluble salts (see Section 11.15), or solutions such as very weak acids or bases (see Section 11.14).

Worked problem 11.1

Question

(i) A conductance cell has electrodes which have an area $6.00\,cm^2$ and are a distance $1.50\,cm$ apart. Calculate the cell constant.

An experiment at $25°C$ gives the following results:

(ii) The conductance of the solvent water is $5.0 \times 10^{-6}\,S$. Calculate the conductivity of the water.

(iii) The conductance of $1.00 \times 10^{-4}\,mol\,dm^{-3}$ KCl solution is $6.42 \times 10^{-5}\,S$. Calculate the conductivity of this solution. What is the conductivity due to the $K^+(aq)$ and $Cl^-(aq)$ (aq) ions?

Answer

(i) The cell constant $= \dfrac{l}{A} = \dfrac{1.50\,cm}{6.00\,cm^2} = 0.250\,cm^{-1}$

(ii) The conductivity for the water as solvent is:

$$\kappa = G \times \text{cell constant} = 5.0 \times 10^{-6} \times 0.250\,S\,cm^{-1} = 1.25 \times 10^{-6}\,S\,cm^{-1}$$

(iii) The total observed conductivity of the solution of KCl(aq) is:

$$\kappa = G \times \text{cell constant} = 6.42 \times 10^{-5} \times 0.250\,S\,cm^{-1} = 1.605 \times 10^{-5}S\,cm^{-1}$$

This is made up from the contribution from the solvent water and the contribution from the $K^+(aq)$ and $Cl^-(aq)$ ions.
    The contribution from the $K^+(aq)$ and $Cl^-(aq)$ ions $= (1.605 - 0.125) \times 10^{-5}\,S\,cm^{-1}$
$$= 1.480 \times 10^{-5}\,S\,cm^{-1}$$
i.e. κ for KCl(aq) due to the $K^+(aq)$ and $Cl^-(aq)$ ions only $= 1.480 \times 10^{-5}\,S\,cm^{-1}$

11.4   Conductivities and molar conductivities: the ideal case

All that is discussed in Sections 11.4 onwards is approximate in so far as it ignores non-ideality due to ionic interactions. Chapter 12 discusses the effects which these ionic interactions have on the conductivity, i.e. they consider the effects of non-ideality.
    The current is carried through a solution by the movement of the ions of the solution. It is likely that the more ions there are present, the lower will be the resistance of the solution to the passage of the current and the greater will be the conductivity, κ. Unsurprisingly, experiment shows that κ varies considerably with concentration, and a further quantity, Λ the molar conductivity is defined.

In the **ideal** situation:

$$\kappa \propto c_{\text{stoich}} \tag{11.17}$$

$$\therefore \kappa = \Lambda c_{\text{stoich}} \tag{11.18}$$

where Λ is a constant of proportionality, called the molar conductivity.

$$\Lambda = \frac{\kappa}{c_{\text{stoich}}} \tag{11.19}$$

Note well as this is important: the concentration which appears in this expression is **always** a **stoichiometric** concentration. All **experimentally determined** molar conductivities are therefore based on stoichiometric concentrations. However, in the **theoretical descriptions** of conductance, the arguments are generally given in terms of **actual** concentrations. Being aware of this distinction is vital in the **theoretical description** of the effect of non-ideality on molar conductivity (see Chapter 12). It is also important in the **theoretical analysis** of the determination of the degree of ionisation of very weak acids, or the degree of protonation of very weak bases, using conductance measurements (see Section 11.14).

Care must always be taken when assigning units:

$$\kappa = \frac{1}{R}\left(\frac{l}{A}\right) = G\left(\frac{l}{A}\right) \text{ and } \Lambda = \frac{\kappa}{c_{\text{stoich}}} \qquad (11.15) \text{ and } (11.19)$$

where the units are:

R, the resistance: ohm, Ω

G, the conductance: siemens, S, and $S = \Omega^{-1}$,

l/A, the cell constant: length^{-1}, generally cm^{-1}

κ, the conductivity: ohm^{-1} cm^{-1} i.e. Ω^{-1} cm^{-1} or siemens cm^{-1} i.e. S cm^{-1}

c has units mol dm^{-3}

and so Λ must have units: $\dfrac{\text{ohm}^{-1}\text{cm}^{-1}}{\text{mol dm}^{-3}} = \dfrac{10^3\text{ohm}^{-1}\text{cm}^{-1}}{\text{mol cm}^{-3}} = 10^3\text{ohm}^{-1}\text{ cm}^2\text{mol}^{-1}$

$$= 10^3 \text{ S cm}^2\text{mol}^{-1}$$

Remember: $1\,\text{dm}^3 = 10^3\,\text{cm}^3$ so that $1/1\,\text{dm}^3 = 1/10^3\,\text{cm}^3$ giving $1\,\text{dm}^{-3} = 10^{-3}\text{cm}^{-3}$

It is important to get the units correct when calculating Λ from κ, i.e. remember the 10^3.

Worked problem 11.2

Question

A conductance cell has platinum electrodes which have a cross sectional area of 4.50 cm^2 and are 1.72 cm apart. Calculate the cell constant. A 0.500 mol dm^{-3} solution of a salt is placed in the conductance cell and the conductance found to be 0.0400 S. Calculate the conductivity, κ, and the molar conductivity, Λ, of the solution.

Answer

The cell constant $= \dfrac{l}{A} = \dfrac{1.72\,\text{cm}}{4.50\,\text{cm}^2} = 0.382\,\text{cm}^{-1}$

The conductance, G, is the reciprocal of the resistance, R.

$$\kappa = \frac{1}{R}\left(\frac{c}{n}\right) = G\left(\frac{l}{A}\right) = 0.04\,005 \times 0.382\,\text{cm}^{-1}$$

$$= 1.528 \times 10^{-2}\,\text{S cm}^{-1}$$
$$\text{or } \kappa = 1.528 \times 10^{-2}\,\text{ohm}^{-1}\text{cm}^{-1}$$

There is no information given as to the conductance of the solvent water, and so it will be ignored. This is likely to be justified as the concentration of the salt is high and so the contribution to the conductance and conductivity of the solution from the water is negligible.

$$\kappa = \Lambda c_{\text{stoich}}$$

$$\therefore \Lambda = \frac{\kappa}{c_{\text{stoich}}} = \frac{1.528 \times 10^{-2}\,\text{S cm}^{-1}}{0.500\,\text{mol dm}^{-3}} = \frac{1.528 \times 10^{-2}\,\text{S cm}^{-1}}{0.500 \times 10^{-3}\,\text{mol cm}^{-3}} = 30.56\,\text{S cm}^2\,\text{mol}^{-1}$$

$$\text{or } \Lambda = 30.56\,\text{ohm}^{-1}\,\text{cm}^2\text{mol}^{-1}$$

Take care: watch the units here. $1\,\text{dm}^3 = 1000\,\text{cm}^3$.

$$\frac{1}{1\,\text{dm}^3} = \frac{1}{1000\,\text{cm}^3}$$

$$\therefore 1\,\text{dm}^{-3} = 10^{-3}\,\text{cm}^{-3}$$

Worked problem 11.3

Question

A conductance cell containing $0.01000\,\text{mol dm}^{-3}$ KCl solution was found to have a resistance of 2573 ohm at 25°C, and a conductivity of $1.409 \times 10^{-3}\,\text{S cm}^{-1}$. The same cell when filled with a $0.2000\,\text{mol dm}^{-3}$ solution of CH_3COOH has a resistance of 5085 ohm. Calculate:

 (i) the cell constant;

 (ii) the conductivity, κ, of the CH_3COOH solution;

(iii) the molar conductivity, Λ, of the solution of KCl, and of the CH_3COOH solution;

(iv) comment on the results.

Assume that the solvent water makes a negligible contribution to the conductance of each solution.

Answer

 (i) The conductivity and the resistance of the KCl solution are given, from which the cell constant can be found.

$$\kappa = \frac{1}{R}\left(\frac{l}{A}\right)$$

$$\therefore \left(\frac{l}{A}\right) = \kappa \times R = 1.409 \times 10^{-3}\text{ohm}^{-1}\,\text{cm}^{-1} \times 2573\,\text{ohm} = 3.625\,\text{cm}^{-1}$$

(ii)

$$\kappa_{CH_3COOH} = \frac{1}{R}\left(\frac{l}{A}\right) = \frac{3.625\,cm^{-1}}{5085\,ohm} = 7.129 \times 10^{-4}\,ohm^{-1}\,cm^{-1} = 7.129 \times 10^{-4}\,S\,cm^{-1}$$

(iii)

$$\Lambda = \frac{\kappa}{c_{stoich}}$$

$$\Lambda_{CH_3COOH} = \frac{7.129 \times 10^{-4}\,ohm^{-1}\,cm^{-1}}{0.2000\,mol\,dm^{-3}} = \frac{7.129 \times 10^{-4}\,S\,cm^{-1}}{0.2000 \times 10^{-3}\,mol\,cm^{-3}} = 3.56\,S\,cm^2\,mol^{-1}$$

$$\Lambda_{KCl} = \frac{1.409 \times 10^{-3}\,ohm^{-1}\,cm^{-1}}{0.01000\,mol\,dm^{-3}} = \frac{1.409 \times 10^{-3}\,S\,cm^{-1}}{0.01000 \times 10^{-3}\,mol\,cm^{-3}} = 140.9\,S\,cm^2\,mol^{-1}$$

(iv) The molar conductivity of the CH_3COOH solution is very much lower than that for the KCl solution, despite the factor of 20 in the concentration. This is a direct reflection of the fact that the KCl is known to be a strong electrolyte, i.e. is fully ionised, whereas the CH_3COOH is a weak electrolyte, i.e. is only slightly ionised (see Sections 11.5 and 11.7).

11.5 The physical significance of the molar conductivity, Λ

The molar conductivity is a very useful quantity when comparing the properties of electrolytes dissolved in water. In particular, it is a quantity which is used to determine whether the **actual** concentration of ions in solution is equal to, or not equal to, the **stoichiometric** concentration of the ions.

If the solution is considered to be **ideal** there will be **no interactions between the ions**. The current is carried by the ions of the solution, and so it would be expected that the current would be strictly proportional to the number of ions per unit volume, i.e. the concentration, in solution. Making the assumption that the solution is **ideal**, then the conductivity, κ, would also be strictly proportional to the number of ions per unit volume, i.e. to the actual concentration of the ions present.

Now, whether or not the actual concentration of ions present in solution is equal to the stoichiometric concentration of the ions depends critically on the type of electrolyte under consideration.

- If the electrolyte is a strong electrolyte, then it is considered to be fully ionised in solution, and:

   the actual concentration of ions = the stoichiometric concentration of the ions
and so for the strong electrolyte the conductivity, κ, would be expected to be strictly proportional to the stoichiometric concentration of the ions, i.e. κ/c_{stoich} is expected to be constant, **provided non-ideality is ignored**.
Since:

$$\Lambda = \frac{\kappa}{c_{stoich}} \tag{11.19}$$

then the observed molar conductivity, Λ, would be expected to be independent of concentration. It will be shown later (Section 11.6 and Chapter 12) that this is not what is found experimentally, and this is one of the main pieces of evidence for the **non-ideality** of aqueous solutions of strong electrolytes.

- If the electrolyte is a weak electrolyte, then it is not fully dissociated in solution and an equilibrium exists between the un-ionised form and the ions:

$$\text{un-ionised form(aq)} \rightleftharpoons \text{ions(aq)}$$

and so:

the actual concentration of ions $\neq$ the stoichiometric concentration of the ions.

In particular, for weak electrolytes the actual concentration of ions only becomes equal to the stoichiometric concentration of the ions in the limiting case of very dilute solutions when the fraction ionised tends to unity, but see Section 4.6 and Section 11.10.

For such electrolytes, since $c_{actual} = \alpha c_{stoich}$, the molar conductivity will depend on the stoichiometric concentration in a precise manner reflecting the fraction ionised.

Because weak electrolytes are often only slightly ionised at low concentrations it is possible to treat the solutions as ideal. However, at higher concentrations corrections will have to be made for non-ideality.

11.6 Dependence of molar conductivity on concentration for a strong electrolyte: the ideal case

For the **ideal** case the conductivity, κ, would be expected to be strictly proportional to the concentration of ions actually present (see Figure 11.4(a)): i.e.

$$\kappa \propto c_{actual} \tag{11.20}$$

The molar conductivity has been defined as:

$$\Lambda = \frac{\kappa}{c_{stoich}} \tag{11.19}$$

and so:

$$\Lambda \propto \frac{c_{actual}}{c_{stoich}} \tag{11.21}$$

But for a strong electrolyte:

$$c_{actual} = c_{stoich} \tag{11.22}$$

$$\therefore \Lambda \propto \frac{c_{stoich}}{c_{stoich}} = \text{constant} \tag{11.23}$$

Therefore, for a strong electrolyte **under ideal conditions** the molar conductance is expected to be independent of concentration (see Figure 11.4(b)). However, when measurements are made on strong electrolyte solutions it is clear that Λ depends on the concentration (see Figure 11.4(c)), with deviations being greater the larger the concentration of the solution. This is interpreted as indicating that such solutions cannot be considered as ideal, and that the variation in Λ with concentration reflects increasing non-ideality as the concentration increases. This non-ideality results from electrostatic interactions between the oppositely charged ions. This will be considered in Chapter 12.

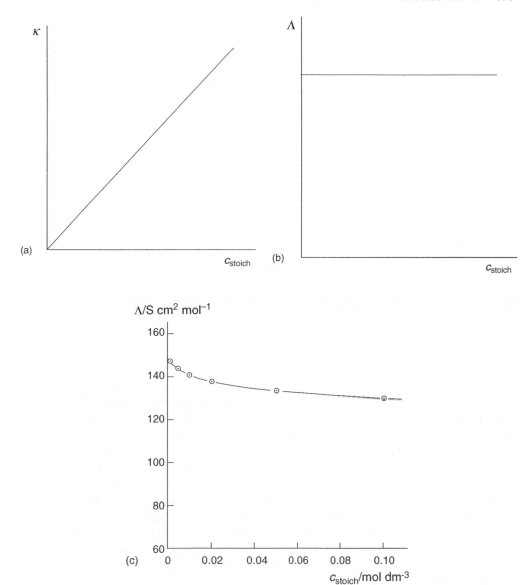

Figure 11.4 **(a)** Ideal case: plot of κ vs. c_{stoich} for a strong electrolyte. **(b)** Idealised behaviour of Λ for a strong electrolyte. **(c)** Experimental plot of Λ vs. c_{stoich} for a strong electrolyte.

11.7 Dependence of molar conductivity on concentration for a weak electrolyte: the ideal case

As with the strong electrolyte, for the **ideal** case, the conductivity, κ, would be expected to be strictly proportional to the concentration of ions **actually** present. For weak electrolytes which are known not to be fully dissociated, experimental data show that κ is not strictly proportional

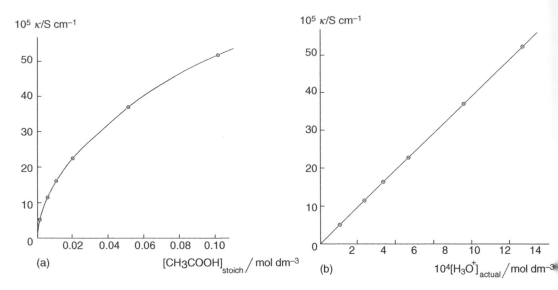

Figure 11.5 **(a)** Experimental plot of κ vs. c_{stoich} for a weak electrolyte. **(b)** Idealised behaviour of κ vs. $[H_3O^+]_{actual}$ for a weak electrolyte.

(i.e. the graph of κ vs. c is curved) to the stoichiometric concentration, see Figure 11.5(a) which shows the dependence of κ on c_{stoich} for ethanoic acid. This should be compared with the proportionality (i.e. linearity) found for the strong electrolyte, KCl(aq) (see Figure 11.4(a)). However, when the **actual** concentrations of H_3O^+(aq) and CH_3COO^-(aq) are calculated from the known K_a and stoichiometric concentrations, and a graph of κ versus $[H_3O^+]_{actual}$ is drawn this is found to be **linear** (see Figure 11.5(b)). This indicates that κ is almost exactly proportional to the **actual** concentration of ions, as expected. A further graph can be drawn, viz, the fraction of the CH_3COOH(aq) ionised, α, versus $[CH_3COOH]_{stoich}$, (see Figure 11.6(a)). The shape of this graph resembles exactly that of Λ vs. c_{stoich}, (see Figure 11.6(b)). For a weak electrolyte it is only in very, very dilute solutions that almost all of the molecules are dissociated to ions, and it is only in such solutions that the actual concentration of ions equals the stoichiometric concentration; see Section 4.1.2, but see also Section 4.6 which deals with very weak acids and bases.

In solutions of weak electrolytes, the concentration of ions is often sufficiently low for the solution to approximate to ideality. However, with moderately weak electrolytes, it becomes possible to have higher concentrations of ions present and deviations from the behaviour described above become apparent. For such electrolytes corrections for non-ideality must be made (see Chapter 12).

Solutions of weak electrolytes are not confined to acids and bases, e.g. solutions of the 1–1 electrolyte, tetrabutylammonium iodide, $N(C_4H_9)_4^+I^-$(aq), the 2–2 electrolyte, $MgSO_4$(aq), and the 3–3 lanthanum hexacyanoferrate(III), $La^{3+}Fe(CN)_6^{3-}$(aq), all form ion pairs in aqueous solution, and so are weak electrolytes.

A similar argument to that carried out for the strong electrolyte can be carried out for the weak electrolyte, but the conclusions are different.

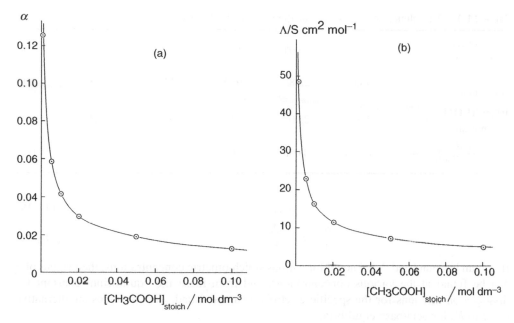

Figure 11.6   (a) Dependence of α on c_{stoich} for a weak electrolyte. **(b)** Dependence of Λ on c_{stoich} for a weak electrolyte.

The conductivity, κ is again proportional to c_{actual} and:

$$\Lambda = \frac{\kappa}{c_{stoich}} \qquad (11.19)$$

and so:

$$\Lambda \propto \frac{c_{actual}}{c_{stoich}} \qquad (11.21)$$

But

$$c_{actual} = \alpha c_{stoich} \qquad (11.24)$$

$$\therefore \Lambda \propto \frac{\alpha c_{stoich}}{c_{stoich}} \propto \alpha \qquad (11.25)$$

Now α is not a constant over a range of concentration for a weak electrolyte (see Table 11.1 and Figure 11.6(a)). In particular, the fraction ionised increases dramatically as the concentration decreases and tends to zero. Hence for a weak electrolyte Λ is expected to vary with the stoichiometric concentration, and in particular to increase dramatically with decrease in concentration to a limiting value when $c_{stoich} \rightarrow 0$ (see Figure 11.6(b)).

Note: the similarity in shape of the graph of α vs. c_{stoich} and Λ vs. c_{stoich} for ethanoic acid.

The limiting value to which Λ approaches as $c_{stoich} \rightarrow 0$ is called the limiting molar conductivity, Λ^0 and:

when

$$\alpha = 1, \quad \Lambda = \Lambda^0 \qquad (11.26 \text{ and } 11.27)$$

Table 11.1   Dependence of α on c_{stoich} for the weak acid CH_3COOH

$\dfrac{[CH_3COOH]_{stoich}}{mol\ dm^{-3}}$	0.001	0.005	0.010	0.020	0.050	0.100
$\dfrac{\Lambda}{S\ cm^2 mol^{-1}}$	48.6	22.8	16.2	11.5	7.36	5.20
$\dfrac{10^4 \times [H_3O^+]_{actual}}{mol\ dm^{-3}}$	1.25	2.90	4.15	5.90	9.40	13.3
$\alpha = \dfrac{[H_3O^+]_{actual}}{[CH_3COOH]_{stoich}}$	0.125	0.058	0.042	0.029	0.019	0.013

Since
$$\alpha \propto \Lambda \tag{11.25}$$

$$\frac{\alpha}{1} = \alpha = \frac{\Lambda}{\Lambda^0} \tag{11.28}$$

If Λ^0 can be found and Λ is known for various stoichiometric concentrations, then values of α can be found at the various concentrations, and from these the equilibrium constant for dissociation into ions for the specific electrolyte can be found. This provides an alternative route to Ks for acid/base equilibria.

However, this argument has to be modified when dealing with solutions of very weak acids and bases where the fraction ionised can never reach unity (see Section 11.10).

11.8   Determination of Λ^0

A look at the graph of Λ vs. c_{stoich} (Figure 11.6(b)) shows just how impossible it is to get an accurate extrapolation to obtain Λ^0 at $c_{stoich} = 0$ for weak electrolytes, and so an alternative route to Λ^0 must be found. This is described in Section 11.13 and in Worked Problems 11.7 to 11.9.

Worked problem 11.4

Question

At $25°C$, the molar conductivities, Λ, of $HCOOH$ and CH_3CH_2COOH at a given known concentration are 4.0 and 2.5 S cm^2 mol^{-1} respectively, and the corresponding limiting molar conductivities, Λ^0, are 404.4 and 385.6 S cm^2 mol^{-1} respectively. Explain how these could be used to determine the fraction of $HCOOH$ and the fraction of CH_3CH_2COOH ionised at the given concentration. In what direction would the molar conductivities of each acid move if the concentration were increased, and what would happen to the fractions ionised with increase in concentration?

Answer

By definition $\kappa = \Lambda c_{stoich}$

$$\therefore \Lambda = \frac{\kappa}{c_{stoich}}$$

But the conductivity is proportional to the actual concentration of ions present.

$$\therefore \Lambda \propto \frac{c_{actual}}{c_{stoich}}$$

where

$$c_{actual} = \alpha c_{stoich}$$

$$\therefore \Lambda \propto \frac{\alpha c_{actual}}{c_{stoich}} \propto \alpha$$

When $\alpha = 1$ then the acid is fully ionised and this will happen in very dilute solution when the molar conductivity will take its limiting value, Λ^0, i.e. $\Lambda = \Lambda^0$.

$$\therefore \alpha = \frac{\Lambda}{\Lambda^0}$$

	HCOOH	CH_3CH_2COOH
$\Lambda / S\,cm^2\,mol^{-1}$	4.0	2.5
$\Lambda^0 / S\,cm^2\,mol^{-1}$	404.4	385.6
$\alpha = \dfrac{\Lambda}{\Lambda^0}$	0.0099	0.0065

If the concentration of the acids were increased, then the number of ions present per unit volume would increase and the conductivity of the solutions would increase **but** the molar conductivity would decrease. The fraction ionised would also decrease.

Worked problem 11.5

Question

At 25°C, the molar conductivity of a 4.00×10^{-3} mol dm^{-3} solution of $Cl_2CHCOOH$ is 359.0 S cm^2 mol^{-1}. If Λ^0, the limiting molar conductivity, for $Cl_2CHCOOH$ is 385.0 S cm^2 mol^{-1}, what is the fraction ionised at this concentration? From this calculate the equilibrium concentrations of H_3O^+(aq), Cl_2CHCOO^-(aq) and $Cl_2CHCOOH$(aq), and from these calculate the ionisation constant for $Cl_2CHCOOH$(aq).

Answer

Following the argument given in Section 11.7 and Worked Problem 11.4:

$$\alpha = \frac{\Lambda}{\Lambda^0} = \frac{359.0\,S\,cm^2\,mol^{-1}}{385.0\,S\,cm^2\,mol^{-1}} = 0.932$$

This is a much higher value than the values of α found for the two acids in Worked Problem 11.4, and reflects the larger value of the ionisation constant for $Cl_2CHCOOH$(aq).

$$Cl_2CHCOOH(aq) \rightleftharpoons H_3O^+(aq) + Cl_2CHCOO^-(aq)$$

$$K_a = \left(\frac{[H_3O^+]_{actual}[Cl_2CHCOO^-]_{actual}}{[Cl_2CHCOOH]_{actual}} \right)_{eq}$$

Following the arguments given in Section 4.1.

$$[Cl_2CHCOOH]_{stoich} = [Cl_2CHCOOH]_{actual} + [Cl_2CHCOO^-]_{actual}$$

$\alpha = 0.932$ and so the acid is 93% ionised and so this full expression **must not be approximated**.

Since there is only one source of the $H_3O^+(aq)$ and $Cl_2CHCOO^-(aq)$ ions and ignoring the self ionisation of water, then:

$$[H_3O^+]_{actual} = [Cl_2CHCOO^-]_{actual}$$

$$\therefore K_a = \left(\frac{[H_3O^+]^2_{actual}}{[Cl_2CHCOOH]_{actual}}\right)_{eq} = \left(\frac{[H_3O^+]^2_{actual}}{[Cl_2CHCOOH]_{actual}}\right)$$

$$[Cl_2CHCOOH]_{stoich} = 4.00 \times 10^{-3} \, mol \, dm^{-3}$$

$$\therefore [Cl_2CHCOO^-]_{actual} = [H_3O^+]_{actual} = \alpha[Cl_2CHCOOH]_{stoich} = 0.932 \times 4.00 \times 10^{-3} \, mol \, dm^{-3}$$
$$= 3.73 \times 10^{-3} \, mol \, dm^{-3}.$$

$$\therefore [Cl_2CHCOOH]_{actual} = [Cl_2CHCOOH]_{stoich} - [Cl_2CHCOO^-]_{actual}$$
$$= (4.00 \times 10^{-3} - 3.73 \times 10^{-3}) \, mol \, dm^{-3}$$
$$= 2.7 \times 10^{-4} \, mol \, dm^{-3}$$

$$\therefore K_a = \left(\frac{[H_3O^+]^2_{actual}}{[Cl_2CHCOOH]_{actual}}\right)_{eq} = \left(\frac{[H_3O^+]^2_{actual}}{[Cl_2CHCOOH]_{actual}}\right) = \frac{(3.73 \times 10^{-3})^2 \, mol^2 \, dm^{-6}}{2.7 \times 10^{-4} \, mol \, dm^{-3}}$$
$$= 5.15 \times 10^{-2} \, mol \, dm^{-3}$$

11.9 Simultaneous determination of K and Λ^0

A weak base B is protonated in the following equilibrium:

$$B(aq) + H_2O(l) \rightleftharpoons BH^+(aq) + OH^-(aq)$$

$$K_b = \frac{[BH^+]_{actual}[OH^-]_{actual}}{[B]_{actual}} = \frac{\alpha^2 c^2_{stoich}}{(1-\alpha)c_{stoich}} = \frac{\alpha^2 c_{stoich}}{(1-\alpha)} \qquad (11.29)$$

But

$$\alpha = \frac{\Lambda}{\Lambda^0} \qquad (11.28)$$

$$\therefore K_b = \frac{\left(\frac{\Lambda}{\Lambda^0}\right)^2 c_{stoich}}{\left(1 - \frac{\Lambda}{\Lambda^0}\right)} \qquad (11.30)$$

$$\therefore K_b - K_b\left(\frac{\Lambda}{\Lambda^0}\right) = \left(\frac{\Lambda}{\Lambda^0}\right)^2 c_{stoich} \qquad (11.31)$$

$$\therefore 1 - \frac{\Lambda}{\Lambda^0} = \left(\frac{\Lambda}{\Lambda^0}\right)^2 \frac{c_{stoich}}{K_b} \qquad (11.32)$$

Divide throughout by Λ

$$\therefore \frac{1}{\Lambda} = \frac{1}{\Lambda^0} + \frac{\Lambda c_{stoich}}{(\Lambda^0)^2 K_b} \qquad (11.33)$$

A graph of $1/\Lambda$ vs. Λc_{stoich} should be linear with intercept $= 1/\Lambda^0$ and slope $= 1/(\Lambda^0)^2 K_b$ from which, knowing Λ^0, K_b can be found.

Worked problem 11.6

Question

Use the following data at 25°C for the ionisation of butanoic acid to find Λ^0 and K_a

$10^3 \times \dfrac{c_{stoich}}{\text{mol dm}^{-3}}$	1.00	2.50	5.00	7.50	10.00
$\dfrac{\Lambda}{\text{S cm}^2\text{mol}^{-1}}$	44.5	28.8	20.6	16.8	14.6

Answer

$\dfrac{10^3 \times \Lambda c_{stoich}}{\text{S cm}^2\text{dm}^{-3}}$	44.5	72.0	103	126	146
$\dfrac{1/\Lambda}{\text{S}^{-1}\text{cm}^{-2}\text{mol}^1}$	0.0225	0.0347	0.0485	0.0595	0.0685

The graph of $1/\Lambda$ against Λc_{stoich} is linear (see Figure 11.7).

From the slope:

$$\frac{1}{(\Lambda^0)^2 K_a} = 0.454\,\text{S}^{-2}\,\text{cm}^{-4}\,\text{mol dm}^3$$

$$\therefore (\Lambda^0)^2 K_a = 2.20\,\text{S}^2\,\text{cm}^4\,\text{mol}^{-1}\,\text{dm}^{-3}$$

The intercept gives:

$$\frac{1}{\Lambda^0} = 0.0023\,\text{S}^{-1}\,\text{cm}^{-2}\,\text{mol}$$

$$\therefore \Lambda^0 = 435\,\text{S cm}^2\,\text{mol}^{-1}$$

which leads to:

$$K_a = \frac{2.20\,\text{S}^2\,\text{cm}^4\,\text{mol}^{-1}\,\text{dm}^{-3}}{(435)^2\,\text{S}^2\,\text{cm}^4\,\text{mol}^{-2}} = 1.2 \times 10^{-5}\,\text{mol dm}^{-2}$$

$$pK_a = 4.93$$

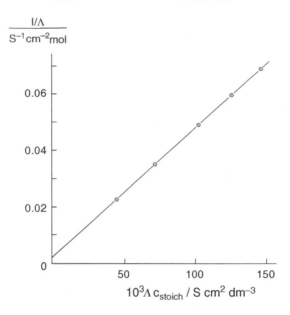

Figure 11.7   Graph of $1/\Lambda$ vs. Λc_{stoich}

Note, however: it is very difficult with this graph to determine the intercept accurately. It is very small and the 3 is decidedly unreliable. If the intercept were taken to be 0.0021, then $\Lambda^0 = 476\,S\ cm^2 mol^{-1}$ but if it is taken to be 0.0025 then $\Lambda^0 = 400\,S\ cm^2\ mol^{-1}$. It is quite clear that when the intercept is so small then the accuracy will be low, and the resulting quantities derived from it will be equally inaccurate and unreliable. This is a problem which one must be aware of when interpreting graphical data. This worked problem highlights the value of determining Λ^0 from Kohlrausch's Law (see Sections 11.11 to 11.13 below). Extrapolation to zero concentration of data for strong electrolytes is highly accurate and by suitable use of Kohlrausch's Law highly accurate values of Λ^0 for weak electrolytes can be obtained (see Worked Problems 11.8 and 11.9).

Using a value of $\Lambda^0 = 382\,S\ cm^2\ mol^{-1}$ calculated from Λ^0 values for $C_3H_7COONa(aq)$, $NaCl(aq)$ and $HCl(aq)$, and using the accurate value for the slope, $(\Lambda^0)^2 K_a = 2.20\,S^2\ cm^4$ $mol^{-1}\ dm^{-3}$ gives a clear-cut value for K_a:

$$K_a = \frac{2.20\,S^2\ cm^4\ mol^{-1}\ dm^{-3}}{(382)^2 S^2\ cm^4\ mol^{-2}} = 1.51 \times 10^{-5}\,mol\ dm^{-3}$$

$$\therefore pK_a = 4.82$$

Also note: $\Lambda^0 = 382\,S\ cm^2\ mol^{-1}$ corresponds to $1/\Lambda^0 = 0.0026\,S^{-1}\ cm^{-2}\ dm^{-3}$ which is within the accuracy of the intercept.

11.10 Problems when an acid or base is so weak that it is never 100% ionised, even in very, very dilute solution

A second read of Section 4.6.1 raises a potentially disturbing question. Here it was shown that the maximum % ionised depends on pH, and also for weak acids with $pK_a > 5$ it becomes impossible to approach 100% dissociation, i.e. α can never approach 1. A similar conclusion holds for weak bases with $pK_b > 5$.

Conductance methods have been used extensively for determining pK_as and pK_bs for weak acids and weak bases. These are based on determination of values of α over a range of concentrations found from measurements of molar conductivities which have been corrected for the contribution from the solvent. The theoretical basis of the equation:

$$\alpha = \Lambda/\Lambda^0 \text{ assumes that } \Lambda \to \Lambda^0 \text{ as } \alpha \to 1 (\text{see Section 11.7}).$$

If α cannot approach unity for very weak acids and bases, then this calls into question the applicability of the use of the ratio Λ/Λ^0 for determining α.

There are many weak acids, particularly phenols, which have pK_as > 5, and some very weak bases, especially substituted anilines which have pK_bs > 5 or whose protonated forms have pK_as < 9. For these substances 100% ionisation or protonation cannot be achieved and the question can be asked: how does this affect the possibility of determining α from conductance studies?

This may seem, at first sight, to be a contrived question in so far as experimentally the conductivities for such solutions would be very low and close to the value for the solvent, water. However, with modern equipment it is possible to obtain highly accurate measurements of the conductivities for both the water and the solutions. The same experimental situation would exist for determinations of solubility products where the conductivity due to the ions will be low. For such determinations there will be no need to correct for non-ideality since the ionic concentrations will be so low.

Analysis of the situation for such acids and bases will be given (Section 11.14), once individual molar ionic conductivities have been discussed (see Section 11.11 below).

11.11 Contributions to the conductivity of an electrolyte solution from the cation and the anion of the electrolyte

The ions in solution are made up of the cations and the anions of the electrolyte, and these move in opposite directions when the current is passed (see Sections 9.1 and 9.2 and Worked Problem 9.1). It is the movement or migration of these charged particles which carries the current through the solution with the cations moving to the cathode and the anions to the anode. This holds even in the case where AC current is used and the cathode and anode keep reversing sign. The total current carried is made up of the contribution from the cations and the contribution from the anions. Likewise the conductivity of the electrolyte solution is made up of contributions from the cation and the anion, i.e.:

$$\kappa \qquad = \qquad \kappa_+ \qquad + \qquad \kappa_- \qquad\qquad (11.34)$$

$$\downarrow \qquad\qquad\qquad \downarrow \qquad\qquad\qquad \downarrow$$

the total conductivity   the contribution   the contribution
from the cation   from the anion

Individual ionic molar conductivities can be also be defined as:

$$\kappa_+ = \lambda_+ c_{+\text{stoich}} \qquad (11.35)$$

$$\kappa_- = \lambda_- c_{-\text{stoich}} \qquad (11.36)$$

$$\therefore \kappa = \kappa_+ + \kappa_- = \lambda_+ c_{+\text{stoich}} + \lambda_- c_{-\text{stoich}} \qquad (11.37)$$

$$\lambda_+ = \frac{\kappa_+}{c_{+\text{stoich}}} \qquad (11.38)$$

$$\lambda_- = \frac{\kappa_-}{c_{-\text{stoich}}} \qquad (11.39)$$

where λ_+ is the molar conductivity of the cation; called an ionic molar conductivity,

$c_{+\text{stoich}}$ is the stoichiometric concentration of the cation which is calculated from the stoichiometric concentration of the electrolyte **assuming complete ionisation**

where λ_- is the molar conductivity of the anion; called an ionic molar conductivity,

$c_{-\text{stoich}}$ is the stoichiometric concentration of the anion which is calculated from the stoichiometric concentration of the electrolyte **assuming complete ionisation**

Each ion present in the solution thus makes a contribution to the total conductivity of the solution. Likewise the total molar conductivity of the solution will be made up of contributions from the cations and anions present in the solution. But care must be taken here; the relation is not a simple additive one, but is one which reflects the stoichiometric formula of the electrolyte, and, in particular, whether the electrolyte is symmetrical or unsymmetrical.

11.12 Contributions to the molar conductivity from the individual ions

• The symmetrical electrolyte

The symmetrical electrolyte AB, e.g. KCl(aq), $MgSO_4$(aq), ionises giving A^{z+}(aq) and B^{z-}(aq). For complete ionisation, 1 mol of AB gives 1 mol of A^{z+}(aq) and 1 mol of B^{z-}(aq):

$$c_{+\text{stoich}} = [A^{z+}]_{\text{stoich}} = [AB]_{\text{stoich}} = c_{\text{stoich}} \text{ for the electrolyte}$$

$$c_{-\text{stoich}} = [B^{z-}]_{\text{stoich}} = [AB]_{\text{stoich}} = c_{\text{stoich}} \text{ for the electrolyte}$$

Each ion will make a contribution to the total conductivity of the solution:

$$\kappa = \kappa_+ + \kappa_- \qquad (11.34)$$

$$\therefore \kappa = \lambda_+ c_{+\text{stoich}} + \lambda_- c_{-\text{stoich}} \qquad (11.37)$$

$$= \lambda_+ c_{\text{stoich}} + \lambda_- c_{\text{stoich}} \qquad (11.40)$$

Note well: the concentration appearing in Equations (11.35) to (11.39) relates to the ionic concentrations, while that in Equation (11.40) is the stoichiometric concentration of the electrolyte. It is important to distinguish between these, especially for unsymmetrical electrolytes, see below.

But $$\kappa = \Lambda c_{\text{stoich}}$$

$$\therefore \Lambda = \lambda_+ + \lambda_- \tag{11.41}$$

• The unsymmetrical electrolyte

The unsymmetrical electrolyte $A_x B_y$, e.g. $K_2SO_4(aq)$, $MgCl_2(aq)$, $Fe_2(SO_4)_3$, ionises giving $A^{y+}(aq)$ and $B^{x-}(aq)$. For complete ionisation, 1 mol of $A_x B_y$ gives x mol of $A^{y+}(aq)$ and y mol of $B^{x-}(aq)$.

Each ion will make a contribution to the total conductivity of the solution:

$$\kappa = \kappa_+ + \kappa_- \tag{11.34}$$

$$\therefore \kappa = \lambda_+ c_{+\text{stoich}} + \lambda_- c_{-\text{stoich}} \tag{11.37}$$

$$= x\lambda_+ c_{\text{stoich}} + y\lambda_- c_{\text{stoich}} \tag{11.42}$$

But $\kappa = \Lambda c_{\text{stoich}}$ $\tag{11.18}$

$$\therefore \Lambda = x\lambda_+ + y\lambda_- \tag{11.43}$$

Watch carefully the subscripts here, i.e. $c_{+\text{stoich}}$, $c_{-\text{stoich}}$ and c_{stoich}

11.13 Kohlrausch's law of independent ionic mobilities

The equations:

$$\Lambda = \lambda_+ + \lambda_- \tag{11.41}$$
$$\Lambda = x\lambda_+ + y\lambda_- \tag{11.43}$$

formulate Kohlrausch's law of independent ionic mobilities or migration which stated verbally is: 'each ion contributes a definite amount to the total molar conductivity of the electrolyte irrespective of the nature of the electrolyte'.

It is used for calculating the **limiting** molar conductivity, Λ^0, of an electrolyte from tabulated individual **limiting** ionic conductivities, λ^0, i.e. for very, very low concentrations. Under such conditions, the law can handle calculations of predicted limiting molar conductivities for both strong and weak electrolytes.

For weak electrolytes this can be extremely useful, since it is very difficult to determine Λ^0 by extrapolation of experimental Λ values at various concentrations (see Sections 11.7 and 11.8, Worked Problem 11.6 and Figure 11.7).

There are two ways of using the law for estimating Λ^0 for a weak electrolyte, e.g. CH_3COOH:

• Tabulated data gives $\lambda_-^0(CH_3COO^-)$ and $\lambda_+^0(H_3O^+)$ from which:

$$\Lambda^0(CH_3COOH) = \lambda_-^0(CH_3COO^-) + \lambda_+^0(H_3O^+)$$

This method relies on data for the individual ions being available. These come from data on strong electrolytes, e.g. $\lambda_-^0(CH_3COO^-)$ can be found from Λ^0 for the strong electrolyte, $CH_3COONa(aq)$, and the transport number for either $Na^+(aq)$ or $CH_3COO^-(aq)$ (see Section 9.21 and Section 11.16 below).

- If the relevant data on the individual ionic molar conductivities is not available, then the calculation can still be made by using data from strong electrolytes. Determination of Λ^0 by extrapolation to zero concentration is very easy and highly accurate provided the electrolyte is strong (see Section 11.6, Figure 11.4 and Chapter 12). Λ^0 for the particular weak electrolyte can be built up by combinations of experimentally determined Λ^0 values for strong electrolytes. If the weak electrolyte CH_3COOH is again chosen as an illustrative example, what is required is:

$$\Lambda^0(HCl) = \lambda^0_{H_3O^+} + \lambda^0_{Cl^-} \tag{11.44}$$

$$\Lambda^0(NaCl) = \lambda^0_{Na^+} + \lambda^0_{Cl^-} \tag{11.45}$$

$$\Lambda^0(CH_3COONa) = \lambda^0_{Na^+} + \lambda^0_{CH_3COO^-} \tag{11.46}$$

$$\Lambda^0(CH_3COOH) = \Lambda^0(CH_3COONa) + \Lambda^0(HCl) - \Lambda^0(NaCl) \tag{11.47}$$

$$= \lambda^0_{Na^+} + \lambda^0_{CH_3COO^-} + \lambda^0_{H_3O^+} + \lambda^0_{Cl^-} - \lambda^0_{Na^+} + \lambda^0_{Cl^-} \tag{11.48}$$

$$= \lambda^0_{CH_3COO^-} + \lambda^0_{H_3O^+} \tag{11.49}$$

Students should be warned that many investigators tabulate individual molar conductivities in the form of λ^0 for $\frac{1}{2} Ca^{2+}$, λ^0 for $\frac{1}{3} La^{3+}$ and so forth, i.e. they would give λ^0 for $\frac{1}{2} Ca^{2+}$ as 59.5 S cm^2 mol^{-1} instead of giving λ^0 for Ca^{2+} as 119.0 S cm^2 mol^{-1}. These investigators also would give a Λ^0 for $\frac{1}{2} CaCl_2$ and Λ^0 for $\frac{1}{3} LaCl_3$ etc. This is most unfortunate and not to be recommended.

Worked problem 11.7

Question

Calculate the limiting molar conductivities, Λ^0, at 25°C for the following electrolytes: $NaNO_3$, $Mg(OH)_2$, $CaSO_4$, K_2SO_4, $(CH_3COO)_2Mg$, $La_2(SO_4)_3$ given the following information on individual ionic conductivities at infinite dilution, i.e. λ^0_+ and λ^0_-.

cation	$\dfrac{\lambda^0_+}{\text{S cm}^2 \text{ mol}^{-1}}$	anion	$\dfrac{\lambda^0_-}{\text{S cm}^2 \text{ mol}^{-1}}$
$Na^+(aq)$	50.1	$NO_3^-(aq)$	71.4
$K^+(aq)$	73.5	$OH^-(aq)$	197.6
$Mg^{2+}(aq)$	106.1	$SO_4^{2-}(aq)$	160.0
$Ca^{2+}(aq)$	119.0	$CH_3COO^-(aq)$	40.9
$La^{3+}(aq)$	209.1		

Answer

For a fully ionised symmetrical electrolyte AB giving $A^{z+}(aq)$ and $B^{z-}(aq)$.
$$\Lambda^0 = \lambda^0_+ + \lambda^0_-$$

For a fully ionised unsymmetrical electrolyte A_xB_y, giving $A^{y+}(aq)$ and $B^{x-}(aq)$
$$\Lambda^0 = x\lambda^0_+ + y\lambda^0_-$$

$$\Lambda^0_{NaNO_3} = \lambda^0_{Na^+(aq)} + \lambda^0_{NO_3^-(aq)} = (50.1 + 71.4)\,S\,cm^2mol^{-1} = 121.5\,S\,cm^2mol^{-1}$$

$$\Lambda^0_{Mg(OH)_2} = \lambda^0_{Mg^{2+}(aq)} + 2\lambda^0_{OH^-(aq)} = (106.1 + 2 \times 197.6)\,S\,cm^2\,mol^{-1} = 501.3\,S\,cm^2\,mol^{-1}$$

$$\Lambda^0_{CaSO_4} = \lambda^0_{Ca^+(aq)} + \lambda^0_{SO_4^{2-}(aq)} = (119.0 + 160.0)\,S\,cm^2\,mol^{-1} = 279.0\,S\,cm^2\,mol^{-1}$$

$$\Lambda^0_{K_2SO_4(aq)} = 2\lambda^0_{K^+(aq)} + \lambda^0_{SO_4^{2-}(aq)} = (2 \times 73.5 + 160.0)\,S\,cm^2\,mol^{-1} = 307.0\,S\,cm^2\,mol^{-1}$$

$$\Lambda^0_{(CH_3COO)_2Mg(aq)} = 2\lambda^0_{CH_3COO^-(aq)} + \lambda^0_{Mg^{2+}(aq)} = (2 \times 40.9 + 106.1)\,S\,cm^2\,mol^{-1}$$
$$= 187.9\,S\,cm^2\,mol^{-1}$$

$$\Lambda^0_{La_2(SO_4)_3(aq)} = 2\lambda^0_{La^{3+}(aq)} + 3\lambda^0_{SO_4^{2-}(aq)} = (2 \times 209.1 + 3 \times 160.0)\,S\,cm^2\,mol^{-1}$$
$$= 898.2\,S\,cm^2\,mol^{-1}$$

Worked problem 11.8

Question

Calculate the limiting molar conductivities at infinite dilution, Λ^0, for the following weak acids at 25°C.

(i) HCOOH, given that the limiting molar conductivities, Λ^0, for HCl(aq), NaCl(aq), and HCOONa(aq) are 426.1, 126.4 and 104.7 S cm^2 mol^{-1} respectively.

(ii) CH_3CH_2COOH, given that the limiting molar conductivities, Λ^0, for HNO$_3$(aq), KNO$_3$(aq) and CH$_3$CH$_2$COOK(aq) are 421.1, 144.9 and 109.3 S cm^2 mol^{-1} respectively.

Answer

This is a question on Kohlrausch's Law:

(i) $\Lambda^0_{HCOOH(aq)} = \lambda^0_{H_3O^+(aq)} + \lambda^0_{HCOO^-(aq)}$

$\Lambda^0_{HCOOH(aq)} = \Lambda^0_{HCl(aq)} - \Lambda^0_{NaCl(aq)} + \Lambda^0_{HCOONa(aq)}$

$\quad = \lambda^0_{H_3O^+(aq)} + \lambda^0_{Cl^-(aq)} - \lambda^0_{Na^+(aq)} - \lambda^0_{Cl^-(aq)} + \lambda^0_{HCOO^-(aq)} + \lambda^0_{Na^+(aq)}$

$\quad = \lambda^0_{H_3O^+(aq)} + \lambda^0_{HCOO^-(aq)}$

$\Lambda^0_{HCOOH(aq)} = (426.1 - 126.4 + 104.7)\,S\,cm^2\,mol^{-1} = 404.4\,S\,cm^2\,mol^{-1}$

(ii) $\Lambda^0_{CH_3CH_2COOH(aq)} = \lambda^0_{H_3O^+(aq)} + \lambda^0_{CH_3CH_2COO^-(aq)}$

$\Lambda^0_{CH_3CH_2COOH(aq)} = \Lambda^0_{HNO_3(aq)} - \Lambda^0_{KNO_3(aq)l} + \Lambda^0_{CH_3CH_2COOK(aq)}$

$\quad = \lambda^0_{H_3O^+(aq)} + \lambda^0_{NO_3^-(aq)} - \lambda^0_{K^+(aq)} - \lambda^0_{NO_3^-(aq)} + \lambda^0_{CH_3CH_2COO^-(aq)} + \lambda^0_{K^+(aq)}$

$\quad = \lambda^0_{H_3O^+(aq)} + \lambda^0_{CH_3CH_2COO^-(aq)}$

$\Lambda^0_{CH_3CH_2COOH(aq)} = (421.1 - 144.9 + 109.3)\,S\,cm^2\,mol^{-1} = 385.5\,S\,cm^2\,mol^{-1}$

Worked problem 11.9

Question

The values of Λ^0 for NaOH(aq), NH$_4$Cl(aq) and NaCl(aq) in aqueous solution at 18°C are 219.96, 129.54, and 108.40 S cm^2 mol^{-1} respectively. The molar conductivity of a 1.00×10^{-3} mol dm^{-3} aqueous solution of NH$_3$ is 29.03 S cm^2 mol^{-1} at 18°C. Find the degree of protonation and the equilibrium constant for the protonation of ammonia in water. From this calculate the pK_a for NH$_4$$^+$(aq) acting as an acid.

Confirm that in the limit of infinite dilution the NH$_3$(aq) would be almost totally protonated.

$$pK_w = 14.24 \text{ at } 18°C$$

Answer

What is first required is the value for the limiting molar conductivity of NH$_3$(aq) in terms of the values given for the strong electrolytes NaOH(aq), NH$_4$Cl(aq) and NaCl(aq).

In a solution of NH$_3$ in water, the following equilibrium representing the protonation of NH$_3$(aq) is set up.

$$NH_3(aq) + H_2O(l) \rightleftharpoons NH_4{}^+(aq) + OH^-(aq)$$

Under normal experimental conditions of concentrations, this equilibrium is set up with **only** a certain fraction of the NH$_3$(aq) protonated to give NH$_4$$^+$(aq) and OH$^-$(aq). However, for the limiting conditions of infinite dilution, weak electrolytes are considered to be fully ionised; and hence for infinite dilution all the NH$_3$(aq) can be considered to be protonated giving NH$_4$$^+$(aq) and OH$^-$(aq). For a solution of x mol dm^{-3} of NH$_3$(aq) present initially there will be present x mol dm^{-3} of NH$_4$$^+$(aq) and x mol dm^{-3} of OH$^-$(aq) **at infinite dilution and only at infinite dilution**. Hence for infinite dilution the limiting molar conductivity for NH$_3$(aq) is made up of a contribution from the NH$_4$$^+$(aq) and OH$^-$(aq) ions.

$$\Lambda^0_{NH_3(aq)} = \lambda^0_{NH_4^+(aq)} + \lambda^0_{OH^-(aq)}$$

Using Kohlrausch's Law:

$$\Lambda^0_{NH_3(aq)} = \lambda^0_{NH_4^+(aq)} + \lambda^0_{OH^-(aq)} = \Lambda^0_{NH_4Cl(aq)} + \Lambda^0_{NaOH(aq)} - \Lambda^0_{NaCl(aq)}$$

$$= \lambda^0_{NH_4^+(aq)} + \lambda^0_{Cl^-(aq)} + \lambda^0_{Na^+(aq)} + \lambda^0_{OH^-(aq)} - \lambda^0_{Na^+(aq)} - \lambda^0_{Cl^-(aq)}$$

$$= (129.54 + 219.96 - 108.40) \text{ S cm}^2 \text{ mol}^{-1}$$

$$= 241.10 \text{ S cm}^2 \text{ mol}^{-1}$$

For weak electrolytes:

$$\alpha = \frac{\Lambda}{\Lambda^0} = \frac{29.03 \text{ S cm}^2 \text{ mol}^{-1}}{241.10 \text{ S cm}^2 \text{ mol}^{-1}} = 0.120$$

$\therefore$ At a concentration of 1.00×10^{-3} mol dm^{-3} NH$_3$(aq) is 12.0 % protonated.

Provided the self ionisation of water can be ignored then there will be present in a solution of $NH_3(aq)$ of concentration 1.00×10^{-3} mol dm^{-3}:

$$\alpha \times 1.00 \times 10^{-3} \text{ mol dm}^{-3} = 0.120 \times 1.00 \times 10^{-3} \text{ mol dm}^{-3} = 1.20 \times 10^{-4} \text{ mol dm}^{-3} \text{ of}$$

both $NH_4^+(aq)$ and $OH^-(aq)$.

$$\therefore [NH_4^+]_{actual} = [OH^-]_{actual} = 1.20 \times 10^{-4} \text{ mol dm}^{-3}$$
$$[NH_4^+]_{stoich} = [NH_4^+]_{actual} + [NH_3]_{actual}$$
$$\therefore [NH_3]_{actual} = [NH_4^+]_{stoich} - [NH_4^+]_{actual}$$
$$= (1.00 \times 10^{-3} - 0.120 \times 10^{-3}) \text{ mol dm}^{-3}$$
$$= 0.88 \times 10^{-3} \text{ mol dm}^{-3}$$

$$K_b = \left(\frac{[NH_4^+]_{actual}[OH^-]_{actual}}{[NH_3]_{actual}}\right) = \frac{0.120 \times 10^{-3} \text{ mol dm}^{-3} \times 0.120 \times 10^{-3} \text{ mol dm}^{-3}}{0.88 \times 10^{-3} \text{ mol dm}^{-3}}$$
$$= 1.65 \times 10^{-5} \text{ mol dm}^{-3}$$
$$\therefore pK_b = 4.78 \text{ for } NH_3(aq) \text{ acting as a base}$$
$$\text{and } pK_a = 9.46 \text{ for } NH_4^+(aq) \text{ acting as an acid}$$

Note well: this question is for experimental data at 18°C, not 25°C, and so $pK_b = 4.78$ must be subtracted from $pK_w = 14.24$ not 14.00. The temperature should thus always be quoted. The limiting value of the degree of protonation (see Section 4.6.1 and Worked Problem 4.7)

$$= \frac{K_b}{\sqrt{K_w} + K_b} = \frac{1.65 \times 10^{-5} \text{ mol dm}^{-3}}{7.6 \times 10^{-8} \text{ mol dm}^{-3} + 1.65 \times 10^{-5} \text{ mol dm}^{-3}} = 0.995$$

i.e. at infinite dilution $NH_3(aq)$ is 99.5% protonated.
$NH_3(aq)$ is a sufficiently strong base for the assumption that there is complete protonation at infinite dilution, i.e. $\alpha = 1$, to be justified. For a weaker base, e.g. urea, this would not be the case (see Worked Problems 4.6 and 4.7).

11.14 Analysis of the use of conductance measurements for determination of pK_as for very weak acids and pK_bs for very weak bases: the basic quantities involved

There are two simultaneous equilibria occurring.

$$HX(aq) + H_2O(l) \rightleftharpoons H_3O^+(aq) + X^-(aq)$$
$$2H_2O(l) \rightleftharpoons H_3O^+(aq) + OH^-(aq)$$

For such weak acids the self ionisation of $H_2O(l)$ cannot be ignored.
By electrical neutrality:

$$[H_3O^+]_{actual} = [X^-]_{actual} + [OH^-]_{actual} \tag{11.50}$$

by stoichiometry:

$$[HX]_{stoich} = [HX]_{actual} + [X^-]_{actual} \tag{11.51}$$

by definition :

$$\alpha = \frac{[X^-]_{actual}}{[HX]_{stoich}} \tag{11.52}$$

and :

$$K_a = \left(\frac{[H_3O^+]_{actual}[X^-]_{actual}}{[HX]_{actual}} \right)_{eq} \tag{11.53}$$

i.e. ignoring non-ideality when activities are approximated to concentrations.

$[H_3O^+]_{actual}$, $[X^-]_{actual}$ and $[OH^-]_{actual}$ will all contribute to the conductivity of the solution. ∴The total conductivity of the solution is made up of:

- a contribution from the ionisation of the weak acid. This contribution must include allowance for the effect which the self ionisation of water has on the ionisation of the weak acid, HX(aq),

- and a contribution from the ions formed by the self ionisation of the solvent water.

It is at this stage that the distinction between the frames of reference used in the experimental description of molar conductivity and the theoretical description must be clearly explained.

- **For the 'experimental description'**, the observed conductivity, κ, the stoichiometric concentration, c_{stoich} and the molar conductivity, Λ , are related by the equation: $\kappa = \Lambda c_{stoich}$.

- **For the 'theoretical description'**, actual concentrations are used, with the relation between κ and c_{actual} being: $\kappa = \Lambda^0 c_{actual}$.

These two descriptions must be clearly distinguished and must never be confused. In particular, in the theoretical description it is Λ^0 which appears **not** Λ. This can be shown by the following argument giving the relation between the two frames of reference.

$$\frac{\Lambda}{\Lambda^0} = \alpha = \frac{c_{actual}}{c_{stoich}} \tag{11.23 – 11.28}$$

$$\therefore \Lambda = \Lambda^0 \frac{c_{actual}}{c_{stoich}} \tag{11.54}$$

But

$$\Lambda = \frac{\kappa}{c_{stoich}} = \Lambda^0 \frac{c_{actual}}{c_{stoich}} \tag{11.55}$$

$$\therefore \kappa = \Lambda^0 c_{actual} \tag{11.56}$$

The same arguments can be given for individual ionic conductivities and individual contributions to the conductivity, i.e.:

For the 'experimental description':

$$\kappa_+ = \lambda_+ c_{+stoich} \quad \text{and} \quad \kappa_- = \lambda_- c_{-stoich} \tag{11.35 and 11.36}$$

For the 'theoretical description':

$$\kappa_+ = \lambda_+^0 c_{+actual} \quad \text{and} \quad \kappa_- = \lambda_-^0 c_{-actual} \tag{11.57 and 11.58}$$

11.14.1 Analysis of the use of conductance measurements for determination of pK_as for very weak acids and pK_bs for very weak bases: the argument

The total conductivity of the solution is given in terms of the theoretical framework, i.e. λ^0's and c_{actual}.

$$\kappa_{total} = \lambda^0_{H_3O^+(aq)}[H_3O^+]_{actual} + \lambda^0_{X^-(aq)}[X^-]_{actual} + \lambda^0_{OH^-(aq)}[OH^-]_{actual} \qquad (11.59)$$

But electrical neutrality requires that

$$[H_3O^+]_{actual} = [X^-]_{actual} + [OH^-]_{actual} \qquad (11.60)$$

$$\therefore \kappa_{total} = \lambda^0_{H_3O^+(aq)}[X^-]_{actual} + \lambda^0_{H_3O^+(aq)}[OH^-]_{actual}$$
$$+ \lambda^0_{X^-(aq)}[X^-]_{actual} + \lambda^0_{OH^-(aq)}[OH^-]_{actual} \qquad (11.61)$$

$$= (\lambda^0_{H_3O^+(aq)} + \lambda^0_{X^-(aq)})[X^-]_{actual} + (\lambda^0_{H_3O^+(aq)} + \lambda^0_{OH^-(aq)})[OH^-]_{actual} \qquad (11.62)$$

This is what the total conductivity will be. However, this includes the contribution from the solvent, and so this must be formulated and subtracted off to be compatible with what is done experimentally.

$$\kappa_{for\ solvent\ on\ its\ own} = \lambda^0_{H_3O^+(aq)}[H_3O^+]_{in\ pure\ water} + \lambda^0_{OH^-(aq)}[OH^-]_{in\ pure\ water} \qquad (11.63)$$

In pure water:

$$[H_3O^+] = [OH^-] = K_w^{1/2} \qquad (11.64)$$

$$\therefore \kappa_{for\ solvent\ on\ its\ own} = (\lambda^0_{H_3O^+(aq)} + \lambda^0_{OH^-(aq)})[H_3O^+]_{in\ pure\ water}$$
$$= (\lambda^0_{H_3O^+(aq)} + \lambda^0_{OH^-(aq)})K_w^{1/2} \qquad (11.65)$$

This must now be subtracted from equation (11.62)

$$\kappa_{corrected\ for\ the\ solvent} = (\lambda^0_{H_3O^+(aq)} + \lambda^0_{X^-(aq)})[X^-]_{actual} + (\lambda^0_{H_3O^+(aq)} + \lambda^0_{OH^-(aq)})[OH^-]_{actual}$$
$$- (\lambda^0_{H_3O^+(aq)} + \lambda^0_{OH^-(aq)})K_w^{1/2} \qquad (11.66)$$

$$\therefore \kappa_{corrected\ for\ the\ solvent} = (\lambda^0_{H_3O^+(aq)} + \lambda^0_{X^-(aq)})[X^-]_{actual}$$
$$- (\lambda^0_{H_3O^+(aq)} + \lambda^0_{OH^-(aq)})(K_w^{1/2} - [OH^-]_{actual}) \qquad (11.67)$$

This is the conductivity which would be expected on the basis of what is thought theoretically to be occurring in the solution. If the arguments are correct, then this conductivity should be the same as that observed experimentally where:

$$\Lambda = \frac{\kappa}{c_{stoich}} \qquad (11.19)$$

$$\Lambda_{HX(aq)} = \frac{(\lambda^0_{H_3O^+(aq)} + \lambda^0_{X^-(aq)})[X^-]_{actual}}{[HX]_{stoich}} - \frac{(\lambda^0_{H_3O^+(aq)} + \lambda^0_{OH^-(aq)})(K_w^{1/2} - [OH^-]_{actual})}{[HX]_{stoich}} \qquad (11.68)$$

But

$$\Lambda^0_{HX(aq)} = \lambda^0_{H_3O^+(aq)} + \lambda^0_{X^-(aq)} \qquad (11.69)$$

and

$$\alpha = \frac{[X^-]_{actual}}{[HX]_{actual}} \qquad (11.70)$$

$$\therefore \frac{\Lambda}{\Lambda^0} = \alpha \times \frac{(\lambda^0_{H_3O^+(aq)} + \lambda^0_{X^-(aq)})}{(\lambda^0_{H_3O^+(aq)} + \lambda^0_{X^-(aq)})} - \frac{(\lambda^0_{H_3O^+(aq)} + \lambda^0_{OH^-(aq)})}{(\lambda^0_{H_3O^+(aq)} + \lambda^0_{X^-(aq)})} \times \left(\frac{K_w^{1/2} - [OH^-]_{actual}}{[HX]_{stoich}} \right) \quad (11.71)$$

$$\therefore \frac{\Lambda}{\Lambda^0} = \alpha - \frac{(\lambda^0_{H_3O^+(aq)} + \lambda^0_{OH^-(aq)})}{(\lambda^0_{H_3O^+(aq)} + \lambda^0_{X^-(aq)})} \times \left(\frac{K_w^{1/2} - [OH^-]_{actual}}{[HX]_{stoich}} \right) \quad (11.72)$$

This analysis shows that for very weak acids and bases, $\alpha \neq \Lambda/\Lambda^0$.

It is, however, still possible to use the experimental data and computer fit to the five equations below to enable α and K_a to be found:

$$\frac{\Lambda}{\Lambda^0} = \alpha - \frac{(\lambda_{H_3O^+(aq)} + \lambda_{OH^-}(aq))}{(\lambda^0_{H_3O^+(aq)} + \lambda^0_{X^-(aq)})} \times \left(\frac{K_w^{1/2} - [OH^-]_{actual}}{[HX]_{stoich}} \right) \quad (11.72)$$

$$[H_3O^+]_{actual} = [X^-]_{actual} + [OH^-]_{actual} \quad (11.60)$$

$$[HX]_{stoich} = [HX]_{actual} + [X^-]_{actual} \quad (11.73)$$

$$K_a = \frac{[H_3O^+]_{actual}[X^-]_{actual}}{[HX]_{actual}} \quad (11.74)$$

$$[H_3O^+]_{actual} \times [OH^-]_{actual} = K_w \quad (11.75)$$

11.14.2 Application of the above analysis to the cases of weak acids and weak bases for which the relation $\alpha = \Lambda/\Lambda^0$ is taken to be valid and $\alpha \to 1$ as $c \to 0$

The total conductivity is again made up from contributions from all the ions present:

$$\kappa_{total} = \lambda^0_{H_3O^+(aq)}[H_3O^+]_{actual} + \lambda^0_{X^-(aq)}[X^-]_{actual} + \lambda^0_{OH^-(aq)}[OH^-]_{actual} \quad (11.59)$$

The analysis given in Sections 4.5 and 4.6 shows that for weak acids with pK_as <5 the fraction ionised can approach unity. For such acids it also can be assumed that the self ionisation of water could be ignored. Under these conditions:

$$[H_3O^+]_{actual} = [X^-]_{actual} \quad (11.76)$$

and $\lambda^0_{OH^-(aq)}[OH^-]_{actual}$ can be dropped out. Under the conditions of this analysis, the solution will be sufficiently acidic for $[OH^-]_{actual} \ll [H_3O^+]_{actual}$ and the total conductivity can be approximated to the simpler expression:

$$\kappa_{total} = (\lambda^0_{H_3O^+(aq)} + \lambda^0_{X^-(aq)})[X^-]_{actual} \quad (11.77)$$

The contribution from the solvent which arises by virtue of the ions arising from the self ionisation of water must again be subtracted, giving:

$$\kappa_{total} = (\lambda^0_{H_3O^+(aq)} + \lambda^0_{X^-(aq)})[X^-]_{actual} - (\lambda^0_{H_3O^+(aq)} + \lambda^0_{OH^-(aq)})K_w^{1/2} \quad (11.78)$$

This must be equal to the experimental κ.

$$\Lambda = \frac{\kappa}{[HX]_{stoich}} \quad (11.19)$$

and so

$$\Lambda_{HX(aq)} = \frac{(\lambda^0_{H_3O^+(aq)} + \lambda^0_{X^-(aq)})[X^-]_{actual}}{[HX]_{stoich}} - \frac{(\lambda^0_{H_3O^+(aq)} + \lambda^0_{OH^-(aq)})K_w^{1/2}}{[HX]_{stoich}} \quad (11.79)$$

and

$$\therefore \frac{\Lambda}{\Lambda^0} = \alpha \times \frac{(\lambda^0_{H_3O^+(aq)} + \lambda^0_{X^-(aq)})}{(\lambda^0_{H_3O^+(aq)} + \lambda^0_{X^-(aq)})} - \frac{(\lambda^0_{H_3O^+(aq)} + \lambda^0_{OH^-(aq)})}{(\lambda^0_{H_3O^+(aq)} + \lambda^0_{X^-(aq)})} \times \left(\frac{K_w^{1/2}}{[HX]_{stoich}} \right) \tag{11.80}$$

$$\therefore \frac{\Lambda}{\Lambda^0} = \alpha - \frac{(\lambda^0_{H_3O^+(aq)} + \lambda^0_{OH^-(aq)})}{(\lambda^0_{H_3O^+(aq)} + \lambda^0_{X^-(aq)})} \times \left(\frac{K_w^{1/2}}{[HX]_{stoich}} \right) \tag{11.81}$$

Unless $[HX]_{stoich}$ becomes very low, then the second term in Equations (11.78) to (11.81) can be ignored with respect to the first and $\Lambda/\Lambda^0 = \alpha$.

If $[HX]_{stoich}$ becomes very low, then it is not valid to ignore the self ionisation of water, and in this case the rigorous expression Equation (11.72) must be used.

11.15   Use of conductance measurements in determining solubility products for sparingly soluble salts

Conductance measurements are an important method for the determination of solubility products. The basis of the method is that for sparingly soluble salts the concentration is sufficiently low that, to a first approximation, non-ideality can be ignored. This means that the molar conductivity of a saturated solution of the sparingly soluble salt can be taken as the molar conductivity for infinite dilution, Λ^0. The limiting molar conductivity, Λ^0, of the electrolyte can be calculated using Kohlrausch's Law. Since $\kappa = \Lambda^0 c_{stoich}$, c_{stoich} can be found and this will give K_s. The method is illustrated in the following two worked problems.

Worked problem 11.10

Question

After correcting for the conductivity of the solvent, the conductivity of a saturated solution of silver bromide in water is 6.99×10^{-8} S cm^{-1} at 18°C. If the molar ionic conductivities of Ag$^+$(aq) and Br$^-$(aq) are 53.5 and 68.0 S cm^2 mol^{-1} respectively, find the solubility of silver bromide in water. From this calculate the solubility product for silver bromide.

Answer

$$\Lambda^0_{AgBr(aq)} = \lambda^0_{Ag^+(aq)} + \lambda^0_{Br^-} = (53.5 + 68.0)\, S\, cm^2\, mol^{-1} = 121.5\, S\, cm^2\, mol^{-1}$$

Since AgBr(s) is a highly insoluble salt, then a saturated solution of AgBr(aq) will be a very dilute solution of Ag$^+$(aq) and Br$^-$(aq), and so the molar conductivity, Λ, for the saturated solution can be taken to be the limiting value of the molar conductivity at infinite dilution, i.e. Λ^0.

$$\therefore \Lambda = \Lambda^0 = 121.5\, S\, cm^2\, mol^{-1}$$

$$\kappa_{for\ very\ dilute\ solution} = \Lambda^0 c_{stoich}$$

where c_{stoich} is here the concentration of the saturated solution of AgBr.

$$c_{stoich} = \frac{\kappa}{\Lambda^0} = \frac{6.99 \times 10^{-8} \, S \, cm^{-1}}{121.5 \, S \, cm^2 \, mol^{-1}} = 5.75 \times 10^{-10} \, mol \, cm^{-3} = 5.75 \times 10^{-10} \times 10^3 \, mol \, dm^{-3}$$

$$\therefore c_{stoich} = 5.75 \times 10^{-7} \, mol \, dm^{-3}$$

$\therefore$ The solubility of AgBr in water $= 5.75 \times 10^{-7} \, mol \, dm^{-3}$

$$AgBr(s) \rightleftharpoons Ag^+(aq) + Br^-(aq)$$

$$K_s = ([Ag^+][Br^-])_{eq}$$

$$= (5.75 \times 10^{-7})^2 \, mol^2 \, dm^{-6}$$

$$= 3.3 \times 10^{-13} \, mol^2 \, dm^{-6}$$

Worked problem 11.11

Question

At 25°C, the conductivity of $8.0 \times 10^{-4} \, mol \, dm^{-3}$ aqueous K_2SO_4 is $2.35 \times 10^{-4} \, S \, cm^{-1}$, and this rises to $3.04 \times 10^{-4} \, S \, cm^{-1}$ if the solution is saturated with $SrSO_4$. The limiting molar ionic conductivities of $K^+(aq)$ and $Sr^{2+}(aq)$ are 73.5 and $119.0 \, S \, cm^{-2} \, mol^{-1}$ respectively. Find the molar conductivity of the K_2SO_4 solution, and then the solubility product for $SrSO_4$.

Answer

This question is more difficult than the previous one, and has to be taken in stages.

- From the conductivity of the K_2SO_4 solution, the molar conductivity of the solution can be found;

- given λ_{K^+} and knowing $\Lambda_{K_2SO_4}$, then $\lambda_{SO_4^{2-}}$ can be found;

- knowing $\lambda_{Sr^{2+}}$ and having found $\lambda_{SO_4^{2-}}$, then Λ_{SrSO_4} can be found;

- the difference between the conductivities of the two solutions allows the contribution to the total conductivity from the ions $Sr^{2+}(aq)$ and $SO_4^{2-}(aq)$ formed from the $SrSO_4$ to be found;

- and this allows the concentration of $SrSO_4(aq)$ dissolved in the solution to be found, from which the solubility product can be calculated.

The calculation is as follows:

- $\kappa_{K_2SO_4(aq)} = 2.35 \times 10^{-4} \, S \, cm^{-1}$

$\kappa = \Lambda c_{stoich}$

$$\therefore \Lambda_{K_2SO_4(aq)} = \frac{\kappa}{c_{stoich}} = \frac{2.35 \times 10^{-4} \, S \, cm^{-1}}{8.00 \times 10^{-4} \, mol \, dm^{-3}} = 2.94 \times 10^{-1} \, S \, cm^{-1} \, mol^{-1} \, dm^3$$

$$= 2.94 \times 10^{-1} \times 10^3 \, S \, cm^2 \, mol^{-1}$$

$$\therefore \Lambda_{K_2SO_4(aq)} = 294 \, S \, cm^2 \, mol^{-1}$$

- $\Lambda_{K_2SO_4(aq)} = 2\lambda_{K^+(aq)} + \lambda_{SO_4^{2-}(aq)}$

Remember: K_2SO_4 is an unsymmetrical electrolyte.

$$\therefore \lambda_{SO_4^{2-}(aq)} = \Lambda_{K_2SO_4(aq)} - 2\lambda_{K^+(aq)} = (294 - 2 \times 73.5)\,S\,cm^2\,mol^{-1} = 147\,S\,cm^2\,mol^{-1}$$

- $\Lambda_{SrSO_4(aq)} = \lambda_{Sr^{2+}(aq)} + \lambda_{SO_4^{2-}(aq)} = (119 + 147)\,S\,cm^2\,mol^{-1} = 266\,S\,cm^2\,mol^{-1}$

Note: $SrSO_4$ is a symmetrical electrolyte.

- The conductivity, κ, of the K_2SO_4 solution saturated with $SrSO_4 = 3.04 \times 10^{-4}\,S\,cm^{-1}$
  The conductivity, κ, of the K_2SO_4 solution on its own $= 2.35 \times 10^{-4}\,S\,cm^{-1}$
  $\therefore$ the conductivity due to the dissolved $SrSO_4 = (3.04 \times 10^{-4} - 2.35 \times 10^{-4})$
  $S\,cm^{-1} = 6.9 \times 10^{-5}\,S\,cm^{-1}$

- $\kappa = \Lambda c_{stoich}$

$$\therefore c_{stoich} = \frac{\kappa}{\Lambda} = \frac{6.9 \times 10^{-5}\,S\,cm^{-1}}{266\,S\,cm^2\,mol^{-1}} = 2.6 \times 10^{-7}\,mol\,cm^{-3} = 2.6 \times 10^{-7} \times 10^3\,mol\,dm^{-3}$$

$$= 2.6 \times 10^{-4}\,mol\,dm^{-3}$$

$\therefore$ solubility of $SrSO_4$ in K_2SO_4 solution $= 2.6 \times 10^{-4}\,mol\,dm^{-3}$

$$SrSO_4(s) \rightleftharpoons Sr^{2+}(aq) + SO_4{}^{2-}(aq)$$

$$K_s = ([Sr^{2+}][SO_4^{2-}])_{eq}$$

Be careful: in this problem there are two sources of $SO_4{}^{2-}(aq)$:

- from the initial K_2SO_4 aqueous solution;

- from the dissolved $SrSO_4$.

$[SO_4^{2-}]_{total} = 8.0 \times 10^{-4}\,mol\,dm^{-3}$ from $K_2SO_4(aq) + 2.6 \times 10^{-4}\,mol\,dm^{-3}$ from the dissolved $SrSO_4$

$$= 1.06 \times 10^{-3}\,mol\,dm^{-3}$$

$[Sr^{2+}] = 2.6 \times 10^{-4}\,mol\,dm^{-3}$

$\therefore K_s = 2.6 \times 10^{-4}\,mol\,dm^{-3} \times 1.06 \times 10^{-3}\,mol\,dm^{-3} = 2.8 \times 10^{-7}\,mol^2\,dm^{-6}$

11.16   Transport numbers

Section 9.1 defines the transport number and Section 9.21 describes how transport numbers can be found from the emf of concentration cells.

The transport number of an ion is the fraction of the total current carried by the ion:

$$t_+ = \frac{I_+}{I_{total}} \quad \text{and} \quad t_- = \frac{I_-}{I_{total}} \qquad (11.82) \text{ and } (11.83)$$

where I_+ and I_- are the current carried by the cation and anion respectively; and I_{total} is the total current carried.

Since the current flowing is proportional to the potential difference across the electrodes:

$$I \propto E \qquad (11.1)$$

$$\therefore I = GE \qquad (11.2)$$

and the conductance is proportional to A/l so that $G = \kappa \dfrac{A}{l}$ $\qquad (11.11)$

$$I = \kappa E \frac{A}{l} \qquad I_+ = \kappa_+ E \frac{A}{l} \qquad I_- = \kappa_- E \frac{A}{l} \qquad (11.84), (11.85), \text{ and } (11.86)$$

$$\kappa = \frac{I}{E}\frac{l}{A} \qquad \kappa_+ = \frac{I_+}{E}\frac{l}{A} \qquad \kappa_- = \frac{I_-}{E}\frac{l}{A} \qquad (11.87), (11.88) \text{ and } (11.90)$$

$$t_+ = \frac{I_+}{I_{total}} \qquad t_- = \frac{I_-}{I_{total}} \qquad (11.82) \text{ and } (11.83)$$

$$\therefore t_+ = \frac{\kappa_+}{\kappa}\left(\frac{EA}{l}\right)\left(\frac{l}{EA}\right) = \frac{\kappa_+}{\kappa} \quad (11.90) \qquad \therefore t_- = \frac{\kappa_-}{\kappa}\left(\frac{EA}{l}\right)\left(\frac{l}{EA}\right) = \frac{\kappa_-}{\kappa} \qquad (11.91)$$

$$= \frac{\lambda_+ c_{+stoich}}{\Lambda c_{stoich}} \quad (11.92) \qquad\qquad = \frac{\lambda_- c_{-stoich}}{\Lambda c_{stoich}} \qquad (11.93)$$

where λ_+, λ_-, $c_{+stoich}$ and $c_{-stoich}$ are as given in Sections 11.11 and 11.12.

- **For the symmetrical electrolyte AB:**

$$c_{+stoich} = [A^{z+}]_{stoich} = [AB]_{stoich} = c_{stoich} \text{ for the electrolyte} \qquad (11.94)$$

$$c_{-stoich} = [B^{z-}]_{stoich} = [AB]_{stoich} = c_{stoich} \text{ for the electrolyte} \qquad (11.95)$$

$$\therefore t_+ = \frac{\lambda_+}{\Lambda} \qquad\qquad t_- = \frac{\lambda_-}{\Lambda} \qquad (11.96) \text{ and } (11.97)$$

- **For the unsymmetrical electrolyte A_xB_y:**

$$c_{+stoich} = [A^{y+}]_{stoich} = x[A_xB_y]_{stoich} = xc_{stoich} \text{ for the electrolyte} \qquad (11.98)$$

$$c_{-stoich} = [B^{x-}]_{stoich} = y[A_xB_y]_{stoich} = yc_{stoich} \text{ for the electrolyte} \qquad (11.99)$$

$$\therefore t_+ = \frac{x\lambda_+}{\Lambda} \qquad\qquad t_- = \frac{y\lambda_-}{\Lambda} \qquad (11.100) \text{ and } (11.101)$$

Since a transport number is a fraction, the sum of the transport numbers for cation and anion will be unity:

$$\therefore t_+ + t_- = \frac{x\lambda_+}{\Lambda} + \frac{y\lambda_-}{\Lambda} \tag{11.102}$$

$$= \frac{x\lambda_+}{x\lambda_+ + y\lambda_-} + \frac{y\lambda_-}{x\lambda_+ + y\lambda_-} = 1 \tag{11.103}$$

Experiment shows that transport numbers depend on the concentration and the temperature. The concentration dependence is a result of non-ideality causing a lack of strict proportionality between κ and the total number of ions present per unit volume. The non-ideality is caused primarily by electrostatic interactions between the ions; these being mainly coulombic in nature. As has been seen in the case of activity coefficients, these deviations are greater the higher the concentration.

Since $t_+ = x\lambda_+/\Lambda$ and $t_- = y\lambda_-/\Lambda$ and Λ is the molar conductivity of the electrolyte in question, then the transport number of a particular cation or anion will depend on the electrolyte, e.g.

$t_{+Mg^{2+}}$ is not the same in $MgSO_4(aq)$ solution as in $Mg(NO_3)_2(aq)$

$$t_{+Mg^{2+}} \text{ in } MgSO_4(aq) = \frac{\lambda_{+Mg^{2+}}}{\Lambda_{MgSO_4}} \tag{11.104}$$

$$t_{+Mg^{2+}} \text{ in } Mg(NO_3)_2(aq) = \frac{\lambda_{+Mg^{2+}}}{\Lambda_{Mg(NO_3)_2}} \tag{11.105}$$

Since Λ_{MgSO_4} and $\Lambda_{Mg(NO_3)_2}$ are different, then $t_{+Mg^{2+}}$ in $MgSO_4(aq)$ is different from $t_{+Mg^{2+}}$ in $Mg(NO_3)_2(aq)$. This can be contrasted with the observation that $\lambda_{+Mg^{2+}}$ in $MgSO_4(aq)$ and $\lambda_{+Mg^{2+}}$ in $Mg(NO_3)_2(aq)$ are the same.

t_+ is the fraction of the total current which is carried by the cation, and since the total current has a contribution from the anion then t_+ will depend on the nature of the anion. A similar argument will apply to the transport number of an anion which will also depend on the nature of the cation, and thus of the electrolyte being studied.

Worked problem 11.12

Question

If the limiting molar conductivity, Λ^0, for $KCl(aq)$ at 25°C is 149.8 S cm^2 mol^{-1} and the limiting molar ionic conductivity for $K^+(aq)$ is 73.5 S cm^2 mol^{-1}, calculate the transport number t_+ for the $K^+(aq)$ ion. What is t_- for $Cl^-(aq)$? Comment on the values.

Answer

In Sections 11.12 and 11.16 it was shown that for a symmetrical electrolyte:

$$\Lambda^0 = \lambda_+^0 + \lambda_-^0$$

$$t_+ = \frac{\lambda_+}{\Lambda} \qquad t_- = \frac{\lambda_-}{\Lambda}$$

and in the limit of infinite dilution:

$$t_+^0 = \frac{\lambda_+^0}{\Lambda^0} \qquad\qquad t_-^0 = \frac{\lambda_-^0}{\Lambda^0}$$

$$t_+^0 = \frac{73.5\,\text{S cm}^2\,\text{mol}^{-1}}{149.8\,\text{S cm}^2\,\text{mol}^{-1}} = 0.491$$

$$\therefore t_-^0 = 1 - t_+^0 = 1 - 0.491 = 0.509$$

Since the transport number for the $K^+(aq)$ is only slightly less than that for $Cl^-(aq)$, then the fraction of the current carried by each is fairly similar. This is reflected in the limiting molar ionic conductivities for each ion: $\lambda_{K^+(aq)}^0 = 73.5\,\text{S cm}^2\,\text{mol}^{-1}$ while $\lambda_{Cl^-(aq)}^0 = (149.8 - 73.5)\,\text{S cm}^2\,\text{mol}^{-1} = 76.3\,\text{S cm}^2\,\text{mol}^{-1}$.

Worked problem 11.13

Question

If the limiting molar conductivity, Λ^0, for $Na_2SO_4(aq)$ at 25°C is 260.2 S cm² mol⁻¹ and the limiting molar ionic conductivity for $Na^+(aq)$ is 50.1 S cm² mol⁻¹, calculate the transport number, t_+, for $Na^+(aq)$. What are $t_{SO_4^{2-}(aq)}^0$ and $\lambda_{SO_4^{2-}(aq)}^0$?

Answer

In Sections 11.12 and 11.16 it was shown that for a unsymmetrical electrolyte, A_xB_y ionising to give $A^{y+}(aq)$ and $B^{x-}(aq)$:

$$\Lambda^0 = x\lambda_+^0 + y\lambda_-^0$$

$$t_+^0 = \frac{x\lambda_+^0}{\Lambda^0} \qquad\qquad t_-^0 = \frac{y\lambda_-^0}{\Lambda^0}$$

$$t_{Na^+(aq)}^0 = \frac{2\lambda_{Na^+(aq)}^0}{\Lambda_{Na_2SO_4(aq)}^0} = \frac{2 \times 50.1\,\text{S cm}^2\,\text{mol}^{-1}}{260.2\,\text{S cm}^2\,\text{mol}^{-1}} = 0.385$$

$$t_{Na^+(aq)}^0 + t_{SO_4^{2-}(aq)}^0 = 1$$

$$\therefore t_{SO_4^{2-}(aq)}^0 = 1 - 0.385 = 0.615$$

$$\Lambda_{Na_2SO_4(aq)}^0 = 2\lambda_{Na^+}^0 + \lambda_{SO_4^{2-}(aq)}^0$$

$$\therefore \lambda_{SO_4^{2-}(aq)}^0 = \Lambda_{Na_2SO_4(aq)}^0 - 2\lambda_{Na^+(aq)}^0 = (260.2 - 2 \times 50.1)\,\text{S cm}^2\,\text{mol}^{-1} = 160.0\,\text{S cm}^2\,\text{mol}^{-1}$$

Worked problem 11.14

Question

The limiting molar ionic conductivities at 25°C for $La^{3+}(aq)$ and $SO_4^{2-}(aq)$ are 209.1 and 160.0 S cm² mol⁻¹ respectively. Find Λ^0, the limiting molar conductivity, for $La_2(SO_4)_3(aq)$. Thence find the transport numbers for $La^{3+}(aq)$ and $SO_4^{2-}(aq)$ in a solution of $La_2(SO_4)_3(aq)$.

Answer

This is like Worked Problem 11.13 where the electrolyte is unsymmetrical.

$$\Lambda^0_{La_2(SO_4)_3} = 2\lambda^0_{La^{3+}(aq)} + 3\lambda^0_{SO_4^{2-}(aq)} = (2 \times 209.1 + 3 \times 160.0)\,S\,cm^2\,mol^{-1} = 898.2\,S\,cm^2\,mol^{-1}$$

$$t^0_{La^{3+}(aq)} = \frac{2\lambda^0_{La^{3+}(aq)}}{\Lambda^0_{La_2(SO_4)_3(aq)}} = \frac{2 \times 209.1\,S\,cm^2\,mol^{-1}}{898.2\,S\,cm^2\,mol^{-1}} = 0.466$$

$$t^0_{SO_4^{2-}(aq)} = \frac{3\lambda^0_{SO_4^{2-}(aq)}}{\Lambda^0_{La_2(SO_4^{2-})_3(aq)}} = \frac{3 \times 160.0\,S\,cm^2\,mol^{-1}}{898.2\,S\,cm\,mol^{-1}} = 0.534$$

A check: $t^0_{La^{3+}(aq)} + t^0_{SO_4^{2-}(aq)} = 0.466 + 0.534 = 1.000$ which is as it should be.

11.17 Ionic mobilities

The ionic mobility, transport number and ionic molar conductivity are all linked. The velocity with which an ion migrates is proportional to the drop in potential over the region in which the ion migrates. Put otherwise, the velocity of migration is proportional to the electric field.

In conductance studies the conductance cell is connected to a source of alternating current. In one half-cycle the negative pole of the source of the current is connected to the negative electrode of the conductance cell which thus acts a cathode, and the positive pole of the source is connected to the positive electrode of the conductance cell which acts as an anode. In the subsequent half-cycle the situation will be reversed. But for the following, the discussion will be restricted to what happens in one half-cycle. The field is taken to be aligned in a direction from the positive electrode to the negative and this is indicated by $^+\!\longrightarrow^-$. The cations will migrate under the influence of the field in the direction of the field, and the anions will move in the opposite direction.

The velocities of migration, v_+ for the cations, and v_- for the anions are different. Since these velocities are proportional to the field, X, then:

$$v_+ \propto X \qquad \text{and} \qquad v_- \propto -X \qquad\qquad (11.106)\text{ and }(11.107)$$

$$\therefore v_+ = u_+ X \qquad \text{and} \qquad \therefore v_- = -u_- X \qquad\qquad (11.108)\text{ and }(11.109)$$

where u_+ and u_- are constants of proportionality for cation and anion respectively, called the ionic mobilities. The minus sign in the equation for the anion appears since it takes cognisance of the fact that the anion is moving in the opposite direction to the cation.

The region between the electrodes of the conductance cell is considered. The area of the electrodes is A, and the distance apart is l (see Section 11.1.1), and let there be N_+ cations and N_- anions in the volume of solution contained between the electrodes (see Figure 11.8). Application of the field, X, will cause the cations to move from left to right. What is of interest is the time that it will take for the cations lying on the left hand anode to reach the right hand cathode. Within that time all the cations in the volume of solution between the electrodes will also have reached the cathode on the right.

The distance travelled $= l$ and the velocity of migration for the cations $= v_+$

The time taken for cations to reach the cathode on the right, is $\tau_+ = \dfrac{l}{v_+}$ $\qquad\qquad (11.110)$

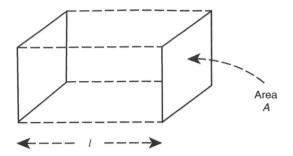

Figure 11.8   Region between the electrodes in a conductance cell.

where τ is used rather than t to distinguish the time from the transport number.
Since all of the cations in the volume of solution between the electrodes reach the right hand
electrode in time, τ_+, this will be a total number $= N_+$.

$$\text{The charge conveyed in time } \tau_+ = N_+z_+e \qquad (11.111)$$

But current $\times$ time $=$ charge
    $\therefore$ current flowing in the time τ_+ is:

$$I_+ = \frac{\text{charge conveyed}}{\text{time taken}} = \frac{N_+z_+e}{\tau_+} = \frac{N_+z_+ev_+}{l} \qquad (11.112)$$

where

$$\tau_+ = \frac{l}{v_+} \qquad (11.110)$$

Substituting Equation (11.108) gives:

$$I_+ = \frac{N_+z_+eu_+X}{l} = \frac{N_+z_+eu_+\Delta E}{l^2} \qquad (11.111)$$

where ΔE is the difference in potential between the two electrode plates separated by the
distance, l. In the case of the battery this would be called the potential difference, or emf, and
the field X is given by the change in potential, ΔE over the distance l.

$$\text{The field } X = -\frac{\partial \psi}{\partial r} \qquad (11.14)$$

$$\therefore -\mathrm{d}\psi = X\mathrm{d}r \qquad (11.112)$$

$$-\int_0^l \mathrm{d}\psi = X\int_0^l \mathrm{d}r \qquad (11.113)$$

$$\therefore -(\psi_l - \psi_0) = \Delta E = Xl \qquad (11.114)$$

$$\therefore X = \frac{\Delta E}{l} \qquad (11.115)$$

Ohm's law gives an expression for $I_+/\Delta E = G_+$ (11.116)

where G_+ is the contribution to the conductance from the cation.

Taking this expression for G_+, and the expression for $I_+/\Delta E$ found from Equation (11.111) gives:

$$\frac{I_+}{\Delta E} = G_+ = \frac{N_+ z_+ e u_+}{l^2}$$ (11.117)

But $$G_+ = \kappa_+ \frac{A}{l}$$ (11.118)

$$\therefore \kappa_+ = \frac{N_+ z_+ e u_+}{l^2} \frac{l}{A} = \frac{N_+ z_+ e u_+}{V}$$ (11.119)

where V = the volume of the region between the electrodes of the conductance cell.

The number of mol of cations in this volume, V, $= \dfrac{N_+}{N}$ (11.120)

where N is Avogadro's number.

$$\therefore c_+ = \frac{N_+}{VN}$$ (11.121)

$$\therefore \frac{N_+}{V} = c_+ N$$ (11.122)

$$\therefore \kappa_+ = N e z_+ u_+ c_+ = F z_+ u_+ c_+$$ (11.123)

$$\lambda_+ = \frac{\kappa_+}{c_+} = F z_+ u_+$$ (11.124)

Since the electrolyte is symmetrical $z_+ = -z_- = |z_-|$ (11.125)

$$\therefore \kappa_- = N e |z_-| u_- c_- = F |z_-| u_- c_-$$ (11.126)

and $$\lambda_- = \frac{\kappa_-}{c_-} = F |z_-| u_-$$ (11.127)

where $|z_-|$ is a number.

$$t_+ = \frac{\lambda_+}{\lambda_+ + \lambda_-} = \frac{F z_+ u_+}{F z_+ u_+ + F |z_-| u_-} = \frac{z_+ u_+}{z_+ u_+ + |z_-| u_-} = \frac{u_+}{u_+ + u_-}$$ (11.128)

and $$t_- = \frac{\lambda_-}{\lambda_+ + \lambda_-} = \frac{F |z_-| u_-}{F z_+ u_+ + F |z_-| u_-} = \frac{|z_-| u_-}{z_+ u_+ + |z_-| u_-} = \frac{u_-}{u_+ + u_-}$$ (11.129)

Remember: $z_+ = -z_- = |z_-|$

For the unsymmetrical electrolyte the final relations are the same, but the argument is more complex, see Section 11.19.3.

Ionic mobilities are strictly constant if the solution is ideal. However, in practice they show a dependence on concentration. This is due to non-ideality which occurs because of the existence of interactions between ions.

Ionic mobilities are important quantities when considering the moving boundary method of measuring transport numbers (see Section 11.19.2) and in the theory of conductance (see Chapter 12 and Appendix 12.1).

Worked problem 11.15

Question

The following are ionic mobilities at infinite dilution at 25°C:

	$\dfrac{10^4 u^0_+}{\text{cm}^2\,\text{s}^{-1}\text{V}^{-1}}$		$\dfrac{10^4 u^0_-}{\text{cm}^2\,\text{s}^{-1}\text{V}^{-1}}$
$Li^+(aq)$	4.01	$Cl^-(aq)$	7.91
$Mg^{2+}(aq)$	5.50	$SO_4^{2-}(aq)$	8.29

(i) Calculate the velocities with which each ion moves:

   (a) in a field, $X = 6.84\ \text{V cm}^{-1}$

   (b) in a field, $X = 50.0\ \text{V cm}^{-1}$

(ii) Calculate the time taken for the $Li^+(aq)$ to migrate a distance of 2.50 cm under the influence of the two fields.

(iii) Calculate the velocity of the $H_3O^+(aq)$ under the action of each field, and comment. Mobility of $H_3O^+(aq)$ is $36.2 \times 10^{-4}\ \text{cm}^2\,\text{s}^{-1}\,\text{V}^{-1}$.

Answer

(i) $v_+ = u_+ X$ and $v_- = u_- X$

For $Li^+(aq)$:

   (a) $v_{Li^+(aq)} = 4.01 \times 10^{-4}\ \text{cm}^2\,\text{s}^{-1}\text{V}^{-1} \times 6.84\ \text{V cm}^{-1} = 2.74 \times 10^{-3}\ \text{cm s}^{-1}$

   (b) $v_{Li^+(aq)} = 4.01 \times 10^{-4}\ \text{cm}^2\,\text{s}^{-1}\,\text{V}^{-1} \times 50.0\ \text{V cm}^{-1} = 2.00 \times 10^{-2}\ \text{cm s}^{-1}$

Proceeding similarly for the other ions gives:

	$\dfrac{X}{\text{V cm}^{-1}}$	$\dfrac{10^2 v}{\text{cm s}^{-1}}$
$Li^+(aq)$	6.84	0.274
	50.0	2.00
$Mg^{2+}(aq)$	6.84	0.376
	50.0	2.75
$Cl^-(aq)$	6.84	0.541
	50.0	3.96
$SO_4^{2-}(aq)$	6.84	0.567
	50.0	4.15

From this data it can be seen that there is no great effect of charge on the velocity with which an ion moves under each field. This is a direct reflection of the small variation in mobilities with charge.

(ii) The velocity, v, is related to the distance, l, covered and the time, t, taken to cover that distance by $v = l/t$, or $t = l/v$.

(a) when the field, $X = 6.84 \text{ V cm}^{-1}$, $t = \dfrac{l}{v} = \dfrac{2.50 \text{ cm}}{2.74 \times 10^{-3} \text{ cm s}^{-1}} = 912 \text{ s}$

(b) when the field, $X = 50.0 \text{ V cm}^{-1}$, $t = \dfrac{l}{v} = \dfrac{2.50 \text{ cm}}{2.00 \times 10^{-2} \text{ cm s}^{-1}} = 125 \text{ s}$

There is a quite dramatic decrease in the time taken to migrate 2.50 cm as the field is increased.

(iii) $v_{H_3O^+(aq)} = 36.2 \times 10^{-4} \text{ cm s}^{-1} \times 6.84 \text{ V cm}^{-1} = 2.48 \times 10^{-2} \text{ cm s}^{-1}$

$$v_{H_3O^+(aq)} = 36.2 \times 10^{-4} \text{ cm s}^{-1} \times 50.0 \text{ V cm}^{-1} = 0.181 \text{ cm s}^{-1}$$

The corresponding times for the ion to move 2.50 cm under the influence of these fields are: 101 s and 13.8 s.

The $H_3O^+(aq)$ has an abnormally high velocity of migration under the influence of an applied field, and this is despite its being highly hydrated (see Section 11.18).

Worked problem 11.16

Question

The following are values of ionic mobilities at infinite dilution at 25°C:

	$\dfrac{10^4 u_+^0}{\text{cm}^2 \text{ s}^{-1} \text{ V}^{-1}}$		$\dfrac{10^4 u_-^0}{\text{cm}^2 \text{ s}^{-1} \text{ V}^{-1}}$
$Li^+(aq)$	4.01	$F^-(aq)$	5.74
$Na^+(aq)$	5.19	$Cl^-(aq)$	7.91
$Mg^{2+}(aq)$	5.50	$Br^-(aq)$	8.10
$Ca^{2+}(aq)$	6.17	$SO_4^{2-}(aq)$	8.29

(i)   Infer the corresponding values for λ_+^0 and λ_-^0.

(ii) From these values calculate the transport numbers for the cation and anion of $NaBr(aq)$, $Na_2SO_4(aq)$ and $CaBr_2(aq)$.

(iii) Compare these with the corresponding values calculated directly from the ionic mobilities given above.

Answer

(i) $\lambda_+^0 = z_+ F u_+^0$ $\qquad\qquad\qquad\qquad\qquad\qquad \lambda_-^0 = |z_-| F u_-^0$

For $Li^+(aq)$:

$\lambda_+^0 = 1 \times 96485 \text{ C mol}^{-1} \times 4.01 \times 10^{-4} \text{ cm}^2 \text{ s}^{-1} \text{ V}^{-1} = 38.7 \text{ C s}^{-1} \text{ V}^{-1} \text{ cm}^2 \text{ mol}^{-1}$

Be careful with the units: $1\,C\,s^{-1}\,V^{-1} = 1\,A\,V^{-1} = 1\,S$

$$\lambda_+^0 = 38.7\,S\,cm^2\,mol^{-1}$$

For $Mg^{2+}(aq)$:

$$\lambda_+^0 = 2 \times 96485\,C\,mol^{-1} \times 5.50 \times 10^{-4}\,cm^2\,s^{-1}\,V^{-1} = 106.1\,C\,s^{-1}\,V^{-1}\,cm^2\,mol^{-1}$$
$$\lambda_+^0 = 106.1\,S\,cm^2\,mol^{-1}$$

Be careful and remember: for negative ions the charge on the ion comes in through $|z_-|$ which is a number only and does not include the sign.
For $F^-(aq)$:

$$\lambda_-^0 = 1 \times 96485\,C\,mol^{-1} \times 5.74 \times 10^{-4}\,cm^2\,s^{-1}\,V^{-1} = 55.4\,C\,s^{-1}\,V^{-1}\,cm^2\,mol^{-1}$$
$$\lambda_-^0 = 55.4\,S\,cm^2\,mol^{-1}$$

For $SO_4{}^{2-}(aq)$:

$$\lambda_-^0 = 2 \times 96485\,C\,mol^{-1} \times 8.29 \times 10^{-4}\,cm^2\,s^{-1}\,V^{-1} = 160.0\,C\,s^{-1}\,V^{-1}\,cm^2\,mol^{-1}$$
$$\lambda_-^0 = 160.0\,S\,cm^2\,mol^{-1}$$

Continuing as above gives the following table:

	$\dfrac{\lambda_+^0}{S\,cm^2\,mol^{-1}}$		$\dfrac{\lambda_-^0}{S\,cm^2\,mol^{-1}}$
$Li^+(aq)$	38.7	$F^-(aq)$	55.4
$Na^+(aq)$	50.1	$Cl^-(aq)$	76.3
$Mg^{2+}(aq)$	106.1	$Br^-(aq)$	78.2
$Ca^{2+}(aq)$	119.0	$SO_4^{2-}(aq)$	160.0

(ii) For the electrolyte A_xB_y ionising to give the ions A^{y+} and B^{x-}:

$$\Lambda = x\lambda_+ + y\lambda_-$$

$$t_+ = \frac{x\lambda_+}{\Lambda} = \frac{x\lambda_+}{x\lambda_+ + y\lambda_-} \quad \text{and} \quad t_- = \frac{y\lambda_-}{\Lambda} = \frac{y\lambda_-}{x\lambda_+ + y\lambda_-}$$

$$t_+ = \frac{u_+}{u_+ + u_-} \quad \text{and} \quad t_- = \frac{u_-}{u_+ + u_-}$$

where t_+ and t_- are transport numbers.

$$NaBr(aq) : \Lambda^0 = \lambda_+^0 + \lambda_-^0 = (50.1 + 78.2)\,S\,cm^2\,mol^{-1} = 128.3\,S\,cm^2\,mol^{-1}$$

$$t_+ = \frac{\lambda_+}{\Lambda} = \frac{\lambda_+}{\lambda_+ + \lambda_-} = \frac{50.1\,S\,cm^2\,mol^{-1}}{128.3\,S\,cm^2\,mol^{-1}} = 0.391$$

$$t_- = \frac{\lambda_-}{\Lambda} = \frac{\lambda_-}{\lambda_+ + \lambda_-} = \frac{78.2\,S\,cm^2\,mol^{-1}}{128.3\,S\,cm^2\,mol^{-1}} = 0.609$$

$$t_+ = \frac{u_+}{u_+ + u_-} = \frac{5.19 \times 10^{-4}\,cm^2\,s^{-1}\,V^{-1}}{(5.19 + 8.10) \times 10^{-4}\,cm^2\,s^{-1}\,V^{-1}} = 0.391$$

$$t_- = \frac{u_-}{u_+ + u_-} = \frac{8.10 \times 10^{-4}\,cm^2\,s^{-1}\,V^{-1}}{(5.19 + 8.10) \times 10^{-4}\,cm^2\,s^{-1}\,V^{-1}} = 0.609$$

For $Na_2SO_4(aq): \Lambda^0 = 2\lambda_+^0 + \lambda_-^0 = (2 \times 50.1 + 160.0) \, S \, cm^2 \, mol^{-1} = 260.2 \, S \, cm^2 \, mol^{-1}$

$$t_+ = \frac{2\lambda_+}{\Lambda} = \frac{2\lambda_+}{2\lambda_+ + \lambda_-} = \frac{2 \times 50.1 \, S \, cm^2 \, mol^{-1}}{260.2 \, S \, cm^2 \, mol^{-1}} = 0.385$$

$$t_- = \frac{\lambda_-}{\Lambda} = \frac{\lambda_-}{2\lambda_+ + \lambda_-} = \frac{160.0 \, S \, cm^2 \, mol^{-1}}{260.2 \, S \, cm^2 \, mol^{-1}} = 0.615$$

$$t_+ = \frac{u_+}{u_+ + u_-} = \frac{5.19 \times 10^{-4} \, cm^2 \, s^{-1} \, V^{-1}}{(5.19 + 8.29) \times 10^{-4} \, cm^2 \, s^{-1} \, V^{-1}} = 0.385$$

$$t_- = \frac{u_-}{u_+ + u_-} = \frac{8.29 \times 10^{-4} \, cm^2 \, s^{-1} \, V^{-1}}{(5.19 + 8.29) \times 10^{-4} \, cm^2 \, s^{-1} \, V^{-1}} = 0.615$$

For $CaBr_2(aq): \Lambda^0 = \lambda_+^0 + 2\lambda_-^0 = (119.0 + 2 \times 78.2) \, S \, cm^2 \, mol^{-1} = 275.4 \, S \, cm^2 \, mol^{-1}$

$$t_+ = \frac{\lambda_+}{\Lambda} = \frac{\lambda_+}{\lambda_+ + 2\lambda_-} = \frac{119.0 \, S \, cm^2 \, mol^{-1}}{275.4 \, S \, cm^2 \, mol^{-1}} = 0.432$$

$$t_- = \frac{2\lambda_-}{\Lambda} = \frac{2\lambda_-}{\lambda_+ + 2\lambda_-} = \frac{2 \times 78.2 \, S \, cm^2 \, mol^{-1}}{275.4 \, S \, cm^2 \, mol^{-1}} = 0.568$$

$$t_+ = \frac{u_+}{u_+ + u_-} = \frac{6.17 \times 10^{-4} \, cm^2 \, s^{-1} \, V^{-1}}{(6.17 + 8.10) \times 10^{-4} \, cm^2 \, s^{-1} \, V^{-1}} = 0.432$$

$$t_- = \frac{u_-}{u_+ + u_-} = \frac{8.10 \times 10^{-4} \, cm^2 \, s^{-1} \, V^{-1}}{(6.17 + 8.10) \times 10^{-4} \, cm^2 \, s^{-1} \, V^{-1}} = 0.568$$

The values for the transport numbers are internally self consistent between the calculation from molar ionic conductivities and from the ionic mobilities. This is as it should be. The sum of the transport numbers for the ions of a given electrolyte is unity. This, again, is as it should be.

11.18 Abnormal mobility and ionic molar conductivity of $H_3O^+(aq)$

At 25°C the mobilities of $H_3O^+(aq)$ and $OH^-(aq)$ are 30.2×10^{-4} and $20.5 \times 20^{-4} \, cm^2 \, s^{-1} \, V^{-1}$ respectively and the corresponding molar ionic conductivities at infinite dilution are 349.82 and $198.0 \, S \, cm^2 \, mol^{-1}$ respectively. These are much larger than typical values for other ions, e.g. for $Mg^{2+}(aq)$ the values are $u_+ = 5.50 \times 10^{-4} \, cm^2 \, s^{-1} \, V^{-1}$ and $\lambda^0 = 106.12 \, S \, cm^2 \, mol^{-1}$. At first sight this may seem surprising. First thoughts might suggest that this is due to the small size of the proton, but this cannot be the explanation since the proton is known to be highly hydrated. The size of the hydrated $H_3O^+(aq)$ would be reasonably comparable to many hydrated ions, so hydration in itself is also not the answer. The abnormally high ionic mobility and ionic conductivity are attributed to the ability of the bare H^+ to jump between successive water molecules as shown in Figure 11.9(a).

For the transference of the proton to continue along the row of water molecules, each water molecule must rotate after the proton has passed to the subsequent water molecule. This will allow the vacated water molecule to be in the right orientation to receive another proton. In this way a proton can be passed quickly from one water molecule to the next. Calculations show that this process will occur much faster than ordinary migration of $H_3O^+(aq)$ through the

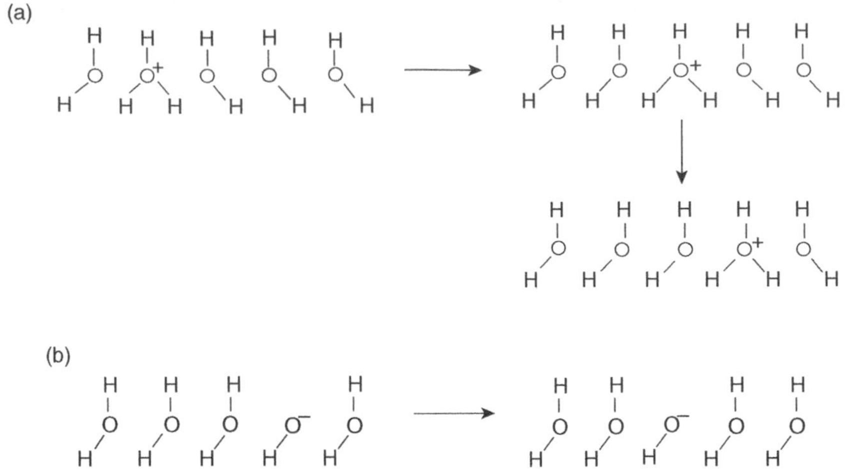

(a)

(b)

Figure 11.9 (a) Proton jumps for $H_3O^+(aq)$. **(b)** Proton jumps for $OH^-(aq)$.

solution. A similar mechanism has been proposed to account for the high mobility and ionic molar conductivity of $OH^-(aq)$ as given in Figure 11.9(b).

11.19 Measurement of transport numbers

It is very important to be able to measure transport numbers over a range of concentrations as this is the only way to determine the dependence of individual molar ionic conductivities as a function of concentration. If this can be done, then it means that observed molar conductivities for any electrolyte at given concentrations can be split up into the contributions from the ions of the electrolyte.

For a symmetrical electrolyte AB ionising to give $A^{z+}(aq)$ and $B^{z-}(aq)$:

$$\Lambda = \lambda_+ + \lambda_- \tag{11.41}$$

$$t_+ = \frac{\lambda_+}{\Lambda} \tag{11.96}$$

If Λ and t_+ are known, then λ_+ can be found. From this, λ_- is found, since $\Lambda = \lambda_+ + \lambda_-$. If Λ and t_+ have been found over a range of concentrations, then λ_+ and λ_- can now be found over that range of concentrations. This is particularly important since these can now be used to test theories of conductance dealing with non-ideality (see Chapter 12).

For the unsymmetrical A_xB_y ionising to give A^{y+} and B^{z-}:

$$\Lambda = x\lambda_+ + y\lambda_- \tag{11.43}$$

$$t_+ = \frac{x\lambda_+}{\Lambda} \tag{11.100}$$

Again, if Λ and t_+ are known, then λ_+ can be found. From this, λ_- is found since $\Lambda = x\lambda_+ + y\lambda_-$, and likewise the dependence of λ_+ and λ_- on concentration can be found.

There are two main methods for determining transport numbers, both of which were developed early in the study of conductance. They are the Hittorf method and the moving boundary method. The emf method has been described in Section 9.21.

11.19.1   The Hittorf method for determining transport numbers

In this method an electrolysis experiment is set up. The passage of direct current causes changes in concentration around each electrode as a result of:

- the chemical reaction occurring at the electrode as a result of passage of the current;

- migration of ions to and from the electrodes, i.e.:

  - cations migrating to the cathode compartment and away from the anode compartment and

  - anions migrating to the anode compartment and away from the cathode compartment.

A typical Hittorf apparatus is illustrated in Figure 11.10.

The Hittorf apparatus is filled with a solution of known concentration. A known current is then passed through the solution for a known time from which the quantity of electricity

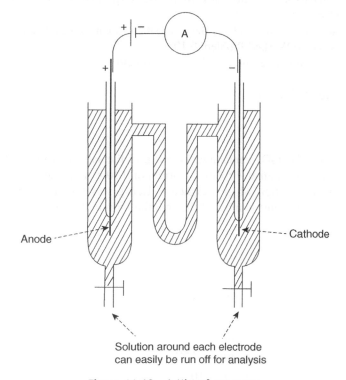

Solution around each electrode
can easily be run off for analysis

Figure 11.10   A Hittorf apparatus.

passed can be calculated. The solution around the electrode in each electrode compartment is run off; each is weighed and then analysed to give the number of gram of electrolyte per given mass of electrolyte solution run off. From this the change in concentration around each electrode can be found. Comparison with the theoretically expected change in concentration enables the transport numbers to be found. In practice, the analysis is often carried out for one compartment only.

The technique can be illustrated by considering what happens when a solution of $AgNO_3$ is electrolysed in a Hittorf apparatus with silver electrodes. The quantity of electricity passed is $It/96485$ faraday where the current is in amp and the time in seconds.

Processes occurring at the anode:

Chemical reaction: $Ag(s) \rightarrow Ag^+(aq) + e$
Migration: $NO_3^-(aq)$ migrates to the anode solution
$\qquad\qquad Ag^+(aq)$ migrates away from the anode solution

If 1 faraday of electricity is passed this will cause 1 mol of $Ag^+(aq)$ to be formed and t_+ mol of $Ag^+(aq)$ to migrate away, giving a net gain of $(1-t_+)$ mol in the anode compartment. If b faraday is passed the net gain in the anode compartment will be $b(1-t_+)$ mol of $Ag^+(aq)$.

Analysis of the anode compartment for $Ag^+(aq)$ allows the number of mol of $Ag^+(aq)$ to be found. If there is a gain of y mol of $Ag^+(aq)$ around the anode, then $y = b(1-t_+)$. Since both y and b have been found experimentally the t_+ for the $Ag^+(aq)$ can be found. t_- for the $NO_3^-(aq)$ can then be found since $t_+ + t_- = 1$.

If the electrolyte is unsymmetrical, e.g $BaCl_2(aq)$ then care must be taken in formulating the migration processes (see Worked Problem 9.10).

Hittorf's method is illustrated in the following problem.

Worked problem 11.17

Question

An aqueous solution of $CdSO_4$ containing 115.60 g of the salt per 1000 g of solution was electrolysed between cadmium electrodes. A current of 0.1000 amp was passed for 5 h. After electrolysis the anode solution which weighed 171.66 g was found to contain 20.72 g of $CdSO_4$. Calculate the transport number of the cadmium ion in the solution.

Answer

Quantity of electricity passed $=$ current $\times$ time for which it is passed

$$= 0.1000 \times 5 \times 3600 \text{ amp s}$$
$$= 1800\,C$$
$$= \frac{1800}{96485} \text{ faraday}$$
$$= 1.866 \times 10^{-2} \text{ faraday}$$

(see Section 9.2.1 and Worked Problem 9.3).
The question gives information as to what happens to the solution around the anode as a result of electrolysis.

Passage of current has two effects:

- Reaction at the anode involves release of electrons: $Cd(s) \rightarrow Cd^{2+}(aq) + 2e$

- $Cd^{2+}(aq)$ migrates from the anode compartment to the cathode compartment.

Hence as a result of passage of the current:

- $Cd^{2+}(aq)$ is **formed** at the anode as a result of the electrode reaction;

- $Cd^{2+}(aq)$ is **lost** from the solution around the anode as a result of migration.

For passage of 1 faraday of electricity:

- electrode reaction: ½ mol of $Cd^{2+}(aq)$ will be formed in the anode compartment;

- ½ t_+ mol of $Cd^{2+}(aq)$ migrates out of the anode compartment;

- net gain in the anode compartment for passage of 1 faraday is $(½ - ½ t_+)$ mol $= ½(1 - t_+)$ mol of $Cd^{2+}(aq)$;

- and for passage of 1.866×10^{-2} faraday there will be a gain in the anode compartment of ½ $(1 - t_+) \times 1.866 \times 10^{-2}$ mol of $Cd^{2+}(aq)$.

- This will be equal to the observed gain in $CdSO_4(aq)$ in the anode compartment.

Initially: there is 115.60 g of $CdSO_4(aq)$ per 1000 g of solution, i.e. 115.60 g of $CdSO_4(aq)$ for 884.40 g of $H_2O(l)$, which corresponds to $115.60/208.47 = 0.5545$ mol of $Cd^{2+}(aq)$ per 884.4 g of $H_2O(l)$.
Finally: there is 20.72 g of $CdSO_4(aq)$ in 171.66 g of solution, i.e. 20.72 g of $CdSO_4(aq)$ in 150.94 g of $H_2O(l)$ which corresponds to $20.72/208.47 = 0.09939$ mol of $Cd^{2+}(aq)$ in 150.94 g of $H_2O(l)$.
Initially 150.94 g of $H_2O(l)$ would contain $0.5545 \times 150.94/884.4 = 0.09464$ mol of $Cd^{2+}(aq)$.

Finally 150.94 g of $H_2O(l)$ contains 0.09939 mol of $Cd^{2+}(aq)$.

∴ net gain in the anode compartment $= (0.09939 - 0.09464) = 0.00475$ mol of $Cd^{2+}(aq)$.
This has been shown to be equal to $½(1 - t_+) \times 1.866 \times 10^{-2}$ mol of $Cd^{2+}(aq)$.

$$\therefore (1 - t_+) = \frac{0.00475 \text{ mol}}{0.5 \times 1.866 \times 10^{-2} \text{mol}} = 0.509$$

$$\therefore t_+ = 0.491$$

i.e. the transport number for $Cd^{2+}(aq) = 0.491$

There are two potential difficulties when working through this problem:

- problems inherent in comparing the anode compartment solution before and after electrolysis. The number of mol of Cd^{2+} (aq) present in the solution (a) before electrolysis and (b) after electrolysis is known. But since these relate to **different** masses of solution and thus to **different** masses of water, it is essential that these be adjusted to relate to the **same** mass of water, i.e. the mass of water in the final solution. It is only after this has been done that the gain in the number of mol of Cd^{2+}(aq) can be estimated by direct subtraction.

- problems in finding the number of mol of Cd^{2+}(aq) migrating as a result of passage of the current. The argument is given in terms of the passage of 1 faraday of electricity. Of this, a fraction, t_+, is conveyed by the Cd^{2+}(aq) and a fraction t_- is conveyed by SO_4^{2-}(aq), where $t_+ + t_- = 1$. A charge of t_+ faraday is conveyed by the migrating Cd^{2+}(aq) per one faraday passed. But 1 mol of Cd^{2+}(aq) has a charge of 2 faraday, and so a charge of t_+ faraday will correspond to $\frac{1}{2} t_+$ mol of Cd^2(aq).

11.19.2   The moving boundary method

This method for determining transport numbers is the preferred method since it is inherently more accurate than the Hittorf method. Even more important is the fact that the transport numbers can be found over a range of concentrations. This, in turn, means that ionic mobilities and individual ionic molar conductivities can also be found over a range of concentrations.

The technique consists of a tube with cathode and anode inserted at the ends. These are connected to a source of current, generally a battery where the potential difference generated by the battery remains constant throughout the experiment. The simplest moving boundary experiment has an arrangement as in Figure 11.11(a). Two electrolyte solutions of known concentrations are used. One of these contains a solution of the cation for which the transport number is being determined, the other is a solution of an electrolyte with the same anion but with a different cation. In the illustrative example the two solutions are KCl(aq) and LiCl(aq). The LiCl(aq) has a lower concentration than the KCl(aq) and consequently has a lower density. The tube is also in a vertical position with the solution of lower density on the top. This cuts down the extent of mixing between the two solutions due to gravity mixing. A distinct boundary appears between the two solutions. This occurs because of the different refractive indices of the two solutions.

A known current is then passed through the solutions for a known length of time and the movement of the boundary is followed with time (see Figure 11.11(b)). After time t the boundary will have moved from the initial position, x, to the final position, x'.

In this time, all of the K^+(aq) at x will have moved to x'; and all of the other K^+(aq) in between will have moved beyond x'. All of the K^+(aq) which started in the volume of the solution of KCl(aq) between x and x' will have moved downward.

$$\text{amount of } K^+(aq) \text{ in that volume} = n_+ = c_+ \times V = c_+ \times A \times l \qquad (11.130)$$

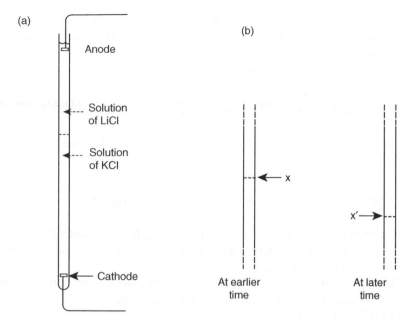

Figure 11.11 (a) A moving boundary experiment showing a schematic apparatus. **(b)** Movement of the boundary with time.

where $l = x' - x$, A is the cross sectional area of the tube and c_+ is the concentration of $K^+(aq)$.

$$\text{number of } K^+(aq) \text{ in that volume} = N \times c_+ \times V = N \times c_+ \times A \times l \qquad (11.131)$$

$$\text{charge on that number of } K^+(aq) = N \times (z_+e) \times c_+ \times A \times l \qquad (11.132)$$

where N is Avogadro's number.

$$\text{But } N \times e = F \qquad (11.133)$$

$\therefore$ charge carried by the $K^+(aq)$ in migrating from x to x'

$$= N \times (z_+e) \times c_+ \times A \times l = z_+c_+ \times F \times A \times l \qquad (11.134)$$

The current and time will give an experimental value of the total charge passed:
total charge passed during the time that it takes for the boundary to move from x to $x' = I \times t$

$$(11.135)$$

The transport number for the $K^+(aq)$, t_+ is:

$$t_+ = \frac{\text{charge carried by } K^+(aq)}{\text{total charge carried}} = \frac{z_+c_+ \times F \times A \times l}{I \times t} \qquad (11.136)$$

All the quantities in this equation are known and so the transport number for $K^+(aq)$ can be found for the given [KCl].

Since $t_+ + t_- = 1$, the transport number for the $Cl^-(aq)$ can be found.

Varying the concentration of the KCl solution will allow determination of the transport numbers for both ions over a range of concentrations. From these the individual molar ionic conductivities for KCl(aq) over a range of concentrations can be found.

$$t_+ = \frac{\lambda_+}{\lambda_+ + \lambda_-} = \frac{\lambda_+}{\Lambda} \quad \text{and} \quad t_- = \frac{\lambda_-}{\lambda_+ + \lambda_-} = \frac{\lambda_-}{\Lambda} \qquad (11.137) \text{ and } (11.138)$$

This is the only way in which the molar conductivity of an electrolyte can be split up into its individual ionic contributions for situations other than infinite dilution.

11.19.3 The argument for the unsymmetrical electrolyte

The argument given above for the symmetrical electrolyte, KCl(aq), will hold for an unsymmetrical electrolyte **provided** c_+ refers to the concentration of the cation and not to the overall concentration of the electrolyte, e.g.

for the electrolyte A_xB_y, $[A^{y+}] = c_+ = xc$ where c is the overall concentration of A_xB_y. A_xB_y ionises to give A^{y+} and B^{x-}, and so $z_+ = y$ and $|z_-| = x$.

Equations (11.124) and (11.127), i.e.:

$$\lambda_+ = Fz_+u_+ \qquad (11.124) \quad \text{and} \qquad \lambda_- = F|z_-|u_- \qquad (11.127)$$

can be translated into the more useful forms for unsymmetrical electrolytes:

$$\lambda_+ = Fyu_+ \qquad (11.139) \quad \text{and} \qquad \lambda_- = Fxu_- \qquad (11.140)$$

$$t_+ = \frac{x\lambda_+}{x\lambda_+ + y\lambda_-} \qquad (11.141) \quad \text{and} \qquad t_- = \frac{y\lambda_-}{x\lambda_+ + y\lambda_-} \qquad (11.142)$$

$$= \frac{xFyu_+}{xFyu_+ + yFxu_-} \qquad (11.143) \quad \text{and} \qquad = \frac{yFxu_-}{xFyu_+ + yFxu_-} \qquad (11.144)$$

$$= \frac{u_+}{u_+ + u_-} \qquad (11.145) \quad \text{and} \qquad = \frac{u_-}{u_+ + u_-} \qquad (11.146)$$

which are the same relations as derived in Section 11.17 for the symmetrical electrolyte (Equations 11.128 and 11.129).

11.19.4 The fundamental basis of the moving boundary method

The migration of the ions under the influence of the electric field, set up by virtue of the potential difference between the electrodes, is governed by the mobilities of the ions, u. The movement of the boundary between the two electrolyte solutions, on the other hand, is governed by the velocities, v, of the two cations, here $K^+(aq)$ and $Li^{2+}(aq)$. In particular, the ratio of the velocities of the two cations must be unity which means that the $K^+(aq)$ and the $Li^+(aq)$ must move at the same speed. This results from the need to maintain electrical neutrality throughout, as will be shown below.

The experiment can be done under conditions such that the potential gradient, i.e. the field, X, under which the $Li^+(aq)$ moves is independent of how far down the tube the solution has reached. The same is true for the K^+(aq). **But watch**: the fields are different for each ion. This is because the field under which each ion moves will be different because the ionic mobilities are different for the two ions. If the Li^+(aq) moved at a slower velocity than the K^+(aq), then the boundary would become diffuse with a region between devoid of cations. This cannot happen because the solution must be electrically neutral throughout. This requires that the Li^+(aq) and K^+(aq) must move with the same velocity, and so a sharp boundary between the two solutions must be maintained.

Note well: it is crucial to distinguish between velocity and mobility. The velocity with which an ion moves under the influence of an applied field is proportional to the applied field:

$$v \propto X \tag{11.147}$$

$$\therefore v = uX \tag{11.148}$$

where u, the mobility, is a constant of proportionality. From which it can be seen that the mobility of an ion and the velocity of the ion **must not be confused**.

Since, in the situation under consideration u and X are constant, then v must also be constant.

Consider the situation at two times t_1 and t_2 when the boundary is at positions x and x': Since electrical neutrality requires that the velocity with which the Li^+(aq) moves and that at which the K^+(aq) must be equal, then it follows from Equation (11.144) that:

$$uX_1(Li^+) = uX_1(K^+) \quad (11.149) \quad \text{and} \quad uX_2(Li^+) = uX_2(K^+) \quad (11.150)$$

$$\frac{X_1(Li^+)}{X_1(K^+)} = \frac{u(K^+)}{u(Li^+)} \quad (11.151) \quad \text{and} \quad \frac{X_2(Li^+)}{X_2(K^+)} = \frac{u(K^+)}{u(Li^+)} \quad (11.152)$$

Since $u(Li^+)$ is a constant, and $u(K^+)$ is another constant, it follows that the ratio of the fields under which each ion moves will remain constant, and it will also be >1 since the mobility of the K^+(aq) is greater than the mobility of the Li^+(aq). Figure 11.12 shows how the field will vary for each solution at two different times, t_1 and t_2.

Worked problem 11.18

Question

Give an interpretation of the observation made directly above that the mobility of the K^+(aq) is greater than the mobility of the Li^+(aq).

Answer

Since potassium lies below lithium in the periodic table, the bare K^+ ion will be larger than the bare Li^+ ion. Large ions would be expected to move more slowly than small ions under the influence of the same electric field, and so the mobility of a large ion would be expected to be smaller than that of a small ion. The fact that the mobility of K^+(aq) is larger than that for Li^+(aq) indicates that Li^+(aq) is larger than K^+(aq), and this would suggest that the lithium ion is more highly hydrated in aqueous solution than is the potassium ion.

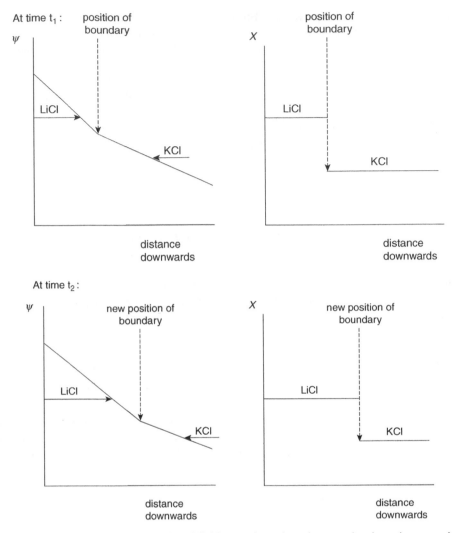

Figure 11.12 Behaviour of potential and field at various times in a moving boundary experiment. Magnitude of field X is the gradient of the graph of ψ vs distance.

Worked problem 11.19

Question

The cations in a moving boundary experiment are $Li^+(aq)$ and $Mg^{2+}(aq)$. The LiCl solution is less dense than the $MgCl_2(aq)$ and the $Li^+(aq)$ is the following ion. The field under which the $Li^+(aq)$ moves is 5.0 V cm^{-1}. Calculate the field under which the $Mg^{2+}(aq)$ is moving. Use this to draw a schematic diagram of the potential difference as a function of distance along the tube as the two solutions migrate along the tube. If the boundary moves a distance of 1.75 cm along the tube during the course of an experiment, calculate the time taken to carry out the experiment.

The mobilities are: $Li^+(aq)$ $4.01 \times 10^{-4}\,cm^2\,s^{-1}\,V^{-1}$ and $Mg^{2+}(aq)$ $5.50 \times 10^{-4}\,cm^2\,s^{-1}\,V^{-1}$.

Answer

$$v_+ = u_+X$$

$$Li^+(aq) : v_{Li^+(aq)} = 4.01 \times 10^{-4}\,cm^2\,s^{-1}\,V^{-1} \times 5.0\,V\,cm^{-1} = 20.05 \times 10^{-4}\,cm\,s^{-1}$$

In a moving boundary experiment electrical neutrality requires that the two cations migrate down the tube at the same velocity and this, in turn, means that a sharp boundary is always maintained. In turn, this implies that the fields under which the ions move are different.

The velocity of migration of the $Mg^{2+}(aq)$ must therefore, be $20.05 \times 10^{-4}\,cm\,s^{-1}$.

$$X = \frac{v}{u} = \frac{20.05 \times 10^{-4}\,cm\,s^{-1}}{5.50 \times 10^{-4}\,cm^2\,s^{-1}\,V^{-1}} = 3.65\,V\,cm^{-1}$$

Figure 11.12 gives a schematic diagram of the potential drop, i.e. the field, with distance for each solution, where $M_gCl_2(aq)$ is substituted for KCl (aq).

The time taken for the boundary to move a distance of 1.75 cm down the tube $= \dfrac{distance}{velocity}$

$$\therefore t = \frac{l}{v} = \frac{1.75\,cm}{20.05 \times 10^{-4}\,cm\,s^{-1}} = 873\,s.$$

The experiment, therefore, lasts 873 s or 14.55 minutes.

11.19.5 Summary of the use made of transport numbers and mobilities

These are both very fundamental quantities in the study of electrolyte solutions. Transport numbers are necessary for the subdivision of molar conductivities into the individual ionic contributions. Because they can be studied over a range of concentrations this allows tabulated data to be drawn up for ionic conductivities for a wide variety of ions over a range of concentrations. This is important in itself, but is also useful when comparing theory with experiment.

Mobilities are important in themselves simply as a property of the hydrated ions. But, more importantly they enable determination of ionic radii to be made for the hydrated ions. This makes use of Stokes' Law (see Section 12.4.2), which requires that for an ion of charge ze and effective radius, a, moving with a velocity, v, in a medium of viscosity, η, under the influence of an external field, X:

$$6\pi a\eta v = zeX \tag{1.153}$$

$$\therefore v = \frac{zeX}{6\pi a\eta} \tag{1.154}$$

$$= uX \tag{1.155}$$

where u is the mobility of the ion and equals $\dfrac{ze}{6\pi a\eta}$.

If u has been determined independently, then a follows directly. Table 11.1 gives a few values determined in this way.

Table 11.1 Values of the effective radius found from measured ionic mobilities and Stokes' Law, taking $\eta = 8.903 \times 10^{-4} \, \text{kg m}^{-1}\text{s}^{-1}$

ion	$\dfrac{a}{\text{nm}}$	crystallographic radii/nm
$Li^+(aq)$	0.238	0.060
$Na^+(aq)$	0.184	0.096
$Mg^{2+}(aq)$	0.347	0.065
$Ca^{2+}(aq)$	0.309	0.099
$F^-(aq)$	0.166	0.134
$Cl^-(aq)$	0.121	0.181
$Br^-(aq)$	0.118	0.195
$SO_4^{2-}(aq)$	0.230	

The effective ionic radius of an ion in solution is an important quantity in the discussion of the behaviour of electrolyte solutions. Mobilities are thus fundamental to both Debye-Hückel theory and conductance theory.

It is clear from Table 11.1 that the effective ionic radius found from Stokes' Law is greater than the crystallographic radius for the cations, suggesting that these ions are probably hydrated in solution. The situation is much less clear for the anions. $F^-(aq)$ is the only anion where the effective radius gives evidence for hydration. Chapter 13 looks at this in more detail.

12

Theories of Conductance:
The Non-ideal Case for
Symmetrical Electrolytes

Chapter 11 focused attention on methods of analysing conductance data where the effects of non-ideality have been ignored, i.e. it has been assumed that there are no ionic interactions. The movement of ions in solution is then a result of motion induced by an applied potential gradient, i.e. an external field superimposed on random Brownian motion. The applied electric field will cause the positive ions to move in the direction of the field and anions to move in the opposite direction. The direction of the field is from the positive pole to the negative pole of the electrical system, and the field is set up by virtue of the potential drop between the two poles.

Non-ideality has been shown to be due to ionic interactions between the ions and consideration of these led to the concept of the ionic atmosphere (see Sections 10.3 and 10.5). These interactions must be taken into account in any theory of conductance. Most of the theories of electrolyte conduction use the Debye-Hückel model, but this model has to be modified to take into account extra features resulting from the movement of the ions in the solvent under the applied field. This has proved to be a very difficult task and most of the modern work has attempted many refinements all of which are mathematically very complex. Most of this work has focused on two effects which the existence of the ionic atmosphere imposes on the movement and velocity of the ions in an electrolyte solution. These are the relaxation and electrophoretic effects.

Section 12.9 on post 1950 modern conductance theories for symmetrical electrolytes and Section 12.10 on Fuoss-Onsager's 1957 conductance equation for symmetrical electrolytes can be omitted until earlier sections are assimilated. These two sections deal with more up to date work which is able to be formulated in a straightforward analytical equation. The development behind these theories is complex and only a brief overview of the ideas behind these theories is given. Nonetheless the Fuoss-Onsager 1957 equation has been much used to analyse experimental data. How this is carried out in practice is given in Sections 12.10.1 to 12.13 and this is now standard procedure and is relevant to anyone analysing conductance work.

An Introduction to Aqueous Electrolyte Solutions. By Margaret Robson Wright
© 2007 John Wiley & Sons Ltd   ISBN 978-0-470-84293-5 (cloth) ISBN 978-0-470-84294-2 (paper)

The whole of Section 12.17 discusses the more recent thoughts on conductance theory. This is given in a qualitative manner, and should be useful in illustrating modern concepts in the microscopic description of electrolyte solutions. These sections, taken in conjunction with Sections 10.14 onwards in Chapter 10 on the theory of electrolyte solutions, and with Chapter 13 on solvation, should give the student a qualitative appreciation of more modern approaches to non-ideality in electrolyte solutions.

Aims

By the end of this chapter you should be able to:

- describe and explain the significance of the relaxation and electrophoretic effects in conductance theory;

- appreciate how the Wien and Debye-Falkenhagen effects support the concept of an ionic atmosphere;

- discuss the models used in the early conductance theories;

- follow through the arguments leading up to the Debye-Hückel-Onsager equation;

- explain how the Debye-Hückel-Onsager equation is modified to account for weak electrolytes and ion pairing;

- follow through the outline of the arguments and modifications introduced by Fuoss and Onsager in their 1957 theory for unassociated electrolytes;

- show how this equation was modified to account for association;

- be aware of how this equation can be used to determine Λ^0, α, $\mathring{a}$ and K_{assoc},

- realise how complex any further conductance theories have to be.

12.1 The relaxation effect

In the absence of an electric field the ionic atmosphere is symmetric about the central ion with the charge distribution fading exponentially to zero with distance from the central ion. When the ion moves under the influence of an applied external field, it will normally still have an ionic atmosphere associated with it. However, at each stage of movement, a new ionic atmosphere must be built up around the moving ion, and it takes time for this to happen. The ionic atmosphere has to build up in front of the moving ion and to decay behind it. Because this process cannot occur instantaneously, the net result is that the ionic atmosphere gets displaced with respect to the moving ion. There is more of the ionic atmosphere behind the ion than in front of the ion, i.e. the ionic atmosphere is asymmetric in contrast to the symmetry obtained in the absence of an externally applied electric field.

There are two processes involved in the build up of this asymmetry:

- diffusive processes of the ions through the solution. These will occur irrespective of the movement under the influence of the applied external field. They are present in the equilibrium situation when the ionic atmosphere is symmetric and are nearly random in nature.

- the movement of the ions under the influence of the external field where the velocity of an ion is proportional to the field, i.e. $v = u X$ where u is the mobility of the ion (see Section 11.17). If the central reference ion is a cation, movement under the influence of the external field will cause this central ion and the cations of the ionic atmosphere to move in the same direction towards the negative pole of the applied potential, i.e. in the direction of the field. The anions will move at a different velocity in the opposite direction. This causes a deficit of negative charge in front of the moving central reference cation and an excess of negative charge behind it. In consequence, the ionic atmosphere becomes asymmetric around the moving cation. A similar process will occur when the moving reference ion is an anion. In this case, there will be a deficit of positive charge in front of the moving anion and an excess of positive charge behind it.

The process giving rise to this asymmetry is called the relaxation effect, and the time taken for the ion to build up its new ionic atmosphere is called the relaxation time. As the ion moves, the build up and decay are occurring continuously. Since the process is a relaxation phenomenon, then the build up of the asymmetry in the ionic atmosphere under the influence of the external field and the decay of the asymmetry to the symmetrical ionic atmosphere once the external field is removed will be first order processes. The overall rate constant for a first order relaxation process can be shown to be given by $k_{overall} = k_{build\,up} + k_{decay}$, and the relaxation time is given by $\tau = 1/(k_{build\,up} + k_{decay})$. This applies even if the rate constants for build up and decay are different. This has the consequence that the same relaxation time is applicable to both build up and decay.

The effect of the asymmetry of the ionic atmosphere when a cation is moving forward in the direction of the applied field, or potential, is to 'pull the cation back'. The velocity of the ion in the forward direction is, therefore, less than it would be if the ionic atmosphere were symmetrical. The same effect will occur when an anion is moving in a direction opposite to the applied field, i.e. the ionic atmosphere 'will pull the anion back'. This results in the velocity of the anion also being less than it would be in the absence of the asymmetry. The ionic conductivity of the ions is therefore less than it would be if the ionic atmosphere were symmetrical.

12.1.1 Approximate estimate of the relaxation time for the ionic atmosphere

In Section 12.1 it was shown that when the central reference ion moves under the influence of an external field there is an excess of negative charge behind the cation and a deficit of negative charge in front of it. If the external field is removed, the asymmetric ionic atmosphere relaxes back to the symmetric distribution. This process can be pictured as if some negative charge

Table 12.1 Approximate τ values at various concentrations of a 1–1 electrolyte

$\dfrac{c}{\text{mol dm}^{-3}}$	0.0010	0.0100	0.100
τ/s	10^{-7}	10^{-8}	10^{-9}

moves from that part of the ionic atmosphere behind the cation to that part in front of the ion, i.e. anions migrate across the ionic atmosphere. A rough estimate of the relaxation time can be found from the time it takes for an anion to move across a distance comparable to the diameter of the ionic atmosphere. This distance can be taken to be of the order of $x(1/\kappa)$ where x is a numerical factor of magnitude probably < 10. A similar situation can be envisaged when the central ion is an anion when the relaxation time back to the symmetric distribution can be visualised as positive charge moving across the ionic atmosphere from behind to in front of the anion. And again the relaxation time could be taken as approximately the time for a cation to move across a distance corresponding to $x(1/\kappa)$. Table 12.1 gives approximate relaxation times calculated in this way for various concentrations of a 1–1 electrolyte.

For 2–2 and 3–3 electrolytes at the same concentration, κ will be larger with the largest increase being for the more highly charged ions. The relaxation times for such electrolytes will be smaller than those quoted above for the 1–1 electrolytes.

For a given concentration, the relaxation time is directly related to $(u_+ + u_-)/u_+u_-$, to $\Lambda/\lambda_+\lambda_-$ and to Λ for the given electrolyte. The relaxation time is thus a fundamental quantity in any theory of conductance. In turn, the relaxation time is a property of the ionic atmosphere which is regarded as the crucial concept in the Debye-Hückel theory of non-ideality. These statements could be summarised as:

- κ is a crucial feature of the ionic atmosphere in equilibrium situations which correspond to the ionic atmosphere being symmetrical;

- while τ is a crucial feature of the non-equilibrium situation which corresponds to the setting up of an asymmetric ionic atmosphere.

12.1.2 Confirmation of the existence of the ionic atmosphere

There are two experiments which beautifully illustrate the correctness of the idea of an ionic atmosphere and its manifestation in terms of the relaxation and electrophoresis which occur when the ion moves under the influence of an external field. These are the Debye-Falkenhagen effect and the Wien effect.

12.1.3 The relaxation time and the Debye-Falkenhagen effect, i.e. the effect of high frequencies

With alternating fields at ordinary frequencies, typical of standard conductance experiments, the time for one oscillation is large compared with the typical times for relaxation of the ionic atmosphere. For example a frequency of 50 Hz corresponds to a time of 2×10^{-2} s for one oscillation and this is very much longer than the times for relaxation given in Table 12.1, e.g.

10^{-7} s for a 1–1 electrolyte at a concentration of $0.0010\,\text{mol dm}^{-3}$. Under these circumstances the ionic atmosphere will always be in its asymmetric state and the relaxation will affect the ionic molar conductivities and the molar conductivity of the electrolyte.

However, for very high frequencies of the external field, such as 10^8 Hz, the time for one oscillation is 10^{-8}s and this is comparable to or even smaller than the relaxation time. Under these conditions the ion is impelled backwards and forwards so rapidly that the build up of the ionic atmosphere to the asymmetric state cannot occur fast enough, and the asymmetric ionic atmosphere will not be fully set up. At even higher frequencies the ionic atmosphere never has time to reach the asymmetric state and the relaxation effect totally disappears. The retarding effect of the asymmetry on the movement of the ions under the influence of the external field is removed. In consequence, the velocity of the ions and their individual ionic molar conductivities are significantly higher than for ordinary frequencies and are much nearer what would they would be expected to be if there were no retarding effect of the ionic atmosphere.

12.1.4 The relaxation time and the Wien effect, i.e. the effect of very large fields

The velocity at which an ion moves under the influence of an external field is proportional to the field, with the constant of proportionality being the mobility of the ion. Provided this strict simple proportionality remains for large fields and, provided the mobility is known, the velocity of an ion can be calculated. $5 \times 10^{-4}\,\text{cm}^2\,\text{s}^{-1}\,\text{V}^{-1}$ is a typical value for an ionic mobility.

Using this and the value of $\kappa = 0.33\,\text{nm}^{-1}$ for a $1.00 \times 10^{-2}\,\text{mol dm}^{-3}$ 1-1 electrolyte, the Table 12.2 can be drawn up, $1/0.33\,\text{nm}^{-1} \approx 3\,\text{nm}$.

$5\,\text{V cm}^{-1}$ is a typical field used in standard conductance experiments. From the table, it can be seen that this will correspond to a time of 1.2×10^{-4} s. This can then be compared with a relaxation time of 10^{-8} s. The relaxation time is short compared with the time taken by the ion to move a distance of the order of the extent of the ionic atmosphere. There is quite clearly ample time for the ionic atmosphere to reach its asymmetric state. As the fields get larger, the velocities become very high and there is progressively less time available for the ionic atmosphere to build up to its asymmetric state, and it is totally destroyed. At very large fields the relaxation effect will disappear. Because of the very high velocities the electrophoretic effect (see Section 12.2) also becomes negligible and so both retarding effects of relaxation and electrophoresis disappear. The molar conductivity of the electrolyte becomes very much larger than at lower field strengths.

The discussion in this section and in the previous one on the Debye-Falkenhagen effect is qualitative, but it does illustrate the physical significance of the Debye-Falkenhagen and the Wien effects.

Table 12.2 Time for an ion to move a distance comparable to the extent of the ionic atmosphere

field $\dfrac{}{\text{V cm}^{-1}}$	5	10	10^3	10^4	10^5
velocity $\dfrac{}{\text{cm s}^{-1}}$	2.5×10^{-3}	5×10^{-3}	0.5	5	50
Time required to move forward by 3 nm	1.2×10^{-4} s	6×10^{-5} s	6×10^{-7} s	6×10^{-8} s	6×10^{-9} s

These two experiments coupled with ordinary conductance and activity coefficient studies demonstrate conclusively the correctness of the basic postulate of the Debye-Hückel theory, that is, the existence of the ionic atmosphere which is itself a manifestation of the inter-ionic interactions occurring in electrolyte solutions. They also suggest that the main properties of the ionic atmosphere which must be taken into consideration in developing a theory of conductance are:

- the relaxation effect, as a consequence of the asymmetry of the ionic atmosphere;

- the electrophoretic effect, i.e. viscous drag on the ionic atmosphere by the solvent.

12.2   The electrophoretic effect

The ions are moving in the solvent, and the effect of the solvent on the movement of the ion and its ionic atmosphere under the applied field must also be considered. This is discussed under the heading of the electrophoretic effect.

When an ion moves it carries its ionic atmosphere with it. This ionic atmosphere is made up, in part, of ions which have the same charge as the central ion and will move in the same direction as the central ion under an external field. It also contains ions which will move in the opposite direction. An ion will drag the solvent molecules in its vicinity along with it. Since cations and anions move in opposite directions as a result of the externally applied electric field, each ion moves against a stream of solvent molecules.

If the central reference ion and an ion of **opposite charge** in the ionic atmosphere are passing each other, solvent will be pulled along with each ion but in opposite directions. So the central reference ion of the pair will, in effect, 'see' solvent streaming past itself in the opposite direction, and this will exert a viscous drag on the central reference ion, slowing it down.

If the central reference ion and an ion of **like charge** in the ionic atmosphere are passing each other, the differing speeds of the two ions will cause streaming of the solvent past the central reference ion. If an ion in the ionic atmosphere is overtaking the central ion, this central reference ion will be 'pulled along by the solvent' associated with the overtaking ion streaming past it in the same direction, resulting in an increase in its velocity. If the central reference ion is overtaking the ion of the ionic atmosphere, it will experience a viscous drag by the solvent around the ionic atmosphere ion. This will make this solvent appear as though it were, in effect, moving in the opposite direction to the central ion and so the ion will be 'pulled back', resulting in a decrease in its velocity.

These effects are covered by the general term 'electrophoretic effect', and their net effect is always to slow the ion down, resulting in a lower ionic conductivity than would be expected if there were no ionic atmosphere.

12.3   Conductance equations for strong electrolytes taking non-ideality into consideration: early conductance theory

If strong electrolytes are fully dissociated and there are no significant interactions between the ions, the molar conductivity should be strictly constant over a range of concentrations (see Section 11.6). But experimentally it is found that Λ depends approximately linearly on $\sqrt{c}$:

$$\Lambda = \Lambda^0 - \text{constant}\sqrt{c} \tag{12.1}$$

where Λ^0 is the limiting molar conductivity found as $c \to 0$, as ideal conditions are approached.

The deviations of Λ from Λ^0 are attributed to non-ideality associated with:

- electrostatic interactions between ions giving rise to the ionic atmosphere;

- the relaxation effect;

- the electrophoretic effect.

Debye and Hückel tackled this non-equilibrium case once they had formulated their equation:

$$\log_{10}\gamma_{\pm} = -\frac{A|z_1 z_2|\sqrt{I}}{1 + B\mathring{a}\sqrt{I}} \tag{12.2}$$

for the equilibrium situation in an electrolyte solution. They were closely followed by Onsager (in 1927) who incorporated corrections into the Debye-Hückel conductance equation and whose equation gave the basis of all modern theories of conductance based on the Debye-Hückel model. This was followed in 1932 by a modified theory by Fuoss and Onsager, but electrochemistry had to wait until 1957 before a full conductance equation of the level of refinement of the activity coefficient expression quoted above was derived (see Section 12.10).

The early conductance theories given by Debye and Hückel in 1926, Onsager in 1927 and Fuoss and Onsager in 1932 used a model which assumed **all** the postulates of the Debye-Hückel theory (see Section 10.3). The factors which have to be considered **in addition** are the effects of the asymmetric ionic atmosphere, i.e. relaxation and electrophoresis, and viscous drag due to the frictional effects of the solvent on the movement of an ion under an applied external field. These effects result in a decreased ionic velocity and decreased ionic molar conductivity and become greater as the concentration increases.

The theory aims to calculate the ionic velocities and thence the molar conductivity as modified by relaxation and electrophoresis. This is done as a function of concentration.

12.3.1 Qualitative aspects of the derivation of the Debye-Hückel-Onsager 1927 equation

The basic equation of the Debye-Hückel theory of non-ideality was deduced in Section 10.6 and is:

$$\psi_j = \frac{z_j e}{4\pi\varepsilon_0\varepsilon_r} \frac{e^{\kappa\mathring{a}}}{1 + \kappa\mathring{a}} \frac{e^{-\kappa r}}{r} \tag{12.3}$$

Under an externally applied electric field, the central reference ion will move with a velocity $v_j = u_j X$ and the force on the ion due to the field will be $z_j eX$, (Section 10.4.4). This force will be opposed by the electrophoretic and relaxation effects, and by the frictional effect of the solvent as the ion moves through it. This latter effect will be present in the absence of the effects of non-ideality, i.e. relaxation and electrophoresis. The calculation reduces to calculating an explicit expression for each of these three effects. Since the ion will move with a constant velocity, the force exerted on it by the applied field will be equal and opposite to the sum of the forces due to the three effects: electrophoretic, relaxation and frictional.

The main assumptions which Debye *et al.* made were:

- The ions are hard spheres of radius $\mathring{a}/2$ and charge $z_j e$, drifting in a continuum of relative permittivity, ε_r, and viscosity, η, under the influence of an external field, X.

- The progressive reduction in mobility as the concentration increases is due to interionic interactions of a coulombic nature. The dominant effects are due to relaxation, electrophoresis and their cross terms.

- Stokes' Law was used to estimate the frictional forces due to the movement of the ion in the solvent.

- In the derivation of the electrophoretic effect Stokes' Law was assumed to hold. There are problems with this assumption, but Onsager showed that it was not necessary for Stokes' Law to be strictly applicable in the immediate vicinity of the ion.

- In the first derivation of the relaxation effect the random Brownian motion of the ions in the solution was ignored, but Onsager modified the theory to take this into account.

- Electrophoresis and relaxation were taken to be totally independent phenomena, whereas they are not. As a result the derivation neglected, (i) the effect of the asymmetry of the ionic atmosphere on the electrophoretic effect, and (ii) the effect of electrophoresis on the movement of the ion in an asymmetrical ionic distribution. These are 'cross terms' described below.

- 'Cross term' due to (i): An applied external field has an indirect effect on the solvent through the ions. The force exerted on the moving central ion by the solvent in between the ions of the ionic atmosphere is dependent on the ionic distribution in the ionic atmosphere. Hence the interaction between the ions and the solvent will be determined by the interactions between the ions themselves. If the symmetrical distribution is perturbed by the externally applied field this will have an effect on the interactions between an ion and the solvent, and this will result in an additional solvent flow about the ion. For a calculation of the electrophoretic effect this asymmetry should be considered, but in this derivation it is not and it is assumed that the symmetrical distribution given by Equation (12.3) can be used. This added effect is considered in more advanced treatments (see Section 12.10).

- 'Cross term' due to (ii): The central ion of the asymmetric distribution is affected by the movement of the solvent around it due to the electrophoretic effect. Since any ion in the ionic atmosphere can be considered to be a central reference ion, the electrophoretic effect will affect all the ions. Because there are interactions between the ions and the solvent molecules, this will alter the asymmetry of the ionic atmosphere which would be set up due to relaxation in the absence of the electrophoretic effect. This has the consequence of an added perturbation on the asymmetry. For a calculation of the relaxation effect this extra perturbation should be considered, but in this derivation it is not though in more advanced theories it is included (see Section 12.10).

- Because of the complex mathematics involved, in the Debye-Hückel-Onsager derivation the asymmetrical distribution is **approximated** to the symmetrical form in the **final** stages of the calculation. This **must not** be taken to mean that the perturbation due to asymmetry has been ignored. This perturbation is considered in more detail in more advanced treatments (see Section 12.10).

- In the derivation, the treatment is simplified by assuming that $\mathring{a}$ is very small so that $(1 + \kappa\mathring{a})$ is very nearly unity and this restricts the resulting conductance equation to very low concentrations, e.g. to $\kappa\mathring{a} \approx 0.03$, which for a 1–1 electrolyte corresponds to a concentration of around $1 \times 10^{-3} \, \text{mol dm}^{-3}$, and to progressively lower concentrations as the charge type increases. Furthermore, the fact that this approximation is made means that $e^{\kappa\mathring{a}}$ is also very nearly unity. This means, in effect, that Equation 10.3 is approximated to $\psi_j = \dfrac{z_j e}{4\pi\varepsilon_0\varepsilon_r} \dfrac{e^{-\kappa r}}{r}$. This approximation is often described as taking the ions to be point charges, a statement which must not be taken literally. If the charges were indeed point charges oppositely charged ions would stick together.

Using these assumptions the derivation can be given in Section 12.4.

12.4 A simple treatment of the derivation of the Debye-Hückel-Onsager equation 1927 for symmetrical electrolytes

12.4.1 Step 1 Stating the problem

The aim is to calculate:

the dependence of the ionic molar conductivities and the molar conductivity as a function of concentration for a symmetrical electrolyte.

This is done by calculating:

- **the contribution at infinite dilution resulting from a balance of the effects of the field and the frictional force of the solvent on the moving ion;**

- **the negative contribution due to the electrophoretic effect;**

- **and the negative contribution due to the relaxation effect.**

12.4.2 Step 2 The contribution at infinite dilution resulting from a balance of the effects of the field and the frictional force on the moving ion

Here the effects of non-ideality can be ignored and the behaviour of the moving central j-ion is not modified by the presence of the other ions. Stokes' Law is used here. This states that a macroscopic sphere of radius a moving with a velocity, v, through a continuous medium experiences a retarding force equal to $6\pi a \eta v$, where η is the viscosity of the medium. It is known that the law is inadequate for small spheres such as ions. Onsager, however, showed that for the derivation it is not necessary for Stokes' Law to be strictly applicable in the immediate vicinity of the moving ion. Also the solvent is not a continuum, and in the immediate vicinity of ions it will be modified by the presence of the ions. This effect will, however, become small at infinite dilution.

When there is an external field, the force acting on the moving central reference j-ion due to this field will be balanced by the frictional force experienced by the ion as it moves through the solvent. Since infinite dilution corresponds to ideality, there will be no extra forces on the ion due to relaxation and electrophoresis.

The frictional force due to the solvent is given by

$$f = 6\pi \frac{\mathring{a}}{2} \eta v_j \tag{12.4}$$

where $\mathring{a}/2$ is the radius of the moving ion.

This frictional force must balance the force due to the external field so that the ion moves with a constant velocity, v_j, i.e.:

$$6\pi \frac{\mathring{a}}{2} \eta v_j = z_j eX \tag{12.5}$$

$$v_j = \frac{z_j eX}{6\pi \frac{\mathring{a}}{2} \eta} \tag{12.6}$$

$$= \text{a constant} \times X \tag{12.7}$$

$$= u_j X \tag{12.8}$$

$$u_j = \frac{z_j e}{6\pi \frac{\mathring{a}}{2} \eta} \tag{12.9}$$

where u_j is the mobility of the ion (see Section 11.17).

At infinite dilution corresponding to the ideal solution where relaxation and electrophoresis are not relevant

$$v_j = u_j^0 X \tag{12.10}$$

where u_j^0 is the mobility of the ion under ideal conditions.

12.4.3 Step 3 The contribution from the electrophoretic effect

This is a contribution operative at finite concentrations where the effects of non-ideality appear. Electrophoresis manifests itself as a result of the existence of the ionic atmosphere. Calculation of the electrophoretic effect, therefore, relates to non-ideal conditions. Because of electrophoresis, solvent molecules are pulled along by the ions of opposite sign to the central reference ion in the opposite direction to that of the moving central ion. This causes a change in the velocity of the central ion from that which results from the movement of the ion through the solvent under ideal conditions (see Step 2). The theory calculates this change in velocity. The change in velocity as a result of electrophoresis will be influenced by all the ions of the ionic atmosphere and this is described by the distribution function for the ionic atmosphere. This has been calculated by the Debye-Hückel theory and is given by Equation 10.28. But for ease of calculation the theory is limited to concentrations where $\kappa \mathring{a}$ is small compared with unity and $1 + \kappa \mathring{a}$ can be approximated to unity and $e^{\kappa \mathring{a}}$ also approximates to 1.

The change in velocity of the ion under the action of the modified force due to electrophoresis is Δv_j and the theory gives this as:

$$\Delta v_j = \frac{X z_j e \kappa}{6 \pi \eta} \tag{12.11}$$

Since the electrophoretic effect manifests itself as a modified viscous drag on the ion by the solvent, the force on the ion due to electrophoresis can be given by Stokes' Law. The change in velocity Δv_j can then be given in terms of the two Stokes' Law terms one for the viscous drag due to the solvent in the absence of electrophoresis, and the other due to the modified viscous drag resulting from electrophoresis.

$$6\pi\eta\frac{\mathring{a}}{2}v_j \qquad - 6\pi\eta\frac{\mathring{a}}{2}v_j' \qquad = 6\pi\eta\frac{\mathring{a}}{2}(v_j - v_j') \ = \ 6\pi\eta\frac{\mathring{a}}{2}\Delta v_j \tag{12.12}$$

$\downarrow$	$\downarrow$	$\downarrow$
involves the velocity, in absence of electrophoresis	involves the velocitywhen electrophoresis is considered	calculated in Step 3

Δv_j has been calculated by theory, and is given by Equation 12.11 and so:

$$\text{the force due to electrophoresis} = 6\pi\eta\frac{\mathring{a}}{2}\Delta v_j = 6\pi\eta\frac{\mathring{a}}{2}\frac{X z_j e \kappa}{6\pi\eta} = z_j e \kappa\frac{\mathring{a}}{2}X \tag{12.13}$$

12.4.4 Step 4 The contribution from the relaxation effect

This involves considerably more intractable mathematics and again only the final result will be given. The calculation again assumes $\kappa \mathring{a}$ to be small compared with unity, and the approximate symmetrical distribution, $\psi_j = \dfrac{z_j e}{4\pi\varepsilon_0\varepsilon_r}\dfrac{e^{-\kappa r}}{r}$, is used to calculate a first approximate perturbed distribution ψ_j'. The change in the field due to the relaxation effect is then $\Delta X_{\text{relaxation}} = -(\partial\psi_j'/\partial x)$ where ψ_j' is the perturbed potential and the component of the field is in the x direction which is taken as the direction in which the central ion is moving. Once this has been done this approximate perturbed distribution is used to calculate the resulting force on the central moving ion.

The main problem lies in calculating the perturbation. This is done by integrating an expression called 'the continuity equation'. This is a fundamental equation involving the velocities with which the ions move under the influence of the perturbed asymmetric field and is a fundamental and important aspect of later theories (see Section 12.10).

Calculations are for symmetrical electrolytes, and so the resulting theory is restricted to symmetrical electrolytes.

$$\text{The relaxation force} = \frac{e^2 z_j e \kappa}{6 \times 4\pi\varepsilon_0\varepsilon_r kT}\frac{z_j^2 X}{1 + \sqrt{1/2}} \tag{12.14}$$

12.4.5   Step 5 The final step in arriving at the conductance equation for symmetrical electrolytes

The force on the moving central j-ion due to the external field under ideal conditions is z_jeX. The effective force on the moving central ion once the electrophoretic and relaxation effects are taken into consideration will be:

$$z_jeX - z_je\kappa\frac{\mathring{a}}{2}X - \frac{z_j^3e^3\kappa}{6\times4\pi\varepsilon_0\varepsilon_rkT}\times\frac{1}{1+\sqrt{1/2}}X \tag{12.15}$$

Since the ion moves at a constant velocity, this force will be balanced by the viscous drag on the ion exerted by the solvent irrespective of the effects of the ionic atmosphere and is given by Stokes Law as

$$6\pi\eta\frac{\mathring{a}}{2}v_j = 6\pi\eta\frac{\mathring{a}}{2}u_jX \tag{12.16}$$

$$\therefore\ 6\pi\eta\frac{\mathring{a}}{2}v_j = z_jeX - z_je\kappa\frac{\mathring{a}}{2}X - \frac{z_j^3e^3\kappa}{6\times4\pi\varepsilon_0\varepsilon_rkT}\times\frac{1}{1+\sqrt{1/2}}X \tag{12.17}$$

Divide by $6\pi\frac{\mathring{a}}{2}\eta = 3\pi\mathring{a}\eta$

$$\tag{12.18}$$

$$\therefore\ v_j = \frac{z_jeX}{3\pi\mathring{a}\eta} - \frac{z_je\kappa X}{6\pi\eta} - \frac{z_j^3e^3\kappa}{6\times4\times3\pi^2\eta\mathring{a}\varepsilon_0\varepsilon_rkT}\frac{X}{(1+\sqrt{1/2})} \tag{12.19}$$

Divide by X and replace v_j/X by u_j

$$\therefore u_j = \frac{z_je}{3\pi\mathring{a}\eta} - \frac{z_je\kappa}{6\pi\eta} - \frac{z_j^3e^3\kappa}{6\times4\times3\pi^2\eta\mathring{a}\varepsilon_0\varepsilon_rkT}\frac{1}{(1+\sqrt{1/2})} \tag{12.20}$$

At infinite dilution the electrophoretic and relaxation effects drop out giving

$$u_j = u_j^0 = \frac{z_je}{3\pi\mathring{a}\eta} \tag{12.21}$$

And for finite concentrations it follows that

$$u_j = u_j^0 - \frac{z_je\kappa}{6\pi\eta} - \frac{z_j^3e^3\kappa}{6\times4\times3\pi^2\eta\mathring{a}\varepsilon_0\varepsilon_rkT}\frac{1}{(1+\sqrt{1/2})} \tag{12.22}$$

The terms in the mobility can be expressed in terms of the ionic molar conductivities, since $\lambda_j = z_jFu_j$ see Section 11.17, and so multiplying throughout by z_jF gives:

$$z_jFu_j = z_jFu_j^0 - \frac{z_j^2Fe\kappa}{6\pi\eta} - \frac{z_j^3e^3z_jF\kappa}{6\times4\times3\pi^2\eta\mathring{a}\varepsilon_0\varepsilon_rkT}\frac{1}{(1+\sqrt{1/2})} \tag{12.23}$$

Replacing the terms $z_je/3\pi\mathring{a}\eta$ in the final term by u_j^0 gives:

$$\therefore\ z_jFu_j = z_jFu_j^0 - \frac{z_j^2Fe\kappa}{6\pi\eta} - \frac{z_j^3e^2F\kappa u_j^0}{6\times4\times\pi\varepsilon_0\varepsilon_rkT}\frac{1}{(1+\sqrt{1/2})} \tag{12.24}$$

$$\lambda_j = \lambda_j^0 - \frac{z_j^2Fe\kappa}{6\pi\eta} - \frac{z_j^2e^2\kappa\lambda_j^0}{24\pi\varepsilon_0\varepsilon_rkT}\frac{1}{(1+\sqrt{1/2})} \tag{12.25}$$

Sections 10.6.10 and 10.6.11 give an expression for κ^2 (Equations 10.36 to 10.39):

$$\kappa^2 = \frac{e^2}{\varepsilon_0 \varepsilon_r kT} \sum_j n_j z_j^2 = \frac{e^2}{\varepsilon_0 \varepsilon_r kT} N \sum_j c_j z_j^2 \tag{12.26}$$

Substituting this expression and carrying out the necessary algebra results in an expression of the form:

$$\lambda_j = \lambda_j^0 - [a' + b' \lambda_j^0] \sqrt{c_j^{\text{actual}}} \text{ for the } j\text{-ion, and} \tag{12.27}$$

$$\lambda_i = \lambda_i^0 - [a' + b' \lambda_i^0] \sqrt{c_i^{\text{actual}}} \text{ for the } i\text{-ion} \tag{12.28}$$

Since the derivation deals with a symmetrical electrolyte only:

$$\Lambda = \lambda_i + \lambda_j = (\lambda_i^0 + \lambda_j^0) - [2a' + b'(\lambda_i^0 + \lambda_j^0)] \sqrt{c_{\text{actual}}} = \Lambda^0 - [a + b\Lambda^0] \sqrt{c_{\text{actual}}} \tag{12.29}$$

This is often written in the form:

$$\Lambda = \Lambda^0 - S\sqrt{c_{\text{actual}}} \tag{12.30}$$

$$\text{where } S = a + b\Lambda^0 \tag{12.31}$$

- a is a constant dependent on temperature, relative permittivity and viscosity, and takes account of electrophoresis;

- b depends on the temperature and the relative permittivity only, and takes account of relaxation.

Equation (12.30) is reminiscent of the empirically determined equation describing the effect of concentration on the molar conductivity where:

$$\Lambda = \Lambda^0 - \text{constant} \sqrt{c_{\text{stoich}}} \tag{12.32}$$

Pay particular attention: the concentration which appears in **theoretical** expressions for the ionic molar conductivities of ions and the molar conductivity of an electrolyte is an **actual** concentration and **not** the **stoichiometric** concentration which is used in the expression for the molar conductivity, $\Lambda = \kappa/c_{\text{stoich}}$ and which is used in experimental work.

For aqueous solutions at 25°C, on the basis that $\varepsilon_r = 78.46$, $\eta = 8.903 \times 10^{-4} \text{ kg m}^{-1} \text{ s}^{-1}$, a and b take the values:

$$a = z^3 \times 60.58 \text{ S cm}^2 \text{ mol}^{-3/2} \text{dm}^{3/2}$$

$$b = z^3 \times 0.2293 \text{ mol}^{-1/2} \text{dm}^{3/2}$$

Although there is a considerable amount of algebra involved in the complete derivation, the basic concepts involved are simple, viz.:
the force on the moving central j-ion due to the field
  = force on the moving central j-ion due to the field at infinite dilution
    − the retarding electrophoretic force − the retarding relaxation force
which reduces to a form giving the molar conductivity of a solution in terms of the molar conductivity at infinite dilution and the square root of the **actual** concentration of the electrolyte.

The expressions used in the Debye-Hückel-Onsager equation for the electrophoretic and relaxation effects are simple in algebraic form, viz.:

$$\text{Electrophoretic force} = z_j e \kappa \frac{\mathring{a}}{2} X \tag{12.13}$$

$$\text{Relaxation force} = \frac{z_j^3 e^3 \kappa}{6 \times 4\pi\varepsilon_0\varepsilon_r kT} \times \frac{1}{1 + \sqrt{1/2}} X \tag{12.14}$$

But the actual derivation of these expressions is very decidedly not simple. They are also very approximate in so far as they involve considerable approximations as outlined in Section 12.3.1.

Because of the assumptions and approximations made in the derivation, and in particular the use of $\psi_j = \dfrac{z_j e}{4\pi\varepsilon_0\varepsilon_r} \dfrac{e^{-\kappa r}}{r}$ rather than $\psi_j = \dfrac{z_j e}{4\pi\varepsilon_0\varepsilon_r} \dfrac{e^{\kappa \mathring{a}}}{1 + \kappa \mathring{a}} \dfrac{e^{-\kappa r}}{r}$, this means that the theory can only be accurate under circumstances where $e^{\kappa \mathring{a}} = 1$ and $1 + \kappa \mathring{a} = 1$ which can only be in very dilute solutions where non-ideality can be ignored.

12.5   The Fuoss-Onsager equation 1932

This treatment is only slightly more physically realistic than the 1927 equation outlined above. The improvement lies in the use of the non-approximate form of the truncated Debye-Hückel expressions in the derivation of the electrophoretic effect, i.e. they used:

$$\psi_j = \frac{z_j e}{4\pi\varepsilon_0\varepsilon_r} \frac{e^{\kappa \mathring{a}}}{1 + \kappa \mathring{a}} \frac{e^{-\kappa r}}{r}$$

rather than the expression

$$\psi_j = \frac{z_j e}{4\pi\varepsilon_0\varepsilon_r} \frac{e^{-\kappa r}}{r}$$

for the symmetrical distribution of the ionic atmosphere.

The relaxation term was calculated as in the earlier theory with electrophoresis and relaxation being treated independently. Other severe approximations were made in the treatments of the electrophoretic and relaxation effects, but this was done for mathematical ease. This resulted in a $(1 + \sqrt{c})$ term in the denominator.

Theoretically this treatment represents a **very primitive** approach, but this merely reflects the complexity of the mathematics which is involved in translating the statement of the physical model into the mathematical framework of the derivation.

The 35 years of work which went into producing the second modified 1957 Fuoss-Onsager equation did not represent any alteration or improvement on the **model**, it merely allowed for a **few** less approximations to be made to the mathematical framework and derivation. However, these did represent a considerable advance on the physically unrealistic approximations in the earlier 1927 and 1932 equations.

12.6   Use of the Debye-Hückel-Onsager equation for symmetrical strong electrolytes which are fully dissociated

This equation only holds at low concentrations, where the solution approximates to ideality. At higher concentrations deviations are expected to appear, and these will become greater the greater the non-ideality. For an unassociated electrolyte where $c_{actual} = c_{stoich}$ the theory predicts that a plot of Λ vs. $\sqrt{c_{stoich}}$ should be linear with slope $= a + b\Lambda^0$ and intercept $= \Lambda^0$. The approximations made in the theory corresponding to taking $e^{\kappa \mathring{a}} = 1$ and to $1 + \kappa \mathring{a} = 1$ limit the range of applicability to $\kappa \mathring{a} <$ around 0.03 which for a 1–1 electrolyte corresponds to an ionic strength of $< 1 \times 10^{-3}$ mol dm^{-3}, and to progressively lower concentrations for higher charge types. The equation thus corresponds to very low concentrations as prescribed by the highly approximate model chosen. It is likely to hold over the same range of concentration as the Debye-Hückel limiting law since both are based on the same approximations regarding the ionic atmosphere, and both are limiting laws.

Three important points must be made:

- when testing the validity of this law **both of the following predictions must be satisfied simultaneously,**

  - the graph must be linear;

  - and it must have slope $= a + b\Lambda^0$ where a and b take specified values.

  - The Debye-Hückel-Onsager equation is a limiting slope only and holds rigorously only at zero concentration – a consequence of the highly approximate treatment of the model chosen. The Debye-Hückel-Onsager equation is a straight **line** and is the limiting tangent to the conductance **curve**. It is **not the equation** of the conductance **curve**. In the same way, the Debye-Hückel limiting law is the limiting tangent to the graph of $\log \gamma_{\pm}$ vs. $\sqrt{I}$.

Care must be taken when testing this theory against experimental data. Experimental data are presented in terms of Λ_{obsvd} values based on the stoichiometric concentration, viz. $\Lambda = \kappa/c_{stoich}$, and these Λ_{obsvd} are then plotted against $\sqrt{c_{stoich}}$. The theoretical equation is based on actual concentrations, being:

$$\Lambda = \Lambda^0 - [a + b\Lambda^0]\sqrt{c_{actual}} = \Lambda^0 - S\sqrt{c_{actual}} \qquad (12.29 \text{ and } 12.30)$$

where $S = a + b\Lambda^0$ \hfill (12.31)

12.6.1 Assessment of experimental results for 1–1, 1–2 and 1–3 electrolytes in concentration ranges where they are expected to be fully dissociated

There are no problems when considering 1–1 electrolytes. These are expected to be fully dissociated into ions over the range of the low concentrations for which the expression is expected to be valid. Under these conditions the actual concentration of free ions is equal to the stoichiometric concentration and so the experimental Λ_{obsvd} vs. $\sqrt{c_{stoich}}$ can be compared directly with the theoretical equation. For the range of concentrations for which the theory is expected to be valid, the Debye-Hückel-Onsager equation is very closely obeyed by 1–1 electrolytes which are fully dissociated. Problems arise for 1–1 strong electrolytes at concentrations above 1×10^{-3} mol dm^{-3} when non-linearity sets in and the Λ_{obsvd} approach the limiting slope from above.

Data for 1–2 and 1–3 electrolytes which are unassociated at concentrations $< 1 \times 10^{-3}$ mol dm^{-3} also appear to fit the Debye-Hückel-Onsager equation up to concentrations $< 1 \times 10^{-3}$ mol dm^{-3} but at higher concentrations some of these electrolytes approach the limiting slope from above, but others approach it from below.

The apparent close fit to the Debye-Hückel-Onsager equations at low concentrations has to be reassessed in the light of the cross-over predicted by the later Fuoss-Onsager 1957 equation (see Section 12.10.2).

When the Λ_{obsvd} approach the limiting slope from above this is due to the approximations made in the derivation of the conductance equation. Empirical corrections (see Section 12.8) have been made which postulate higher order terms to be necessary, viz. terms in c, $c\log c$ and c^2 with the coefficients of these terms being determined experimentally. But an explanation of this behaviour had to wait until the Fuoss-Onsager equation of 1957 had been formulated (see Section 12.10).

Other workers have looked at typical 2–2 electrolytes at concentrations as low as 1×10^{-6} mol dm^{-3} and found that the data fit the Debye-Hückel-Onsager slope up to around 1×10^{-5} mol dm^{-3}. At higher concentrations the observed conductance curve approaches the limiting slope from below, clear evidence of the expected ion association found in typical 2–2 electrolytes (see below).

12.7   Electrolytes showing ion pairing and weak electrolytes which are not fully dissociated

For electrolytes which are not fully dissociated highly accurate data should be able to detect the onset of ion pair formation and, as the concentration increases, significant association is expected. The higher the charge type the lower the concentration at which association will be observed. Ion association results in Λ_{obsvd} values being lower than expected. For solutions of weak electrolytes where undissociated molecules are present similar behaviour is observed, viz. Λ_{obsvd} values will approach the limiting law slope from below.

For these electrolytes typical experimental graphs of Λ_{obsvd} vs. $\sqrt{c_{stoich}}$ are shown in Figure 12.1 and demonstrate that the slopes at various points on the graph are greater than

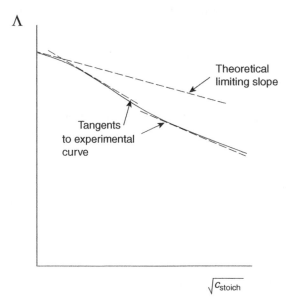

Figure 12.1   A schematic graph of Λ_{obsvd} vs. $\sqrt{c_{stoich}}$ for an associated electrolyte, with the limiting Debye-Hückel-Onsager slope drawn in.

expected theoretically. The net effect of ion association for the strong electrolytes and incomplete dissociation for typical weak electrolytes is that there are fewer conducting species present in the solution than would be expected in terms of the stoichiometric concentration and the assumption of complete dissociation. Where the weak electrolyte is concerned there will be molecular species present which cannot conduct the current. In symmetrical strong electrolytes the overall charge of the ion pair will be zero and so the ion pair is not expected to conduct the current. However, the ion pair will be a dipolar charge separated species which could still be weakly conducting by virtue of the charge separation. In unsymmetrical electrolytes the ion pair will be charged, but of lower charge than the sum of the charges of the individual ions. It will be less conducting, and the molar conductivity would thus be less than would be expected for no ion association. For both situations, i.e. weak electrolytes and ion pairing, $c_{actual} = \alpha c_{stoich}$ and so the theoretical expression must be converted from one in terms of actual concentrations, into one in terms of stoichiometric concentrations for a direct comparison of the experimental graph of Λ_{obsvd} vs. $\sqrt{c_{stoich}}$ to be made, i.e. it is reformulated in terms of the degree of dissociation, α.

The Debye-Hückel-Onsager equation is:

$$\Lambda_{theoretical} = \Lambda^0 - (a + b\Lambda^0)\sqrt{c_{actual}} \tag{12.29}$$

$$\text{where } \Lambda_{theoretical} = \frac{\kappa}{c_{actual}} = \frac{\kappa}{\alpha c_{stoich}} = \frac{1}{\alpha}\Lambda_{observed} \tag{12.33}$$

$$\therefore \Lambda_{theoretical} = \frac{1}{\alpha}\Lambda_{observed} \tag{12.34}$$

$$\text{and } \Lambda_{observed} = \alpha\Lambda_{theoretical} \tag{12.35}$$

$$\text{and the term}(a + b\Lambda^0)\sqrt{c_{actual}} = (a + b\Lambda^0)\sqrt{\alpha c_{stoich}} \tag{12.36}$$

From this it can be seen that the correct form of the Debye-Hückel-Onsager equation to be used to compare with the experimental results is one in which stoichiometric concentrations appear.

$$\Lambda_{observed} = \alpha[\Lambda^0 - (a + b\Lambda^0)\sqrt{\alpha c_{stoich}}] = \alpha\Lambda^0 - \alpha^{3/2}(a + b\Lambda^0)\sqrt{c_{stoich}} \tag{12.37}$$

This is often written in the abbreviated form:

$$\Lambda_{observed} = \alpha\Lambda' \tag{12.38}$$

$$\text{where } \Lambda' = \Lambda^0 - (a + b\Lambda^0)\sqrt{\alpha c_{stoich}} \tag{12.39}$$

$$\text{and } \alpha = \frac{\Lambda_{observed}}{\Lambda'} \tag{12.40}$$

The value of α deduced from this equation can be compared with the simple expression assuming ideality given in Section 11.7, Equation 11.28

$$\alpha = \frac{\Lambda_{observed}}{\Lambda^0} \tag{12.41}$$

But from Equation (12.39), $\Lambda' < \Lambda^0$

$$\therefore \frac{\Lambda_{obsvd}}{\Lambda'} > \frac{\Lambda_{obsvd}}{\Lambda^0} \tag{12.42}$$

Thus the value of α from the equation which makes some attempt to take into account the asymmetry of the ionic atmosphere in a conducting solution via primitive corrections for the electrophoretic and relaxation effects will be larger than that from the equation assuming ideality.

Equation (12.37) also shows that the greater is the value of α, i.e. the more fully dissociated either a weak electrolyte or an ion pair is, the greater will be the difference between $\Lambda_{observed}$ and $\alpha\Lambda^0$. And so calculations determining the degree of dissociation from the conductance ratio Λ/Λ^0 will always give a value which is too small, with this discrepancy being greater the more highly ionised the electrolyte. Not surprisingly, this means that the K_{dissoc} found from the ratio, Λ/Λ^0, will be smaller when ideality is assumed than when an attempt is made to account for non-ideality by using the ratio Λ/Λ'.

12.7.1   Calculation of Λ^0, α and K_{dissoc} for weak electrolytes and ion pairs using the Debye-Hückel-Onsager equation

If Λ^0 is known independently, α can be obtained for each experimental stoichiometric concentration by a series of successive approximations using the Debye-Hückel-Onsager equation and the known value of Λ^0:

$$\begin{aligned}
\Lambda_{observed} &= \alpha[\Lambda^0 - (a + b\Lambda^0)\sqrt{\alpha c_{stoich}}\,] \\
&= \alpha\Lambda^0 - \alpha a\sqrt{\alpha c_{stoich}} - \alpha b\Lambda^0\sqrt{\alpha c_{stoich}} \\
&= \alpha\Lambda^0 - \alpha^{3/2}a\sqrt{c_{stoich}} - \alpha^{3/2}b\Lambda^0\sqrt{c_{stoich}}
\end{aligned} \tag{12.43}$$

A series of successive approximations is required since this equation is a cubic equation in $\alpha^{1/2}$.

If Λ^0 is not known, α and Λ^0 can both be found from computer fitting to the observed conductance curve.

Computer fitting to Equation (12.37) is now the standard procedure.

The equilibrium constant is then found from values of α and c_{stoich} and values of the mean activity coefficient of the electrolyte, making sure that $\gamma_{\pm}$ is calculated for each concentration using an ionic strength based on the **actual** concentrations. Alternatively a non-ideal K can be found which is then corrected for non-ideality by plotting $\log_{10}K$ vs. $\sqrt{I}/(1 + \sqrt{I})$.

12.8   Empirical extensions to the Debye-Hückel-Onsager 1927 equation

When data on electrolytes believed to be fully dissociated at moderate concentrations are studied, the molar conductivities are too large even though they **appear** to fit the equation at low concentrations, i.e. the conductance curve approaches the limiting slope from above.

These anomalies are taken to be a result of the very considerable approximations made in the theory. Allowances for this were made by empirical extensions such as:

$$\Lambda_{obsvd} = \Lambda^0 - Sc_{actual}^{1/2} + A'c_{actual} - B'c_{actual}^{3/2} + C'c_{actual}^2 - D'c_{actual}^{3/2} \tag{12.44}$$

Another empirical extension suggested that there should be a term such as $Dc_{actual}\log_e c_{actual}$. This term will come into prominence in the later theories of Fuoss and Onsager (1957 and later) and other workers (see Sections 12.10 and 12.17).

Although the experimental work suggests terms which should be incorporated into the theoretical equation, it was a further 30 or more years before any detailed theoretical work was produced. This was a direct result of the complexity of the mathematical problems in setting up less primitive expressions for the electrophoretic and relaxation effects.

12.9   Modern conductance theories for symmetrical electrolytes – post 1950

In the early 1950s several workers suggested modifications to the 1927 Debye-Hückel-Onsager equation and to the 1932 Fuoss-Onsager equation. All of these modifications allowed for $\kappa\overset{\circ}{a}$ being comparable to unity in the electrophoretic term, i.e. $\psi_j = \dfrac{z_j e}{4\pi\varepsilon_0\varepsilon_r} \dfrac{e^{\kappa\overset{\circ}{a}}}{1 + \kappa\overset{\circ}{a}} \dfrac{e^{-\kappa r}}{r}$ was used in the derivation of the electrophoretic term. This resulted in a $(1 + \sqrt{c})$ term in the denominator, as in the Fuoss-Onsager 1932 equation. The main thrust of the modifications, however, lay in developments to the relaxation term. Several of these modified theories fitted with the empirically determined modifications described in Section 12.8. These will be summarised below:

- Falkenhagen considered a modified distribution function which took account of the finite size of the ions by recognising that the total space available to the ions is less than the total volume of the solution. This implicitly means that $(1 + \kappa\overset{\circ}{a})$ is not approximated to unity. A modified Poisson-Boltzmann equation was thus used in the derivation of the relaxation effect, but the solution of this modified Poisson-Boltzmann equation was approximated to the first two terms. These modifications gave higher order terms in c_{actual} of the type which had been empirically observed.

- Pitts used an approach in which the Poisson-Boltzmann exponential term was expressed as a series, but where only a few terms in the exponential were retained though more than in Debye-Hückel. This also gave higher order terms in concentration.

- Mirtskhulava also used a similar approach. This involved a complex series expansion which also included a term of the type $\int_x^\infty (e^{-x}/x)\mathrm{d}x$ (called a transcendental term) which gives rise to a $c\log c$ term in the conductance equation, as well as the higher order terms in powers of c and $c^{1/2}$. This confirmed the postulate made empirically that any conductance equation must include a $c\log c$ term as well as higher order terms in powers of c and $c^{1/2}$.

In all of these modifications no account was taken of the need to consider cross terms arising from the effect of relaxation on the electrophoretic effect, and from the effect of electrophoresis on relaxation, but they did hint at the form of the conductance theory put forward later by Fuoss and Onsager.

12.10   Fuoss-Onsager 1957: Conductance equation for symmetrical electrolytes

There are very good reasons why there was the long delay of 30 years between formulation of the Debye-Hückel model, recognition of the necessity to consider the effect of relaxation and electrophoresis, and formulation of a more exact theory relating molar conductivity with concentration.

The physical model is simple, but the mathematics involved is formidable and almost intractable. The problem does not lie in the numerical aspects of the calculation, but in the **analytical** developments of the **types** of equations which are required to describe the electrophoretic and relaxation effects and their cross terms. Computing technology will not help here, unless a model were developed to which molecular dynamics could be applied.

Because of the mathematical intractability of this conductance theory, it is not easy, and perhaps impossible, to demonstrate in a short account the cross-linking between the model and the mathematics. Nor is it possible to give an account of the theory at a level and in the same detail as that given for the Debye-Hückel theory in Chapter 10. And so only a brief description of the theoretical treatment and the results of conductance theory, post 1950, are given.

For theories later than the Fuoss-Onsager 1932 treatment it is useful to express the effects of electrophoresis, relaxation and other contributions in a form showing how they modify the external field under which the ions are migrating.

If X is the external field, then this is modified by changes, ΔX, due to the electrophoretic, relaxation and other effects which are operating. These changes in X, reduce the effective field so that the net effective field on the migrating ions is $X - \Delta X$. Carrying through the argument given in Appendix 12.1 results in the following equations:

$$\lambda_+^{\text{actual}} = \lambda_+^{\text{ideal}} \frac{X - \Delta X}{X} = \lambda_+^0 \left(1 - \frac{\Delta X}{X} \right) \tag{12.45}$$

$$\lambda_-^{\text{actual}} = \lambda_-^{\text{ideal}} \frac{X - \Delta X}{X} = \lambda_-^0 \left(1 - \frac{\Delta X}{X} \right) \tag{12.46}$$

where the ideal ionic molar conductivities are written in the conventional manner, λ^0.

The actual ionic molar conductivity which is a value for situations where non-ideality must be considered is equal to the ideal molar ionic conductivity at infinite dilution modified by the term $1 - \Delta X/X$. ΔX is a sum of all the terms which are taken to lead to non-ideality.

If the effects of viscosity on the mobilities of the ions are included, this results in a term, $(1 + Fc_{\text{stoich}})$, where F is a constant involving the viscosity. The equations are then modified as:

$$\lambda_+^{\text{actual}} (1 + Fc_{\text{stoich}}) = \lambda_+^{\text{ideal}} \frac{X - \Delta X}{X} = \lambda_+^0 \left(1 - \frac{\Delta X}{X} \right) \tag{12.47}$$

$$\lambda_-^{\text{actual}} (1 + Fc_{\text{stoich}}) = \lambda_-^{\text{ideal}} \frac{X - \Delta X}{X} = \lambda_-^0 \left(1 - \frac{\Delta X}{X} \right) \tag{12.48}$$

$$\lambda_+ (1 + Fc_{\text{stoich}}) = \lambda_+^0 \left[1 - \frac{\Delta X_{\text{relaxation}}}{X} - \frac{\Delta X_{\text{electrophoretic}}}{X} - \frac{\Delta X_{\text{Brownian}}}{X} \right.$$

$$\left. - \frac{\Delta X_{\substack{\text{cross term for effect of} \\ \text{relaxation on electrophoresis}}}}{X} - \frac{\Delta X_{\substack{\text{cross term for effect of} \\ \text{electrophoresis on relaxation}}}}{X} \right] \tag{12.49}$$

A similar equation describes the effects of non-ideality on the limiting ionic molar conductivity, λ_-^0, for the case where the central reference ion is an anion.

$$\lambda_-(1 + Fc_{\text{stoich}}) = \lambda_-^0 \left[1 - \frac{\Delta X_{\text{relaxation}}}{X} - \frac{\Delta X_{\text{electrophoretic}}}{X} - \frac{\Delta X_{\text{Brownian}}}{X} \right.$$
$$\left. - \frac{\Delta X_{\text{cross term for effect of relaxation on electrophoresis}}}{X} - \frac{\Delta X_{\text{cross term for effect of electrophoresis on relaxation}}}{X} \right] \quad (12.50)$$

The equation for the molar conductivity for a symmetrical electrolyte is:

$$\Lambda(1 + Fc_{\text{stoich}}) = \Lambda^0 \left[1 - \frac{\Delta X_{\text{relaxation}}}{X} - \frac{\Delta X_{\text{electrophoretic}}}{X} - \frac{\Delta X_{\text{Brownian}}}{X} \right.$$
$$\left. - \frac{\Delta X_{\text{cross term for effect of relaxation on electrophoresis}}}{X} - \frac{\Delta X_{\text{cross term for effect of electrophoresis on relaxation}}}{X} \right] \quad (12.51)$$

- $\Delta X_{\text{relaxation}}$

In the calculation, the term $1 + \kappa \mathring{a}$ is retained and so the finite size of the ions is taken account of, in contrast to the previous theories where this term was approximated to unity. The perturbed distribution of the ionic atmosphere is made up of the unperturbed distribution given by the Debye-Hückel distribution truncated to two terms, i.e. $\psi_j = \frac{z_j e}{4\pi\varepsilon_0\varepsilon_r} \frac{e^{\kappa \mathring{a}}}{1 + \kappa \mathring{a}} \frac{e^{-\kappa r}}{r}$ and a term for the perturbation. This perturbation can be described by a complex equation involving integrals (called the equation of continuity), which can be solved by a series of successive approximations to obtain the best possible approximation to the perturbation. This contributes to the $c^{1/2}$ term and also leads to higher order terms in the concentration, viz., c, $c^{3/2}$, c^2, $c^{5/2}$ etc. and to transcendental terms, i.e. $\int_x^\infty (e^{-x}/x)\mathrm{d}x$. These transcendental terms give rise to logarithmic terms in the conductance equation. Once the perturbation of the ionic atmosphere has been found, the force under which the central ion moves can be found and this, in turn, will lead to $\Delta X_{\text{relaxation}}$.

- $\Delta X_{\text{electrophoresis}}$

Again the $1 + \kappa \mathring{a}$ term is retained and the truncated form of the Debye-Hückel distribution in its **unperturbed form** is used throughout the calculation. From this and a knowledge of the velocity of the ions in the ionic atmosphere, the resulting viscous force on the central reference ion can be calculated. However, in this derivation the integrations involved are taken further than the leading term, in contrast to the earlier theories where only the leading term was used. This again contributes to the $c^{1/2}$ term and also leads to higher order terms and transcendental terms. This viscous force has an effect which is equivalent to a modification of the field.

- $\Delta X_{\text{Brownian}}$

This is a modification of the Brownian diffusive motion of the ions in the solution. Because of non-ideality and its description in terms of the ionic atmosphere it has been shown (see Section 12.1), that, if the central reference ion is a cation, then there is a deficit of negative ions in front of the ion and an excess of negative ions behind. This means that the probability of the central reference cation colliding with a negative ion is greater behind the cation than in front. This effect is equivalent to a small force which acts in the direction of the external field and in the direction in which the cation is moving. This will result in a small increase in the velocity of the central reference cation. A similar effect would occur if the central ion were an anion.

- ΔX cross terms for effect of relaxation on electrophoresis

The $\Delta X_{\text{electrophoresis}}$ was worked out using an unperturbed distribution, but it should have involved a perturbed distribution. When this is done it gives rise to transcendental terms and to higher order terms in c, but starting with c, $c^{3/2}$, c^2, $c^{3/2}$ etc. It does not contribute to the term in $c^{1/2}$.

- ΔX cross terms for effect of electrophoresis on relaxation

The fact that a correction has to be made to account for the effect of asymmetry of the ionic atmosphere on electrophoresis means that this correction will imply an alteration of the perturbed field used in $\Delta X_{\text{relaxation}}$. When this is done it gives rise to transcendental terms and to higher order terms in c, but starting with c, $c^{3/2}$, c^2, $c^{5/2}$ etc. It does not contribute to the term in $c^{1/2}$.

Pay particular note: the direct effect of considering both relaxation, $\Delta X_{\text{relaxation}}$, and electrophoresis, $\Delta X_{\text{electrophoresis}}$, contributes a term in $c^{1/2}$, i.e. they give the $S\sqrt{c}$ term in the Debye-Hückel-Onsager equation. When the effects of the cross terms are considered they have no term in $c^{1/2}$ and thus do not contribute to the $S\sqrt{c}$ term. But all five corrections contribute to the other terms in the final Fuoss-Onsager equation.

The formulation of these modifications to the external field are complex and lengthy and lie in the field of highly advanced mathematics, but eventually they are resolved into the Fuoss-Onsager equation for unassociated symmetrical electrolytes:

$$\Lambda = \Lambda^0 - S\sqrt{c_{\text{actual}}} + (E_1\Lambda^0 - 2E_2)c_{\text{actual}}\log c_{\text{actual}} + J_1 c_{\text{actual}} + J_2 c_{\text{actual}}^{3/2} - F\Lambda^0 c_{\text{actual}}$$

$$(12.52)$$

in which one subsequent correction to the equation of 1957 has been incorporated, viz., this equation has the term $2E_2$ whereas the original published version had the term E_2. This is mentioned so as to avoid confusion for students if this equation is compared with the original Fuoss-Onsager equation where the logarithmic term was wrongly given as $(E_1\Lambda^0 - E_2)c_{\text{actual}}\log c_{\text{actual}}$.

Also take note: the base for the logarithmic term in Equation (12.52) is not specified, This is because the Fuoss-Onsager equation (Equation 12.52) can be quoted in terms of $\log_{10}$ or $\log_e$. This will affect the expression for E_1, E_2, J_1, J_2, and their values (see Appendix 2, Table 12.3).

The first two terms on the right hand side of the equation give the Debye-Hückel-Onsager limiting law equation:

- S is a function of ε_r, η, T and results from terms of the order of $c_{\text{actual}}^{1/2}$ in the electrophoretic and relaxation terms.

- E_1 is a function of ε_r, T, and it arises from terms to the order of $c_{\text{actual}}\log c_{\text{actual}}$ in the expression for the relaxation effect and the cross terms for the effect of relaxation on electrophoresis and the effect of electrophoresis on relaxation. The term $2E_2$ arises from electrophoresis and the cross terms. These include all terms up to the order of $c_{actual} \log c_{actual}$.

- The term, $J_1 c_{actual}$, and the higher order terms, $J_2 c_{actual}^{3/2}$, arise from higher terms in electrophoresis and for relaxation and are complex functions of $\varepsilon_r, \eta, T, \mathring{a}$.

- The term in $F\Lambda^0 c_{actual}$ is the effect of the viscosity on the ionic mobilities. It can either be calculated, or can be found from studies on the viscosity of solutions of the electrolyte at various concentrations.

- Throughout bulk ε_r, η, are used even though this is an approximation.

The equation is often written in the form where the viscosity term is subsumed into Λ:

$$\Lambda = \Lambda^0 - S\sqrt{c_{actual}} + (E_1\Lambda^0 - 2E_2)c_{actual}\log c_{actual} + J_1 c_{actual} + J_2 c_{actual}^{3/2} \qquad (12.53)$$

though, if viscosity data are available which will give a value of the hydrodynamic radius, R, which appears in the expression, $F = $ a constant $\times R$, and if Λ^0 is known, this term can be included.

The Fuoss-Onsager equation is also often used in the more approximate form:

$$\Lambda = \Lambda^0 - S\sqrt{c_{actual}} + (E_1\Lambda^0 - 2E_2)c_{actual}\log c_{actual} + J c_{actual} \qquad (12.54)$$

The constants which appear are of the form given above, and the numerical values of the constants for $25°$ C are given in Appendix 2, Table 12.3.

It can be seen that this treatment gives a theoretical justification for the empirical corrections given in Section 12.8 to the 1927 Debye-Hückel-Onsager and the 1932 Fuoss-Onsager equations. The denominator $(1 + \sqrt{c_{actual}})$ in the 1932 Fuoss-Onsager equation is now incorporated in higher order terms in c, i.e. in the terms:

$$(E_1\Lambda^0 - 2E_2)c_{actual}\log c_{actual} + J c_{actual} + \text{higher terms in } c_{actual} - F\Lambda^0 c_{actual} \qquad (12.55)$$

If the treatment is restricted to very low concentrations these higher terms drop out and the equation reduces to the simple Debye-Hückel-Onsager equation:

$$\Lambda = \Lambda^0 - S\sqrt{c_{actual}} \qquad (12.30)$$

The concentration which appears in the equation is the actual concentration of the ions. If the electrolyte is assumed to be completely unassociated, e.g. many 1–1 electrolytes, then the actual concentration can be replaced by the stoichiometric concentration.

The Fuoss-Onsager equation provides a good description of the effects of non-ideality for concentrations up to ca. 4×10^{-2} mol dm^{-3} for 1–1 electrolytes, to ca. 1×10^{-2} mol dm^{-3} for 2–2 electrolytes and to ca. 4×10^{-3} mol dm^{-3} for 3–3 electrolytes, **provided that there is no association**. If there is association considerable negative deviations are observed. One of the main difficulties in testing the Fuoss-Onsager equation for the higher charge types has been the lack of 2–2 and 3–3 unassociated electrolytes. Most 2–2 electrolytes are associated even at low concentrations, though divalent cation salts of disulphonic acids and trivalent cation salts of naphthalenetrisulphonic acid have been shown to be unassociated over a considerable range of concentrations and these have been used to test the general validity of the equation for these higher charge types.

Fuoss and Onsager subsequently modified their treatment to account for association (see Section 12.12).

For a given solvent at a fixed temperature and making the approximations for ε_r and η given above, the equation reduces to one in terms of the concentration, Λ^0 and $\mathring{a}$ and so is an equation in two unknowns. However, it is not possible to solve the equation as a simultaneous equation in the two unknowns using values of Λ at two concentrations. This is because of the complexity of the functional form of the Fuoss-Onsager equation.

12.10.1 Use of the Fuoss-Onsager equation to determine Λ^0 and $\mathring{a}$

Modern studies use computer fitting of experimental data to the 1957 Fuoss-Onsager equation, from which values of Λ^0, S and $E_1 \Lambda^0 - 2E_2$ and J can be found.

The value of J can be then compared with the theoretical values of J calculated for various assumed values of $\mathring{a}$ using the expression for J in terms of $\mathring{a}$ given previously. The value of $\mathring{a}$ which gives the calculated value of J closest to the experimentally determined J is taken to be the 'true' value of $\mathring{a}$. This value can then be compared with the value found from Λ^0 using Stokes' Law arguments, the Debye-Hückel theory and with the appropriate crystallographic radii.

Data given at the time of publication of the Fuoss-Onsager equation showed that, for small ion 1–1 electrolytes which are unassociated, the theory copes extremely well even if the higher terms in c, e.g. $c^{3/2}$, are omitted. The values of $\mathring{a}$ found in this way suggest that, at the distance of closest approach, the bare ions are in contact with all solvent having been expelled. However, this was for simple alkali halides, and for bigger and more complex ions this need not necessarily be the case. This is to be contrasted with diffraction data on 1–1 electrolytes which indicates 6 H_2O as nearest neighbours, and suggests that different methods measure different things (see Section 13.9 and 13.14).

The conclusion from the early work was that the equation was as successful as the Debye-Hückel theory relating to mean ionic activity coefficients was in coping with the effects of non-ideality in solutions of electrolytes.

12.10.2 Implications of the Fuoss-Onsager equation for unassociated symmetrical electrolytes

For 2–2 and 3–3 electrolytes in which there is no ion association, the graphs of Λ_{obsvd} vs. $\sqrt{c}$ lie above the Debye-Hückel-Onsager limiting slope at sufficiently high c_{stoich} but then cross over this line as c_{stoich} is decreased. Thereafter the graph approaches the limiting slope from below. This is a direct consequence of the $(E_1 \Lambda^0 - 2E_2)c_{stoich}\log c_{stoich}$ and Jc_{stoich} terms in the Fuoss-Onsager equation. Over the range of concentrations for which the equation is valid $(E_1 \Lambda^0 - 2E_2)c_{stoich}\log c_{stoich}$ is always negative while Jc_{stoich} is positive. For these higher charge types, the $(E_1 \Lambda^0 - 2E_2)c_{stoich}\log c_{stoich}$ term is dominant at very low values of c_{stoich} but is more than compensated by the Jc_{stoich} term at higher c_{stoich}. If the Fuoss-Onsager equation is taken to be a good description of the conductance data, then the graph of Λ_{obsvd} vs. $\sqrt{c}$ will cross over the limiting slope when:

$$(E_1\Lambda^0 - 2E_2)c_{stoich}\log c_{stoich} + Jc_{stoich} = 0 \tag{12.56}$$

and the value of the concentration at which this cross-over occurs is given by:

$$\log c_{stoich} = -\frac{J}{E_1\Lambda^0 - 2E_2} \tag{12.57}$$

The value of c_{stoich} at which this occurs is characteristic of the electrolyte. This is because both $E_1\Lambda^0 - 2E_2$ and J depend on Λ^0 while J is a function of $\mathring{a}$, the ion size parameter (see Table 12.3). Both Λ^0 and $\mathring{a}$ are characteristic of the electrolyte.

For 1–1 electrolytes, the concentration at which the cross-over occurs is sufficiently low that, even with very high precision measurements, such behaviour would probably just be barely detectable.

Typical values at which the cross-over would be expected to occur are:

1–1 electrolytes: concentration of the order of $10^{-4}\,\mathrm{mol\,dm^{-3}}$ and would require an experimental accuracy of 0.003% to detect.

2–2 electrolytes: concentration of the order of $10^{-3}\,\mathrm{mol\,dm^{-3}}$.

3–3 electrolytes: concentration of the order of $10^{-3}\,\mathrm{mol\,dm^{-3}}$.

Cross-over points for both 2–2 and 3–3 electrolytes should be easily and precisely detected with accurate experimental work.

The cross-over can be masked by the onset of ion association which causes the molar conductivity to become progressively lower than that predicted by an equation which is only valid for unassociated electrolytes. This explains why even highly accurate data on 2–2 and 3–3 electrolytes rarely show a cross-over point (see Figure 12.2). Only a very few 2–2

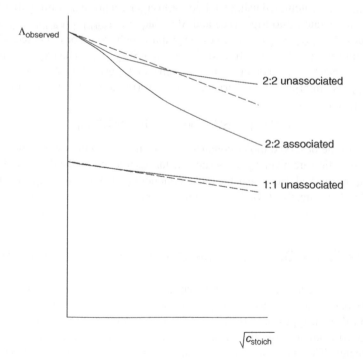

Figure 12.2   Cross-over diagram for a 1–1 and a 2–2 unassociated electrolyte; and a 2–2 associated electrolyte.

electrolytes, such as salts of disulphonic acids, have been shown to be unassociated even at moderate concentrations, and these show the expected cross-over point at the predicted concentration. Salts of trisulphonic acids are among the few known unassociated 3–3 electrolytes and these also show the predicted cross-over.

Within the range of concentrations for which the Fuoss-Onsager equation is expected to be valid, this equation accounts well for the effects of non-ideality in solutions of symmetrical electrolytes in which there is no ion association. It can thus be taken as a base-line for non-associated electrolytes and any deviations from this predicted behaviour can be taken as evidence of ion association (see Section 12.12).

12.11   A simple illustration of the effects of ion association on experimental conductance curves

The Fuoss-Onsager equation gives a good approximation to the experimental graphs of Λ vs. $\sqrt{c_{\text{stoich}}}$ for non-associated electrolytes and as such can serve as a base line for the behaviour of unassociated electrolytes. Any deviations from this predicted behaviour can be taken as evidence of ion association. This can be illustrated by experiments on the conductance of aqueous solutions of $MgSO_4$.

Λ^0 is known accurately for $MgSO_4(aq)$ and is equal to $266.1\ \text{S cm}^2\text{mol}^{-1}$.

$$\Lambda = \Lambda^0 - S\sqrt{c_{\text{actual}}} + (E_1\Lambda^0 - 2E_2)c_{\text{actual}}\log c_{\text{actual}} + Jc_{\text{actual}} \tag{12.54}$$

Values of $S, E = (E_1\Lambda^0 - 2E_2)$ and J can be calculated for $MgSO_4$ solutions and predicted values of $\Lambda_{\text{predicted}}$ are calculated using the Fuoss-Onsager equation assuming that $MgSO_4(aq)$ is completely dissociated into $Mg^{2+}(aq)$ and $SO_4^{2-}(aq)$ i.e. $c_{\text{actual}} = c_{\text{stoich}}$. A graph of these predicted values, $\Lambda_{\text{predicted}}$ vs. $\sqrt{c_{\text{stoich}}}$, gives the graph which would be expected if $MgSO_4(aq)$ were present entirely in solution as free $Mg^{2+}(aq)$ and $SO_4^{2-}(aq)$. The experimental graph of Λ_{obsvd} vs. $\sqrt{c_{\text{stoich}}}$ can be drawn on the same diagram (see Figure 12.3). Any deviation of Λ_{obsvd} from $\Lambda_{\text{predicted}}$ is interpreted as arising from formation of the overall uncharged $Mg^{2+}SO_4^{2-}(aq)$ ion pairs.

$$Mg^{2+}(aq) + SO_4^{2-}(aq) \rightleftharpoons Mg^{2+}SO_4^{2-}(aq)$$

The deviations of the observed molar conductivities from those which would be found if there were no ion association are given by the vertical distances marked on the diagram. It can be seen that there is a quite marked decrease in the molar conductivity from values expected for complete dissociation. Furthermore, the deviations become larger as the concentration increases. This is in keeping with the increase in the fraction associated as the concentration increases.

12.12   The Fuoss-Onsager equation for associated electrolytes

The 1957 Fuoss-Onsager equation can be adapted to take account of association of ions to form ion pairs and to account for incomplete dissociation of weak electrolytes. Chemically these are two different types of situation, but physically they are the same, viz. some of the ions are removed from solution by formation of ion pairs, or by formation of undissociated molecular species. The physical manifestation is that not all of the solute will be able to conduct the current, and so the observed conductance will be lower than that predicted by the

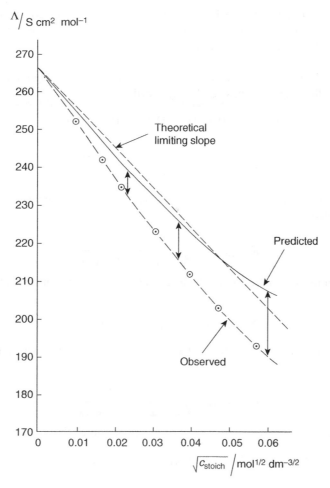

Figure 12.3 Schematic graphs of Λ_{obsvd} vs. $\sqrt{c_{stoich}}$ and $\Lambda_{predicted}$ vs. $\sqrt{c_{stoich}}$ for $MgSO_4(aq)$. Vertical distances between the curves show deviations from predicted behaviour due to ion association.

Fuoss-Onsager 1957 equation. Fuoss later modified his treatment to include the possibility that some ion pairs are conducting (see Section 12.17).

The equation for fully dissociated electrolytes in compact form and ignoring higher order terms is:

$$\Lambda_{theoretical} = \Lambda^0 - S\sqrt{c_{actual}} + (E_1\Lambda^0 - 2E_2)c_{actual}\log c_{actual} + Jc_{actual} \qquad (12.54)$$

If removal of ions from the solution of electrolyte at a stoichiometric concentration, c_{stoich} occurs then:

$$\Lambda_{observed} = \alpha\Lambda_{theoretical} = \alpha[\Lambda^0 - S\sqrt{c_{actual}} + (E_1\Lambda^0 - E_2)c_{actual}\log c_{actual} + Jc_{actual}] \quad (12.55)$$

where α is the fraction of the electrolyte **present as ions** and able to conduct the current. This equation can also be written in terms of $K_{association}$ rather than α. The equilibrium constant appearing in the equation for an associated electrolyte is an association constant for removal of the ions from solution rather than the conventional dissociation constant describing the formation of ions from either a weak electrolyte or an ion pair, e.g.

- For the weak acid, HA, for which the self ionisation of water can be ignored, the fraction of the weak acid dissociated to ions can be described as:

$$HA(aq) + H_2O(l) \rightleftharpoons H_3O^+(aq) + A^-(aq)$$

$$K_{dissociation} = \frac{\alpha^2 c_{stoich}^2}{(1-\alpha)c_{stoich}} = \frac{\alpha^2 c_{stoich}}{(1-\alpha)} \tag{12.56}$$

The expression for formation of the weak acid, $K_{association}$, can also be written in terms of the fraction dissociated, α.

$$K_{association} = \frac{1}{K_{dissociation}} = \frac{1-\alpha}{\alpha^2 c_{stoich}} \tag{12.57}$$

- For the case of ion association again the fraction of the ion pairs converted to free ions can be described as:

$$MX(aq) \rightleftharpoons M^{z+}(aq) + X^{z-}(aq)$$

and again the expression for association of the ions to the ion pair can be written in terms of the fraction associated, α.

$$K_{association} = \frac{1}{K_{dissociation}} = \frac{(1-\alpha)}{\alpha^2 c_{stoich}} \tag{12.58}$$

- From these definitions and Equation (12.55), it can be shown that:

$$\Lambda_{observed} = \Lambda^0 - S\sqrt{c_{actual}} + Ec_{actual}\log c_{actual} + Jc_{actual} - K_{association}c_{actual}\Lambda_{observed} \cdots \cdots \tag{12.59}$$

Although this form of the conductance equation is very useful for theoreticians, it is not useful for the experimentalist because α, $K_{association}$ and c_{actual} are unknown. It is, therefore, better to express the equation in terms of the known stoichiometric concentrations making it then more directly comparable to the experimental data. Carrying through some complex algebra and dropping some higher order terms leads to:

$$\Lambda_{observed} = \Lambda^0 - S\sqrt{c_{stoich}} + Ec_{stoich}\log c_{stoich} + Jc_{stoich} - K_{association}c_{stoich}\Lambda_{observed} \cdots \cdots \tag{12.60}$$

It may seem surprising that these two forms of the conductance equation for associated electrolytes are so similar. There is, however, no mistake here. Neither equation is exact, and the difference between the two equations is of the order of terms such as $c^{3/2}$ or $c^{3/2}\log c$ which are omitted in both. Because of this, the equation in terms of the stoichiometric concentration should only be used to obtain first approximate values of Λ^0, $K_{association}$ or its equivalent values of α at various stoichiometric concentrations and J which will allow determination of $\mathring{a}$. Thereafter the equation in terms of c_{actual} should be used to determine Λ^0, J, $K_{association}$ and $\mathring{a}$.

This is an equation in three unknowns Λ^0, $K_{association}$ and $\mathring{a}$. The distance of closest approach comes in through the J terms which include the term $\mathring{a}$. And so, experiment can determine all three quantities from a series of experimental observations of $\Lambda_{observed}$ at various stoichiometric concentrations.

12.12.1   Determination of Λ^0, $K_{association}$ and $\mathring{a}$ using the Fuoss-Onsager equation for associated electrolytes

Modern methods use computer fitting of experimental data to the conductance curve. However, there are hidden dangers here for the student since computer fitting does not lend itself to developing an understanding of the physical principles involved in the setting up of the conductance equation and its solution. Furthermore, more than one set of Λ^0, $K_{association}$ and $\mathring{a}$ may fit the data.

12.13   Range of applicability of Fuoss-Onsager 1957 conductance equation for symmetrical electrolytes

Fuoss has stressed that his equations can only be legitimately applied for concentration ranges where $\kappa\mathring{a} < 0.2$. This corresponds to approximately $4 \times 10^{-2}\,mol\,dm^{-3}$ for 1–1 electrolytes, $1 \times 10^{-2}\,mol\,dm^{-3}$ for 2–2 electrolytes and $4 \times 10^{-3}\,mol\,dm^{-3}$ for 3–3 electrolytes. The theory also only applies to symmetrical electrolytes, though later workers were able to obtain expressions for unsymmetrical electrolytes. These are much more complex than that of Fuoss and Onsager.

The restriction to these concentrations is partly mathematical and partly physical.

- The mathematical restriction results from the complicated functions which result from the integrations involved in setting up expressions for the various $\Delta X/X$ used to account for non-ideality. These functions were given as a series expansion where, because of mathematical complexity, it was necessary to neglect terms in concentration of order $c^{3/2}$ or higher. The errors involved become significant for $\kappa\mathring{a} > 0.2$. This gives an upper limit beyond which these approximations would lead to systematic deviations.

- The physical restriction arises from the concept of the ionic atmosphere which is the basic postulate of the Debye-Hückel theory. The ions in the ionic atmosphere are not treated as discrete charges, but are described in terms of a smeared out cloud of charge density. Such a description is only valid if $1/\kappa$ is **significantly greater** then the diameter of an ion, $\mathring{a}$. The model of a continuous charge fails when $1/\kappa$ becomes of the order of the distance of closest approach, $\mathring{a}$ (see Table 12.4).

The data in Table 12.4 is given in terms of increasing concentrations, where $\kappa\mathring{a} = B\mathring{a}\sqrt{I} = B\mathring{a}z\sqrt{c_{stoich}}$ for an unassociated symmetrical electrolyte. From the table it

Table 12.4   Values of $1/\kappa$ as a function of $\kappa\mathring{a}$

$\kappa\mathring{a}$	0.05	0.1	0.2	0.3	0.5
$1/\kappa$	$20\mathring{a}$	$10\mathring{a}$	$5\mathring{a}$	$3\mathring{a}$	$2\mathring{a}$

can be seen that $1/\kappa$ is significantly greater than $\mathring{a}$ only for concentrations less than those corresponding to $\kappa\mathring{a} \approx 0.2$. Hence Fuoss's statement that the theory is only valid for concentration regions corresponding to $\kappa\mathring{a} < 0.2$.

Furthermore if the concept of replacing the discrete ion surrounding a central reference ion by a continuous charge cloud fails when $\kappa\mathring{a} > 0.2$, then use of $\psi_j = \dfrac{z_j e}{4\pi\varepsilon_0\varepsilon_r} \dfrac{e^{\kappa\mathring{a}}}{1+\kappa\mathring{a}} \dfrac{e^{-\kappa r}}{r}$ in the various integrals of the various $\Delta X/X$ terms in the theory will also fail. If, however, the free ions were assumed to be solvated and **this** dictated the distance of closest approach, then this restriction would disappear. More recent work allows this by using the Gurney co-sphere approach (see Section 12.17).

12.14 Limitations of the treatment given by the 1957 Fuoss-Onsager conductance equation for symmetrical electrolytes

In many respects it is perverse and verging on the impertinent to criticise this equation when the following points are considered:

- the complexity of the derivation (a concise abbreviated version runs to around 100 pages);

- the enormous amount of mathematical skills required to solve the equations and the massive expenditure of time and energy which was required;

- the problems lying in the complexity of the mathematics and in the setting up of the analytical forms necessary: the whole treatment is an exercise in abstract pure mathematics.

However, despite the sophistication of the mathematics, it is necessary to realise that this is only a very primitive treatment at the microscopic level.

The properties of the ions and the solvent which are ignored are similar to those ignored in the Debye-Hückel treatment. These are very important properties at the microscopic level, but it would be a thankless task to try to incorporate them into the treatment used in the 1957 equation. Furthermore, Stokes' Law is used in the equations describing the movement of the ions. This law applies to the motion of a macroscopic sphere through a structureless continuous medium. But the ions are microscopic species and the solvent is not structureless and use of Stokes' Law is approximate in the extreme. Likewise, the equations describing the motion also involve the viscosity which is a macroscopic property of the solvent and does not include any of the important microscopic details of the solvent structure. The macroscopic relative permittivity also appears in the equation. This is certainly not valid in the vicinity of an ion because the intense electrical field due to an ion will cause dielectric saturation of the solvent immediately around the ion. In addition, alteration of the solvent structure by the ion is an important feature of electrolyte solutions (see Section 13.16). However, solvation is ignored. As in the Debye-Hückel treatment the physical meaning of the distance of closest approach, i.e. $\mathring{a}$ is also open to debate.

It is quite obvious that further modification to this conductance theory to take account of the shortcomings of the Debye-Hückel model as outlined would be even less fruitful than similar attempts on the Debye-Hückel theory itself.

However, this must be seen in the context of the considerable impetus and stimulus which the Fuoss-Onsager treatment of conductance has given to the experimentalist who has striven to find more and more precise methods with which to test the various theories outlined. This has resulted in very considerable improvements being made to conductance apparatus. It has also placed a very detailed emphasis on obtaining precision and accuracy of the measurements themselves. This has been of considerable import when making measurements at very low concentrations where the experimental difficulties are greatest, but where it is important to test the theory in regions where it is expected to be valid. Such expectations have been vindicated by precision low concentration work where confidence can be placed in the accuracy of the conductance equation. This is reminiscent of the impetus to experimentalists after the Debye-Hückel equation had been put forward.

Nonetheless, the Fuoss-Onsager 1957 equations for unassociated and associated electrolytes cope with non-ideality extremely well, **provided** the concentration range is limited to up to:

- 4×10^{-2} mol dm^{-3} for 1–1 electrolytes,

- up to 1×10^{-2} mol dm^{-3} for 2–2 electrolytes

- and up to 4×10^{-3} mol dm^{-3} for 3–3 electrolytes.

i.e. if the condition is $\kappa\mathring{a} < 0.2$ and $\mathring{a} \approx 0.3$ nm

12.15 Manipulation of the 1957 Fuoss-Onsager equation, and later modifications by Fuoss and other workers

In the period between 1957 and 1978 various modifications and extensions were made to the Fuoss-Onsager equations for unassociated and associated electrolytes, but there were no major changes in the model. All that these studies had done was to produce modified conductance equations.

Comparisons of the various modifications with the Fuoss-Onsager 1957 equation showed that:

- Λ^0 appears to be insensitive to the conductance equation being considered and, probably more significantly, it appeared to be insensitive to the value of K_{assoc}. This is not surprising considering the status of Λ^0 as a quantity pertaining to infinite dilution, i.e. ideal conditions.

- However, the value of K_{assoc} not only depended on the conductance equation to which the data was fitted, but also on the value of $\mathring{a}$.

- This suggests that fitting experimental data to the conductance equation reduces in effect to finding best fit values of K_{assoc} and $\mathring{a}$.

- Experiment can thus give the $(K_{assoc}, \mathring{a})$ best fit. However, it was found that there was a range in the values which would give the best fit. This is a problem inherent in the use of computing techniques for obtaining best fit parameters, and one which the experimenter must be fully aware of.

- The experimenter must use chemical judgement to decide on a given pair of $(K_{assoc}, \mathring{a})$ values.

Once this is done, it is then possible to use the value of $\mathring{a}$ to find a **calculated** value of K_{assoc} using the equation describing

- Bjerrum's ion pair model (see Section 10.12);

- Fuoss's contact ion pair model (see Section 10.12.6);

- or other modified ion pair models (see Section 10.12.6).

and to see which fits best with the experimental K_{assoc}. Here the best fit could give clues as to which ion pair model best fits the data, but see Section 12.17.

12.16 Conductance studies over a range of relative permittivities

One of the most important studies involves the dependence of the molar conductivity on the solvent. The solvent is characterised by its chemical nature, its macroscopic relative permittivity and viscosity. Studies carried out in non-aqueous solvents and/or mixed solvents one of whose components is often water, allow conductance studies to be made over a range of relative permittivities and viscosities. The dependence of Λ^0, K_{assoc} and $\mathring{a}$ on relative permittivity has furnished a vast amount of data which is of fundamental importance in conductance studies. In particular, the relation between K_{assoc} and ε_r has given considerable insight into the correctness or otherwise of the various models developed to describe the ion pair.

Studies in mixed solvents and non-aqueous solvents are outside the scope of this book. But they indicate conclusively that the primitive model used by Fuoss and Onsager up to 1978 is wrong, and that a new model needs to be developed.

12.17 Fuoss *et al.* 1978 and later

By the 1970s it became clear that the model on which current conductance theories were based was wrong. The discussion of the limitations of the theory to concentrations corresponding to values of $\kappa\mathring{a} < 0.2$ (see Sections 12.13 and 12.14) showed that the approximations made in the theory which led to higher order terms being dropped out would not be valid at moderate concentrations. Furthermore experiment showed unambiguously that ion association had to be considered explicitly, especially for 2–2 and higher charge type electrolytes, i.e. some form of short range coulombic ion–ion interaction must be considered.

The model and theory, like that of the Debye-Hückel treatment of non-ideality, were based on consideration of long range electrostatic coulombic interactions only. The model was most likely to be inadequate because it did not take into account specific short range interactions corresponding to ion–ion, ion–solvent, and solvent–solvent interactions.

For the ion–ion interactions, Fuoss indicated that a fruitful line of enquiry would be that of working downwards from

- the crystal lattice structure of the solid electrolyte;

- to the fused electrolyte structure;

- to the highly concentrated solution;

which it was hoped would link up with the satisfactory theory for dilute solutions. This is discussed in Section 10.21.

Fuoss also indicated that a second line of approach could be to consider specific ion–solvent interactions. All the previous theories had considered the solvent to be a continuum which did not recognise the discrete molecular structure of the solvent, here water. The discrete structure should, however, be considered for at least the first few nearest neighbour layers of water around an ion. In this region the macroscopic relative permittivity and viscosity would not be applicable because the solvent structure would be modified by the intense field of the ion, and induced dipoles would be generated (see Sections 1.3 to 1.6 and Section 13.16, on solvation). However, there will be a region of the solvent beyond a certain distance from the ion in which the solvent structure remains unmodified from its macroscopic structure and its properties are thus those of the bulk solvent. It is to this region of unmodified solvent that the conductance theories based on the original Fuoss-Onsager model apply.

Fuoss in fact took neither of these two routes, and developments along these lines were left to other workers. Instead, he approached the problem by basing his new theory on the concept of the Gurney co-sphere.

The Gurney co-sphere defines a region around the ion which has solvent molecules whose structure has been modified by the field of the ion (see Section 10.16). Outside this region, the solvent has its macroscopic bulk structure. The diameter of the Gurney co-sphere takes a value, R. The distance between the centre of ions with such co-spheres where the co-spheres just touch is also R (see Figure 12.4). This assumes that the ions are spherical, and the situation could well prove to be different if the ions are non-spherical. In addition, ion association was taken to be an integral part of the model and theory rather than an 'added-on' factor used to explain deviations from a conductance equation based on the concept of complete dissociation. All ion association takes place within the Gurney co-sphere.

The distance of closest approach was no longer used, being replaced by the distance, R, which defines the distance from the centre of one ion to the centre of another ion, both having a co-sphere with the co-spheres just touching (see Figure 12.4). For a given ion, the co-sphere defines the region of solvent which is modified from its bulk state, i.e. in the region, $0 < r < R/2$, and the distance $R/2$ defines the distance beyond which the solvent behaves as a continuum. This means that it is possible to take all the short range interactions, ion–ion

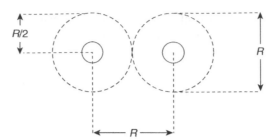

Figure 12.4   A Gurney type diagram showing two ions with co-spheres just touching. R is the distance between the centres of the ions and is the Gurney diameter.

and ion–solvent, and place them within the Gurney co-sphere and hence be able to avoid specifying them. This has considerable significance in the development of the theory, as will be seen later.

The theory also saw a major departure from the more traditional concepts of an ion pair which had stated that an ion pair consisted of one of the following:

- two ions in contact at a distance, $\overset{\circ}{a}$, between the charges at the centre of the ions;

- two ions lying within the Bjerrum distance, q, defined as a distance where the electrostatic energy of interaction of the two ions was $2kT$;

- two ions at a distance apart given by Fuoss's earlier concept of an ion pair.

Fuoss now allowed two types of ion pair to occur within the Gurney co-spheres of the central ion and an ion of opposite sign, i.e. within the distance, $R/2$ from the centre of the ion. The new description now involves:

- contact ion pairs: these have an ion of opposite charge to the central ion in the first shell of solvent neighbours, i.e. its nearest neighbours. These are characterised by a distance between the centres of the two ions falling within a range from $\overset{\circ}{a}$ to $\overset{\circ}{a} + s$, where s is the diameter of a solvent molecule (see Figure 12.5(a) and 12(b)). In these ion pairs some of the solvent molecules in the Gurney co-spheres have been squeezed out and replaced by the second ion of opposite charge.

- solvent separated ion pairs: these are those which are further apart than the contact ion pair, but with distances apart of the ions which lie in the range between $\overset{\circ}{a} + s$ and R (see Figure 12.5(a) and 12.5(c)). Here the maximum extent of the region which would define a solvent separated ion pair corresponds to the two co-spheres of each ion just touching. The region thus includes regions where the co-spheres overlap, but the overlap is restricted to regions with more than one solvent molecule.

- free ions: these are at distances apart greater than R.

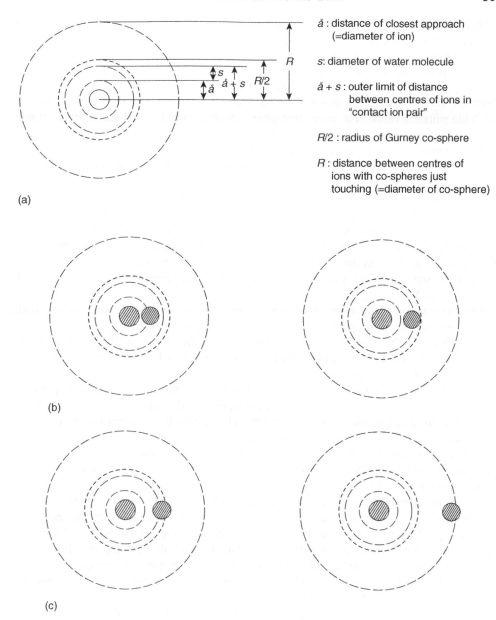

å : distance of closest approach
(=diameter of ion)

s: diameter of water molecule

å + s : outer limit of distance
between centres of ions in
"contact ion pair"

R/2 : radius of Gurney co-sphere

R : distance between centres of
ions with co-spheres just
touching (=diameter of co-sphere)

(a)

(b)

(c)

Figure 12.5 **(a)** Distances involved in specifying contact ion pairs and solvent-separated ion pairs; **(b)** Fuoss contact ion pairs – solvent molecules totally squeezed out; **(c)** Fuoss solvent-separated ion pairs.

There is also a further crucial change to the model. Up to this time the free ions were taken to be the only species able to conduct the current. Contact ion pairs, Bjerrum ion pairs and early Fuoss ion pairs do not conduct the current. But now Fuoss allows solvent separated ion pairs as well as free ions to conduct the current.

This has repercussions on how the theory is developed. Calculations of activity coefficients, relaxation and electrophoresis and their cross terms are carried out by assuming that only long range coulombic forces operate in regions where the macroscopic relative permittivity is applicable. Calculations of the quantities, $\gamma_{\pm}, \Delta X_{\text{relaxation}}/X, \Delta X_{\text{electrophoresis}}/X$, $\Delta X_{\text{cross terms}}/X$, will, therefore, relate only to distances from the centre of the central ion which are greater than R, the diameter of the Gurney co-sphere and the region outside of which ion pairs can reside. The ionic atmosphere which encapsulates all the elements of non-ideality will now start at a distance R rather than at $\mathring{a}$ as in the Debye-Hückel theory and in the former theories of conductance. The substitution of R for $\mathring{a}$ is not a simple substitution. Instead it has major implications for the development of the theory and radically alters the equations involved in the mathematical derivation. It also has the consequence that for these purposes contact ion pairs and solvent separated ion pairs do not appear in the calculations; the only forces allowed are the long range interactions between free ions. These are the only interactions which are relevant, and will only apply at distances greater than R. It is in this way that, by using the Gurney co-sphere model, Fuoss can ignore the discrete structure of the solvent and its modification by the ions, as well as the presence of the contact ion pair and the solvent separated ion pairs in the calculations pertaining to $\gamma_{\pm}, \Delta X_{\text{relaxation}}/X$, $\Delta X_{\text{electrophoresis}}/X, \Delta X_{\text{cross terms}}/X$.

When it comes to what is allowed to carry the current a further crucial change is also made. Now, solvent separated ion pairs as well as free ions can conduct. The theory must then be expressed in terms of the fraction of ions which can conduct the current and this fraction contains a contribution from the fraction of the electrolyte converted to solvent separated ion pairs.

His rationale for this is as follows.

Normal Brownian motion is a result of solvent molecules impacting on the solute particles, and these give both translational and rotational movement to the solute. An ion on its own will execute this Brownian motion. Since an ion has a charge which can interact with an external electric field, this interaction will perturb the translational Brownian motion, with a cation moving in the direction of the field while an anion will move in the opposite direction. The field will have a minor effect on the rotational Brownian motion, but this will not contribute to the translational mobility.

Ions in contact will behave as an uncharged entity. They will also be surrounded by a cage of solvent molecules. Many collisions will be required before the ions can separate and escape from the cage. They will execute Brownian motion, but since the ion pair will exist for a time which is much longer than the interval between collisions it will behave during that time as though it were uncharged, and so the translational Brownian motion will not be perturbed by the external field. The only effect which the external field will have on a contact ion pair will be to perturb the rotational part of the Brownian motion by virtue of its dipolar nature. Since it is only translational motion which contributes to the ionic mobility, this effect can be ignored.

Solvent separated ion pairs will also be overall uncharged and will execute Brownian motion. They will also be enclosed in a cage of solvent molecules, but since the interactions between the ions will be considerably smaller than those between ions in contact, they will separate and escape from the cage much sooner than the contact ion pairs. On the time scale of collisions they can be considered as separating and thus not moving as a single entity. The translational Brownian motion could then be perturbed by an external field, so that, on average, the motion of a cation could be in the direction of the external field and so be able to conduct the current, and if the ion is an anion it will move in the opposite direction.

But is this reasonable? There are two possible sources of doubt as to the physical correctness of this argument:

- If the solvent separated ion pair behaves in this manner, can it then be counted as an ion pair?

- If the solvent separated ion pair behaves in this manner, should it make the same contribution to carrying the current as does a free ion?

12.17.1   The fraction of ions which are free to conduct the current

The equilibria involved for the symmetrical electrolyte, MX are:

$$M^{z+}(aq) + X^{z-}(aq) \quad \rightleftarrows \quad M^{z+}(aq)-----X^{z-}(aq) \quad \rightleftarrows \quad M^{z+}X^{z-}(aq)$$
$$\text{free ions} \qquad\qquad\qquad \text{solvent separated} \qquad\qquad \text{contact}$$
$$\text{ion pair} \qquad\qquad\qquad \text{ion pair}$$

- Let γ be the fraction of ions which are free and unassociated, and have a concentration given by γc_{stoich}.

- $\therefore 1 - \gamma$ is the fraction of ions associated in some way, and these will be the contact ion pairs and solvent separated ion pairs with a total concentration given by $(1 - \gamma)c_{\text{stoich}}$. This has to be split up into a fraction associated to solvent separated ion pairs and the fraction of these which are associated to contact ion pairs.

- Define α as the fraction of the associated ion pairs which are contact ion pairs. The concentration of contact ion pairs is, therefore, $\alpha(1 - \gamma)c_{\text{stoich}}$.

- The total concentration of associated ion pairs
  = the concentration of solvent separated ion pairs
  + the concentration of contact ion pairs

i.e. $(1 - \gamma)c_{\text{stoich}} = $ concentration of solvent separated ion pairs $+ \alpha(1 - \gamma)c_{\text{stoich}}$ (12.61)

$\therefore$ concentration of solvent separated ion pairs $= (1 - \gamma)c_{\text{stoich}} - \alpha(1 - \gamma)c_{\text{stoich}}$ (12.62)

$$= (1 - \gamma - \alpha + \alpha\gamma)c_{\text{stoich}} \qquad\qquad (12.63)$$

$$= (1 - \alpha)(1 - \gamma)c_{\text{stoich}} \qquad\qquad (12.64)$$

- The free ions and the solvent separated ion pairs are those which, on this model, can conduct the current. The total fraction of ions able to conduct the current is, therefore,

$$\gamma + (1 - \alpha)(1 - \gamma) = \gamma + 1 - \gamma - \alpha + \alpha\gamma = 1 - \alpha(1 - \gamma) \qquad\qquad (12.65)$$

- The overall equilibrium constant, $K_{assoc} = \dfrac{[\text{ion pairs}]_{actual}}{[\text{free ion}]^2_{actual}} = \dfrac{(1-\gamma)c_{stoich}}{\gamma^2 c^2_{stoich} f^2_\pm}$ (12.66)

where $f_\pm$ is used to denote the mean activity coefficient of the free ions, − the standard symbol, $\gamma_\pm$, is not used to avoid confusion with the fraction of ions, γ, which are free.

The Debye-Hückel expression is used as : $\log_{10} f_\pm = -\dfrac{A|z_1 z_2|\sqrt{I}}{1 + BR\sqrt{I}}$ (12.67)

Note: the distance of closest approach is now R not $\mathring{a}$.

12.17.2 The Fuoss 1978 equation for associated symmetrical electrolytes

The basic conductance expression is:

$$\Lambda(1 + Fc_{stoich}) = \Lambda^0 \left[1 - \frac{\Delta X_{relaxation}}{X} - \frac{\Lambda X_{electrophoresis}}{X} - \frac{\Delta X_{Brownian}}{X} \right.$$
$$\left. - \frac{\Delta X \underset{\text{relaxation on electrophoresisn}}{\text{cross term for effect of}}}{X} - \frac{\Delta X \underset{\text{electrophoresis on relaxation}}{\text{cross term for effect of}}}{X} \right]$$ (12.51)

when corrected for the fraction of ions which can conduct the current is then:

$$\Lambda(1 + Fc_{stoich}) = [1 - \alpha(1-\gamma)]\Lambda^0 \left[1 - \frac{\Delta X_{relaxation}}{X} - \frac{\Lambda X_{electrophoresis}}{X} - \frac{\Delta X_{Brownian}}{X} \right.$$
$$\left. - \frac{\Delta X \underset{\text{relaxation on electrophoresisn}}{\text{cross term for effect of}}}{X} - \frac{\Delta X \underset{\text{electrophoresis on relaxation}}{\text{cross term for effect of}}}{X} \right]$$ (12.68)

α and γ can then be expressed in terms of the K_{assoc} for each relevant process.

As mentioned, the move from a distance of closest approach of $\mathring{a}$ to R in the theoretical expressions is not a simple substitution. Instead there are complex changes in the integrals appearing in the $\Delta X/X$ terms. The theory now becomes so complex that it is impossible to express Λ in the simple form of the 1957 equations. Instead the equations are left in their basic form as integrals, or sometimes in approximate form as a series expansion in polynomials. Whichever is done, it becomes necessary to computer fit the experimental data to the integral form, or to the approximate polynomial form. Computer programs are available for doing this.

Other modern approaches use distribution functions and computer simulations analogous to those described in the statistical mechanical procedures in Sections 10.17 to 10.19. These will be very complex, but may well be the most fruitful developments of the future.

Appendix 1

The velocity, v, with which an ion moves under an external field, X, is proportional to the external field (see Section 11.17).

$$v = uX$$ (1)

where u is a constant of proportionality.

Under ideal conditions, this can be written as:

$$v^{\text{ideal}} = u^{\text{ideal}}X = u^0 X \tag{2}$$

where u^0 is the mobility under ideal conditions where $c \to 0$.

Under non-ideal conditions, the external field is modified by electrophoresis and relaxation, and this modified field is $(X - \Delta X)$. As the concentration decreases, ΔX also decreases, and in the limit as $c \to 0$, i.e. ideal conditions, $\Delta X \to 0$. **But**, the velocity with which the ion migrates under this modified field is **still defined** in terms of the external field, **even** for the non-ideal case where the **actual effective field** under which the ion migrates is $(X - \Delta X)$, i.e.:

$$v^{\text{actual}} \propto X = u^{\text{actual}}X \tag{3}$$

where u^{actual} is the actual mobility, not the ideal mobility.

However, it could **also be written** in terms of the modified field rather than in terms of the externally applied field:

$$v^{\text{actual}} \propto (X - \Delta X) = u(X - \Delta X) \tag{4}$$

where u is a mobility whose value has yet to be determined.

These two equations for v^{actual} can be equated, giving:

$$u(X - \Delta X) = u^{\text{actual}}X \tag{5}$$

But as $\Delta X \to 0$, i.e. $(X - \Delta X) \to X$ when $c \to 0$ and ideal conditions are approached, then:

$$u^{\text{actual}} = u^0 \tag{6}$$

$$\text{i.e. } uX = u^{\text{actual}}X = u^0 X \text{ as } c \to 0 \tag{7}$$

$$\therefore u = u^0 \text{ for ideal conditions} \tag{8}$$

Hence:

$$v^{\text{ideal}} = u^0 X \tag{9}$$

$$\text{and } v^{\text{actual}} = u^0(X - \Delta X) \tag{10}$$

The velocities of migration, v_+ for the cations, and v_- for the anions are different. Applying the above relations, Equations (9) and (10), to the individual ions gives:

- **The ideal case**:

$$v_+^{\text{ideal}} = u_+^0 X \qquad \text{and} \qquad v_-^{\text{ideal}} = -u_-^0 X \tag{11 and 12}$$

where u_+^0 and u_-^0 are ideal mobilities corresponding to infinite dilution.

The minus sign in the equation for the anion appears since it takes cognisance of the fact that the anion is moving in the opposite direction to the cation.

- **The non-ideal case**:

$$v_+^{\text{actual}} = u_+^0(X - \Delta X) \qquad \text{and} \qquad v_-^{\text{actual}} = -u_-^0(X - \Delta X) \tag{13 and 14}$$

where u_+^0 and u_-^0 are again the constants of proportionality for cation and anion deduced in the argument given in the Equation (10).

Using these relations:

$$\therefore \quad \frac{v_+^{\text{actual}}}{v_+^{\text{ideal}}} = \frac{X - \Delta X}{X} \qquad \text{and} \qquad \frac{v_-^{\text{actual}}}{v_-^{\text{ideal}}} = \frac{X - \Delta X}{X} \qquad (15) \text{ and } (16)$$

Using Equations (11.13) and (11.14) applied to each ion:

$$u_+^{\text{actual}} X = u_+^0 (X - \Delta X) \qquad \text{and} \qquad u_-^{\text{actual}} X = u_-^0 (X - \Delta X) \qquad (17) \text{ and } (18)$$

$$\therefore \quad \frac{u_+^{\text{actual}}}{u_+^0} = \frac{X - \Delta X}{X} \qquad \text{and} \qquad \frac{u_-^{\text{actual}}}{u_-^0} = \frac{X - \Delta X}{X} \qquad (19) \text{ and } (20)$$

In Section 11.17 (Equations 11.124 and 11.127) it was shown that

$$\lambda_+ = F z_+ u_+ \qquad \text{and} \qquad \lambda_- = F |z_-| u_-$$

$$\therefore \quad \frac{\lambda_+^{\text{actual}}}{\lambda_+^{\text{ideal}}} = \frac{u_+^{\text{actual}}}{u_+^{\text{ideal}}} = \frac{X - \Delta X}{X} \qquad \text{and} \qquad \frac{\lambda_-^{\text{actual}}}{\lambda_-^{\text{ideal}}} = \frac{u_-^{\text{actual}}}{u_-^{\text{ideal}}} = \frac{X - \Delta X}{X} \qquad (21) \text{ and } (22)$$

$$\text{and} \qquad \lambda_+^{\text{actual}} = \lambda_+^{\text{ideal}} \frac{X - \Delta X}{X} = \lambda_+^0 \left(1 - \frac{\Delta X}{X} \right) \qquad (23)$$

$$\lambda_-^{\text{actual}} = \lambda_-^{\text{ideal}} \frac{X - \Delta X}{X} = \lambda_-^0 \left(1 - \frac{\Delta X}{X} \right) \qquad (24)$$

where the ideal ionic molar conductivities are now written in the conventional manner, λ^0.

The actual ionic molar conductivity which is a value for situations where non-ideality must be considered is equal to the ideal molar ionic conductivity at infinite dilution modified by the term $1 - \Delta X/X$. ΔX is a sum of all the terms which are taken to lead to non-ideality.

If the effects of viscosity on the mobilities of the ions are included, this results in a term, $(1 + F c_{\text{stoich}})$, where F is a constant involving the viscosity. The equations are then modified as:

$$\lambda_+^{\text{actual}} (1 + F c_{\text{stoich}}) = \lambda_+^{\text{ideal}} \frac{X - \Delta X}{X} = \lambda_+^0 \left(1 - \frac{\Delta X}{X} \right) \qquad (25)$$

$$\lambda_-^{\text{actual}} (1 + F c_{\text{stoich}}) = \lambda_-^{\text{ideal}} \frac{X - \Delta X}{X} = \lambda_-^0 \left(1 - \frac{\Delta X}{X} \right) \qquad (26)$$

$$\lambda_+ (1 + F c_{\text{stoich}}) = \lambda_+^0 \left[1 - \frac{\Delta X_{\text{relaxation}}}{X} - \frac{\Delta X_{\text{electrophoresis}}}{X} - \frac{\Delta X_{\text{Brownian}}}{X} \right.$$

$$\left. - \frac{\Delta X \; \substack{\text{cross term for effect of} \\ \text{relaxation on electrophoresisn}}}{X} - \frac{\Delta X \; \substack{\text{cross term for effect of} \\ \text{electrophoresis on relaxation}}}{X} \right] \qquad (27)$$

A similar equation describes the effects of non-ideality on the limiting ionic molar conductivity, λ_-^0, for the case where the central reference ion is an anion.

$$\lambda_-(1 + Fc_{\text{stoich}}) = \lambda_-^0\left[1 - \frac{\Delta X_{\text{relaxation}}}{X} - \frac{\Delta X_{\text{electrophoresis}}}{X} - \frac{\Delta X_{\text{Brownian}}}{X}\right.$$
$$\left. - \frac{\Delta X \underset{\text{relaxation on electrophoresisn}}{\text{cross term for effect of}}}{X} - \frac{\Delta X \underset{\text{electrophoresis on relaxation}}{\text{cross term for effect of}}}{X}\right] \tag{28}$$

The equation for the molar ionic conductivity for a symmetrical electrolyte is:

$$\Lambda(1 + Fc_{\text{stoich}}) = \Lambda^0\left[1 - \frac{\Delta X_{\text{relaxation}}}{X} - \frac{\Delta X_{\text{electrophoresis}}}{X} - \frac{\Delta X_{\text{Brownian}}}{X}\right.$$
$$\left. - \frac{\Delta X \underset{\text{relaxation on electrophoresisn}}{\text{cross term for effect of}}}{X} - \frac{\Delta X \underset{\text{electrophoresis on relaxation}}{\text{cross term for effect of}}}{X}\right]. \tag{12.29}$$

Appendix 2

Table 12.3 Expression for the constants in the Fuoss-Onsager 1957 equation along with formulae for calculating their values

$S = a + b\Lambda^0$

with $a = \dfrac{z^3 eFB}{3\pi\eta}$

$b = \dfrac{z^3 e^2 B}{24\pi\eta\varepsilon_0\varepsilon_r kT(1 + \sqrt{1/2})}$

$B = \left(\dfrac{2e^2 N}{\varepsilon_0\varepsilon_r kT}\right)^{1/2}$

For aqueous solutions at 25°C, on the basis that $\varepsilon_r = 78.46$, $\eta = 8.903 \times 10^{-4}\,\text{kg m}^{-1}\text{s}^{-1}$, a and b take the values:

$a = z^3 \times 60.58\,\text{S cm}^2\,\text{mol}^{-3/2}\text{dm}^{3/2}$

$b = z^3 \times 0.2293\,\text{mol}^{-1/2}\text{dm}^{3/2}$

$B = 3.288\,\text{nm}^{-1}\,\text{mol}^{-1/2}\text{dm}^{3/2}$

But note: B is incorporated into the values of a and b given immediately above.

$E_1 = \dfrac{z^6 B^2}{24}\left(\dfrac{e^2}{4\pi\varepsilon_0\varepsilon_r kT}\right)^2$

$2E_2 = \dfrac{z^6 B^2 eF}{24\pi\eta}\dfrac{e^2}{4\pi\varepsilon_0\varepsilon_r kT}$

These formulae for E_1 and $2E_2$ are given for the case of the Fuoss-Onsager equation written in terms of logarithms to base e. They should be multiplied by 2.303 if base 10 is used. For aqueous solutions at 25°C, on the basis that $\varepsilon_r = 78.46$, $\eta = 8.903 \times 10^{-4}\,\mathrm{kg\,m^{-1}s^{-1}}$, E_1 and $2E_2$ take the values:

$$E_1 = z^6 \times 0.2299\,\mathrm{mol^{-1}dm^3}\,\text{for logarithms to base e}$$

$$E_1 = z^6 \times 0.5293\,\mathrm{mol^{-1}dm^3}\,\text{ for logarithms to base 10}$$

$$2E_2 = z^6 \times 17.78\,\mathrm{S\,cm^2\,mol^{-2}dm^3}\,\text{ for logarithms to base e}$$

$$2E_2 = z^6 \times 40.95\,\mathrm{S\,cm^2\,mol^{-2}dm^3}\,\text{ for logarithms to base 10.}$$

$$J = (\sigma_1\Lambda^0 + \sigma_2)\,\mathrm{cm^2 S\,mol^{-2}dm^3}$$

$$\sigma_1 = 0.4582z^6[h(b) + \log_e \mathring{a} + \log_e z - 0.0941]$$

$$\sigma_2 = 15.48z^2 + 18.15z^4\mathring{a} - 17.66z^6[\log_e \mathring{a} + \log_e z]$$

$$h(b) = \frac{2b^2 + 2b - 1}{b^3}$$

$$b = 7.135 \times 10^{-10}\mathrm{m} \times \frac{z^2}{\mathring{a}}$$

$\mathring{a} = $ the distance of closest approach.

The expression for J is based on the 1957 un-amended equation, i.e. the equation where the logarithmic term is given as $E_1\Lambda^0 - E_2$. Fuoss has not given any corrections for J in the corrected version which uses $E_1\Lambda^0 - 2E_2$.

13

Solvation

Solvation occurs when a solute is dissolved in a solvent and has come to be seen as a crucial and fundamental feature in determining the behaviour and properties of solutes and of the solution itself. In this chapter the discussion will be restricted to water as a solvent. Water molecules are dipolar, and as a consequence liquid water has a definite microstructure due to H-bonding throughout the bulk liquid. Ions have charges and these will interact with the dipoles of the water. As a result they will also have an effect on the structure of water, and this is now considered to be a very important feature in solvation.

Aims

By the end of this chapter you should be able to:

- describe the structure of water and its importance in solvation of ions;

- explain how diffraction studies and computer simulations give the details of this structure;

- recognise the terms used to describe solvation;

- discuss traditional methods for studying solvation and assess their limitations;

- understand the value of NMR in studying solvation;

- show how diffraction studies give radial distribution functions for the ions and the water molecule in an electrolyte solution;

- and explain how these can be used to obtain hydration numbers;

- describe the three region model and its significance in defining 'structure making' and 'structure breaking' ions;

An Introduction to Aqueous Electrolyte Solutions.   By Margaret Robson Wright
© 2007 John Wiley & Sons Ltd   ISBN 978-0-470-84293-5 (cloth) ISBN 978-0-470-84294-2 (paper)

- use standard partial molar entropies of ions and standard molar volumes of hydration as an aid to classifying ions as 'structure makers' and 'structure breakers';

- define hydrophobic hydration and discuss how it is studied;

- define thermodynamic transfer functions and use them to compare the behaviour of apolar solutes and ions in solution.

13.1   Classification of solutes: a resumé

- **Ions:** these are charged atoms or groups of atoms with positive or negative charges, or both in charge separated ions.

- **Polar molecules:** these are uncharged neutral molecules with an overall dipole moment which may be the result of one individual polar bond within the molecule, or the result of a vector summation of several individual bond dipoles.

- **Non-polar molecules:** these are uncharged neutral molecules with a zero dipole moment, but containing bonds which are polar. Vector summation of individual bond dipoles within the molecule cancel to give a zero dipole moment by cancellation.

- **Apolar molecules:** these are molecules which are neutral with an overall zero dipole moment. The crucial feature is that the bonds in the molecule are all completely non-polar, or virtually non-polar. There is no cancellation of any individual bond dipoles.

- **Large molecules which are biologically important:** such molecules are often very large and include both polymer and non-polymer molecules such as proteins and phospholipids. Depending on their environment they can be charged or uncharged, and they contain regions which are very decidedly polar and regions which are apolar. This particular aspect of biologically important molecules is often associated with their overall structure and biological function. Solvation of the polar regions is of crucial importance and the alteration of the water structure around these molecules, by virtue of both the polar and apolar regions, is crucial in determining their structure and behaviour.

13.2   Classification of solvents

Solvents can be classified as polar, non-polar and apolar, with the definitions being as above. However, a more general classification uses the relative permittivity. There are high relative permittivity solvents such as (at 25°C):

$$H_2O : 78.46 \quad HF : 83.6 \quad H_2SO_4 : 101$$
$$HCN : 106.8 \quad \text{formamide} : 109.5$$

Other common solvents have much lower relative permittivities:

acetamide : 59 at 83°C   methanol 32.6 at 25°C

hydrazine : 25 at 25°C   ammonia : 22 at −34°C

Non-polar liquids have relative permittivities of around 2 at 25°C such as:

benzene : 2.27   dioxan : 2.21

The range of permittivities can be extended by using mixed solvents. Many studies of the behaviour of electrolyte solutions have been carried out in mixed solvents. But it is important to realise that there may be preferential solvation by one of the solvents under these conditions, and this could affect the correctness or otherwise of the interpretations which can be made.

13.3   Solvent structure

The structure of a crystalline solid is an extensive repeating fixed and regular pattern in three dimensions. A liquid, however, is characterised by random Brownian motion with no permanent large-scale repeating periodic structure throughout the liquid. Nonetheless, liquids do have a certain degree of transient structure. If one molecule is chosen to be a central reference molecule there is a regular arrangement, rather than a random arrangement, of other molecules around it. This arrangement is not on the large scale and is constantly being broken down and re-formed, but its lifetime is long on a Brownian motion time-scale. Structure in this sense cannot be defined by a set of unchanging atomic coordinates as in a crystal, but X-ray and neutron diffraction show conclusively that diffraction patterns characteristic of a repeating short range structure are obtained from liquids.

13.3.1   Liquid water as a solvent

A prerequisite to the study of ionic solvation is an understanding of the structure of the pure liquid solvent. This is because introduction of a solute molecule into a solvent can alter or disrupt the solvent's molecular structure and behaviour. For the purposes of this book on aqueous electrolyte studies this means focusing attention onto liquid water. However, the techniques used to study the structure of water can be carried over to the determination of the structure of other solvents.

The bulk properties of liquid water are well characterised and quantified, e.g. values for important properties such as dipole moment, relative permittivity and viscosity are known with considerable accuracy.

It is also well known that water is anomalous in some of its bulk properties and this has been interpreted in terms of H-bonding. The existence of the H-bond has a major influence on the structure of both liquid water and its solid phase, ice.

13.3.2   H-bonding in water

A 'partial structure' exists in water as a result of H-bonding between water molecules. The charge distribution in the H-bond is given in Figure 13.1, and the most stable form is believed

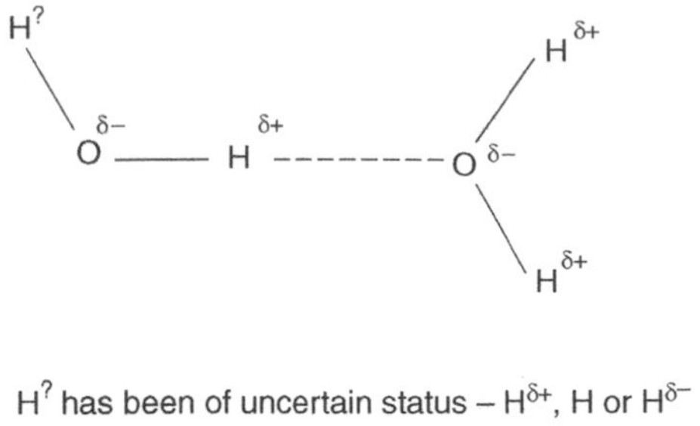

$$H^? \diagdown$$

$$\diagdown \underset{\delta-}{O} \underline{\qquad} \underset{\delta+}{H} \text{--------} \underset{\delta-}{O} \diagup \overset{\diagup \overset{\delta+}{H}}{} \diagdown \underset{\delta+}{H}$$

$H^?$ has been of uncertain status $-$ $H^{\delta+}$, H or $H^{\delta-}$

Figure 13.1 Charge distribution in the H-bond.

to be the linear H-bond, though this does not preclude the possibility that bent H-bonds could exist within the overall structure of water.

In the isolated water molecule the oxygen atom is joined to the two H-atoms by covalent bonds and these with the two lone pairs on the oxygen are approximately tetrahedrally disposed around the O-atom (see Figure 13.2). The bonds are polar with a partial negative charge on the O-atom and a partial positive charge on the H-atoms. Because of the tetrahedral arrangement there is an overall dipole moment. When a H-bond is formed between two water molecules it is primarily a result of dipole–dipole interactions with a certain degree of covalent bonding.

The water molecule can donate two H-atoms and a set of two lone pairs on the O-atom to form four H-bonds arranged tetrahedrally around the central O-atom. This O-atom, in turn, has four O-atoms arranged tetrahedrally around it. These are the O-atoms of the four H-bonds made with four other water molecules (see Figure 13.3). This structure can be extended outwards from the oxygens. As a result, a very open three-dimensional network array of H-bonds characterises the structure of water, both in the solid and the liquid phases. Each water molecule is surrounded by only four other water molecules and it is this factor which gives the open network array, in contrast to the close array which would result from a hexagonal close packed structure where 12 molecules would be arranged around a central molecule.

In the normal polymorph of ice the array is strictly tetrahedral, and the H-O-H angle is very nearly the tetrahedral angle of $109°28'$. When ice melts, this regular tetrahedral array partially

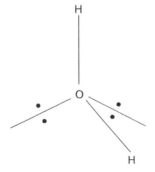

Figure 13.2 H_2O molecule showing the lone pairs on the O atom tetrahedrally disposed.

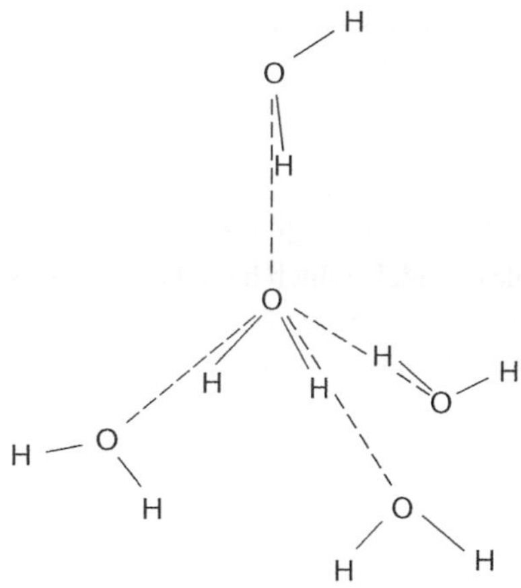

Figure 13.3 The pentameric unit: H_2O molecule surrounded by four H_2O molecules tetrahedrally disposed.

collapses into itself to give a more closely packed and more dense structure which is a distorted tetrahedral array. However, the resulting structure is still an open network array, and this persists, though to a progressively decreasing extent, as the temperature is raised. It is this structure of water which is of interest to the physical chemist studying solvation.

The H-bonding extends through the liquid but this does not mean that the H-bonded structure is fixed and static. Rather it is a dynamic structure with H-bonds constantly being broken and new ones formed. The structure and, on the average, each H-bond exists for a relatively long time of the order of 3×10^{-12} s. This is a longer time than that for Brownian motion. This dynamic picture of the arrangement of the structure of liquid water was developed to account for the anomalous properties of bulk water, such as the dependence on temperature and pressure of the thermal expansion, compressibility, viscosity and C_p for liquid water.

As a result it is now believed that water is made up of clusters of water molecules linked together by H-bonds in an irregular tetrahedral array. These are often described as 'ice-like'. In between the clusters the water is still H-bonded, but in a much less ordered way giving a less open structure. The distribution of water molecules between clusters and the intervening regions will depend on the temperature and pressure. As a rough guide, near the melting point the clusters are expected to be predominant, but near the boiling point the clusters would contain a smaller proportion of the total number of water molecules present. The clusters could be defined as the 'less dense' region, the intervening regions the 'more dense' region. It is the existence of these different regions which gives rise to the anomalous macroscopic properties of water.

The water molecules in these regions are constantly interchanging, with some moving out of the 'less dense' region and some moving into the 'more dense' region. The mechanism for this is a cooperative breaking and making of H-bonds throughout the whole structure. The H-bonds do not break, or form, one at a time and independently of each other. It is believed that once one bond breaks or forms it sets off a chain of events in which many bonds are broken or formed. And so whole clusters may be broken or formed. This is called the 'flickering cluster' model. The regions in which there are clusters of 'normal ice-like' water are constantly disappearing and appearing, and so appear to move around the bulk of the solvent. It is possible that there could be a wide range in the sizes of the clusters. Some authors believe that up to 70% of the water molecules exist as clusters of 50 molecules, or more, and that the mean lifetime of a cluster is of the order of 10^{-10} or 10^{-11} s. However, the sizes of the clusters and their lifetimes are still a subject for investigation. What is clear is that the lifetime of an individual H-bond is considerably less than that of a cluster.

There are other more complex models which have been proposed for the structure of liquid water but these are beyond the scope of this book. However, what all of the models do indicate is that it is likely that the introduction of an ion into water must have considerable effects on the structure of water.

This microscopic structure can be probed in two ways:

• experimentally;

• theoretically.

13.4   The experimental study of the structure of water

The structure of the isolated molecule in the gas phase can be determined spectroscopically using UV, IR, Raman and microwave analysis. This gives information about bond lengths, bond angles, bond strengths and the dipole moment.

Recent far IR spectroscopy on liquid water indicates that the basic H-bonded unit is indeed the pentameric unit described above. The clusters are then built up by further H-bonding from this unit. The structure which is relevant to hydration is that of bulk water where what is primarily studied is the relation of the individual molecules with respect to each other throughout the water in bulk. Diffraction and NMR are the major techniques for probing this structure.

13.5   Diffraction studies

X-ray and neutron diffraction are probably the most important and the most profitable techniques for studying the structure of liquid water. Electron diffraction is much less useful for the liquid state and it is used primarily for studies of isolated molecules in the gas phase.

The electrons of atoms cause X-rays to be scattered. This scattering is difficult to detect and analyse for light atoms, and for water it is very difficult to detect the H atoms though the scattering due to the oxygen atoms is easier to detect. However, modern technological advances, including synchrotron-generated X-rays, give vastly superior X-ray data leading to greater scope for X-ray diffraction techniques in solution.

Neutron diffraction has contributed enormously to the knowledge of the molecular structure of liquids. This is because it is much more sensitive and can pinpoint light atoms more accurately than X-rays can.

Detection of D in D_2O is even more accurately determined and structural studies using neutron diffraction are, more often than not, carried out in D_2O. The overall structure of liquid D_2O very closely resembles that of H_2O.

Diffraction studies gives atom–atom distances and bond angles between the various atoms. But more significantly, such experiments give the overall structure of a substance in terms of radial distribution functions $g(r)$ (see Section 10.17). If, say, the O atom in an H_2O molecule is taken as a central reference point and placed at the origin of the coordinate system used to define positions, then $g_{O-O}(r)$ gives the probability of finding a second O atom in a given small region lying at a distance r from the O atom at the origin. Similarly, $g_{O-H}(r)$ gives the probability of finding an H atom in a given small region lying at a distance r from the O atom at the origin. If H were chosen as the central reference atom at the origin, then $g_{H-H}(r)$ gives the probability of finding a second H atom in a given small region distance r from the origin. $g_{H-O}(r)$ is the same as $g_{O-H}(r)$. These three radial distribution functions are all that is required to define the overall structure of bulk water. Since there is no preferred direction for finding other atoms around the central reference atom, the distribution is spherically symmetrical, hence the term **radial** distribution function.

Processing the diffraction data for liquids is more difficult than for solids, though it is greatly facilitated by comparing the scattering patterns for $H_2O(l)$ and the isotopically substituted $D_2O(l)$. Ultimately this information will give the radial distributions, $g_{O-O}(r)$, $g_{O-H}(r)$, $g_{H-H}(r)$, and these can be summarised in an overall total distribution function $g(r)$. This represents the probability of finding a second H_2O molecule in a given small region of the bulk water at any distance, r (over $360°$), from a central reference H_2O molecule situated at the origin. Figure 13.4 gives a qualitative representation of what comes out of a diffraction experiment.

This is probably the most important basic and characteristic property of a liquid, here water, and it gives a lot of information about the arrangements of the molecules in a liquid. A peak indicates that there is a good chance of finding a molecule at a distance from the central molecule given by the position of the peak on the r-axis. The successive maxima show that there is structure in the liquid, and that the central molecule 'sees' another molecule at distances given by the peaks. When the peaks die away the central molecule no longer 'sees' a definite arrangement of molecules around it, and thereafter the arrangement appears random. The structure above extends to about three shells of solvent molecules around the central molecule. The arrangement of the molecules around the central molecule can be inferred from the details of the radial distribution function found from experiment, for instance, it could be tetrahedral or octahedral.

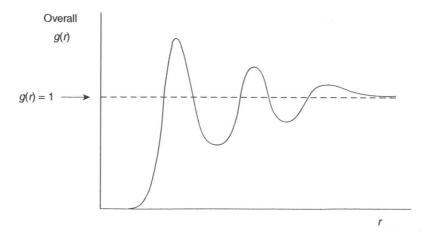

Figure 13.4   Diffraction data results for the liquid water demonstrating short range structure.

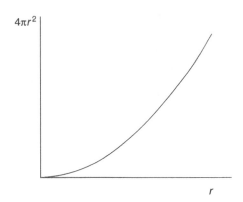

Figure 13.5 $4\pi r^2$ vs. r, showing how the size of each spherical shell around an atom increases with r.

13.5.1   Determination of the number of H_2O molecules corresponding to each peak and the three-dimensional arrangement corresponding to each peak

As the distance from the central reference molecule increases, the number of molecules capable of being packed around the central molecule will increase. Since the distribution functions are spherically symmetrical, each spherical shell around the atom will have an increasing larger volume given by $4\pi r^2 \times$ thickness of the shell (see Figure 13.5). If the value of the total $g(r)$ at a given distance r form the origin is multiplied by the volume of a spherical shell at the same distance, r, then a graph of $4\pi r^2 g(r)$ vs. r results. This graph approaches the graph of $4\pi r^2$ vs. r asymptotically from below (see Figure 13.6). The number of molecules in the region defined by a peak in the distribution function, $g(r)$ can now be found by integration. $\int_0^r g(r)4\pi r^2 dr$ gives the number of molecules disposed around the central molecule between the origin and r. Between 0 and a distance given by σ, $g(r)$ is very nearly zero. σ is taken to be a measure of the effective size of the water molecule situated at the origin.

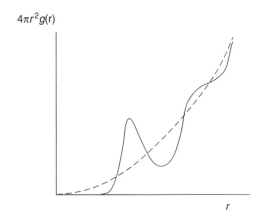

Figure 13.6   Graph of $4\pi r^2 g(r)$ vs. r, showing how this function approaches the graph of $4\pi r^2$ vs. r asymptotically from below.

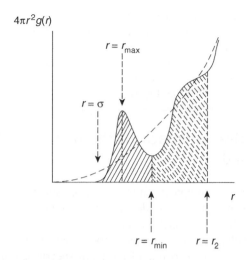

Figure 13.7   Diagram showing how the number of H_2O molecules in each shell of nearest neighbours can be estimated from the area of each peak.

Figure 13.7 shows how the number of water molecules corresponding to each peak can be estimated. The area under the curve between $r = \sigma$ and $r = r_{min}$ is generally asymmetric about the vertical line defined by r_{max}. The area under the curve between $r = \sigma$ and $r = r_{min}$ given by the cross hatched area gives the shell of nearest neighbours around the central molecule. The area between $r = r_{min}$ and $r = r_2$ gives the number of molecules surrounding the central molecule in the second shell of next nearest neighbours, and so forth. Generally the short range order is limited in extent, limiting the number of shells to three, or at the most four.

13.5.2 Results of diffraction studies: the structure of liquid water

The linear H-bond imposes a very open tetrahedral structure on the water molecules throughout the liquid. Diffraction experiments show each water molecule to be tetrahedrally surrounded by about four other water molecules at a distance of 0.28 nm with a next layer at 0.485 nm and a third layer at 0.58 nm. At distances further than this the molecule no longer 'sees' the H-bonded arrangement as such, but 'sees' only an apparent random distribution. Nonetheless, this is totally compatible with an extended H-bonded array throughout the liquid since each central molecule is part of the 'structure' around another molecule which could, instead, have been chosen as the central reference molecule.

However, this is a simplistic description. Modern work has shown that the structure is more complex than that inferred from the diffraction experiments. Unfortunately details about clusters and other possible structures given in Section 13.3.2 cannot be inferred from diffraction results. What is needed is information about many more shells from the central reference molecule, e.g. information regarding the structure over say, 10 shells deep. At present the only probe to this structure is the theoretical work using computer simulation.

Most of the recent work has been focused on studying the effect of temperature and pressure on the H-bonding in water. As has been shown the H-bonding is defined by $g_{OO}(r)\, g_{OH}(r)$ and

$g_{HH}(r)$ and these have been studied over a wide range of pressures and temperatures, and indicate that the H-bonding alters drastically with changes in these parameters. Modern work is also likely to focus on structures other than the normal H-bonding described here. These are the possible cage structures which can be formed by H-bonding and are believed to be crucial in the behaviour of solutions of non-polar and apolar molecules and long chain quaternary ammonium salts (see Sections 13.24 to 13.26).

13.6 The theoretical approach to the radial distribution function for a liquid

With the advent of modern fast computers and the development of Monte Carlo and molecular dynamics techniques, it is now possible to deduce from computer simulation experiments the form of the predicted distribution functions based on assumed molecular potential energy functions. This is a development of supreme importance, well advanced for liquids and progressing rapidly for solutions. It is one of the main spearheads of the modern attack on molecular interactions in liquids and solutions.

The neutron diffraction functions, $g_{O-O}(r)$, $g_{O-H}(r)$, $g_{H-H}(r)$ for a wide range of conditions can be compared with those calculated by computer simulations. For each simulation a given model for water can be chosen and the potential energy functions for the interactions which have been assumed to occur can be fed into the computer. The resultant calculated $g_{O-O}(r)$, $g_{O-H}(r)$, $g_{H-H}(r)$, can be compared with the experimentally determined values. These comparisons can indicate which assumptions could best be modified.

Computer simulations are an ideal way to study the structure of liquids. As indicated above, all possible molecular interactions can be considered and the expressions for the potential energy for each interaction can be fed into the computer. In these methods several hundred water molecules can be considered. Either Monte Carlo or molecular dynamics techniques can be used, though if the time dependent distributions are required molecular dynamics techniques would be necessary. These techniques are limited in principle by the accuracy of the potentials used, and this aspect is probably the main limiting feature of the technique. However, techniques are improving year by year and considerable detail has emerged and will continue to emerge. Present results support the extended H-bonded structure for liquid water, and the concept of clusters and cages.

A further application of computer simulations lies in the calculation of simulated spectra. This has been done for water, with the comparison being between the simulated spectrum and that for the far-infra-red experimental spectrum. This study has shown that a calculated spectrum which is based on the assumption of the pentameric H-bonded structure closely reproduces the observed spectrum. However, this simulation does not reproduce certain dynamic aspects of the water structure.

13.7 Aqueous solutions of electrolytes

In the past the emphasis in the study of solvation has been on what happens to the solute when it changes from the pure state to the dissolved state in the solvent. Obviously the chemist has been very much aware of the role of the solvent in solution, but the focus has always been on the solvated solute species.

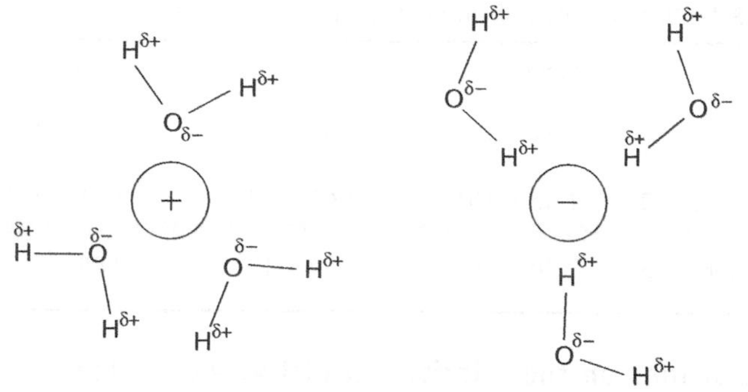

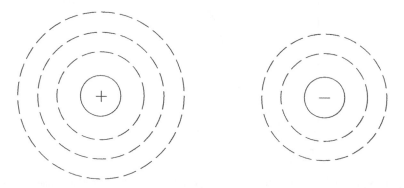

Figure 13.8 Ion–dipole interactions for (i) a cation and (ii) an anion.

However, chemists gradually became aware that much more attention should be given to the change in solvent structure from the pure solvent to that involved in the solution around the solute. Because of the charge on ions and the dipolar nature of water, electrostatic interactions will occur between the ion and the dipole. A cation will be attracted to the partial charge on the oxygen atom, and this interaction will be greater the charge on the ion and the smaller the size. An anion will be attracted to the partial positive charge on the hydrogen atom, and this will be greater the larger the charge and the smaller the size of the anion. These interactions are represented in Figure 13.8. This simple picture of what happens on dissolution of an electrolyte shows that some reorganisation of the structure of water must occur when hydration occurs.

This suggests that ions in water are unlikely to be bare, and hydrated cations and anions are expected (see Figure 13.9). The size of the ion is, therefore, expected to be larger than the crystallographic radius. However, it may well be that the experimental determination of the size of an ion will depend on the method used to determine it. For instance can it be assumed that estimates of the size of an ion from equilibrium properties such as the dependence of the mean ionic activity coefficient on ionic strength will be the same as the size of the moving entity in conductance studies?

Experimental techniques for studying solvation include direct methods such as spectroscopic methods, diffraction techniques and light scattering. Use can also be made of indirect methods such as thermodynamic properties, conductance and activity coefficient studies, and diffusion. Computer simulation experiments have increasingly been playing a major role.

Figure 13.9 A solvated cation and anion.

Table 13.1   Dependence of field on distance from the centre of an ion

$\dfrac{r}{\text{nm}}$	0.500	1.000	1.500
$\dfrac{\text{field}}{\text{V cm}^{-1}}$ for $z = 1$	7.34×10^5	1.83×10^5	8.14×10^4
$\dfrac{\text{field}}{\text{V cm}^{-1}}$ for $z = 2$	1.46×10^6	3.67×10^5	1.63×10^5
$\dfrac{\text{field}}{\text{V cm}^{-1}}$ for $z = 3$	2.20×10^6	5.50×10^5	2.45×10^5

13.7.1   Effect of ions on the relative permittivity of water

The electric field in the vicinity of ions is large, of the order of 10^5 V cm^{-1} and depends on the distance from the ion, being larger the smaller the distance. Table 13.1 gives calculated values of the field for various distances from the ion and for ionic charges of 1, 2 and 3.

These have been worked out using the bulk value of ε_r. But because of dielectric saturation these values may be underestimates – at least for $r = 0.500$ nm. As a result the ion will orientate the water molecules in its vicinity. Close to the ion there will be a region where all the permanent dipoles of water are completely aligned corresponding to a region of dielectric saturation. The relative permittivity in this region will be in the range 2–4. The bulk relative permittivity is 78.46 at 25°C, and this applies in the region where virtually no alignment occurs and the ion has only a small effect on the water dipoles. In between these two regions there will be a region where partial alignment occurs, and in this region the situation changes from complete alignment to near non-alignment. This is shown qualitatively in Figure 13.10. The exact dependence of ε_r on r will depend on the charge on the ion and its size. Qualitatively, the smaller the ion and the larger its charge the greater will be the effect of the ion on the water molecule at small distances apart.

13.8   Terms used in describing hydration

There are several terms commonly used to discuss hydration.

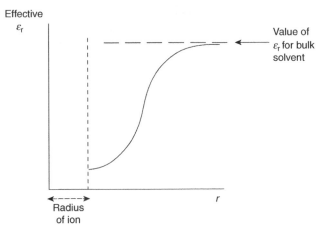

Figure 13.10   Schematic diagram showing the dependence of the effective value of the relative permittivity on distance from the centre of an ion.

13.8.1 Solvation shell

The term solvation shell only describes a region of solvent around the ion. Simply being part of the shell does not imply anything about the interactions between the ion and the water molecules of the shell, or the strength of these interactions. Likewise it says nothing about the state and structure of the water.

The first solvation shell or sheath

This is the region of the solvent in which the molecules are immediate nearest neighbours of the ion, and it is almost certain that they will have properties which are distinct from the molecules of bulk water.

The second solvation shell or sheath

The next region of the solvent which lies further out is described as consisting of next nearest neighbours. This shell is also likely to be distinct from bulk water.

The third or subsequent shells or sheaths

These are water molecules further out from the ion, and they may or may not have properties different from bulk water.

Bulk water

This constitutes regions of the water which are totally unmodified by the ions of the electrolyte.

13.8.2 Bound and non-bound water

Within each of these regions described as solvation shells it is of vital importance to distinguish between:

(a) Regions of the solvent where solvent molecules are actually bound to the ion, are immobilised and move with the ion. These solvent molecules have lost their own translational degrees of freedom.

(b) Regions of the solvent which are affected by the ion but are not bound to it and so are not immobilised. They do not move with the ion and have their own translational degrees of freedom.

13.8.3 Further terms: primary and secondary solvation

There are further terms which are used to describe the strength of the interaction between the ion and the solvent molecules and the region in which strong interaction takes place.

Primary solvation

If this occurs it takes place in the first solvation shell. These are water molecules with which the ion interacts strongly and are coordinated to the ion. They thus move with the ion and could be termed bound water. Often all of the first solvation shell water molecules can be considered to be primary solvation, but some workers feel that this need not necessarily be the case. Some water molecules may be in the first solvation shell but are not bound, and do not move with the ion. It becomes necessary to define a criterion for distinguishing bound from unbound molecules, i.e. primary solvation or not. The criterion used will be determined by the experimental technique which is being used to measure it. Unfortunately, different experiments may actually measure different properties.

Secondary solvation

This is made up of molecules which are affected by the ion, but are not coordinated to them and often do not move with the ion. They generally are found in the second solvation shell or subsequent shells if the water molecules in them have different properties from bulk water. It may also include molecules from the first solvation shell which are not included in the primary solvation.

13.9 Traditional methods for measuring solvation numbers

These are mainly spectroscopic, but also include activity measurements, transport methods such as conductance and diffusion and thermodynamic studies.

13.9.1 Vibrational spectra (IR and Raman) and electronic spectra (UV, visible and in some cases IR)

These are older techniques but are still important, especially Raman.

UV and visible studies tend not to give much direct information about the interatomic structure since the details of the structure tend to require a particular model for solvation to be assumed first. However, they are excellent for probing the possible symmetries of the chosen ion in solution. These techniques are of particular importance and use for transition metal ions.

Infrared and Raman spectroscopy have very wide applications and have proved an important source of information on ion–solvent interactions and hydration numbers. In general, the position of an absorption enables a guess to be made for the hydration numbers, h_+ and h_-. Since spectroscopic techniques deal with frequencies associated with coordinative bonding between the ion and the solvent, such techniques furnish information about the strength of the ion-solvent interactions in the primary solvation.

Ultra violet and visible spectra

This is the spectroscopic technique of preference for transition metal ions. Water molecules can be arranged around the metal cation in a variety of geometries, e.g. four as a tetrahedron, four as square planar, six as an octahedron and eight as a cubic array. Other geometries are possible along with distorted forms. However, the important feature is that the ligand field of the solvent molecules causes splitting of the d-orbitals in a manner characteristic of the geometry and of the number of d-electrons in the ion. The absorption peaks in the spectrum of the cation in solution reflect this splitting and ways of analysing actual spectra, although complex, have been developed. This along with peak areas lead ultimately to solvation numbers, arrangement of the solvent molecule around the ion and strength of the interaction.

Raman or infrared spectra

These are vibrational spectra and lead to fundamental frequencies of vibration and the strength of ion–solvent interactions. When a salt is dissolved in water it will affect the frequencies of the water absorption and in favourable cases it will give rise to new peaks due to ion–solvent interactions. Alteration in the vibrational spectrum of the water due to the presence of the ion gives information regarding the effect of the ion on the water structure. The really useful information, however, comes from a study of the new lines due to the actual bonding of the ion to the solvent molecules. The frequencies and intensities of the vibrational lines give a measure of the strength of the bond between ion and solvent, and the peak areas can in favourable cases lead to hydration numbers.

13.9.2 Transport phenomena

These deal with the properties of ions moving in solution. The results of such experiments indicate that the bound water molecules move with the ion.

Conductance studies and transport number experiments study the ion and its solvation shell moving under the influence of an applied external field, while viscosity and diffusion experiments study the movement of the ion and bound solvent through the solvent.

(a) Conductance data

The molar conductivities of alkali halides in water vary in an order:

$$\Lambda(\text{LiCl}) < \Lambda(\text{NaCl}) < \Lambda(\text{KCl})$$

which is the reverse of what would be expected if ions were unhydrated (see Worked Problem 11.18). The smallest bare ion, Li^+, would be expected to move fastest in an applied field and so be the most highly conducting. This is not so, and it is possible that the moving ions are really Li^+(several H_2O) as against K^+(bare) or K^+(few H_2O). Molar conductivities coupled with transport numbers and Stokes' Law give individual hydration numbers for each ion.

Results suggest that if Cs^+ has one layer of water molecules around it, then smaller ions on a crystallographic radius basis must have more than one layer around them, for instance one to three layers. Cl^- is believed to have one layer, F^- probably has more than one layer, while complex anions such as SO_4^{2-} and NO_3^- have probably less than one layer.

Analysis of the shapes of conductance curves based on the various Fuoss-Onsager conductance equations leads to experimental values for the distance of closest approach, $\mathring{a}$. But since these studies are often carried out at concentrations where ion association can be significant what is obtained is generally a range of values of $\mathring{a}$ and K_{assoc} as best fit parameters.

(b) Viscosity and diffusion coefficients

These are older techniques and are no longer much used. A solute alters the viscosity of the solvent. The ion and its solvent molecules exert a viscous drag on the rest of the solvent and the change in viscosity can be used to calculate a total hydration number which is then split up into its individual components. However, viscosity measurements are normally used as an aid to developing a structural interpretation of solvation (see Section 13.8), and to give correction factors in conductance equations (see Section 12.10).

Diffusion coefficient studies again study the movement of the ion through the solvent, and a large variety of techniques can be used. Total hydration numbers are found.

13.9.3   Relative permittivity of the solution

The relative permittivity measures the alignment of the solvent dipoles and production of induced dipoles by an electric field. An ion produces an intense field on bound solvent molecules, and will cause partial, if not complete, alignment of the dipoles of the solvent molecules affected by the ion. This results in a drop in the observed relative permittivity of the solution relative to the pure solvent. This drop is related to the number of bound solvent molecules. Controversy exists as to whether the effect is restricted to bound molecules only, or whether other solvent molecules are involved. Both theoretical and experimental studies have been carried out. The dependence of the relative permittivity on the distance from a given ion is of fundamental importance in theories of electrolyte solutions where generally the bulk relative permittivity is used in the theoretical expressions. But it is more likely that a varying relative permittivity should be used.

13.9.4   Activity measurements

This has always been an important, though very indirect, source of hydration numbers. Colligative properties and emfs give high precision experimental data from which the activity of the solute can be calculated and stoichiometric mean ionic activity coefficients found. If it is assumed that solvent molecules are bound to the ions, a relation between

- the observed stoichiometric $\gamma_{\pm(obsvd)}$;

- the corresponding $\gamma_{\pm(calc)}$ calculated from the Debye-Hückel theory;

- and the hydration number for the electrolyte as a whole;

can be derived. This hydration number can then be split up into h_+ and h_-.

Deviations from predicted behaviour are here interpreted in terms of solvation, but other factors such as ion association may also be involved. Ion association leads to deviations in the opposite direction and so compensating effects of solvation and ion association may come into play. The deviations may also be absorbing inadequacies of the Debye-Hückel model and theory, and so no great reliance can be placed on the actual numerical value of the values emerging. This major method has now been superseded by X-ray diffraction, neutron diffraction, NMR and computer simulation methods. The importance of activity measurements may lie more in the way in which they can point to fundamental difficulties in the theoretical studies on activity coefficients and conductance. The estimates of ion size and hydration studies could well provide a basis for another interpretation of conductance and activity data, or to modify the theoretical equations for mean activity coefficients and molar conductivities.

13.10 Modern techniques for studying hydration: NMR

This is probably the most powerful spectroscopic technique, and with X-ray and neutron diffraction is now the technique of choice. A shift in the proton resonance frequency and the intensity of the signal tells how many water molecules are responsible. Proton relaxation shifts have proved to be a major advance, and are progressively being applied to solutions containing complex ions. For simple ions they suggest six water molecules around a cation are fairly typical. In favourable cases individual hydration numbers are obtained using this technique. In this respect they are superior to the more traditional methods which on the whole only measure overall hydration numbers and require some arbitrary way of splitting these into cation and anion contributions. Diffraction studies also furnish individual hydration numbers.

Only nuclei with a net nuclear spin will give a signal. However, this is no problem for hydration studies since it is the chemical shift of the proton, ^1H, signal of H_2O which is being measured. In certain cases the signal from ^{17}O in $H_2^{17}O$ is used. In a solution the O atom is the atom closest to the cation and the H atom is the one closest to an anion. Essentially the experiment looks at the environment of the atom whose resonance is being studied. In an electrolyte solution there are three species of interest: the solvated cation, the solvated anion and bulk water. Coordinated water will exchange with secondary solvation of the ion under study, with the primary and secondary solvation of the counter ion and with the bulk water.

If a signal is to be observed then the proton must remain in the same environment for at least 10^{-4} s to be picked up as a distinct entity in that environment. This is a limitation imposed by the nature of NMR spectroscopy and is not an instrumental limitation. This has significance for the study of hydration using NMR. If the proton does not remain in one environment, e.g. coordinated to a cation, for at least 10^{-4} s then the signal found in the spectrum will be the average of that for all the environments open to the proton.

Cations and anions can be studied using NMR, but the focus has predominantly been on the cations. Signals from both can be picked up and it is relatively easy to assign particular signal(s) to the cation and other(s) to the anion. If a series of metal chlorides are used, the proton signal for water associated with the cation will depend on the cation, the chloride signal will not vary in the same way.

13.10.1 Limitations of NMR

- There are a few problems associated with salts of electronically diamagnetic and paramagnetic cations. In general, if the exchange of water molecules is slow, two signals are detected. If the cation is diamagnetic these proton resonances for water molecules are easy to pick out and analyse. These are generally around 100 Hz apart. However, paramagnetic ions, in general, cause much larger differences in the chemical shifts for the two signals, but there is so much broadening in the signals that analysis is problematic. However, in such cases, use of the ^{17}O signal in isotopically enriched H_2O resolves this problem and two signals can be obtained. The use of D in D_2O is also helpful here.

It is best if hydration can be studied using the two peaks corresponding to coordinated water and the rest of the water which comprises secondary solvation of the cation and primary and secondary solvation of the anion and the bulk water. When two signals are observed it is possible to determine individual hydration numbers.

- It is possible to convert a single signal into two signals by lowering the temperature. This slows down the rate of exchange of water between coordinated water and the rest. Because high concentrations are required for NMR signals to be picked up, this means that temperatures below $0°C$ can be attained in the liquid phase. The rate of exchange can also be slowed down by adding propanone (acetone) to the solution. It is essential that any added solvent must not significantly solvate with the cation or anion.

- NMR experiments have to be carried out at very high electrolyte concentrations to be able to obtain a measurable signal. These are way above concentrations corresponding to non-ideality, and so the hydration numbers found in slow exchange experiments in particular will reflect this non-ideality. What is more serious is that the concentrations are in ranges where ion association is expected to be significant, and an anion could well replace a water molecule in the primary solvation shell.

13.10.2 Slow exchanges of water molecules

This is normally found for Be^{2+}, Mg^{2+} and higher charge cations including some of the electronically diamagnetic transition metal ions, lanthanides and actinides. Paramagnetic transition metal cations can be studied using ^{17}O or D resonance.

If there are two peaks:

- the chemical shift gives some indication of cation–solvent interaction;

- peak areas give primary solvation numbers;

- but be aware that these may have to be corrected for ion pairing.

For slow exchange the positions of the chemical shifts do not alter with concentration, though of course the areas under the peaks do. The areas under the peak for coordinated water and the peak for bulk water can be easily measured and the ratio of these areas along with a knowledge of the number of mol of ion and water will lead directly to the hydration number.

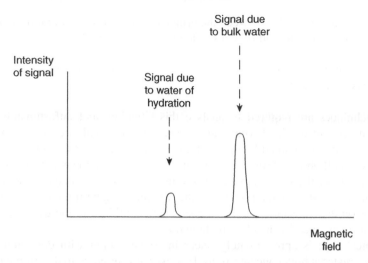

Figure 13.11 Proton chemical shifts for solutions showing how these occur at lower field strengths than for bulk water.

Paramagnetic ions such as Co^{2+}, Ni^{2+} and Fe^{2+} give only one proton resonance, and so cannot be studied in H_2O. However, if ^{17}O enriched water is used, then two peaks are found and the areas under the two peaks will give the hydration number.

A fairly large number of both diamagnetic and paramagnetic ions which have slow exchange of water between coordinated water and the rest can be studied if both H_2O and $H_2^{17}O$ are used.

The electric field of the ion polarises the water molecules and causes a reduction in the electron density around the H-atoms. This causes the resonances for the protons in the coordinated waters to occur at a lower field than that for the bulk water (see Figure 13.11). The magnitude of the differences in these chemical shifts gives some indication of the cation–solvent interaction. These differences show that the main factor governing the shifts is the electric field of the cation. This is larger the larger the charge and the smaller the size of the cation. Minor effects appear if there is a change in geometry of the coordinated water.

Data from slow exchange experiments show that the most common hydration number is six, indicating that the primary solvation shell has an octahedral symmetry. This shell will thus have approximately spherical symmetry.

13.10.3 Fast exchanges of water molecules

For some ions, the water molecules which are bound to the ion exchange rapidly with the water molecules outside the primary solvation shell. If this is the case only one proton signal will be detected. This will represent an average of the environments of the proton in the primary solvation shell and in the rest of the water, i.e. in the secondary shell of the cation, the primary and secondary shells of the anion and the bulk water. Situations where fast exchange is found and one signal observed show that:

- exchange of water molecules between the primary solvation shell of cations and all the rest of the water is fast for alkali metal cations, some alkaline earth cations and a large number of transition metal cations;

- exchange of water molecules between the primary solvation shell of anions and all the rest of the water is always fast and only a single resonance is found;

- exchange of water molecules between both cation and anion secondary solvation shells and bulk water is always fast.

Special techniques are required to analyse this situation, and unfortunately only total hydration numbers are obtained when only one resonance signal is observed. Methods for splitting up these into individual hydration numbers are necessary (see below).

The chemical shift observed for a given electrolyte depends on the nature of both the cation and the anion. Unlike the slow exchange where the chemical shift is the same whatever the concentration of the solution, the position of the resonance signal if the exchange of water is rapid depends on the concentration. This dependency is utilised in the determination of the overall hydration number for the electrolyte in question.

The chemical shift is approximately linear in concentration with deviations becoming progressively greater at high concentrations. However, it is possible to determine the quantity $(\partial \delta / \partial c)_{c=0}$ from the graph of δ, the observed chemical shift, as a function of concentration. This quantity can then be split up into chemical shifts for cation and anion using one or other of the assumptions, e.g. δ_+ for K^+ is equal to δ_- for Cl^-, or by assuming δ_+ for NR_4^+ to be zero. The values of the shifts found using such procedures for splitting up the overall signal into individual components, $h_+\delta_+$ and $h_-\delta_-$ have been used to discuss the structure breaking or making properties of the ions which can be studied in this manner. These individual chemical shifts have also been used for a study of the cation–solvent interaction. Various methods have been tried in an attempt to obtain hydration numbers from fast exchange data, but the values so obtained are to be treated with caution.

13.10.4   Results of NMR studies of hydration

NMR results are limited in so far as individual hydration numbers can only be found for ions for which the two peaks corresponding to coordinated water and the rest can be observed. This limits the study to ions for which there is slow exchange and means that alkali metal cations, and most of the alkaline earth metal cations are excluded. Nonetheless a considerable number of cations have been studied with unambiguous results. The most common coordination number is six, but cations known to favour tetrahedral or square planar configurations show a coordination number of four. It is only with large cations such as the lanthanides and actinides that numbers greater than six are found.

When the exchange is fast only overall hydration numbers can be found. But these, in conjunction with diffraction techniques which do give individual hydration numbers, can extend the range of cations for which individual hydration numbers can be found.

13.10.5   Residence times from NMR and ultrasonic relaxation

The residence time found from NMR data is the average lifetime of a water molecule in the primary solvation shell. Estimates of mean lifetimes can be found from the line widths of the signals obtained for both fast and slow exchange. The resonance signal which is required is that for water coordinated to the cation. If the lifetime of a water molecule in the primary hydration shell is long then a sharply defined peak is found, while shorter lifetimes will give a broader peak. A detailed analysis gives relations between the width of the line, the lifetime of the water molecule in its environment and the rate constant for exchange.

The signals obtained in the NMR study of hydration can be either from proton resonances or from ^{17}O resonances. The latter is preferable for cations since the O is the atom which is coordinated to the cation. Observation of the line width of the ^{17}O signal is easy to detect for paramagnetic ions, not so for diamagnetic ions. Consequently ^{17}O studies are limited to paramagnetic ions and ^1H is used for diamagnetic cations.

There is a fundamental difference in the processes occurring in the two experiments using ^1H and ^{17}O. Exchange of a water molecule from the coordinated water molecules means splitting the bond between the cation and the oxygen atom of the exchanging water molecule. In the proton resonance it is only the proton which is flipping between its two environments, not the whole water molecule. The rate of exchange of a proton of a coordinated water molecule with other water molecules not coordinated to the cation is likely to be faster than exchange of the whole water molecule which is what happens in the ^{17}O experiments.

If anions are studied it is the H end of the O-H dipole which is attracted to the anion. Proton resonance will thus correspond to exchange of the water molecule.

If the resonances are studied over a range of temperatures, then activation parameters can be found for the exchange of water between the solvation shell and the rest of the water.

A wide variety of lifetimes are found ranging from 25000 ps for a water molecule near Mn^{2+} to 10 ps for Na^+, 5 ps for Cl^- and 4 ps for I^-. These can be compared with 3 ps for the average lifetime of a water molecule at a given site in the pure liquid.

Ultrasonic relaxation can also be used. Here relaxation times are measured and these are related to the lifetime of a water molecule in a given environment. The relaxation time can only be defined for a first order process and, like the half-life, is a fractional lifetime. It is the time taken for the concentration to fall to $1/e$ of the initial concentration.

The rate of release of a water molecule is often studied as part of ion association studies, e.g. the following consecutive equilibria are often found in formation of contact ion pairs.

$$M^{z+}(H_2O)_n + X^z - (H_2O)_n \rightleftharpoons (H_2O)_{n-1}M^{z+}(H_2O)(H_2O)X^{z-}(H_2O)_{n-1}$$

free ions   solvent separated ion pair

$$\rightleftharpoons (H_2O)_{n-1}M^{z+}(H_2O)X^{z-}(H_2O)_{n-1} + H_2O$$

solvent separated ion pair

$$\rightleftharpoons (H_2O)_{n-1}M^{z+}X^{z-}(H_2O)_{n-1} + H_2O$$

contact ion pair

The contact ion pair could also be written as $M^{z+}X^{z-}(H_2O)_{2n-2}$.

The rates of release of the water molecules from the two solvent separated ion pairs should enable the lifetime of a water molecule in the two environments to be found.

The basic principle of the techniques can be illustrated by the following hypothetical scheme:

$$M^{z+}(H_2O)_n \rightleftharpoons M^{z+}(H_2O)_{n-1} + H_2O$$

If a sound wave is passed through the solution it will perturb the equilibrium slightly and a wave of new equilibrium positions is called for. If the concentrations can alter rapidly enough to enable each new equilibrium position to be set up as the wave is passed, then the equilibrium can keep in phase with the periodic displacement of the sound wave, and there will be no associated absorption of sound. If the concentrations cannot alter rapidly enough, they will not keep in phase with the sound wave and they will remain approximately constant throughout.

These are two extremes. An intermediate situation occurs when the period of the sound wave is comparable to the relaxation time. Here reaction occurs fast enough to allow the concentrations to alter with time, but the change in concentration is out of phase with the sound wave. This shows up as a point of inflection in the graph of (velocity of sound)^2 vs. frequency and as a maximum in the absorption of sound (see Figures 1.2(a) and 1.2(b)). The relaxation time can be found from the point of inflection, or from the position of the maximum. For the single equilibrium used to illustrate the technique:

$$\tau = \frac{1}{k_1 + k_{-1}} \tag{13.1}$$

If the equilibrium constant is known:

$$K = \frac{k_1}{k_{-1}} \tag{13.2}$$

and so k_1 and k_{-1} can be found independently.

For more than one equilibrium, such as the scheme involving solvent separated and contact ion pairs, several points of inflection or maxima may be observed and a relaxation time for each equilibrium can be found (see Figures 1.2(c) and 1.2(d)). If the relaxation processes are not resolved, best fit computer analysis is used to determine the relaxation times and the rate constants for the desolvation processes.

Table 13.2 gives typical values found for hydration numbers from NMR.

13.11   Modern techniques of studying hydration: neutron and X-ray diffraction

Diffraction patterns formed by scattering of X-rays or neutrons can give the following:

- the solvation or hydration number of the ion;

- an estimate of how the solvent molecules can be arranged around the ion;

Table 13.2   Typical values for hydration numbers found from areas of peaks in NMR

Ion and hydration number		Ion and hydration number		Ion and hydration number	
Be^{2+}	4	V^{2+}	6	Pd^{2+}	4
Mg^{2+}	6	Mn^{2+}	6	Pt^{2+}	4
Al^{3+}	6	Fe^{2+}	6		
Ga^{3+}	6	Co^{2+}	6		
		Ni^{2+}	6		
		Zn^{2+}	6		

- possibly an indication of the numbers of layers bound to the ion; and

- an indication of the stability of the solvation layers, but only if the strong source modern second and third generation synchroton-generated X-rays are used.

The main problem for the X-ray method is that H atoms do not scatter X-rays sufficiently for clear scattering patterns to result. Since what is of interest is the interaction of ions with water, this is a clear drawback. However, modern second and third generation synchroton generated X-rays give a much more intense beam and this has partially compensated for the poor scattering capacity of H atoms. Neutron diffraction is not limited in this way. In modern work, isotopically substituted ions are used, and here accessibility and cost can be a problem. Furthermore, if neutron diffraction with isotope substitution is used it is essential that the isotopes show significantly different scattering. For instance ^{16}O, ^{17}O and ^{18}O show very similar scattering patterns and so cannot be used in this way (see Section 13.11.1).

Despite this, diffraction and computer simulations are easily the most searching tools for studying solvation. The main aim of the diffraction studies has been to determine the details of the atomic environment of the solvent and the solute particles. Other than NMR probes, the best way to determine such environments is through determination of the pair radial distribution function, $g_{ab}(r)$. This gives the probability of finding atom, b, at a distance, r, from atom a. In the present context of an electrolyte solution of X_aY_b in $H_2O(l)$ there are 10 pairs of types of atoms described by 10 pairwise radial distribution functions, $g_{ab}(r)$. They are:

$$\text{Solute}: g_{XX}(r), g_{XY}(r) \text{ and } g_{YY}(r).$$
$$\text{Solvent}: g_{OO}(r), g_{HH}(r) \text{ and } g_{OH}(r)$$
$$\text{Solute/solvent}: g_{XO}(r), g_{XH}(r), g_{YO}(r) \text{ and } g_{YH}(r)$$

The total radial distribution function $g(r)$ is a linear combination of these 10 individual pairwise radial distribution functions, and this can be found from the overall scattering in the diffraction experiment. This is a useful quantity to have, but the more informative pairwise functions are essential in determining the complete environment of the ions in solution. However, it is difficult to split up the observed overall $g(r)$ determined from straightforward analysis of the scattering into its components. It would be much easier to generate $g(r)$ from the individual components if they were available.

Scattering patterns can be translated by standard diffraction techniques to give these pairwise distribution functions. This is difficult to do with X-ray scattering patterns, although modern techniques such as anomalous X-ray scattering have helped. It is relatively easy to determine these using the nuclear diffraction isotope substitution technique.

13.11.1 Neutron diffraction with isotope substitution

The electrolyte X_aY_b is studied in solution. What is wanted is to find out about the hydration around the cation and the anion, and to find out if the water structure is altered. If the cation is substituted by an isotope and if the scattering abilities of both are different, then a different scattering pattern will be found. If there is more than one isotope of the ion then different scattering patterns are found for each isotope compared with that for the unsubstituted electrolyte. This makes the subsequent analysis much easier. Likewise, different scattering patterns are found for isotopically substituted anions. If the H is replaced by its isotope, D, this will also cause a difference in scattering pattern. ^{16}O, ^{17}O and ^{18}O show very similar scattering patterns and so cannot be used in this way.

- (A) The electrolyte solution of X_aY_b in $H_2O(l)$ is studied and its scattering pattern determined. The following 10 radial distribution functions are relevant:

  $g_{XX}(r), g_{XY}(r), g_{YY}(r), g_{OO}(r), g_{HH}(r), g_{OH}(r), g_{XO}(r), g_{XH}(r), g_{YO}(r)$ and $g_{YH}(r)$.

  The total radial distribution function $g(r)$ is a linear combination of these 10 $g_{ab}(r)$.

- (B) A solution of isotopically substituted $X_a^*Y_b$ in water is taken and its scattering pattern is found. The radial distribution functions which are then relevant to the difference in the two scattering patterns of (A) and (B) are:

  $g_{XX}(r), g_{XY}(r), g_{XO}(r)$, and $g_{XH}(r)$.

- (C) Likewise substitution of Y by its isotope Y* gives a further different scattering pattern, and the functions which are now relevant to the difference in the two scattering patterns (A) and (C) are:

  $g_{XY}(r), g_{YY}(r), g_{YO}(r)$ and $g_{YH}(r)$.

- (D) Substitution of X by X*, Y by Y* and H by D will give a further different pattern, and the relevant functions are now:

  $g_{XX}(r), g_{XY}(r), g_{YY}(r), g_{HH}(r), g_{OH}(r), g_{XO}(r), g_{XH}(r), g_{YO}(r)$ and $g_{YH}(r)$.

- By studying the differences between all of the different scattering patterns it becomes possible to deconvolute the information to give all the individual pairwise radial distribution functions.

The important $g(r)$s are $g_{XO}(r), g_{XH}(r), g_{YO}(r)$ and $g_{YH}(r)$. From these it is possible to determine ion–water coordination and conformation in the first shell of nearest neighbours. With suitable data, it has been possible in favourable situations to obtain information about the second hydration shell.

The average hydration number for an ion in solution can be obtained as the integral:

$$\overline{n}_{\text{ion}}^{\text{H}_2\text{O}} = \text{constant} \int_{r_i}^{r_2} g_{\text{ion}-\text{O}}(r) r^2 dr \qquad (13.3)$$

or as $\qquad \overline{n}_{\text{ion}}^{\text{H}_2\text{O}} = \text{constant} \int_{r_i}^{r_2} g_{\text{ion}-\text{H}}(r) r^2 dr \qquad (13.4)$

depending on which function is more readily available. r_1 is generally taken as the radius of the species around which the hydration number is being found, and r_2 is the distance beyond the origin at which the outer boundary of the solvation sphere is assumed to lie (see Figure 13.12).

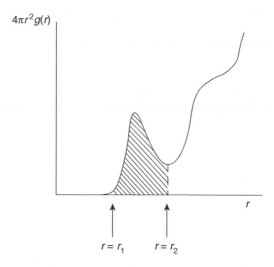

Figure 13.12 Radial distribution function showing how to estimate the average hydration number for an ion from the peak area.

The radial distribution functions for cation–anion can likewise give information on ion-pairing, and in general the coordination number of atom β around atom α can be found from a similar integral:

$$\bar{n}_\alpha^\beta = \text{constant} \int_{r_i}^{r_2} g_{\alpha\beta}(r) r^2 dr \qquad (13.5)$$

where r_1 is again taken as the radius of the species, α, with which β is associated, and r_2 is the outer limit to the region where the ions can be considered to be associated.

Typical values of hydration numbers are given alongside values determined by other methods in Table 13.3.

Six water molecules are found to be arranged octahedrally around such ions as Ni^{2+}, Na^{+}, and Ca^{2+}, with bound water possibly being confined to one layer only. A similar result has been found for Cl^{-}.

13.12 Modern techniques of studying solvation: AXD diffraction and EXAFS

These are two X-ray techniques using X-ray scattering which have become increasingly important. AXD is an acronym for anomalous X-ray diffraction, and is a purely diffraction technique, whereas EXAFS, extended X-ray absorption fine structure, is basically a spectroscopic technique. Both these techniques have become possible because of the advances made in the generation of high intensity X-ray beams using modern synchrotron, second and third generation, equipment.

Table 13.3 Typical hydration numbers found by various methods.
*denotes finding a negative value

Ion	NMR	X-ray or neutron diffraction and related methods	Computer simulation	Stokes' Law radii from ionic conductivities	From ionic conductivities with corrections to Stokes' Law
Li^+	—	4 to 6	6	2.1	7
Na^+	—	ca. 6	6	0.8	5
K^+	—	ca. 6	6	*	>1.5
Be^{2+}	4	—	—	11	14
Mg^{2+}	6	6	—	6.5	12
Ca^{2+}	—	6 to 10	—	4.5	10
Al^{3+}	6	6	—	—	—
Fe^{2+}	6	6	—	—	—
Ni^{2+}	6	6	—	—	—
La^{3+}	—	8	—	9.5	13
Ag^+	—	4	—	0.2	>2.5
F^-	—	6	6 to 8	0.35	4
Cl^-	—	6	6 to 8	*	>0.7
Br^-	—	6 to 7	—	*	>0.3
I^-	—	6 to 9	7	*	≥ 0

In the EXAFS technique the incident X-rays are absorbed by the atom which is being studied. In doing so they interact directly with the electrons of the atom and the excited electrons then probe the immediate environment of the atom. As a result X-rays are emitted and this appears as fine structure in an absorption. From this it is possible to deduce coordination numbers, geometry and mean distances of the atom under study to its nearest neighbours, i.e. to give information about the primary solvation of an ion. Recently it has proved useful in probing the secondary solvation shell.

In the AXD technique, X-rays of a given energy close to a region where the X-rays are absorbed are used, and the resulting scattering pattern is found. Another experiment uses X-rays of a slightly different energy and a further scattering pattern is obtained. By studying the differences between these patterns considerable structural information can be found. This gives the structure around the ion and the information derived fully complements and adds to that derived by more conventional techniques. However, at present it is limited to atoms with atomic number greater than 28, i.e. for Cu and beyond.

13.13 Modern techniques of studying solvation: computer simulations

Computer simulations have proved to be an exciting advance for electrolyte studies; small to moderately large solute molecules are now routinely studied. They are also being developed for the study of large molecules such as those of biological importance, and could possibly enable detail at the molecular level to be correlated with details of the structure of macro-molecules in solution and their biological function.

They are a major tool in the study of electrolyte solutions, and this technique coupled with diffraction and NMR studies may give a means to unravelling the intricacies of solvation.

The technique has been described in Section 10.15. In summary, a model of solvation is decided upon. The computer uses Monte Carlo or molecular dynamics methods and a simulation of the solvation pattern emerges. The beauty of the method lies in the capacity to vary the model at will by varying the type and number of interactions considered for each model. This will give a simulation for each set of conditions which can then be compared with each other and with results from all the experimental methods described earlier. In effect the computer is used to help find a model which fits experiment.

13.14 Cautionary remarks on the significance of the numerical values of solvation numbers

Table 13.3 lists typical values of hydration numbers found by various methods.

Gross discrepancies in the values of h, h_+ and h_- for any given electrolyte are found for the various methods. These may be a result of some or all of the following factors:

- Different methods may measure different things: diffraction and spectroscopic methods deal with an equilibrium situation whereas transport properties deal with the moving ion and its bound solvent molecules. There are obvious discrepancies between:

    (a) the diffraction and NMR results and

    (b) conductance results.

The values of $\mathring{a}$ found from conductance studies are more like crystallographic radii while the diffraction and NMR results indicate bound water.

- Some experiments could measure bound molecules only; some could measure bound plus affected molecules.

- Some of the discrepancies may be absorbing inadequacies in the theoretical model and the equations used in the calculations, for example

    - inadequacies of the Debye-Hückel model and theory;

    - inapplicability of Stokes' Law (Sections 12.4.2 and 12.4.3) in transport methods, and of assumptions that the ion and its bound solvent molecules are incompressible, and electrostriction is ignored;

    - lack of knowledge about the varying relative permittivity as the distance from the ion is increased.

    - NMR, diffraction and computer simulations give individual hydration numbers. These methods are superior to all the other methods quoted and are the most likely to be able to give a comprehensive and accurate study of hydration.

13.15   Sizes of ions

Deciding on the actual size of an ion in solution is one of the fundamental problems in electrolyte solutions. This is because there are several ways in which the actual size could be defined. And so when talking about the actual size and giving it a magnitude, it is necessary to state to what entity this is referring, i.e. is it:

- the bare ion;

- the bare ion plus its primary solvation sphere;

- the moving entity, i.e. ion plus water molecules moving with it;

- or is it the secondary solvation shell?

13.16   A first model of solvation – the three region model for aqueous electrolyte solutions

Around the ion is a region of bound or strongly affected water molecules forming layers or shells which are highly orientated by the ion–dipole interactions between the ion and the polar water molecules. The arrangement of these bound or strongly affected water molecules is considerably distorted from that of the pure solvent. Next to this region, and separating it from the unmodified solvent, is an intervening region of water molecules whose structural arrangement gradually changes from the highly orientated water to that of unmodified water. This is a region of 'misfit' between bound and unbound water, and is expected to be highly disordered because of this misfit.

A comparison can be made between the degree of ordering in the pure solvent and the degree of ordering in the electrolyte solution in terms of:

- (a) the highly ordered region around the ion;

- (b) the disordered intervening region of misfit;

- (c) the order of the unmodified solvent.

Whether an ion will cause an overall increase in order in the solution compared with the pure solvent depends critically on the balance between the order/disorder introduced by (a) and (b).

- If on balance

    region (a) is **dominant**, there will be an **overall** increase in order in the solution relative to the pure solvent, the degree of 'structure' will be enhanced and the ion is termed a '**structure maker**'.

- If, on the other hand, on balance

    region (b) is **dominant**, there will be an **overall** decrease in order in the solution relative to the pure solvent, the degree of 'structure' will be decreased and the ion is termed a '**structure breaker**'.

A cautionary word is necessary here. The structure which is being discussed, i.e. in the strongly orientated and the misfit regions, is not the structure of ordinary water. The structure which the ions enforce on the strongly orientated water is often an octahedral arrangement which is approximately spherically symmetrical around the ion, and this is totally incompatible with the tetrahedral arrangement dictated by the H-bonding array found in pure water. In the intervening 'misfit' region the arrangement is neither a spherically symmetrical nor a tetrahedral one, but something which has to change from one extreme to another, and this could occur over several layers of water molecules. In these layers the molecules must be structurally and dynamically perturbed.

Since this 3-region model was proposed, more precise experimental techniques and theoretical approaches have been developed, and these have led to the possibility of a more detailed description of the three regions.

13.16.1 Structure making and structure breaking ions

- For a **structure maker** the ordering of the bound water molecules must outweigh the disorder induced in the 'misfit' region. Such ions will be highly polarising, and so are expected to be smaller and highly charged as exemplified by Li^+, Mg^{2+}, Al^{3+} and F^-.

- **Structure breaker** ions bind water molecules around them sufficiently to cause a 'mismatch' between the structure of the bound water and that of unmodified water typical of the pure solvent. But the order induced is not great enough to outweigh the disorder created in the 'misfit' region, and this will occur with the less strongly polarising ions such as Rb^+, Cs^{+}, Br^- and I^-.

- Polyatomic ions are more difficult to classify without recourse to experimental clues. But all indications suggest that, possibly because of charge effects,

  - SO_4^{2-} and PO_4^{3-} are powerful structure makers; and

  - NO_3^-, ClO_4^-, and SCN^- are structure breakers.

- Tetraalkylammonium ions, long chain fatty acid anions, detergents and many big biologically important ions are in a class by themselves. They appear generally to be structure makers, but modern work has shown that their structure making behaviour is not a result of the molecular processes proposed for simple ions, but is a result of a totally different physical process called hydrophobic hydration.

13.16.2 Evidence for structure making/structure breaking

Thermodynamic studies are the main source of evidence for these phenomena.

13.16.3 A first attempt at classification

The entropy of solvation of the electrolyte can be compared with that for the corresponding number of noble gas atoms or non-polar molecules. For instance, KCl is compared with two argon atoms while $CaCl_2$ could be compared with three CH_4 molecules.

The processes for the noble gas or non-polar molecule are:

$$\text{noble gas(g)} \rightarrow \text{noble gas(aq)}$$
$$\text{non-polar molecule(g)} \rightarrow \text{non-polar molecule(aq)}$$

It might be expected that because there will be no electrostatic interactions between the noble gas atoms and the water molecules there will be only small changes in entropy. However, they have large negative entropies of solvation. This is surprising since it means that the presence of a non-polar inert solute in water causes a considerable increase in order and must therefore alter the structure of water. It also suggests that **any** solute in water will alter its structure. These processes are, therefore, used as a base-line for comparison with solvation of electrolytes, with the following being used as a 'rule of thumb':

- if the entropy of solvation of an electrolyte has a more negative value than that expected, then it is a structure-maker;

- if the value is less negative than expected it is a structure breaker.

The expected values are computed as follows:

- if the electrolyte is KCl then the value of the entropy of hydration for KCl should be compared with that for 2 Ar;

- if $SrCl_2$ it should be compared with the entropy of hydration for $Kr + 2Ar$;

If hydration entropies of individual ions are considered, the comparisons required can be illustrated by:

- K^+ with Ar;

- I^- with Xe;

- Mg^{2+} with Ne.

However, as will be shown later, the physical processes involved in the solvation of the inert gases, apolar and non-polar molecules are totally distinct from those involved in the solvation of ions. It could be queried as to whether it is valid to use the entropies of solvation of these substances as the basis of a classification of ions in solution.

13.16.4 Thermodynamic evidence

All of the thermodynamic evidence pertaining to hydration is given in terms of standard values which correspond to ideality where the solute concentration $\rightarrow 0$. Under these conditions the only interactions will be those between ion and solvent molecules and between solvent and solvent molecules. Ion–ion interactions can then be ignored.

The thermodynamic procedure is carried out in a series of steps as given in Sections 13.16.5 and 13.16.6.

13.16.5 Determination of partial molar entropies for individual ions

Note well: when data on partial molar entropies are given they are usually in terms of the hydrogen ion written as $H^+(aq)$. This section uses this convention so as not to be in conflict with tabulated data. In all other parts of the book, the standard procedure of writing the H^+ in terms of $H_3O^+(aq)$ is used.

- **From standard emfs of cells**:

  The overall reaction occurring in the cell, $Pt(s)|H_2(g)|HCl(aq) \vdots AgNO_3(aq)|Ag(s)$ is:

$$\tfrac{1}{2}H_2(g) + Ag^+(aq) \rightarrow H^+(aq) + Ag(s)$$

$$\text{and } \Delta G^\theta = \mu^\theta_{H^+} + \mu^\theta_{Ag(s)} - \tfrac{1}{2}\mu^\theta_{H_2(g)} - \mu^\theta_{Ag^+(aq)} \tag{13.6}$$

$$\text{and } \Delta S^\theta = \overline{S}^\theta_{H^+(aq)} + \overline{S}^\theta_{Ag(s)} - \tfrac{1}{2}\overline{S}^\theta_{H_2(g)} - \overline{S}^\theta_{Ag^+(aq)} \tag{13.7}$$

where the $\overline{S}^\theta$'s are standard partial molar entropies.

The partial molar entropy of substance B in a solution relates to the dependence of the entropy of the **solution** on the amount of substance B in it. More precisely, it is the rate of change of S for the solution with change in amount of B present, with the change in amount of B being carried out at constant temperature, constant pressure and constant amount of every other substance.

ΔS^θ can be found from the temperature dependence of the emf of the cell (see Worked Problem 8.11, Section 9.12.2 and Worked Problem 9.16).

$\overline{S}^\theta_{Ag(s)}$ can be calculated from third law entropies,

$\overline{S}^\theta_{H_2(g)}$ can be calculated from statistical mechanics,

$\overline{S}^\theta_{H^+(aq)}$ is allocated specific values using various conventions, e.g. $\overline{S}^\theta_{H^+(aq)} = 0$, or $\overline{S}^\theta_{H^+(aq)} = \tfrac{1}{2}\overline{S}^\theta_{H_2(g)}$. It is vital when quoting partial molar entropies and entropies of hydration to be aware of what convention has been used in deriving the values. In this book the convention used is that $\overline{S}^\theta_{H^+(g)} = 0$.

All the quantities in Equation (13.7) are known except $\overline{S}^\theta_{Ag^+(aq)}$ which can thus be found. If other cells are used further standard partial molar entropies can be found.

- **From solubility products**:

  This can be illustrated by the solubility of the sparingly soluble $AgCl(s)$

$$AgCl(s) \longrightarrow Ag^+(aq) + Cl^-(aq)$$

$$\Delta G^\theta = \mu^\theta_{Ag^+(aq)} + \mu^\theta_{Cl^-(aq)} - \mu^\theta_{AgCl(s)} \tag{13.8}$$

$$\Delta S^\theta = \overline{S}^\theta_{Ag^+(aq)} + \overline{S}^\theta_{Cl^-(aq)} - \overline{S}^\theta_{AgCl(s)} \tag{13.9}$$

ΔS^θ can be calculated from ΔG^θ found from the solubility product, K_s, and ΔH^θ found from the temperature dependence of the solubility product, as illustrated in, for example, Worked Problems 8.6 and 8.7, or from the temperature dependence of ΔG^θ (see Section 8.19).

$\overline{S}^\theta_{AgCl(s)}$ can be found from third law entropies,

$\overline{S}^\theta_{Ag^+(aq)}$ can be found as above from the emf of the cell $Pt(s)|H_2(g)|HCl(aq) \vdots AgNO_3(aq)|Ag(s)$

All the quantities in Equation (13.9) are known except $\overline{S}^\theta_{Cl^-(aq)}$ which can thus be found. Using these methods and others, tables of individual partial molar entropies for many ions have been built up.

But remember: the actual value for any ion will depend on the convention used. Other conventions may be used and they need not necessarily relate to $H^+(aq)$. Care must, therefore, be taken when using such tables.

Similar tables can be found for determining standard partial molar entropies for ions in a variety of solvents. These are important in determining thermodynamic transfer data (see Sections 13.23 and 13.25.5).

13.16.6 Standard entropies of hydration

These relate to the process of transferring the ion from the gas phase to the aqueous phase. They can be calculated from the tabular data of partial molar entropies of the ion in solution and statistical mechanical calculations on the ion in the gas phase.

$$K^+(g) \rightarrow K^+(aq)$$

$$\Delta G^\theta_{hydration} = \mu^\theta_{K^+(aq)} - \mu^\theta_{K^+(g)} \tag{13.10}$$

$$\Delta S^\theta_{hydration} = \overline{S}^\theta_{K^+(aq)} - \overline{S}^\theta_{K^+(g)} \tag{13.11}$$

$\overline{S}^\theta_{K^+(g)}$ can be found from statistical mechanics
$\overline{S}^\theta_{K^+(aq)}$ can be found as given above (Section 13.16.5).
and so $\Delta S^\theta_{hydration}$ can be found.
Tables of entropies of hydration for individual ions can thus be drawn up.

But take care: these will also be dependent on the convention used in the determination of the individual partial molar entropies of ions in aqueous solution.

Table 13.4 lists some values for $\Delta S^\theta_{ion\,hydration}$. It will be seen that they are negative corresponding to a decrease in entropy and to an increase in order on forming the solution. This is not surprising as solvation will occur. To be able to classify ions into being 'structure breakers' or 'structure makers', these $\Delta S^\theta_{ion\,hydration}$ are compared with the entropies of hydration for the noble gases. A value of $\Delta S^\theta_{ion\,hydration} = -100\,J\,mol^{-1}\,mol^{-1}$ is taken as a rough and ready division between 'structure breakers' and 'structure makers'. A value of $\Delta S^\theta_{ion\,hydration}$ which is more negative than this indicates a 'structure maker', a value which is less negative indicates a 'structure breaker'. Obviously there is quite a degree of latitude as to where the dividing line is drawn, but overall most ions are classified correctly using this highly approximate method. However, the values of $\Delta S^\theta_{ion\,structural}$ deduced in Section 13.16.7 are more meaningful.

Table 13.4 Values of $\Delta S^{\theta}_{ion\ hydration}$ for various ions

Cations

Ion	$\dfrac{\Delta S^{\theta}}{J\,mol^{-1}K^{-1}}$			Comment
	(a)	(b)	(c)	
H^+	−109	−131	−104	structure making
Na^+	−89	−111	−84	structure making
K^+	−52	−74	−47	structure breaking
Mg^{2+}	−287	−331	−304	structure making
Ba^{2+}	−161	−205	−178	structure making
Al^{3+}	−472	−538	−511	structure making
La^{3+}	−388	−454	−427	structure making

Anions

Ion	$\dfrac{\Delta S^{\theta}}{J\,mol^{-1}K^{-1}}$			Comment
	(a)	(b)	(c)	
F^-	−159	−137	−110	structure making
Cl^-	−96	−74	−47	structure breaking
Br^-	−81	−59	−32	structure breaking
I^-	−58	−36	−9	structure breaking

(a) based on standard pressure for gaseous ions, ordinary concentrations for aqueous ions; convention as given in Section 13.6.5 for $H^+(aq)$;

(b) based on standard pressure for gaseous ions, ordinary concentrations for aqueous ions; different convention – one which gives the same values of $\Delta S^{\theta}_{hydration}$ for $Cl^-(aq)$ as for $K^+(aq)$;

(c) based on a standard concentration $(1\ mol\ dm^{-3})$ for gaseous ions, ordinary concentrations for aqueous ions; convention – one which gives the same values of $\Delta S^{\theta}_{hydration}$ for $Cl^-(aq)$ as for $K^+(aq)$, i.e. as in (b).

13.16.7 Significance of entropies of hydration for structure making/breaking

The entropy of hydration is the difference between the partial molar entropy of the ion in its standard state in solution and that of the gaseous ion in its standard state. There will be several contributions to this change in entropy, the most significant being:

- (i) long range electrostatic effects due to the interaction of the charge on the ion with the permanent and induced dipoles of the water molecules. This can be calculated using an argument developed by Born. However, there are hidden problems here. The argument used involves an integral over distances from the ion and also the relative permittivity of the solvent. The upper limit in the distance is always ∞. The value for the lower limit is the problem. It could be taken to be one of the following:

- if the radius of the bare ion is used then the relative permittivity close to the ion cannot be the bulk value. What would be needed is a relative permittivity whose value varies with the distance from the ion. It would also include an effect due to immobilisation of the water which is generally treated separately.

- if the lower limit is taken as the radius of the primary solvation sheath, then the entropy which is calculated will include a contribution from the misfit region. This would really be a 'structural' effect which is normally calculated from the entropy of hydration minus each of the contributions (i) and (ii).

- The best procedure is probably to use the Born equation with the lower limit being the distance from the origin to the start of the region corresponding to bulk water. This allows (ii) and (iii) to be assessed separately.

- (ii) immobilisation of solvent molecules in the near environment of the ion. This is likely to be the primary solvation shell for singly charged ions, but for higher charged cations it might include some of the secondary solvation.

- (iii) other effects which are a consequence of the disruption of the water structure in the misfit region.

When the entropy contributions from (i) and (ii) are added up and subtracted from the entropies of hydration, there is a finite value left which is assigned as a change of entropy consequent on the disruption of the water structure in the misfit region, often labelled $\Delta S^{\theta}_{\text{ion structural}}$ Table 13.5 lists some typical values.

Table 13.5   Values of $\Delta S^{\theta}_{\text{ion structural}}$ for various ions

Ion	$\Delta S^{\theta}_{\text{hydration}}$/J mol^{-1} K^{-1}	Approximate 'Born' electrostatic contribution/J mol^{-1} K^{-1}	Approximate contribution from restriction of motion of nearest neighbours of ion/J mol^{-1} K^{-1}	Estimated 'structural' contribution/ J mol^{-1} K^{-1}
Na$^+$	-84	-15	-50	-19
K$^+$	-47	-15	-50	$+18$
Mg^{2+}	-304	-60	-50 or -120	-194 or -124
Ba^{2+}	-178	-60	-50 or -120	-68 or $+2$
Al^{3+}	-511	-135	-50 or -150	-326 or -226
La^{3+}	-427	-135	-50 or -150	-242 or -142
F$^-$	-110	-15	-50	-45
Cl$^-$	-47	-15	-50	$+18$
Br$^-$	-32	-15	-50	$+33$
I$^-$	-9	-15	-50	$+56$

(i)   The values of $\Delta S^\theta_{\text{hydration}}$ listed are based on a standard **concentration** for the gaseous ions, as distinct from the normal values based on a standard **pressure**; and on ordinary concentrations for aqueous ions. They correspond to the convention which takes $\Delta S^\theta_{\text{hydration}}$ for $Cl^-(\text{aq})$ to be the same as for $K^+(\text{aq})$.

(ii) The 'Born' contribution is estimated by taking it to relate to the region beyond 0.270 nm from the centre of the ion. This gives a contribution of $(z^2 \times 15)\,\text{J mol}^{-1}\,\text{K}^{-1}$.

(iii) The contribution from the restriction of motion of H_2O molecules which are nearest neighbours is taken to be **either**:

- $-50\,\text{J mol}^{-1}\,\text{K}^{-1}$ for all monatomic ions;

or

- $-50\,\text{J mol}^{-1}\,\text{K}^{-1}$ (moderate restriction) for singly charged ions;
  $-120\,\text{J mol}^{-1}\,\text{K}^{-1}$ (extensive restriction) for doubly charged ions;
  $-150\,\text{J mol}^{-1}\,\text{K}^{-1}$ (immobilisation) for triply charged ions.

(iv) The 'structural' contribution is estimated as:

$$\Delta S^\theta_{\text{hydration}} - \{\text{'Born' contribution}\} - $$
$$\{\text{contribution from restricted motion of nearest neighbours}\}$$

For some ions $\Delta S^\theta_{\text{ion structural}}$ is negative, i.e. the effect is a loss of entropy corresponding to an increase in order of the solvent. These are called 'structure making ions'. Those for which $\Delta S^\theta_{\text{ion structural}}$ is positive correspond to an increase in entropy and an increase in disorder for the solvent and are 'structure breaking ions'.

Confirmation of this classification comes from studies on the partial molar volumes of ions, and from the compressibility and viscosity of electrolyte solutions and from the activation parameters for water exchange found from NMR and ultrasonic methods.

13.17 Volume changes on solvation

When a solute is added to a solvent, e.g. an electrolyte to water, it might be expected that the final volume of the solution would be greater then the volume of the solvent used. However, this is not always the case, and large negative changes in volume are often found experimentally. These are found by measuring the density of solutions as a function of concentration of the electrolyte.

Analogous interpretations to the entropy data can be given for changes in volume on solution of an electrolyte, though the effects are less definite than with the entropies.

- If the decrease in volume is greater than expected, then the ion is a structure-maker – constriction of the solvent in the bound shell more than outweighs the effect of the 'misfit' region.

- If the decrease in volume is less than expected, then the ion is a structure-breaker – the constriction of the loosely bound water molecules in the bound region is not sufficiently great to outweigh the effects of the 'misfit' region.

13.18 Viscosity data

This is a less clear-cut situation:

- An ion with a strongly bound inner region will give a more ordered solution and hence an increase in viscosity greater than expected. This is found for structure-makers.

- An ion with a loosely bound inner region will give a less ordered solution and hence a lower viscosity than expected. This is found for structure breakers.

13.19 Concluding comment

The models described are expected to be an oversimplified description of what happens in solvation. But this has focused attention on the molecular description and microscopic detail of what happens around an ion in solution, and has thrown into perspective the relevance and importance of diffraction and computer simulation experiments.

13.20 Determination of $\Delta G^{\theta}_{hydration}$

This requires a more complex argument than that used for determining $\Delta S^{\theta}_{hydration}$, and can be illustrated by the process of hydration for $Ag^+(aq)$.

$$Ag^+(g) \rightarrow Ag^+(aq) \tag{1}$$

There are five stages to the argument:

- Use of the cell, $Pt(s)|H_2(g)|HCl(aq) \vdots AgNO_3(aq)|Ag(s)$ whose overall cell reaction is:

$$\tfrac{1}{2} H_2(g) + Ag^+(aq) \rightarrow H^+(aq) + Ag(s) \tag{2}$$

and whose ΔG^{θ}_2 can be found from the emf of the cell.

- The reaction $\qquad H^+(aq) + Ag(s) \rightarrow \tfrac{1}{2} H_2(g) + Ag^+(aq) \tag{3}$

is the reverse of reaction (2) and so:

$$\Delta G^{\theta}_3 = -\Delta G^{\theta}_2 \tag{13.12}$$

Hence ΔG^{θ}_3 is known.

- The reaction: $\qquad \tfrac{1}{2} H_2(g) \rightarrow H^+(aq) + e \tag{4}$

is taken by convention to have $\Delta G^{\theta}_4 = 0$ $\tag{13.13}$
If reactions (3) and (4) are added this will give:

$$Ag(s) \rightarrow Ag^+(aq) + e \tag{5}$$

with $\qquad \Delta G^{\theta}_5 = \Delta G^{\theta}_3 + \Delta G^{\theta}_4 = \Delta G^{\theta}_3 + 0 = \Delta G^{\theta}_3 = -\Delta G^{\theta}_2 \tag{13.14}$

- The reaction which is required is:

$$Ag^+(g) \rightarrow Ag^+(aq) \tag{1}$$

and this is made up from the two reactions:

$$Ag(s) \rightarrow Ag^+(aq) + e \tag{5}$$

and $\qquad\qquad Ag(s) \rightarrow Ag^+(g) + e \tag{6}$

i.e. $(1) = (5) - (6)$

Now $\Delta G_5^\theta = \Delta G_3^\theta = -\Delta G_2^\theta$ see Equation (13.14)

$$\therefore \Delta G_1^\theta = \Delta G_5^\theta - \Delta G_6^\theta = -\Delta G_2^\theta - \Delta G_6^\theta \tag{13.15}$$

- It is now necessary to find (6) which is the sum of:

$$Ag(s) \rightarrow Ag(g) \tag{7}$$

which corresponds to sublimation, and for which the free energy of sublimation can be found, and

$$Ag(g) \rightarrow Ag^+(g) + e \tag{8}$$

which corresponds to ionisation, and for which the free energy of ionisation can be found and so $(6) = (7) + (8)$

and hence $\Delta G_1^\theta = \Delta G_5^\theta - \Delta G_6^\theta = \Delta G_5^\theta - \Delta G_7^\theta - \Delta G_8^\theta = -\Delta G_2^\theta - \Delta G_7^\theta - \Delta G_8^\theta \quad (13.16)$

Since ΔG_2^θ, ΔG_7^θ and ΔG_8^θ are all easily determined experimentally, ΔG_1^θ which is the standard free energy of hydration for Ag^+ can be found.

Using arguments such as these, $\Delta G_{\text{hydration}}^\theta$ for a whole variety of ions can be tabulated.

13.21   Determination of $\Delta H_{\text{hydration}}^\theta$

An exactly analogous argument will result in $\Delta H_{\text{hydration}}^\theta$.

- ΔH_2^θ for reaction (2) can be found from the temperature dependence of the emf of the cell, $Pt(s)|H_2(g)|HCl(aq) \vdots AgNO_3(aq)| Ag(s)$ from which $\Delta H_3^\theta = -\Delta H_2^\theta$

- By convention $\Delta H_4^\theta = 0$, and so $\Delta H_5^\theta = \Delta H_3^\theta = -\Delta H_2^\theta \qquad \text{(13.17) and (13.18)}$

- Standard enthalpies of sublimation and ionisation are easily determined experimentally,

  and so $\Delta H_1^\theta = \Delta H_5^\theta - \Delta H_6^\theta = \Delta H_5^\theta - \Delta H_7^\theta - \Delta H_8^\theta = -\Delta H_2^\theta - \Delta H_7^\theta - \Delta H_8^\theta \qquad (13.19)$

Using arguments such as these, $\Delta H_{\text{hydration}}^\theta$ for a whole variety of ions can be tabulated.

13.22 Compilation of entropies of hydration from $\Delta G^{\theta}_{hydration}$ and $\Delta H^{\theta}_{hydration}$

The values of $\Delta H^{\theta}_{hydration}$ and $\Delta G^{\theta}_{hydration}$ found above can be used to give values of $\Delta S^{\theta}_{hydration}$ for the series of ions for which $\Delta H^{\theta}_{hydration}$ and $\Delta G^{\theta}_{hydration}$ have been found.

Now take very great care: These values for $\Delta S^{\theta}_{hydration}$ have been found using the convention that for the reaction:

$$\tfrac{1}{2}\,H_2(g) \rightarrow H^+(aq) + e \tag{4}$$

$\Delta G^{\theta}_4 = 0$, $\Delta H^{\theta}_4 = 0$ and therefore $\Delta S^{\theta}_4 = 0$. This is a different convention to that used in Section 13.16.6 where $\Delta S^{\theta}_{hydration}$ values have been calculated from standard partial molar entropies for the individual hydrated ions which were based on the convention that $\overline{S}^{\theta}_{H^+(aq)} = 0$. The two sets of values for $\Delta S^{\theta}_{hydration}$ will not be the same.

It is absolutely vital to be aware of these different conventions and any other which may have been used. When compounding standard values for thermodynamic quantities they must all be for the same convention.

13.23 Thermodynamic transfer functions

The three thermodynamic quantities $\Delta G^{\theta}_{solvation}$, $\Delta H^{\theta}_{solvation}$ and $\Delta S^{\theta}_{solvation}$ can be compiled for ions in a variety of solvents. For a given ion each corresponding thermodynamic quantity can be converted into a quantity called a thermodynamic transfer function for transferring the ion from the solvent to water. For instance $\Delta H^{\theta}_{hydration} - \Delta H^{\theta}_{solvation}$ gives the transfer function, $\Delta H^{\theta}_{transfer}(s \rightarrow w)$. These can give some interesting insights into structural aspects of solvation, and will be discussed in Section 13.25.5, and compared with the transfer functions for apolar molecules.

This is a book on aqueous electrolytes, but it is imperative that something is said about the solvation of non-polar solutes. This is a topic of prime concern for biochemists and biologists, but it is also very important for chemists because many of the phenomena observed for non-polar and apolar solutes bear a superficial similarity to those for electrolytes. However, it is important to point out that, despite this similarity, the physical and molecular details are totally different for ions and apolar solutes. A study of the reasons for the difference in molecular behaviour involved in the solvation of ions and apolar molecules has resulted in an increased understanding of solvation phenomena in general. It has also led to an upsurge in the number of investigations, both experimental and theoretical.

13.24 Solvation of non-polar and apolar molecules – hydrophobic effects

The behaviour of noble gases and non-polar molecules has been used as a base-line for discussion of solvation of electrolytes. However, the molecular description of what happens on solvation is different in the two cases with hydrophobic interactions being the predominant factor in the solvation of the apolar or non-polar molecule and interactions of the ion with the dipole of water being the predominant factor in the solvation of ions. If this is the case, it becomes necessary to study hydrophobic interactions and to query the validity of using the behaviour of apolar and non-polar molecules as a base-line for interpreting data on electrolytes.

When an apolar or non-polar solute is dissolved in water, intuitively it is not expected that there would be any interactions of the molecule with the dipoles of the water molecule. Yet, dissolution is accompanied by a large negative change in entropy which can only be explained by saying that the solute is doing something to the solvent structure which results in a large overall increase in order.

This observation demands that the question should be asked:

what large change can an inert molecule impose on the structural behaviour and properties of pure liquid water?

The only feasible answer is that the normal extensive H-bonded structure of water responds to the presence of the solute molecule by spatial and orientational rearrangements which result in an overall increase in order of the solution. The loose H-bonded structure of ordinary water is lost locally near the solute molecule, and in its place a new H-bonded structure is set up where the water molecules form a cage-like structure around the solute molecule. A large number of cage structures can be formed. These differ in size and shape, and can therefore accommodate different sizes of solute molecules, but all cage structures are made up of H-bonded water molecules. The term hydrophobic hydration is used to describe the situation when a non-polar or apolar solute is solvated in this way by having a cage of solvent molecules around it. Hydrophobic hydration is only possible because liquid water can exist in many spatial H-bonded arrangements over and above the open and extensive H-bonded array of ordinary water. Hydrophobic hydration generally occurs only with non-polar or apolar solutes, though alcohols, ethers and amines must be included in an exhaustive study. However, for the electrolyte chemist it becomes important as an aspect of the solvation encountered when alkyl ammonium ions and other large organic ions are dissolved in water (see Section.13.26). Polar solutes, on the other hand, break down the H-bonded arrangement of water and replace it by a spherically symmetrical non-H-bonded shell of water molecules.

Hydrophobic hydration involves relatively rigid cages around molecules of solute. In this way, a structure is formed which is more ordered than ordinary water. The observed large negative ΔS^θ for hydration is thought to be a consequence of this.

It is important to realise that hydrophobic hydration is a physical consequence of rearrangement of the water structure around the solute, and is not primarily a result of direct attractive interactions between solute and water molecules. It is a hydration phenomenon totally distinct from the sort of attractive interaction hydration which occurs when a polar or charged solute is dissolved in water.

It is for this reason that the comparison of thermodynamic quantities for apolar and non-polar solutes with those for charged solutes must be queried. However, it could be asked 'what other comparison can be made?'

13.25   Experimental techniques for studying hydrophobic hydration

These can be broadly classified into three main categories:

(i)   spectroscopic, NMR and ultrasonic relaxation;

(ii) computer simulations;

(iii) thermodynamic studies.

13.25.1   Results of methods (i) and (ii)

Spectroscopic studies strongly support cage-like structures, and show strong similarities to the solid clathrates which are well characterised as being crystalline cage-structures around an inert molecule. NMR relaxation studies show that rotational diffusion motions of the water molecules of the cage are inhibited, but the motion of the solute is not. This would be expected if water molecules are 'fixed' in the cage, but the solute is still free to move around in the space inside.

As in the case of aqueous solutions of electrolytes, computer studies have shed much light on the behaviour of aqueous solutions of non-polar and apolar molecules. They can give information on solute–solvent and solvent–solvent interactions in such solutions. This is a powerful tool for studying hydrophobic phenomena and is limited only by the accuracy of the assumed model and the quantities relevant to this model which are fed into the computer simulations. Simulation is of particular importance in the solution chemistry of large macromolecules and polymers which are extremely difficult to study experimentally, especially in dilute solutions. They are likely to be a dominant feature of the future study of hydrophobic hydration.

This work indicates:

• Agreement with other techniques which suggest that there is disruption of the normal H-bonded water structure on introduction of an apolar or non-polar solute. This is replaced by a cage-like H-bonded structure with the solute molecule sitting in the cavity at the centre of the cage.

• Cages of different sizes can result from different H-bonded arrangements of water molecules.

• The H-bonds around the centre of the cage may be stronger than those in bulk water.

• The rotational motion of the water molecules in the cage is impeded, and this affects viscosity and diffusion rates.

• The thermodynamic functions calculated from computer simulations can be fitted to experimental values.

These findings should be compared with the results of computer simulations on aqueous solutions of electrolytes.

13.25.2   Results of thermodynamic studies

These have been the backbone of studies on hydrophobic hydration. Thermodynamic functions such as ΔG, ΔH and ΔS extrapolated to infinite dilution give information about solute–solvent interactions whereas the same functions studied over a range of concentrations give information on solute–solute interactions (see Sections 1.7.1 and 1.9). Results at infinite dilution will give information on hydrophobic hydration.

13.25.3   Entropies of hydration at infinite dilution, ΔS^{θ}

ΔS^{θ} is always large and negative for hydrophobic hydration. This supports the belief that dissolution must increase the order of the solution, which, in turn, suggests the alteration in structure which would result from cage formation.

13.25.4 ΔV^{θ} of solvation at infinite dilution

This is a change in volume at infinite dilution when the solute and solvent are mixed to give the solution.

Such studies show found that $\Delta V^{\theta}_{\text{solution}}$ is always negative for hydrophobic hydration. If the solute is small then it can fit into the spaces of the normal tetrahedral H-bonded structure of ordinary water, giving a decrease in volume. But if cage structures result it could be expected that the water structure is even more compacted around the solute, resulting in a further decrease in volume.

13.25.5   Thermodynamic transfer functions

When an apolar or non-polar solute is dissolved in a variety of solvents water is always anomalous. So it is useful to compare properties in each solvent with those in water. And so ordinary thermodynamic functions like $\Delta H^{\theta}_{\text{solution}}$ are converted into transfer functions which represent the change, for instance, in enthalpy for the process of transferring the solute from the given solvent to water (in all cases) at infinite dilution. It is interesting and of value to compare the behaviour of apolar solutes with that of ions in solution. The results can be summarised as follows.

- **Ions in solution**

  - When ions are dissolved in water and in non-aqueous solvents, they are more soluble in the water, and $\Delta G^{\theta}(\text{transfer}, s \rightarrow w)$ is negative.

    i.e. $\Delta G^{\theta}_{\text{solv}}(w) < \Delta G^{\theta}_{\text{solv}}(s)$
    and therefore the equilibrium constant for the process gas $\rightleftharpoons$ aqueous solution is larger than that for the process gas $\rightleftharpoons$ non-aqueous solution.

- ΔH^{θ}(transfer, $s \rightarrow w$) is positive.

  i.e. $\Delta H^{\theta}_{solv}(w) > \Delta H^{\theta}_{solv}(s)$
  so that if the only factor affecting dissolution were ΔH^{θ}_{solv}, then the process of forming solvated ions would be more energetically favourable in the non-aqueous solvent than in water, contrary to what might be expected.

- The only way in which ΔG^{θ}(transfer, $s \rightarrow w$) can be **negative** when ΔH^{θ}(transfer, $s \rightarrow w$) is **positive** is for $T\Delta S^{\theta}$(transfer, $s \rightarrow w$) to be **positive** and large enough.

Remember: $\Delta G^{\theta} = \Delta H^{\theta} - T\Delta S^{\theta}$ or $T\Delta S^{\theta} = \Delta H^{\theta} - \Delta G^{\theta}$
  In the present context both ΔH^{θ}(transfer, $s \rightarrow w$) and $\{-\Delta G^{\theta}$(transfer, $s \rightarrow w)\}$ are positive. This conclusion supports the belief that the aqueous solution is more disordered than the non-aqueous solution. Put otherwise, the process of dissolution is energetically unfavourable in water compared with the non-aqueous solution, but this is compensated by an entropically favoured process for the aqueous solution compared with the non-aqueous solution.

- **Apolar solutes in solution**

- Most non-polar and apolar molecules are less soluble in water than in other solvents. Therefore, ΔG^{θ}(transfer, $s \rightarrow w$) is positive.

  i.e. $\Delta G^{\theta}_{solv}(w) > \Delta G^{\theta}_{solv}(s)$
  and therefore the equilibrium constant for the process gas $\rightleftharpoons$ aqueous solution is smaller than that for the process gas $\rightleftharpoons$ non-aqueous solution.

- ΔH^{θ}(transfer, $s \rightarrow w$) is often negative and fairly large which means that the process of forming the solvated solute is more energetically favourable in water than in the non-aqueous solvent, contrary to what might be expected.

- The only way in which ΔG^{θ}(transfer, $s \rightarrow w$) can be **positive** when ΔH^{θ}(transfer, $s \rightarrow w$) is **negative** is for $T\Delta S^{\theta}$(transfer, $s \rightarrow w$) to be **negative** and large enough.

This suggests that the solute must be passing from a less ordered solution to a more ordered solution in going from the solvent to water.

This is very difficult to explain other than in terms of hydrophobic hydration, that is, cage formation of tetrahedrally H-bonded water molecules around the solute. Forming the cage in water with the apolar molecule inside corresponds to an increase in entropy compared with the situation of the apolar molecule in other solvents.

The further conclusion which has to be drawn is that dissolution of ions in water is structurally totally distinct from that for dissolution of apolar solutes in water. The way in which the structure of pure water is modified on forming aqueous solutions of these two types of solutes is crucially different, and raises the question as to whether the changes of entropy on dissolving an apolar or non-polar solute in water can be used as a base-line for interpreting changes in entropy for dissolution of ions in water (see Section 13.16.3).

To summarise:

Ions in solution	Hydrophobic hydration
$\Delta G^{\theta}_{\text{transfer}}(\text{s} \rightarrow w)$ is negative	$\Delta G^{\theta}_{\text{transfer}}(\text{s} \rightarrow w)$ is positive
$\Delta H^{\theta}_{\text{transfer}}(\text{s} \rightarrow w)$ is often positive	$\Delta H^{\theta}_{\text{transfer}}(\text{s} \rightarrow w)$ is often negative
$\Delta S^{\theta}_{\text{transfer}}(\text{s} \rightarrow w)$ is positive	$\Delta S^{\theta}_{\text{transfer}}(\text{s} \rightarrow w)$ is negative
and	
ΔV^{θ} hydration is negative	ΔV^{θ} hydration is negative
ΔS^{θ} hydration is usually large and negative	ΔS^{θ} hydration is large and negative

Note: the distinction in the effect of ions and apolar solutes on the structure of water only becomes apparent in terms of ΔS^{θ} values when thermodynamic transfer ΔS^{θ} values are used.

13.26 Hydrophobic hydration for large charged ions

Tetraalkylammonium ions are interesting. At first sight they appear to be structure-making ions but, in fact, their properties are due to hydrophobic hydration as shown by thermodynamic, spectroscopic, conductance and viscosity data. Computer simulation experiments also fit in with the ideas of hydrophobic interaction. The long alkyl chain acts as a non-polar residue which induces the orientational changes in the water structure characteristic of hydrophobic hydration. It is likely that ions which combine a charge with a substantial alkyl or aryl residue, for example detergents, phospholipids and other large ions will show the same pattern of behaviour. These ions must be thought of in a different way from simple approximately spherically symmetrical ions. They may well form a bridge between the microscopic behaviour of simple electrolytes exhibiting one type of molecular behaviour to large polyelectrolytes and large charged molecules and macromolecules exhibiting an entirely different type of behaviour, even though their macroscopic properties may be similar.

Again it is worthwhile to emphasise the crucial distinction between:

hydration of ionic solutes

and

hydration of non-polar and apolar solutes

Ionic solutes are hydrated as a consequence of attractive electrostatic interactions between the charges on the ions and the dipoles of the water molecules. These interactions break down the tetrahedral arrangements of H-bonded water molecules near the ion to form a spherically symmetrical hydration shell which is not H-bonded in the pattern characteristic of the pure solvent.

Non-polar and apolar solutes are hydrated not because of attractive interactions between the solute molecule and the dipoles of the water molecules, but because the presence of the solute molecule induces rearrangements of the H-bonds. Hydration of these solutes is possible because water can form another type of tetrahedral structure – cage arrays. But in these cages the solute molecule is surrounded by water molecules which are still H-bonded and tetrahedrally disposed.

13.27 Hydrophobic interaction

When finite concentrations of non-polar and apolar molecules are considered, it would be expected that there would be no attractive interactions between such solute molecules. However, when the concentration dependence of various thermodynamic functions is looked at, the behaviour observed is not consistent with that expected for non-interacting solute particles.

Remember: concentration dependence gives information on solute–solute interactions (see Sections 1.7.1 and 1.9).

For instance, the entropy of solution is found to be less negative than would be predicted on the basis of the limiting ΔS^θ (discussed in hydrophobic hydration, Sections 13.24 and 13.25) and the increasing number of particles present in solution. The presence of the supposedly non-interacting solute molecules appears to be reducing the large negative entropy change associated with hydrophobic hydration, and is partially reversing the entropically unfavourable hydrophobic hydration.

Observations on other thermodynamic functions collected at finite concentrations support the idea that something is happening to partially reverse the hydrophobic hydration, and this effect manifests itself more obviously as the concentration increases. This effect is called the hydrophobic effect or interaction. The simplest interpretation pictures two cages each with a solute molecule in its cavity coming close enough to squeeze all the water molecules out from between the two cages, and to enable the solute molecules to touch. This results in increased disorder corresponding to an increase in entropy which fits with the observation that the entropy of solution is less negative than expected. More detailed work shows that this picture is oversimplified.

A lot of evidence supporting the idea of a hydrophobic effect comes from detailed thermodynamic studies over a range of concentrations.

A similar conclusion of too simple a picture of hydrophobic interaction is reached when the physical properties of apolar solutions are looked at in more detail.

However it is again crucial to realise that the hydrophobic effect, like hydrophobic hydration, is physically a consequence of rearrangement of the water structure around the solute molecules. It is not a result of a direct attractive interaction between two solute particles. This becomes important when biologically important solutes are considered. Here both hydrophobic hydration and hydrophobic interaction play a crucial role in determining the structure of biological molecules *in vivo* and in determining their biological function. For instance, when two apolar parts of a protein appear to be 'attracted' to each other it is not because of an attraction between these two parts of the protein molecule. Rather it is because overlap of the cage structures around the apolar residues results in a change of the water structures between the apolar residues of the protein.

13.28 Computer simulations of the hydrophobic effect

Computer simulation experiments suggest that the hydrophobic effect is more complex than previously thought. They fit in with the idea of two sorts of 'interaction', one corresponding to solute–solute contact with no solvent between the solute molecules, the other corresponding to a solvent-separated hydrophobic effect with the two solute molecules still solvent separated.

Subject matter of worked problems

Acids, bases, salts and buffers – a classification: 3.1–3.3

ampholytes, intermediate species: 6.2–6.4

Calculation of pH: 2.4, 4.1–4.6, 4.8, 5.1–5.4, 6.1, 8.20, 9.19, 9.23, 9.24

Cells:

assignment of polarity: 9.10, 9.11, 9.15

cells with liquid junction (salt bridge): 9.13, 9.17, 9.18, 9.20–9.23, 9.25–9.28

cells with liquid junction (tube): 9.12, 9.30

cells without liquid junction: 9.19, 9.24, 9.29

concentration cells: 9.21–9.23, 9.25, 9.26–9.28, 9.30

dependence of emf on ionic strength: 9.17, 9.18, 9.21, 9.23, 9.28, 9.29

determination of E^0: 9.17–9.20, 9.29

electrode reactions and processes in cell: 9.4–9.30

formulating emf from μ's: 9.17–9.30

migration processes: 9.4, 9.5, 9.10–9.13, 9.30

transport number: 9.4, 9.10–9.13, 9.30

use to determine thermodynamic quantities: 9.16, 9.18, 9.19, 9.20, 9.23–9.28

Conductance:

cell constants: 11.1–11.3

equilibrium constants: 11.5, 11.6, 11.9–11.11

Kohlrausch's Law: 11.8–11.11, 11.14

mobilities: 11.15, 11.16, 11.18, 11.19

transport numbers: 11.12–11.14, 11.16, 11.17

Λ: 11.2, 11.3

Λ^0: 11.6–11.11, 11.14

$\alpha = \Lambda/\Lambda^0$: 11.4, 11.5, 11.9

λ_i: 11.7, 11.8, 11.10–11.14, 11.16

Dependence of thermodynamic quantities on T and p:

K: 2.4, 2.5, 8.5–8.9:

ΔH^θ: 8.4, 8.6

emf's: 8.11, 9.16

Direction of reaction: 2.2, 2.3, 7.5, 7.6, 8.1, 8.3, 8.16

possibility of precipitation: 2.3, 7.5, 7.6, 8.3, 8.16

Electrolysis: 9.1–9.3

Electrical neutrality, use of: 4.3, 4.6 (Section 5.4.1), 7.1, 7.7, 9.13

Equilibrium calculations:

buffers: 5.3–5.5

effect of complexing and ion pairing on solubility: 7.8–7.10

formulating ΔG from μ's: 9.17–9.30

effect of ionic strength: 8.15, 8.18–8.24, 9.17–9.19, 9.21, 9.23, 9.29

experimental determination of K: 7.2, 7.3, 7.10, 8.11, 8.15, 8.18–8.22

from conductance: 11.5, 11.6, 11.9–11.11

from emf's 9.16, 9.20, 9.23–9.28

fraction ionised/protonated (α) 4.1–4.8, 5.1, (Section 5.4.1), 11.5, 11.9

dependence of α on concentration: 4.4, 4.8, 5.1, (Section 5.4.1)

maximum ionisation/protonation: 4.7, 4.8, 11.9

An introduction to Aqueous Electrolyte Solutions. By Margaret Robson Wright
© 2007 John Wiley & Sons Ltd   ISBN 978-0-470-84293-5 (HB) ISBN 978-0-470-84294-2 (PB)

Index

see also, where marked *, the appropriate section in "subject matter of worked problems".

An introduction to Aqueous Electrolyte Solutions. By Margaret Robson Wright
© 2007 John Wiley & Sons Ltd   ISBN 978-0-470-84293-5 (HB) ISBN 978-0-470-84294-2 (PB)

Printed and bound by CPI Group (UK) Ltd, Croydon, CR0 4YY

27/10/2024

14580153-0002